Design
to Reduce
Technical Risk

Practical Engineering Guides for Managing Risk

by AT&T and the Department of the Navy

Design's Impact on Logistics
Moving a Design into Production
Testing to Verify Design and Manufacturing Readiness

Design
to Reduce
Technical Risk

AT&T

McGraw-Hill, Inc.

New York   San Francisco   Washington, D.C.   Auckland   Bogotá
Caracas   Lisbon   London   Madrid   Mexico City   Milan
Montreal   New Delhi   San Juan   Singapore
Sydney   Tokyo   Toronto

Library of Congress Cataloging-in-Publication Data

Design to reduce technical risk / AT&T.
    p.   cm. — (Practical engineering guides for managing risk)
    Includes bibliographical references.
    ISBN 0-07-002561-4
    1. Engineering design. 2. Computer-aided design. I. American
Telephone and Telegraph Company. II. Series.
TA174.D485 1993
620'.0042—dc20   92-42750
                                              CIP

1 2 3 4 5 6 7 8 9 0 DOC/DOC 9 9 8 7 6 5 4 3

ISBN 0-07-002561-4

The sponsoring editor for this book was Robert W. Hauserman, the editing supervisor was Kimberly A. Goff, and the production supervisor was Donald F. Schmidt. This book was set in Century Schoolbook. It was composed by McGraw-Hill's Professional Book Group composition unit.

Printed and bound by R. R. Donnelley & Sons Company.

Contents

Part 4. Software Design and Software Test

Part 5. Computer-Assisted Technology

Part 6. Design for Testing

Part 7. Configuration Control

Part 8. Design Reviews

Part 9. Design Release

How This Book Is Organized

This book is divided into five major sections:

Introduction: Explains which templates are covered, contains keys to success of risks to avoid and best practices to consider, and gives definitions for important terms.

Procedures: Presents the steps for reducing the technical risks for each template covered in the book.

Application: Gives examples of the principles and procedures discussed in the Procedures chapter.

Summary: Outlines information from earlier chapters.

References: Gives an annotated bibliography of sources for more information.

Acknowledgments

The following individuals and their respective organizations are hereby recognized for their contributions to the development of the texts on Practical Engineering Guides for Managing Risk.

Bell Labs, Holmdel, New Jersey

Dr. Robert D. Lake

Dr. Margaret Judith Doran

George J. Hudak

David A. Britman

Ann D. Wright

Gus de los Reyes

Upendra Chivukula

Dr. Blake Patterson

Dr. Howard H. Helms

Dr. Behrooz Khorramian

Julie Strachie

David B. Demyan

Valerie Mehlig

Federal Systems, Greensboro, North Carolina

James H. Everett

Robert S. Doar

David L. Hall

Teresa B. Tucker

Clydy D. Gann

Doug H. Weeks

Russell M. Pennington

J. Gil Jasso

Albert T. Mankowski

**Office of Assistant Secretary of the Navy
(Research, Development & Acquisition)
Product Integrity Directorate, Washington, DC**

Willis J. Willoughby, Jr.

Douglas O. Patterson

Edward L. Smith, Jr.

George E. Maccubbin

Louis C. Gills

Joseph G. Cady

General Electric, Arlington, Virginia

William A. Finn

Elwood P. Padgett, Jr.

Front End Process

1

Introduction

To the Reader

The Front End Process is a process to conceive, define, and evaluate the mission and to develop requirements for a new or updated system. It involves three key Transition from Development to Production templates:[1][2] Design Reference Mission Profile, Trade Studies, and Design Requirements. These templates cover activities prior to design. Tasks associated with these templates occur iteratively throughout the Front End Process.

The templates, which reflect engineering fundamentals as well as industry and government expertise, were first proposed in the early 1980s by the Defense Science Board, under the chairmanship of Willis J. Willoughby, Jr. Their intent was to encourage everyone who is involved in the acquisition process to become aware of these templates and actually use them on the job.

This part on the Front End Process is one of a series of parts in this book written to help defense contractor engineers, government program managers, and contract administrators use the templates most effectively. This book is meant to be a stand-alone reference and textbook for related courses.

Clustering several templates makes sense when their topics are closely related. For example, the templates in this part interrelate and occur iteratively within the Front End Process. Other templates, such

---

[1]*Transition from Development to Production.* DoD 4245.7-M, September 1985.

[2]Best Practices: *How to Avoid Surprises in the World's Most Complicated Technical Process.* Department of the Navy (NAVSO P-6071), March 1986.

as Design Reviews, relate to many other templates and are thus dealt with in different parts.

The nineteenth century ironclad vessels, the *Monitor* and the *Merrimack*, and the modern twentieth century SSN688 submarines and the B-1 Bomber faced the same problem: How to develop and deliver sophisticated military systems using unproven technologies and sluggish procurement and acquisition systems. Defining the need, developing requirements, and then developing and producing the system may take 15 years. When the systems are finally delivered, they are already outdated. The cycle repeats with little hope of improvement.

Improvements in the Front End Process will pay off. Most systems acquire their life cycle costs from decisions made early. Some studies show that two-thirds of the projected life cycle costs are built in at the planning and design stages. An industry executive recently stated by the time a system has completed the design stage, 85% of the projected life cycle costs have been built in. These early decisions must consider producibility, installation, deployment, and maintenance in order to keep the life cycle costs at a minimum.

Historically, the conceptual, demonstration, and development phases of a new or updated system focused almost entirely on the development of the system, not on manufacture, deployment, and maintenance. Only later were estimates made about how much the new system would cost to manufacture and support.

Keys to Success

Table 1 gives risks and best practices for the design reference mission profile, design requirements, and trade studies.[3] These risks and best practices will be described more fully in the Procedures chapter.

The Packard Commission

A panel of experts, chaired by David Packard, was convened by President Reagan to identify critical issues in defense management. The Packard Commission report[4] identified several problems for management focus in controlling costs and schedules and ensuring perfor-

---

[3]*Best Practices: How to Avoid Surprises in the World's Most Complicated Technical Process.*

[4]Packard. *A Quest for Excellence. Final Report to the President by the President's Blue Ribbon Commission on Defense Management.* Washington, D.C.: June 1986.

TABLE 1   Front End Process Risks and Best Practices

Risk	Best Practice
Mission environment is defined using military specifications only	Mission environments are defined according to conditions of operating, handling, storing, transporting, and training
Contractor works in vacuum to interpret the system performance requirements	Contractor and government agree on complete mission profiles for the system's life cycle
Mission profile emphasizes just a few key tactical functions	Design reference mission profile is a composite of all functions and environments
Mission profile defines operational requirements only	Mission functional and environmental profiles are included in request for proposals; system functional and environmental profiles are the basis for design requirements
Schedule pressure prevents flow down of design requirements to lowest technical level	Requirements are allocated and verified at the lowest technical level
Design requirements do not flow down to subcontractors and suppliers	Prime contractor communicates design requirements early to subcontractors and suppliers
Design requirements are based on unproven state-of-the-art technology	State-of-the-art technology is used only when trade studies show it is cost-justified
Design requirements change as the design evolves	Design requirements are evaluated, approved, and baselined before full scale design begins
Trade studies are done only once	Trade studies occur iteratively to evaluate requirements and alternative configurations
Trade studies are poorly planned, conducted, and reported	Trade studies are systematically planned, conducted, and reported

mance, quality, and reliability. Many of the problems they identified occur during concept exploration in the Front End Process; similarly many of their recommended solutions take place during the Front End Process. To overcome these problems, the Packard Commission recommended changes in the acquisition process based on their studies of successful commercial programs from IBM, Boeing, AT&T, and Hughes. Although these commercial programs are as big and complex as major weapon systems, they took about half as long to develop and cost much less than major weapons systems. Below are some of the problems and suggested changes.

Problems in defining the mission profile

Two of the problems the Packard Commission identified concerned unsatisfactory methods for defining the mission profile for a new or upgraded system. The commission called these unsatisfactory methods: *user push* and *technology pull*.

User push. In this method, the Armed Services describe what new systems they need to overcome the present inadequacies or meet a new threat. But they often are not actual users and often lack knowledge of the cost and schedule implications. As a result, "goldplating" often occurs—features are included whose cost exceeds their operational value. Also, tradeoffs between performance and cost are made late in the process, if at all.

Systems benefit from more input from users in the planning of new systems. Customer feedback must be built into the Front End Process. Otherwise, delays will occur through time spent defining the wrong or incomplete system.

All too often the contractor develops a weapons system with too little regard for user requirements. When the system is deployed, the users find the system is late, does not perform as required, is difficult to maintain, and lacks training manuals or programs. These problems can be eliminated if, during concept exploration, actual users help define the system requirements. Early user involvement helps minimize "nice-to-have" features that add nothing to the system. The Ford Motor Company asked car buyers what features and requirements they wanted. The results were the well-received Sable and Taurus automobiles.

Technology pull. In this method, available or promised technology is the impetus for describing a need for new or upgraded systems. This method may lead to significant improvements, e.g., radar and jet engines, but it often leads to goldplated features and unproven technology.

Unproven technology and components should be used only when there is a cost advantage and benefits clearly offset the associated risks. One way to prove the advantages is to build and test prototypes using the new technology early in the development phase. Operational tests must be combined with developmental tests to uncover deficiencies before full scale development is started.

Competition to refine the need and the design requirements

Defense contractors assemble proposal teams to describe the functions and environments for a system which meets the customer's needs.

Competition has many good aspects, but it also encourages unrealistic estimates of cost and schedule.

Along with performance (i.e., speed and endurance), cost and schedule are key aspects to emphasize. Cost overruns may occur because of overstated specifications, untested and unreliable technologies, and high maintenance and supportability costs. Cost, schedule, and performance should be included in the requirements.

Competition to evaluate the requirements and propose alternative configurations

If the requirements are frozen before they are fully evaluated and refined, defense contractors are discouraged from carrying out trade studies to uncover effective tradeoffs between performance and cost. The use of trade studies and risk analysis is crucial in today's complex systems. Each requirement or parameter must be weighed with respect to the function it performs in the complex system with key items such as cost, schedule, reliability, producibility, and supportability examined early in the concept or development phases of the project. Risk analysis is a form of cost-benefit analysis to optimize performance, reliability, producibility, and supportability.

Often, contracts are awarded to the lowest bidder, who may not be the most technically sound. Also, many contractors underbid, hoping to negotiate performance tradeoffs or recover understated costs through engineering change orders. To prevent these problems, fixed-price contracts are more common today than previous years in which cost-plus contracts were typical. But fixed-price contracts make it all the more important to describe the mission profile and the design requirements appropriately. Otherwise, there is little flexibility to make effective tradeoffs.

Early prototyping

Early prototyping with extensive testing will help reduce the costs of a system. In many corporations, the development, testing, and production organizations do not communicate with each other until the product is ready to be handed off to the next organization. The testing of a system must begin early in the development phase using prototypes. Prototypes minimize risks, reduce costs, and improve reliability and maintainability.

Today's trends toward concurrent engineering (also called simultaneous engineering or concurrent development) contrast with the slow and costly methods of the past. With concurrent engineering, design and development work with manufacturing, deployment, and maintenance early in the concept phase to address the customer's requirements without sacrificing quality or cost. Companies that tackle downstream issues early have been more successful than those that use the traditional approach.

Small staffs with clear command channels

The successful commercial programs, which the Packard Commission studied, had small, experienced staffs, with a short, defined chain of command, and limited reporting from the program manager to the chief executive officer. In addition to clear objectives, successful commercial program managers usually had clear authority and responsibility to resolve conflicts and responsibility for producing a high-quality system within cost and schedule.

Having small, experienced staffs on both the contractor and military sides of the acquisition process, with total project responsibility for the system, will shorten the acquisition cycles. Government and corporate team members should have enough collective experience in all facets of project and system management to make informed decisions quickly.

Short and stable schedules

Short and stable schedules keep projects on track with minimum risk to the program. Baselines established for requirements and designs lend stability to programs. Short schedules provide better tracking. The systems engineer or systems designer can make required changes immediately, rather than one or two years into the program, when many changes may be required to adjust for one item's failure.

Short schedules must be realistic, with recognition for meeting milestones, especially the early ones. Overly optimistic schedules with punishment for lateness encourage managers to hide problems with costly consequences. Project managers must ensure the project meets deadlines, stays within budget, and achieves the desired performance and quality. Project management requires skills in integration, organization, planning, implementation, cost accounting, and decision-making.

Use of quality principles

The recent explosion of interest in quality has fostered new management theories. Corporations have adopted new principles to improve productivity and quality:

- analyze what is being done correctly to build on successes
- remove barriers that prevent people from doing their jobs right the first time
- use cross-organizational teams to develop systems cost-effectively

Effective teams can be more productive with fewer people, shorter schedules, and more reliable systems. Cross-organizational teams can yield significant results if they are properly managed and the team members are properly rewarded.

During the Front End Process, design and development must work with manufacturing early in the concept phase. This approach addresses the customer's requirements without sacrificing attention to quality or cost. Companies that tackle manufacturing issues early have done better than those that use the traditional approach. Corporations that use concurrent engineering and multi-disciplinary teams have been able to reduce costs, shorten intervals, and improve the system quality.

Benefits

Improving the Front End Process will pay off throughout the system's life cycle. The following benefits can be expected:

- design requirements developed from the strong foundation of a well-defined mission profile
- fewer engineering changes as a result of accurate and complete design requirements
- reduced risks with use of proven technology, defensive designs, and simulations at the concept stage
- realistic configurations chosen as a result of useful and timely trade studies
- costs reduced with early involvement of manufacturing, testing, and deployment in design for assembly, manufacturability, testability, etc.

A more effective Front End Process reduces the development interval, encourages early detection of incorrect engineering and management decisions, and leads to better customer acceptance.

Chapter

2

Procedures

The three key topics in the Front End Process are the Design Reference Mission Profile, Trade Studies, and Design Requirements. Tasks associated with these areas occur iteratively and continuously throughout the Front End Process. (Figure 1 shows the steps in the Front End Process.)

FRONT END PROCESS

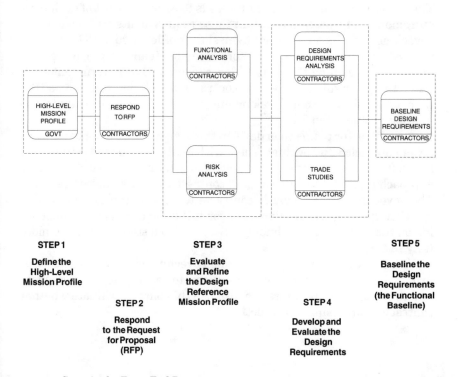

STEP 1

Define the
High-Level
Mission Profile

STEP 2

Respond
to the Request
for Proposal
(RFP)

STEP 3

Evaluate
and Refine
the Design
Reference
Mission Profile

STEP 4

Develop and
Evaluate the
Design
Requirements

STEP 5

Baseline the
Design
Requirements
(the Functional
Baseline)

Figure 1 Steps in the Front End Process.

The Front End Process begins with a need described in the mission profile and ends with a set of design requirements. As with other processes, a successful Front End Process requires

- a manager who has authority and responsibility
- a project plan, with measurable and observable objectives
- periodic reviews to make sure the output is of high quality

Stable (i.e., baselined) profiles and requirements are the basis of good practice. This chapter describes good practices during the Front End Process which are done by the government or by contractors. This part is meant to be useful to those who implement the practices and others who must recognize whether the organizations they deal with actually follow the best practices during the Front End Process.

Chronological Steps

The objective of the Front End Process is to establish a set of design requirements. The process to obtain these requirements starts when the government defines the specific mission profile (Step 1). The process proceeds when potential contractors assemble teams to propose an approach to developing the requirements (Step 2). The government evaluates the proposals and awards contracts to selected companies who then assemble internal development teams which will evaluate the mission profile (Step 3).

The contractor teams then develop the design requirements for their design approach and evaluate them with analyses and trade studies (or tradeoff studies) (Step 4). The contractors document (baseline) their approach and conduct internal and contract-mandated design reviews. The government then awards contracts to develop the concept (called the Concept Exploration Phase) or for full-scale development depending on the status of the technology (Step 5). Each step is discussed more fully in this chapter.

With a few exceptions that describe government activities, the steps primarily describe contractors' activities. However, program managers and contract administrators can also use this part to evaluate whether contractors are using best practices.

How the Steps Relate to Three Key Templates

Design reference mission profile

The specific mission profile is defined in Step 1 and evaluated in Step 3 with trade studies, risk assessment, and functional analyses to evaluate the design reference mission profile.

Trade studies

Trade studies are performed in Step 4 to evaluate the design requirements.

Design requirements

The specific design requirements are developed and evaluated in Step 4 with trade studies and design requirements analysis and then baselined in Step 5.

Step 1: Define the High-Level Mission Profile

The government must describe a need before requesting funding to respond to the threat or opportunity. A mission is selected to counter the threat or take advantage of the opportunity. The mission may be a one-time event such as the launching of a missile or may be a series of events which a submarine would encounter on a six-month voyage at sea.

Sources

The high-level mission profile is generated through studies conducted by:

- the government
- a consulting firm or a contractor, group, or consortium of contractors
- the government and a contractor jointly

Occasionally, the high-level mission profile must be constructed from a number of documents, e.g., planning documents, study reports, contractual supplements. The length and degree of detail grows as the system evolves through the life cycle phases. The mission profile should be a dynamic, rather than a static, document.

Key elements of a mission profile

Key elements of a mission profile include:

- define the mission: what, where, and when
- define potential mission deterrents
- characterize the performance, maintainability, reliability, safety, etc., and how these may affect the primary mission
- give reasons for the functional requirements and constraints
- describe the environments for operating, storing, handling, and transporting

Objectives and operational requirements

The mission profile is first proposed in response to a need. The need may be reactive to a perceived weakness or inability to counter an enemy threat. The need may also be proactive—a vision or strategy for exploring outer space, for example. Operational requirements are defined which ultimately are translated into quantitative design requirements that meet the mission needs. (See Step 4 for details on design requirements). The mission profile, which may be incomplete at first, must be described in detail to correspond to conditions in all operational scenarios. It is left to the contractor to specify how a system will react to the threat. The output of Step 1 is a set of mission objectives and operational requirements that are clear, concise, and focused.

The operational requirements should be based on the best available data. The requirements should specify functions and constraints to be met, rather than actual values, in order to allow flexibility in choosing the best configuration.

Examples.   Table 2 shows illustrative mission objectives and operational requirements for a hypothetical solar probe mission.

Detail in a mission profile.   The length and degree of detail increases as the mission profile is developed. The profile gives the reasons for the constraints and functional requirements. It also discusses various considerations that may affect the design concept. For example, the payload experiment precluded the use of a nuclear power source in the solar probe mission.

Quantitative values, at least maximum and minimum values, are often given in a matrix which lists the operational requirements for various times during the mission. Table 3 illustrates a matrix for the

TABLE 2   Mission Objectives or Operational Requirements

Mission objective or operational requirement	Example
Objective	Explore the Sun and its atmosphere, measuring solar wind, magnetic fields, corona electron densities, solar flares, and cosmic rays
Destination	Approach within 0.3 astronomical units (AU) of the Sun
Reliability	Spacecraft, instruments, and payload perform according to specifications at a 95% probability for the first six months and a 90% probability for the second six months
Instrumentation requirements	The corona meter can distinguish brightnesses 10^{-13} times that of the solar disc
Trajectories	Launched on a heliocentric orbit with nominal perihelion of 0.3 astronomical units
Launch site	Cape Canaveral
Weight limits	Spacecraft should weigh approximately 400 pounds
Configuration limits	Use existing tracking and ground communications facilities
Power sources	Payload experiments preclude use of nuclear power sources (capabilities will be decided in trade studies based on experiment requirements and other functions and constraints)
Storing, transporting, and handling requirements	Must withstand vibration during transit in an over-the-road vehicle; must withstand storage and handling in temperatures which may reach 60°C

solar probe mission. Table 3 has more detail than Table 2. Mission events and operational requirements are described for prelaunch, three stages of initial ascent and transsolar insertion, and data collection while in solar orbit in front of and behind the Sun.

Table 3 also shows that operational requirements may or may not differ in various phases. For example, the acoustic level in decibels relative to a stated reference sound level varies from 100 decibels (db) in prelaunch to 140 db at launch and then to 77 db during the mission. Some operational requirements do not differ from phase to phase. For example, the controlled temperature is 14° to 26° Celsius throughout.

From the operational requirements, a set of performance values can be defined for the system. Later, these system functions are allocated to subsystem functions and components in an iterative derivation process. Step 3 describes the iterative process more fully.

TABLE 3   Matrix of Environments and Phases (Solar Probe)

Item	Prelaunch	Launch & Stage I	Stage II	Stage III	In front of sun	Behind sun
Start Time		0	5.9 min	9.2 min	2 hr	113 day
Max. lateral shock (g in 10 ms)	free drop 1 in.	2.5	3.0	0		
Temperature (°C)	14° to 26°					
Acoustic (db)	100	140	92	77		
Sinusoidal vibration (g. rms)		10	8	4		
Random vibration (g²/cps)		.15		.10		
Longitudinal load (g)	0	6.9	2.3	23	0	
Thermal radiation (BTU/hr-ft²)			540		4930 max	
Solar wind (plasma flux, max.)					Flux 1.2×10^8 particles/sq cm sec; Velocity 500 km sec; Energy .5 to 1.0 keV	

Mission Profile Matrix. The mission profile matrix in Table 3 describes the environments and phases of the hypothetical solar probe mission.[5]

Step 2: Respond to the Request for Proposals(RFP)

Issue request for proposals

After the high-level mission profile has been refined, the government issues a request for proposals to explore the feasibility of the concept. Typically, many contractors prepare a response; the number depends on the size of the program.

---

[5]National Aeronautics and Space Administration. *Introduction to the Derivation of Mission Requirements Profiles for System Elements.* Washington, D.C.: NASA, 1967, p. 9.

Form the proposal team

When a company decides to respond to the RFP, management selects a program manager who then selects the rest of the team. Teams are usually selected well in advance of the formal RFP in order to plan and prepare to respond within the typical 45-day or 60-day period. The primary objective of this team is to propose a plan that meets the performance, cost, and schedule requirements of the mission within the defined constraints.

Proposal team members. Team members typically include the Program Manager, Proposal Manager, Marketing Manager, Pricing Manager, Contract Administration, Subcontractor Administration, Legal Representation, Technical Manager, Systems Engineers, Design Engineers, Manufacturing Representative, Quality Assurance Representative, Logistics Support Representative, etc.

The core team, made up of the contractor's Program Manager, Proposal Manager, Marketing Manager, Technical Manager, and Pricing Manager, schedules all major activities, including the drafting of the proposal, pricing, costing, design reviews, management reviews, and customer contacts.

Develop the proposal. The proposal team develops and evaluates system requirements and proposed designs using trade studies and analyses (which will be discussed in Step 3 and Step 4). The team develops high-level hardware and software requirements (which is discussed in Part 4, Software Design and Software Test). In addition, the team develops plans for reliability, maintainability, system safety, manufacturing, quality assurance, configuration management, subcontractor control, etc.

Proposals often follow a *storyboard* format, rather than a text format. That is, themes, sketches, graphics, and bullet items are used to describe the proposed system.

Receive a short-term contract for concept exploration

Of the many contractors who respond with a proposal, usually a small number (perhaps two) receive a short-term contract to analyze the design requirements and perform trade studies. (The RFP usually says how many winners will be selected.) Companies that develop the design reference mission profile and the design requirements usually do not receive an additional contract to develop or implement the system, according to the conflict of interest clause of the Federal Acquisitions Regulations (FAR). This clause aims to prevent non-competitive bidding. If the company that defines the problem can also design the solution, that company may unfairly define the problem in its favor, with other companies at a disadvantage. (With different companies responsible for the concept exploration, development, and production of a system, documentation, and transition plans become all the more important. For details, see AT&T's *Transition Plan Reference Guide.*)

Before awarding the short-term contract, the government may ask some respondents who are in the competitive range to answer questions in writing, arrange a site survey, or have an oral presentation of their method of analysis with discussion of the written answers to the questions.

Contractors usually request informal debriefings from the customer to discuss proposal strengths and weaknesses and any changes in cost, schedule, performance, or risk. Customer meetings are then held often to make sure the customer and contractor agree on priorities and objectives.

Form the development team

The development team has representatives from design, manufacturing, maintenance, and other key organizations who have been associated with the proposal team. The team develops the project plan, and sets the project's goals, strategies, and cost objectives.

The team emphasizes *what* should be done, not how to do it. Table 4

TABLE 4   Responsibilities: for a Development Team

Team Member	Responsibility
Project Team Leader	▪ Plan and manage the project ▪ Select the project team ▪ Coordinate activities and interfaces to achieve the project's objectives
Systems Engineering	▪ Evaluate mission profile, using analyses and trade studies ▪ Develop and evaluate design requirements, using feasibiity studies, functional analyses, and trade studies ▪ Develop high-level system and subsystem architectures ▪ Develop hardware, software, and firmware requirements ▪ Manage baselining and changes ▪ Investigate technologies and model high-risk areas
Product Design	▪ Plan and manage all aspects of the design effort ▪ Translate high-level architectures into detailed data, drawings, and documents ▪ Help manufacturing optimize manufacturing processes
Product Manufacture	▪ Plan and manage all aspects of manufacturing ▪ Coordinate and integrate activities with design and deployment ▪ Procure materials, equipment, and resources for manufacturing
Logistics	▪ Plan and manage all aspects of the deployment effort ▪ Procure materials, equipment, and resources for operation and maintenance

is a sample of team members and tasks which may be done during the Front End Process or in some cases early in the design process. The team's structure will vary with different environments.

Step 3: Evaluate and Refine the Design Reference Mission Profile

The mission profile is composed of two parts:

- The *functional mission profile* provides a time-series description of the functions the system will perform. The functional mission profile emphasizes what the system must do in every potential situation in the total envelope of environments. Examples of functions include accelerating to maximum speed or hitting the target with a 90% probability.

- The *environmental mission profile* describes the conditions to which the system will be exposed, including being transported and stored for varying durations and at varying temperatures and humidities, loaded and unloaded, stored again, and so on. It is important to describe these environments because transporting and storing may actually be more stressful than the operational environment.

The *design reference mission profile* is a composite or a reasonable sample of all the potential environments and functions a system will encounter during its total life cycle. It does not attempt to match any particular environment.

The objective of this step is to describe how to use the mission profile to refine *system* functional and environmental profiles prior to the development of requirements. These profiles are then used to develop design requirements and alternative configurations to satisfy mission objectives (as discussed in Step 4).

Usually, the government includes high-level functional and environmental profiles in the RFP. The proposal team uses them to establish *system* functional and environmental profiles that are the basis for the system and subsystem design requirements which will then be compatible with the operational requirements.

Examples of system profiles

A system functional profile might include:

- structural support
- navigation and guidance
- command and sequencing

- communications and telemetry
- measurements and operations

A system environmental profile might include:

- pressure, e.g., gravitational force at various mission times
- temperature and humidity, e.g., values for actual use as well as those for transporting and storing
- vibration and acoustic noise, e.g., maximum values during use, transporting, and storing
- trajectories and altitudes, e.g., launch and flight altitudes to deliver payload

System architecture

Refining the system functional and environmental profiles requires a disciplined methodology which will be outlined in this chapter. The methodology is used to arrive at a system architecture that meets the mission profile within cost and schedule constraints. An architecture is a description of the system, its parts, and how they interrelate. The system is developed first at a high level, and then at progressively finer levels of detail for the subsystems and components. This top-down approach sets a framework for developing the lower-level designs and interfaces between the various modules in the system.

Advantages.   The advantages of this structured approach are:

- design, verification, and testing are addressed from the system level to the component level
- each level is integrated from the top down
- requirements are integrated
- interfaces are considered at the outset, not as an afterthought

See Figure 2 for an illustration.

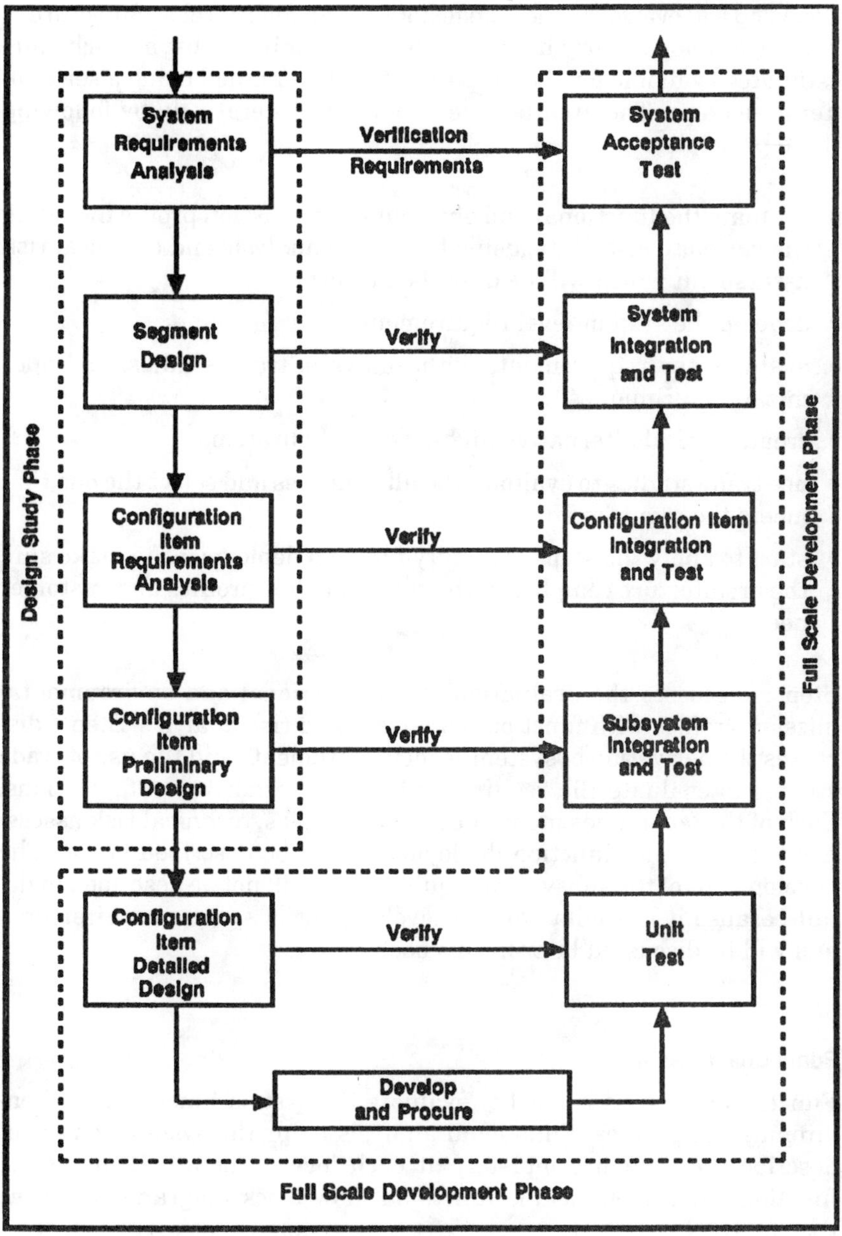

Figure 2 Top-down design and bottom-up verification.

Choosing and evaluating the architecture.   The high-level architecture is usually developed early in order to have preliminary cost and schedule estimates available for the proposal. It is then refined using a series of iterative steps. The architecture is developed iteratively by following these principles:

- evaluate the functional and environmental mission profile including the various constraints, using functional analysis and technical risk assessment which will be described below
- develop the system-level requirements
- evaluate the requirements with analyses, trade studies, and optimization techniques
- create several alternative high-level configurations
- use trade studies to evaluate the alternatives and select the one that meets the requirements
- start to apply the steps of quality function deployment to make sure the architecture complies with the established profiles and customer needs

Step 3 discusses the evaluation of the functional and environmental mission profiles with functional analyses and risk analysis. Step 4 discusses the derivation of system-level requirements and the use of trade studies to evaluate the requirements and alternative configurations. Each of the techniques mentioned (trade studies, technical risk assessment, and quality function deployment) will be described below. The development of the subsystem requirements will not be described in detail because it is similar to the development of system requirements and will be discussed later in this book.

Functional analysis

Functional analysis is used to evaluate the system functional and environmental profiles. Functional analysis helps the systems engineer describe the system completely at each level of detail. Examples of functional analyses include functional flow block diagrams and time line analysis.

Examples of analyses.   A functional flow block diagram displays the primary system functions and breaks them down into subfunctions at

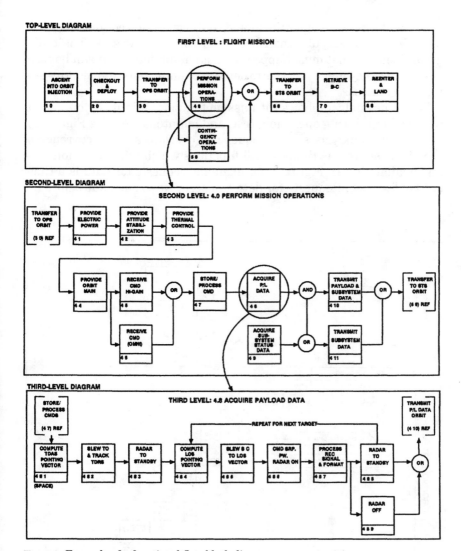

Figure 3   Example of a functional flow block diagram.

finer and finer levels of detail until discrete tasks are identified.(See Figure 3.)[6]

Functional flow block diagrams show the time sequences of all system functions, illustrating which occur in series, in parallel, or on alternative

[6]Defense Systems Management College. *Systems Engineering Management Guide.* Contract Number MDA 903-82-C-0339. Fort Belvoir, VA: DSMC, October 3, 1986, p. 6-4.

paths, and which are operational and contingency procedures. They also identify inputs, outputs, interfaces, and control processes. These diagrams identify what must happen, without assuming how it will happen. In doing so, they make it easier to develop requirements and identify useful trade studies.

Time line analysis shows the mission functions minute by minute, or hour by hour during operation, test, and maintenance. (See Figure 4.)[7]

Time line analysis shows sequences, overlap, and concurrency of functions, as well as time-critical functions which affect reaction time and availability. In Figure 4, for example, requests for data is a time-critical function. Requests for data can occur up to one hour before the

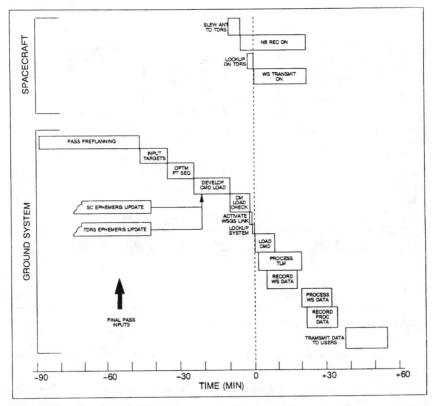

Figure 4 Time line for a flight mission.

---

[7]Defense Systems Management College. p. 6-5.

start of data collection. Data must be processed and given to the users within one hour of acquisition.

In addition to operational functions, time line analysis is useful in describing the key sequences, overlap, and concurrency of maintenance and support tasks. For periodic maintenance of a distiller, for example, the blowdown pump should be lubricated before the oil and filter are replaced; the oil and filter can be replaced concurrently if necessary by two crew members.

Time line analysis can also point out the need for a trade study, for example, which alternative control system will give the precise timing the system requires.

Technical risk assessment

Technical risk assessment combines risk analysis (estimates of chance are assigned to the areas of potential failure) and risk management (the actions taken to minimize the risks). Technical risk assessment must consider technical and financial risks.

Risk analysis must start at the beginning of the project. The scope of this activity is proportional to the complexity of the program. Risks fall into four areas:

- programmatic
- technical
- production-oriented
- engineering

In the early stages of a program, the risks tend to be programmatic and technical. The technical risks are described from an operational standpoint in order to test them in a prototype. Risk analysis is an iterative process that attempts to identify potential problem areas, assess the effect of the risks, and generate alternative solutions to reduce these risks.

Steps in technical risk assessment.   The steps in technical risk assessment are:

- identify and assess the risks
- rank the risks
- develop plans to deal with the risks

With so many computer modeling packages, risk modeling can be done in the concept exploration phase or demonstration/validation phase to detect and minimize risks early. Early detection is important because it is easier and cheaper to make changes and correct errors when the system is being defined and explored than when prototypes or the actual system are tested.

Risk analysis programs range from short personal computer programs to large mainframe programs that simulate many alternatives. The choice of a program should be consistent with a company's design policy.

The risk models can be developed based on the results of trade studies, or conversely, can point out the need for a trade study. Technical risk assessment can be performed to decide whether to start new programs, to evaluate alternative technologies, or to decide whether to make or buy a component. (Technical risk assessment will be described in more detail later in this book.)

Step 4: Develop and Evaluate the Design Requirements

Design requirements are developed from the design reference mission profile. Contractors refine the design requirements to make them detailed and specific. The prime contractor ensures subcontractors and suppliers have complete design requirements when they are baselined.

The design requirements are evaluated with analyses and trade studies. Systems engineering acts as a liaison between the concept originators and the design and development organizations. Systems engineers have a variety of skills and backgrounds (e.g., electrical engineering, mechanical engineering, operations research, computer science). The systems engineers expand the high-level goals into technical requirements by describing the functional requirements and how the system will operate. In writing requirements, systems engineers consider technology constraints, size, quality, thermal aspects, reliability, supportability, producibility, legal and regulatory issues, and human factors. The systems engineers describe functional requirements and the interfaces, often using block diagrams.

The output of this step is a detailed set of requirements and specifications that satisfy the customer needs specified in the mission profile. These requirements and specifications are the key input for design and development.

Analyzing design requirements

Requirements must be defined before selecting the functional or physical configurations to make sure the configurations meet the requirements.

Design requirements are derived from the operational requirements in the mission profile. Well-defined requirements must be used throughout the development process to prevent missed schedules, cost overruns, and inadequate performance.

The contractor defines functional requirements for the system, including interfaces between subsystems. The contractor also defines requirements and constraints for each environment. Ideally, specific values or ranges of values are given, which are quantitative goals against which performance is measured.

Explicit vs. implicit requirements.   Some requirements are explicit in that they are directly traceable to the mission profile, e.g., reliability, cost, performance, and thermal requirements. Other requirements are implicit in that they come from good engineering practices which are usually formulated in design policies or guidelines, e.g., producibility, testability, and component use. Each requirement must be stated, verified, and evaluated for completeness, accuracy, and consistency with the definitions in the mission profile.

Attributes of well-defined requirements.   Below are some attributes of a well-defined requirement.

Specific, clear, and unambiguous.   Contains no vague terms; states what, where, when, and why (but not how).

Understandable.   Stated in everyday language with acronyms, assumptions, and prerequisites clarified; documented to the level of detail required for understanding without elaboration.

Concise.   Contains no unnecessary words.

Consistent.   Usage of terms is identical; conforms to standards; design requirements flow down from the prime contractors to subcontractors to achieve top-to-bottom consistency.

Stable.   Baselined and under change control.

Traceable.   Derived from the mission profile or from a company's design policies or guidelines; may be grouped into classes according to product families, physical realization (e.g., hardware, software, firmware), or duration (when adherence is measured); may use a numerical system to show relationships among requirements (e.g., requirement 4.1.1 was derived from requirement 4.1 which in turn was derived from requirement 4.0 in a top-down manner).

Testable.   Refers to operational events or data and test cases, which will be used to evaluate whether the product is satisfying the requirement (e.g., reliability is stated in terms of mean times between failures and mean time to repair).

Feasible.   Can achieve, produce, and maintain the requirement using the available resources and technology.

Examples of analyses

Requirements allocation analyses help identify how to decompose and allocate the system-level requirements to subsystems and components, e.g., hardware, software, personnel, technical manuals or facilities. Allocatable requirements include weight, power consumption, and reliability. Subsystems and components receive budgets which together add up to the total system requirement. The weight of the subsystems added together, for example, should be equal to or less than the system requirement for weight. (Some requirements, e.g., standards, are not-allocated. Each component must comply with the requirement as stated.)

The allocation should point out areas that are technically difficult or risky. The Requirements Allocation Sheet (RAS) is the primary documentation for requirements identification, especially for lower-level, non-time-critical functions which do not need functional flow block diagrams and time line analyses.

As the requirements are decomposed and allocated, they are stated in increasingly greater detail. At the system level, for example, a requirement for an orbiting spacecraft may be to "*transmit collected data in real time to a remote ground site.*" The requirement for the subsystem is much more detailed: "*Provide 10 MHz link at 17.0 GHz with 10 W effective radiated power for 20 minutes maximum per orbital revolution.*"

Thermal analyses determine the ranges of temperatures which could occur in the various mission scenarios. Techniques include simulations, thermal profiles, and finite element analysis. At this early stage, simulations are more timely and versatile than physical models, but may not be as reliable. The results are used to make sure the requirements are realistic for each key scenario.

Analyses are also done of the stresses in transporting, storing, and handling. Temperature and humidity changes, for example, in transporting, handling, and storing may be more stressful than the primary mission environment. Also, the system may spend more time in storage and in transit than on the primary mission.

Shock and vibration analyses are done for the various profiles. Simulations and mathematical modeling are the main techniques.[8]

---

[8]Don M. Ingels. *What Every Engineer Should Know about Computer Modeling and Simulation.* New York: Marcel Dekker, 1985.

Part 2, Design Policy, Design Process, and Design Analysis, describes procedures and rationale for useful design analyses, including simulation and modeling.

Using trade studies

A trade study is a formal decision-making method that can be used to solve any complex problem. Trade studies (also called tradeoff studies) are used to evaluate the design requirements and alternative designs. Trade studies are used throughout the development process to make cost-effective decisions.

Trade studies are used to:

- rank user needs in order of importance
- evaluate specifications for fuel, target, or defense data
- develop cost models
- identify realistic configurations that meet mission needs
- make designs producible, testable, reliable, supportable
- find manufacturable, testable, and maintainable configurations, with quality, cost, and reliability at the required levels

During the Front End Process, trade studies are used to arrive at the combination of requirements that will result in high quality, high functionality, and low cost. The design requirements may not necessarily be ideal, but they should give the best fit with the mission profile, the acceptable risk levels, and the available resources and technology. Proposed tradeoffs that commit the government to additional investments should not be considered unless there is a long-term benefit to the government. Cost-benefit analyses must be done to justify additional funding.

Iterative process of analyses and trade studies. Analyses and trade studies occur iteratively, rather than just once. After requirements are defined, alternative strategies and technologies are evaluated with analyses and trade studies. These evaluations refine requirements and show if alternative strategies violate any requirements. If analyses point out violations, trade studies can find root causes and possible solutions. The cycle continues to fine tune and balance needs with constraints.

Trade study methods. Trade studies occur at the system level and later at the subsystem and component level. Some subsystem functions (e.g., power sources) may be handled at the system level because they affect many other subsystems. Trade studies at the subsystem level study interactions among requirements. For example, the launch site for a solar probe is a major input into the choice of the thermal control subsystem. To make sure all critical interactions are found, each subsystem expert lists the major functional and environmental profiles for his or her area. These subsystem values are then checked to make sure they are individually and collectively within the range set for the system parameters.

Applying optimization theory in trade studies. Optimization theory is a branch of applied mathematics that helps the systems engineer find the "best" values or range of values for the requirements according to several criteria. Optimization methods take advantage of the mathematical structure of the problem to find the best values efficiently. One example is sensitivity analysis, which identifies "robust" design parameters that remain optimal even if the values of other parameters change. Other methods include linear programming (if the variables are linear) and multivariable search methods (if the variables are nonlinear). These methods have become more popular with the proliferation of high-speed computers. Several recent textbooks are good references for principles and examples of optimization theory.[9][10]

Identify a range of acceptable values. A range of acceptable values for each requirement is provided so that tradeoffs can be made. These values usually come from a synthesis of existing, modified, and original requirements. Alternatives are prescreened to eliminate those obviously unsuitable.

Selection criteria. The criteria for selecting alternative values come from the mission analysis and high-level requirements. The criteria should be quantitative, objective, and useful in comparing alternatives. Examples include effectiveness, cost, producibility, and technical risk. The selection criteria relate to the dependent variables in an experimental trade study and correlated variables in a correlational trade study.

The selection criteria may be weighted according to importance. For example, safety and speed at maximum power may be weighted above

---

[9]Ralph W. Pike. *Optimization for Engineering Systems*. New York: Van Nostrand Reinhold, 1986.

[10]Wolfram Stadler (Ed.). *Multicriteria Optimization in Engineering and in the Sciences*. New York: Plenum Press, 1988.

endurance and cargo capacity. Objective techniques such as paired comparisons are useful. (In paired comparisons, each criterion is compared to every other criterion. Examples: Is safety more important than speed? Is safety more important than agility? Is speed more important than agility?)

Experimental vs. correlational trade studies.   Ideally, trade studies are conducted according to the principles of good experimental design. In the trade study, two or more implementation strategies with different values of an independent variable such as size, shape, or weight are compared using a dependent variable such as speed, fuel use, or reliability. Tradeoff decisions are more meaningful when extraneous variables are kept constant or otherwise controlled.

A correlational study may be done instead of an experimental study due to cost, technology, or time constraints. Correlations do not give information about causes, but strong correlations allow conclusions about associated variables (e.g., as size increases, fuel use and thus costs increase). The cause of the cost increase, however, may be a third variable that actually leads to increased fuel usage (e.g., weight).

The sample size must be large enough to draw conclusions at the desired confidence level. Confidence levels are measures of the accuracy or believability of an estimate, expressed as a percentage. For example, the 95% confidence level means a 5% probability of a wrong decision. When an estimate's accuracy is questionable, the analysis must be refined to reduce uncertainty.

Choosing the appropriate confidence level means making a tradeoff between level of certainty and the cost to attain it. Costs tend to rise as the confidence level rises. The benefits of going from a 98% level to a 99% confidence level, for example, may not justify the added costs.

The results of the trade study can be used to evaluate the design requirements. The tradeoff analysis process allows the engineer to make reasonable decisions taking into account the rigidity of the requirement goals, the confidence levels of the trade studies, and the interdependencies among the requirements.

Trade study reports.   Government offices use a standard format to report the trade study results. During the Front End Process, the trade study reports concentrate on feasibility, system effectiveness, and life cycle cost. Later, the trade studies focus on alternative designs, components, and equipment. These reports are used as input to internal and contract-mandated design reviews. The trade study report summarizes the characteristics of alternative configurations and documents the rationale for selecting among the alternatives.

Limitations of trade studies.   The trade study's quality depends on the quality of the input data. The results will be unreliable if the input data comes only from peoples' memories, estimates, or "best guesses."

The number of alternatives that can be evaluated is limited by available resources, time, and technology. Configurations with many alternatives become very complex with many interactions. If the items studied are independent (e.g., if snaps are better fasteners than screws regardless of whether the material is aluminum or fiberglass), many smaller studies are preferable to one large study. However, if the interactions are significant (e.g., if snaps are better fasteners with fiberglass and screws are better fasteners with aluminum), one large study that compares the interactions is preferable to many smaller studies.

Use of a trade tree.   *Trade trees* can be used to manage tradeoff analyses with many options. Trade trees are similar to decision trees in that decisions are made in a step-by-step order based on risks, costs, availability, etc.

For example, the trade tree in Figure 5 shows the decisions to be made in choosing lifeboats for 200 people. The first decision is the capacity—how many people each lifeboat should carry. A small trade study is done first to decide among five capacities: should a lifeboat handle 20, 40, 50, 100, or 200 people.

The result of the first trade study in which option 3 (50-person capacity) is best is then used in the second trade study which decides which lifeboat type is best. Aluminum and fiberglass were among the 20 lifeboat types studied.

Then, the second study's result (option 8: aluminum) is used in the third trade study which decides which of 10 attachment mechanisms is best.

Finally, the best attachment mechanism (option 4) is used in the fourth trade study to decide among four mounting locations on the ship.

Thus, a trade tree divides a large study into many smaller studies. Figure 6[11] shows that doing four small trade studies instead of one large study reduces the comparisons from 4,000 to 39, which is better than a 100 to 1 reduction. With the large study, each interaction is evaluated in a complete factorial design; thus the number of comparisons is the *product*: $5 \times 20 \times 10 \times 4 = 4,000$. With the trade tree, interactions between alternatives on different branches are not evaluated; the number of comparisons is the *sum*: $5 + 20 + 10 + 4 = 39$.

---

[11]Defense Systems Management College. p. 8-15.

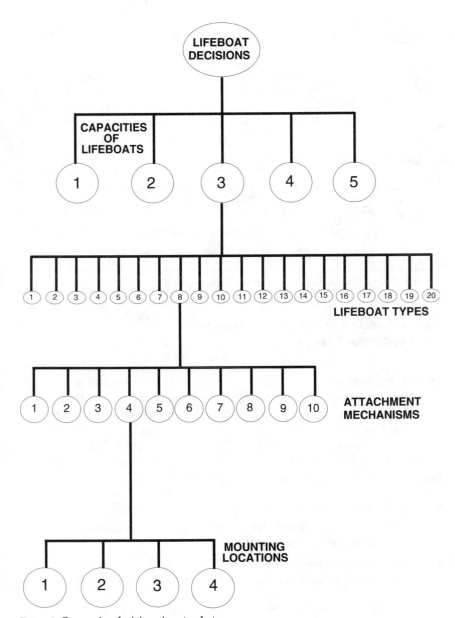

Figure 5 Successive decisions in a trade tree.

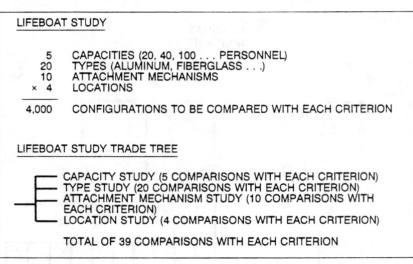

Figure 6 A trade tree reduces the number of comparisons.

Step 5: Baseline the Design Requirements (the Functional Baseline)

A *baseline* is the starting point from which all future changes to the system requirements are controlled. Baselining is a formal process for agreeing on a requirement, design, objective, or plan and then documenting the agreement at a specific time. Any change must formally alter the baseline.

Complex systems need standard, formal change procedures to reduce costs, errors, and rework, and to improve productivity. An effective baselining and change procedure must have:

- a formal tracking system

- a disciplined methodology for controlling and documenting the configuration of the system, subsystems, etc.

- reviews to obtain agreement on modifications to functional and environmental specifications, internally and with the customer

Many organizations "freeze the requirements," allowing no additional modifications while costs are determined, trade studies are conducted, or the design process starts.[12] Freezing the requirements works well for

---

[12]J. M. Juran and F. M. Gryna (Eds.). *Juran's Quality Control Handbook*. 4th Ed. New York: McGraw-Hill, 1988, p. 13-66.

some organizations, minimizing schedule slips, cost overruns, and missed targets. The best procedure is to set a goal of minimal changes after baselining. These relatively few changes are made to adapt to changes in customer needs or evolving technology.

The output of this step is a *functional baseline*—a set of documents defining what the system is expected to accomplish during the system life cycle and a first cut at a high-level system architecture. The functional baseline represents the first step in optimizing the system and building an integrated quality system.

Design reviews

After the needs and mission profile have been established, representatives from systems engineering, electrical design, physical design, software, firmware, manufacturing, testing, and maintenance participate in internal design reviews to ensure customer needs have been identified and to prepare for the contract-mandated Systems Requirements Review with the customer. Part 8, Design Reviews, describes procedures, data packages, analyses, and metrics for effective design reviews.

Internal design reviews are held to obtain internal agreement within the project team on functional and environmental design requirements and high-level allocations. These internal design reviews precede the contract-mandated design reviews.

Obtain customer approval

The contractor and customers should interact continuously throughout the Front End Process. One technique to ensure the functional and environmental specifications agree with high-level mission requirements is Quality Function Deployment (QFD), which appears to be effective. In QFD, both the contractor and the customer participate to make sure the design requirements actually meet customer needs. The QFD methodology is explained in *Better Designs in Half the Time.*[13]

QFD requires the contractor to focus on the mission profile and the design requirements. The contractor and the customer must reach consensus on the requirements. The QFD steps are:

- identify customer needs ("listen to the customer's voice")
- rank the needs in order of importance (as perceived by the contractor and verified by the customer)

---

[13]R. King. *Better Designs in Half the Time.* Methuen, MA: GOAL/QPC, 1987.

- identify technical requirements that meet customer needs
- map the technical characteristics to the needs
- obtain customer agreement on functional and environmental requirements, first in informal reviews and then in formal, contract-mandated reviews

Measures of effectiveness

As with other processes, the Front End Process benefits from efforts to achieve continual improvements in quality, timeliness, cost-effectiveness and other goals. To do this, organizations need to allocate resources to review the outcome of each key milestone and the process under which the outcome was achieved.

The key outcomes in the Front End Process are documents describing the design reference mission profile and the design requirements. Possible *outcome metrics* that measure quality, performance, timely availability, and customer satisfaction are:

- results of customer satisfaction surveys
- results of performance tests conducted in later phases
- timeliness measured by documents available on time

Process metrics are related to outcome metrics, but give immediate feedback on the quality of the development process. They can be used to track progress and identify areas for improvement. Examples of process metrics include:

- percentage of requirements that meet criteria for accuracy, completeness, quality, quantitative values, and timeliness
- staff months required to make major changes to baselined requirements
- percentage of design reviews that follow best practices (e.g., entry and exit criteria, timely delivery of data packages, sufficient preparation time)

Outcome and process metrics should be refined based on actual use. As the project team collects historical data on outcome and process, they can set guidelines and objectives based on their own situation and past performance.

**Receive long-term contract for
demonstration/validation or full scale
development**

After the customer has approved the functional and environmental design requirements, a long-term contract is awarded to one or two contractors. This contract may be for additional demonstration and validation on a complex system or for full scale development on a less complex system. During demonstration and validation, the contractor conducts trade studies and analyses to verify the preliminary configurations.For more details, see Part 2, Design Policy, Design Process, and Design Analysis.

3

Application

In this chapter are examples to illustrate some of the procedures and principles discussed in the Procedures chapter. Examples and mini-case studies show how the design reference mission profile and the design requirements are developed and evaluated with analyses and trade studies.

What Prompts a Mission Profile

Below are examples of how a mission profile arises—from a threat, an opportunity, or user needs.

Mission profile based on analysis of an enemy threat

Early in 1989, a congressional panel of experts urged that the U.S. antisubmarine warfare effort counter a new Soviet threat.[14] Soviet submarines built in the last two decades could dive deeply and move quickly, but were very noisy. These noisy submarines were easy to detect with passive sonar devices which measure the sounds emitted from the submarine's engine and propeller. The new Soviet attack submarines, however, are very quiet and much harder to detect. The panel concluded that the quieter Soviet submarines threaten the survivability of the carrier task forces which carry troops to Europe or other areas in the event of war. New approaches to antisubmarine warfare are needed because enhancing the sensitivity of the existing passive sonar system will result in diminishing returns.

---

[14]C. Norman. "Quiet Subs Prompt Concern." *Science,* Vol. 243, March 31, 1989, pp. 1653–1654.

Experts are exploring active sonar which has disadvantages (e.g., alerting the target submarine and being effective only at relatively short ranges) and nonacoustic detection methods (e.g., magnetic anomaly detection in which sensitive magnetometers on aircraft detect slight distortions in the earth's magnetic field caused by a submarine's steel hull). Magnetic anomaly detection is also only effective at short ranges.

The panel concluded that improved antisubmarine warfare should be a top priority activity in the Department of Defense. Existing technology should be improved and new technologies introduced as soon as they become available.

Mission profile based on analysis of a unique opportunity

In the late 1960s and early 1970s, experts recommended that NASA take advantage of a unique opportunity to explore the outer planets. With the once-in-176-year alignment of the planets, a spacecraft could use the gravity of each planet to bend its path, speed it up, and propel it towards the next planet. Thus it could arrive at Neptune in 12 years rather than the 32 years it would take otherwise.[15]

Congress, however, ignored the experts' advice and approved money for Voyager 1 and 2 to visit only two of the four outer planets: Jupiter and Saturn, but not Uranus and Neptune—a five-year, rather than the original twelve-year program. With the support of high-placed officials, the Voyager program survived attempts to end it. Then, the Jet Propulsion Lab figured out how to send Voyager 2 to the two additional planets for a total cost of less than the annual cost of the shuttle program. The program has been a tremendous success. The mission to Neptune, planned 12 years ago, was almost flawless in execution, sending photos and data 2.8 billion miles to earth.

Mission profile based on user needs

The F-16 fighter plane was initially conceived in the late 1960s by a group of Air Force pilots. They proposed a lighter, cheaper plane in contrast to the expensive, high-technology fighter planes like the F-14 and

---

[15]S. Begley and M. Hager. "A Fantastic Voyage to Neptune." *Newsweek,* September 4, 1989, pp. 50–56.

F-15 with their performance and reliability problems. The pilots expressed their need for a more maneuverable fighter plane for fights at close range. In their experience, close-range fights with enemy planes occur much more often than long-range fights. The F-16 benefited from user input from its conception and throughout its development and deployment. Requirements were initially stated as goals rather than as rigid specifications. The users who ultimately flew the plane gave it excellent ratings.[16]

Recently, there has been interest in a mobile satellite communications system capable of providing both voice and data services to a large number of users in the Continental United States, Canada, and Alaska.[17] The mobile station is similar to existing terrestrial systems providing mobile-to-mobile, mobile-to-fixed-station, and fixed-to-mobile modes of communications. The advantages of the mobile system are:

- multiple services to a large geographical area (e.g., mobile and rural telephone, radio dispatch, electronic mail, emergency communication for disaster relief)

- a link to existing communications networks such as Public Switched Telephone Network

- future expanded use to thousands of voice channels

With growing interest in satellite communications, NASA wants to accelerate the mobile satellite program to test its satellite, with its proposed space station in the 1990s. NASA, working with Hughes Aircraft, Ford Aerospace, RCA Astro Electronics, Lockheed Missile, General Electric, and the Harris Corporation, has developed plans for three generations of a mobile satellite system which progressively increase in capability and complexity.

Example of mission profile.   Excerpts from mission profiles for three generations of the mobile satellite system[18] are given in Table 5.

---

[16]J. A. Adam (Ed.). "Special Report: Military Systems Procurement—The Price for Might." *IEEE Spectrum,* November 1988, pp. 42–43.

[17]M. K. Sue and Y. H. Park. *Second Generation Mobile Satellite System: A Conceptual Design and Tradeoff Study.* Jet Propulsion Laboratory Publication 85-85, June 1, 1985.

[18]M. K. Sue and Y. H. Park. pp. 11–12.

TABLE 5　Example of Mission Profile Feature Charts

Features	1st Generation	2nd Generation	3rd Generation
Number of Satellites	2	2	1
Antenna Size	5–8 m	20 m	55 m
Spacecraft Weight	1500 lbs	6200 lbs	8857 lbs
Coverage Area	Continental US, Alaska, Canada	Continental US, with acceptable performance in low elevation areas	Continental US
Power Requirements	1 kW	4 kW	10 kW
Satellite to Terminal Link	UHF	UHF	UHF
Satellite to Base Station Link	Ku Band	Ku Band	S-Band
Number of Beams	10 or fewer	21–24	87
Channels Needed to Meet Demand	3,663	5,302	7,759
Projected to be Operating by	1990	1995	2000

Examples of Evaluating the Mission Profile with Analyses

Below are examples of analyses to evaluate and refine the design reference mission profile.

Example of an incomplete mission profile

This example shows how even with a seemingly complete mission profile, requirements can be misinterpreted. A tank was designed to provide infantry support in offense and defense against an attacking army which fires anti-personnel shells at the tank. The mission profile defined the threat and the size and weight requirements for the tank. For example, the tank had to fit with the standard Bern Tunnel railway car and weigh between 50 and 55 tons. To survive the variety of threats specified in the mission profile, the armor that is distributed over the tank must be a specified thickness in order to deflect threats from the front, back, and sides.

This thickness specification was related to a requirement written in the mission profile which was that the tank not be stopped by shell fragments falling in different locations on the battlefield. Analyses of this design requirement studied the "worst case," i.e., the location where the most damage would be expected, which was directly over the tank at normal fusing height. Analyses showed there was a 95% probability that a shell fragment falling from this normal fusing height wouldn't stop the tank. Thus this part of the mission profile was met. But when a shell explodes above the tank, a grapefruit-sized brass fuse drops straight down and will damage anything it falls on, thus stopping the tank.

This example illustrates a case where the requirements, as stated, were met (the thickness of the armor protected the engine from fragments), but the mission profile requirements did not take into account the damage caused by the brass fuse dropping from a shell exploding directly above the tank.

Example of effective tradeoffs

The mission of Voyager 2 included visits to the outer planets Jupiter, Saturn, Uranus, and Neptune, which are so far from the sun that solar-electric cells would be almost useless. Instead, an unusual type of nuclear generator was used that turns the heat of decaying plutonium-238 into electric power to run the many items of electronic equipment, computers, and radio transmitters that beam the data and photos to earth. Three units produce about 375 watts of electricity to power all the spacecraft's systems.[19]

This type of nuclear battery is controversial because of the risk of an accident, especially during launch. An explosion would spread plutonium, which can cause cancer. This risk was regarded as slight, but to minimize the danger, the plutonium was surrounded by multiple layers of shielding, including thick graphite shells to withstand the extreme impact of temperatures if the launch had to be aborted. This technology made the mission possible. Its benefits outweighed the risks.

Example of noneffective tradeoffs

In the SSN 688 class of submarines, first developed in the late 1960s, the original mission profile had goals for speed, stealth or quietness, and depth—three critical features for effective attack. The 688 class would have a much larger propulsion plant than the previous 637 class, and

---

[19]W. J. Broad. "Voyager's Heartbeat is Nuclear 'Battery'." *The New York Times,* August 26, 1989, p. 8.

so would be almost twice as heavy as the earlier submarines. The weight made it hard to meet the speed requirement of thirty knots, even with more horsepower. To get the desired speed, an expert panel of submarine commanders decreased the hull's thickness. The increase in speed came at the expense of the submarine's diving ability. With the thinner hull, the submarine could go at thirty knots, but could not dive deeper than 950 feet.[20] For the past seven years, submarines could dive to thirteen hundred feet to avoid detection by sonar beams and to avoid enemy torpedoes and depth charges. Sacrificing depth to speed was viewed as a temporary tradeoff to be remedied in a few years with a new class of attack submarine. But this 688 class is still being produced in the 1980s.

Examples of Analyses and Trade Studies to Evaluate Design Requirements

Below are examples of analyses and trade studies to develop and evaluate design requirements.

Example of an incorrectly stated requirement

In the requirements documents for an airborne camera, the Air Force specified a 2.5-foot cable to connect the control unit to the camera head. In designing the camera, however, the contractor found that a 19-inch cable was sufficient. The contractor and the Air Force then had to decide whether the government should require the change or whether the contractor should request a waiver. Both options added cost. The Air Force chief of engineering documents said that the error was in specifying a 2.5-foot cable, rather than a cable with adequate strain relief and safety margins, without excessive slack.[21]. Requirements should specify goals, not the actual values which will vary with the ultimate design chosen.

Example of an incomplete requirements analysis

The example of another recently designed tank shows how a configuration might be cost-justified and still not be as good as another configuration. This tank was called "the tank that squats." It had a

---

[20]P. Tyler. *Running Critical: The Silent War, Rickover and General Dynamics.* New York: Harper & Row, 1986, pp. 65–67.

[21]J. A. Adam (Ed.). "Special Report: Military Systems Procurement—the Price for Might." *IEEE Spectrum,* November 1988, p. 56.

suspension system that could raise or lower the tank. When the tank was attacking, the tank could be high off the ground and thus travel better. When the tank was defending against an enemy attack, the tank could squat down to lower its silhouette and make it harder to hit. This benefit justified the added cost of the complex suspension system which was prone to mechanical failure and added about 10% to the cost of the tank. Computer simulation models demonstrated that the potential savings in tanks outweighed the additional cost of the suspension system.

The simulation models, however, did not consider a less expensive strategy to avoid losing a tank. The tank crew could be taught to take advantage of the terrain and topography. By moving to lower ground or ravines on the right or left, the crew could lower the tank's silhouette without an expensive, failure-prone suspension system. The cost of additional training for the crew was less than the cost of the suspension system and provided nearly the same benefit.

Examples of tradeoffs to refine the design requirements

A goal of the mobile satellite system is to maximize the number of satellite channels, using existing commercial satellite bus configurations with minimum modifications. (See the mission profile feature charts described earlier.) To accomplish the goal, tradeoffs were performed on the number of satellites required, antenna size, feed configurations, interference modes, and operating frequencies. A cost model was used to weigh the merits of various alternatives.

The tradeoffs required to achieve the objectives for the three generations of satellites were very complex. To facilitate the tradeoff problem, a computer program was developed to simulate various alternatives. The assumptions were based on the best available information or best estimates from the data of previous studies. However, because of some uncertainty in estimating the parameters, subject matter expertise was used to validate results.

This example will focus on the weight constraints of the total system at lift off and how it affects the tradeoffs. It will concentrate on the second-generation system that must be optimized to meet the weight constraint of 6,200 pounds, the maximum payload for the second-generation system. The tradeoff parameters considered to meet the weight constraint are described below.

Market demands. Studies conducted on the projected market demands for the next two decades show the actual number of channels must be two times higher than the first-generation system for coverage in the

TABLE 6   Mobile Satellite Configurations Ranked According to Monthly Service Charge (1 is Most Economical; 8 is Least)

Antenna size	22 m	24 m	31 m	35 m	36 m	37 m	51 m
# of Satellites	2	2	1	1	2	2	1
Rank	1	2	3	4	5–6	7	8

continental United States. The satellite size, weight, and power required for the additional channels require a third-generation satellite with a customized bus. More studies must be conducted to verify the number of users before a final decision is made.

System cost studies. To reduce the total system costs while maintaining the same total system capacity, additional satellites with smaller antenna may be used. For example, a system using a 31 m antenna and one satellite will have a 10% rate of return; a system with the same capacity using a 22 m antenna and two satellites would have an 8.75% return. The rate of return is lower because the spare satellite accounts for a smaller portion of the total system cost. Table 6 shows using smaller antennas with more satellites (keeping all other variables constant) is more economical.

Bus configuration trade study. The projected market demand showed the need for a two-satellite system with a bus capable of operating 7,000 to 15,000 channels. A survey of satellite manufacturers indicated this type of bus would be available for the second-generation satellite in the 1990s. Studies showed this commercially-available bus would meet the power requirements and would be more cost effective than a customized bus.

Antenna trade study. The key feature of the mobile satellite is the communications antenna. Operating at UHF or "L" band frequencies dictates a large antenna with a high gain to reduce the power requirements and a small beam size to reuse the limited frequency band. Due to the complexities of large antennas, optimizing the antenna was a major task. To minimize the payload weight and maintain the mechanical and electrical performance characteristics, trade studies were conducted, which are described below.

Mobile satellite trade studies

Advantages and disadvantages of planar arrays and lenses. The advantages of arrays and lenses are:

- superior scan performance
- higher aperture efficiency
- no aperture blockage
- high reliability

The drawbacks of arrays and lenses are:

- excessively heavy
- no deployment concepts for large arrays
- simultaneous coverage of the entire area not feasible

Center feed vs. offset feed reflectors.   The center feed system is easier to control and structurally simpler. The offset configuration has lower radio frequency (RF) blockage and less weight.

Accuracy requirements.   Pointing accuracy depends on focal length, reflective surface, beam footprints, and lateral panel focusing. Reflector-scattering programs can compute deterministic errors; computer programs can optimize parameters.

The most challenging aspect was to maintain the satellite in the proper orientation and to point the 20 m reflector within the required accuracy. Computer models were used to design the attitude and orbital control subsystems; however, these models involved risk because of the uncertainty of the estimates.

Deployment technologies.   These items were optimized before the system was baselined (see Step 5 for details on baselining):

- transponder design
- antenna feed design
- power
- attitude control
- thermal control
- networking with other systems

Configurations meeting the mission profile.   The antenna-supporting structure has a direct bearing on the weight of the satellite. If the overall length of the satellite could be reduced, then the cost of each satellite may be reduced by as much as 36%, resulting in a potential savings of $60 million for deploying three satellites.

After all the trade studies were run, two configurations were found

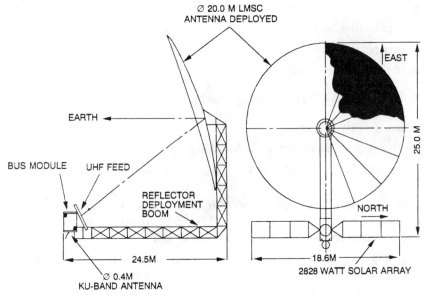

Figure 7 Configuration meeting mobile satellite profile.

to meet the mission profile requirements. One system would have the antenna support beam deployed in an anti-earth direction with the electrical boresight pointed northward toward the center of the Continental United States, Canada, and Alaska for proper coverage. See Figure 7.[22] The other configuration would be similar, except it would have solar panels mounted on a 10 m beam to prevent shadowing of the solar arrays.

Example of a well-designed trade study

The design of sonar domes is a key factor in the detection of enemy submarines which is a primary mission of frigates and destroyers in anti-submarine warfare. The sonar system is housed in a dome in the bow of the submarine. A longer than usual dome allows a larger sonar array and the use of lower frequencies with an increase in the detection range.

As part of the Trident Scholar program, Culbertson[23] conducted a

---

[22]M. K. Sue and Y. H. Park. p. 20.

[23]J. Culbertson. *A Tradeoff Study of Sonar Performance and Powering Requirements for Unconventional Sonar Domes.* U.S. Naval Academy Trident Scholar Project Report. Annapolis, MD: USNA, May 19, 1987.

trade study to investigate designs for a longer dome to house a larger sonar system. One disadvantage of a longer dome is more frictional resistance due to greater surface area, which would require more power. But if the increase in frictional resistance were offset by a decrease in wave-making resistance, power requirements would not increase.

Culbertson's trade study used computer analysis (the Fastship computer-aided hull form design program and wave-resistance flow code) to investigate how different cross-sectional shapes of a sonar dome affect resistance. The trade study also varied the dome's length to compare changes in power requirements, using a fuel-consumption analysis program and measures of effective horsepower at various speeds. (Effective horsepower is a function of velocity and total resistance; total resistance is a function of velocity and wetted surface area. Total wetted surface area was held constant for each hull design.)

The effective horsepower of each alternative design was compared to the effective horsepower of a baseline hull form. A ratio of 1.00 would mean the new design and the baseline design required the same horsepower to achieve a particular speed. A ratio of 1.20 would mean that the new design needed 20% more horsepower than the baseline to achieve the same speed. The effective horsepower ratios were also used to compute changes in fuel used by each new alternative hull form. Culbertson's trade study was composed of several parts, each of which is described below.

Baseline study.   The resistance and powering characteristics were evaluated for the standard destroyer or frigate hull and dome design to which the alternative designs were to be compared.

Shape study.   Four different cross-sectional shapes for sonar domes were designed and fit to the baseline hull form using the Fastship computer simulation program. The wave-making resistance and the effective horsepower ratios were compared to the baseline hull and dome design. For each of these four cross-sectional shapes, three domes varying in area were designed. For each dome shape, Culbertson studied the effect of changes in area on resistance and power requirements.

The same trends were apparent for each shape. For these designs as compared to the baseline design, more power was needed at speeds below 26 knots; less power was needed above 26 knots. The domes with the greatest area had the greatest increase in effective horsepower at speeds below 26 knots and the greatest reduction in effective horsepower at speeds above 26 knots. Hull forms in two of the shapes stood out. For these shapes, the maximum increase in power requirements at low speeds was less than 20% and the maximum reduction at high speeds was less than 4%. One of these shapes was chosen for further study because it had more volume and could more easily house sonar arrays.

Figure 8 shows the trends from the shape trade study. The x-axis gives velocity or speed in knots (KT). The y-axis gives the ratio of effective horsepower of the alternative designs compared to the baseline sonar dome design. Compared to the baseline design, the alternative dome shapes (B1, B2, and B3) require more power at low speeds and less power at high speeds.[24]

Length study. Using the data from the shape study, the three cross-sectional designs with the best power requirements (shapes B1, B2,

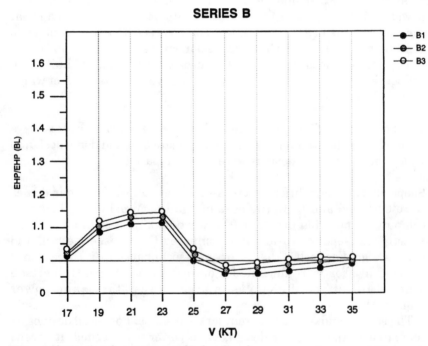

Figure 8 Alternative shapes compared to baseline dome.

---

[24]J. Culbertson. p. 37.

B3), were used to develop three series of domes in which the length of the sonar array housed in the dome varied systematically (10, 20, 40, 80, and 160 feet—lengths were chosen using previously available data on sonar performance).

Each design was evaluated against the standard design for resistance and power requirements. Domes housing a given length gave similar results regardless of transverse area. Domes with greater length generally showed a greater increase in effective horsepower at low speeds. At speeds above 26 knots, the differences were less striking.

Fuel consumption study.   Using the resistance and power data from the length study, Culbertson measured how varying dome length changes fuel consumption on an illustrative mission profile for a destroyer at various speeds. The fuel consumption analysis showed how much more fuel would be needed with each length and each type of cross-sectional shape to obtain the gain in overall sonar performance with the longer domes. The amount of fuel a ship burns is directly proportional to the amount of horsepower the propulsion system needs to run the ship at that speed.

Figure 9 shows an operation profile for a destroyer, which shows it operates at high speeds (above 26 knots) 3% of the time.[25] For every 1,000 hours under way, this destroyer would require 2,914 long tons of fuel to meet its operational profile.

The amount of fuel required by designs with domes housing 160 foot long arrays was less than the amount of fuel for designs with sonar domes housing 80 foot arrays. See the top panel of Figure 10.[26] The bottom panel shows the 80 foot array dome design (the worst case) would require about 10% more fuel than the baseline hull. The 20 foot and 40 foot array dome design would use about 2% more fuel; the 160 foot array about 9% more fuel.

Tradeoff analysis.   Using the results from the fuel consumption analysis, Culbertson computed the costs associated with the changes in fuel requirements with the domes of various shapes and lengths. The cost results were then compared with the gains in sonar performance. The optimal dome design emerged from this cost-benefit analysis. Gains in sonar performance were measured with the directivity index or improvement in signal strength, which measures the ability to

---

[25]J. Culbertson. p. 50.

[26]J. Culbertson. p. 57.

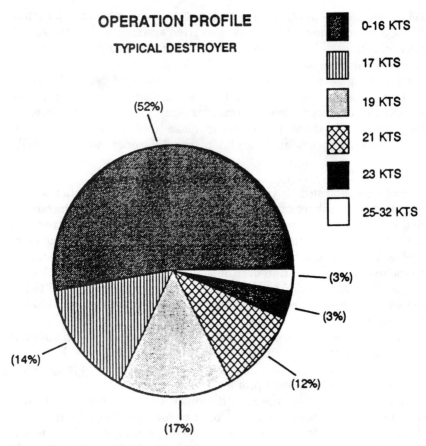

Figure 9 Operation profile for a typical destroyer.

discriminate between the target signal and unwanted noise, expressed in decibels. A 160 foot array would have a signal strength 16 times that of a 10 foot array.

Table 7[27] shows the B3 shape dome housing an 80 foot array is the worst case in terms of fuel costs, with an additional $55 an hour; but even this worst case gives a gain of 9 decibels compared to a 10 foot array. The B1 dome housing an 160 foot array would give the greatest gain (12 decibels), with an additional fuel cost per hour of $38, about the same cost as the 80 foot size.

---

[27]J. Culbertson. p. 64.

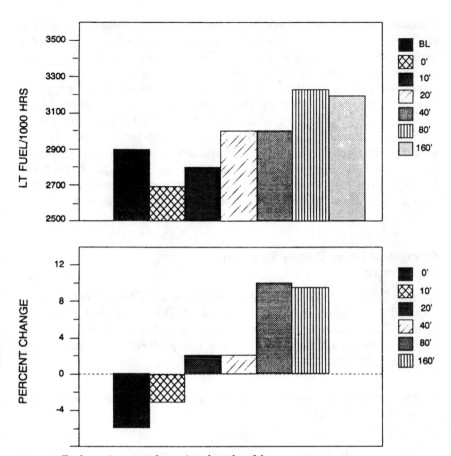

Figure 10 Fuel requirements for various lengths of domes.

TABLE 7   Penalty in Dollars Per Hour for Associated Gains in Detection Measured in Decibels (db)

Array length (ft)	Gain in Detection (db)	Penalty in dollars per hour of additional fuel costs		
		Dome Shape B1	Dome Shape B2	Dome Shape B3
20	+3	$13	$17	$21
40	+6	$14	$18	$21
80	+9	$37	$46	$55
160	+12	$38	$45	$52

Conclusions. The systematic investigations in this trade study showed changing the length of the dome housing a sonar array produces proportional increases in effective horsepower at low speeds and proportional decreases at high speeds. Changing the shape and area affects only the magnitude of the changes in horsepower for a given speed. Long sonar arrays (80 and 160 foot) give significant improvements in detection, at a relatively small additional fuel cost.

Culbertson was able to arrive at these conclusions quickly and cheaply using the Fastship computer-analysis program and the resistance flow code described above. The systematic design of the trade study allowed her to eliminate insignificant variables and concentrate on the variables with the greatest potential for improvement.

Example of Using Quality Function Deployment

Quality function deployment (QFD) is used to obtain approval of design requirements before they are baselined. QFD is a tool that focuses on coordination, planning, communication, and integration to help teams deliver systems on time and within the performance and cost restraints. QFD focuses on:

- translating vague, nonmeasurable requirements into specific requirements
- prioritizing features and functions
- specifying which features contribute to quality
- helping to make tradeoffs necessary to meet conflicting requirements

Digital Equipment Corporation (DEC) used QFD in developing high-level product definitions and strategy for a high-end computer. DEC found the key benefits of QFD are that the customer and the developer agreed on:[28]

- the mission and mission environment
- the need for more information early in the process

---

[28]L. Cohen. "Quality Function Deployment: An Application Perspective from Digital Equipment Corporation." *National Productivity Review,* Summer 1988, pp. 197–208.

- design requirements that actually meet the needs
- tradeoff decisions that optimize all the requirements
- tradeoff decisions to be reviewed at a later date, if necessary

QFD's primary objective is to detect and solve problems early in the concept or demonstration phase, therefore making it easier to manufacture the system with fewer engineering changes.

4

Summary

The Front End Process is a process to conceive, define, and evaluate the mission and to develop and evaluate requirements for a new or updated system. Improvements in the Front End Process will pay off. Most systems acquire their life cycle costs from decisions made early. Some studies show that two-thirds of the projected life cycle costs are built in at the planning and design stages. Historically, the conceptual, demonstration, and development phases of a new or updated system focused almost entirely on the development of the system, not on manufacture, deployment, and maintenance. Only later were estimates made about how much the new system would cost to manufacture and support.

The Packard Commission

A panel of experts, chaired by David Packard, was convened by President Reagan to identify critical issues in defense management. The Packard Commission report identified several problems for management focus in controlling costs and schedules and ensuring performance, quality, and reliability. Many of the problems they identified occur during concept exploration in the Front End Process; similarly many of their recommended solutions take place during the Front End Process.

Keys to Success

Table 8 gives risks and best practices for the design reference mission profile, design requirements, and trade studies.[29]

---

[29]*Best Practices: How to Avoid Surprises in the World's Most Complicated Technical Process.*

TABLE 8   Front End Process Risks and Best Practices

Risk	Best Practice
Mission environment is defined using only military specifications	Mission environments are defined according to conditions of operating, handling, storing, transporting, and training
Contractor works in vacuum to interpret the system performance requirements	Contractor and government agree on complete mission profiles for the system's life cycle
Mission profile emphasizes just a few key tactical functions	Design reference mission profile is a composite of all functions and environments
Mission profile defines operational requirements only	Mission functional and environmental profiles are included in request for proposals; system functional and environmental profiles are the basis for design requirements
Schedule pressure prevents flow down of design requirements to lowest technical level	Requirements are allocated and verified at the lowest technical level
Design requirements do not flow down to subcontractors and suppliers	Prime contractor communicates design requirements early to subcontractors and suppliers
Design requirements are based on unproven, state-of-the-art technology	State-of-the-art technology is used only when trade studies show it is cost-justified
Design requirements change as the design evolves	Design requirements are evaluated, approved, and baselined before full scale design begins
Trade studies are done only once	Trade studies occur iteratively to evaluate requirements and alternative configurations
Trade studies are poorly planned, conducted, and reported	Trade studies are systematically planned, conducted, and reported

Benefits

Improving the Front End Process will pay off throughout the system's life cycle. These benefits can be expected:

- design requirements developed from the strong foundation of a well-defined mission profile
- fewer engineering changes as a result of accurate and complete design requirements
- reduced risks with use of proven technology, defensive designs, and simulations at the concept stage

- realistic configurations chosen as a result of useful and timely trade studies
- costs reduced with early involvement of manufacturing, testing, and deployment in design for assembly, manufacturability, testability, etc.

A more effective Front End Process reduces the development interval, encourages early detection of incorrect engineering and management decisions, and leads to better customer acceptance.

Summary of Procedures

Accurate, specific, and complete design reference mission profiles and design requirements are the result of careful and well-conceived analyses and trade studies. This reference guide has presented step-by-step procedures for achieving these results. Table 9 is a summary of those steps.

TABLE 9   Summary of the Front End Process Procedures

Step 1—Define the High-Level Mission Profile	
Procedure	Supporting Activities
Include key elements	■ Define mission including performance, maintainability, reliability, safety, environments, etc.
Define objectives and operational requirements	■ In operational requirements, specify functions and constraints to be met, rather than actual values, in order to allow flexibility in choosing best configuration
	■ Increase details as the mission profile is developed

Step 2—Respond to the Request for Proposals (RFP)	
Procedure	Supporting Activities
Issue request for proposals	■ Ask contractors to explore feasibility of concept
Form the proposal team	■ Include representatives from each organization
	■ Develop proposal
Receive a short-term contract for concept exploration	■ At informal briefing, discuss proposal strengths and weaknesses and any changes
Form the development team	■ Include representatives from design, manufacturing, maintenance, and other key organizations on the development team
	■ Develop project plan, emphasizing *what* should be done, not how to do it

TABLE 9   **Summary of the Front End Process Procedures (*Continued*)**

Step 3—Evaluate and Refine the Design Reference Mission Profile

Procedure	Supporting Activities
Refine the design reference mission profile, including the system functional mission profile and the system environmental profile	■ Describe a composite or reasonable sample of all the environments and functions in the system's life cycle
	■ Describe in a time series all the functions the system will perform
	■ Describe the conditions to which the system will be exposed: transporting, handling, storage, training, etc.
Choose and evaluate a system architecture that meets the mission profile and constraints	■ Use an iterative methodology that emphasizes top-down design and bottom-up verification
Use functional analyses to evaluate the system functional and environmental profiles	■ Use functional flow block diagrams to display the primary system functions and break them into subfunctions
	■ Use time line analysis to show mission functions minute by minute or hour by hour during operation, test, and maintenance
	■ Identify, assess, and rank risks
Use technical risk assessment to consider technical and financial risks	■ Develop plans to deal with the risks

Step 4—Develop and Evaluate the Design Requirements

Procedure	Supporting Activities
Analyze the design requirements	■ Derive design requirements from the operational requirements in the mission profile
	■ Describe explicit and implicit requirements
	■ Ensure requirements are specific, clear, unambiguous, understandable, concise, consistent, stable, traceable, testable, and feasible
	■ Allocate requirements to subsystems and components, with more detailed requirements
	■ Analyze effects of extreme shock and vibration, temperature and humidity stresses in transporting, handling, and storage
Use trade studies to refine the design requirements	■ Develop cost models
	■ Identify configurations meeting mission needs that are producible, testable, and maintainable
	■ Identify a range of acceptable values and objective selection criteria
	■ Conduct trade studies using principles of experimental design and optimization theory
	■ Perform trade studies and analyses iteratively
	■ Use a standard format for trade study reports
	■ Keep in mind the limitations of trade studies
	■ Use "trade trees" to minimize the number of comparisons

TABLE 9   Summary of the Front End Process Procedures (*Continued*)

Step 5—Baseline the Design Requirements (the Functional Baseline)

Procedure	Supporting Activities
Use a formal process to obtain approval and then document the agreements	■ Use a method of baselining that controls and documents functional and environmental specifications, design requirements, and the system and subsystem configurations
Conduct design reviews internally and externally with the customer	■ Obtain agreement on functional and environmental design requirements and high-level allocations
Obtain customer approval	■ Identify and rank needs
	■ Identify technical requirements that meet needs
	■ State a requirement for each need
	■ Obtain customer approval informally and in design reviews
Use measures of effectiveness	■ Use outcome metrics to measure customer satisfaction and other key goals
	■ Use process metrics to track progress and improve the development processes
Receive long-term contract	■ For additional demonstration and validation, conduct trade studies and analyses to verify preliminary configurations

5

References

The reference list is an annotated list of the references used to develop the material.

Adam, J. A. (Ed.) "Special Report: Military Systems Procurement—The Price for Might." *IEEE Spectrum,* November 1988, pp. 23–67. Describes procedures and examples of the Pentagon's procurement. Discusses problems in procurement and deployment and suggests possible improvements.

Begley, S. and Hager, M. "A Fantastic Voyage to Neptune." *Newsweek,* September 4, 1989, pp. 50–56. Describes the mission to take advantage of the once-in-176-years opportunity to explore the four outer planets.

Belev, G. C. "Guidelines for Specification Development." *Proceedings of the Annual IEEE Symposium on Reliability and Maintainability,* 1989. Describes the process for developing specifications; gives suggested formats and checklists to minimize ambiguity and cost and meet cost and schedule constraints; mentions strategies to distinguish truly needed requirements from not-needed requirements.

Bell, T. (Ed.). "Special Report: Managing Risk in Large Complex Systems." *IEEE Spectrum,* June 1989, pp. 21–52. Discusses principles and procedures of risk analysis and risk management, with illustrations from telephone networks, space shuttle, nuclear weapons reactors, and a pesticide plant.

Best Practices: How to Avoid Surprises in the World's Most Complicated Technical Process. Department of the Navy (NAVSO P-6071), March 1986. Discusses how to avoid traps and risks by implementing best practices for 47 areas or templates, including topics in funding, design, test, production, facilities, and management. Discusses best practices for design reference mission profile, trade studies, design requirements, and technical risk assessment.

Boeing Aerospace. *New Remotely Piloted Vehicle Launch and Recovery Concepts.* Air Force Flight Dynamics Laboratory Technical Report-79-3069, June 1979. Describes dynamic analysis, preliminary design, and performance and cost trade studies of air bag skid and air cushion concepts for launch and recovery of remotely piloted vehicles. Includes dynamic simulations of steady state flight, landing, and takeoff. Describes performance and cost trade study of complexity, fuel requirements, adverse weather capability, ground equipment and facility requirements, survivability, reliability, maintainability, and acquisition and life cycle costs.

Broad, W. J. "Voyager's Heartbeat is Nuclear 'Battery'." *The New York Times,* August 26, 1989, p. 8. Describes the risks and benefits of the plutonium-powered radio isotype thermionic generators used in the Voyager 2 spacecraft.

Cohen, L. "Quality Function Deployment: An Application Perspective from Digital Equipment Corporation." *National Productivity Review,* Summer 1988, pp. 197–208. Discusses principles and procedures of Quality Function Deployment, a technique used to help define and design systems that actually meet customer needs. Describes how this technique has been applied at Digital Equipment Corporation.

Crawford, C. C. "Reduction Cost Avoidance for Full Scale Development Begins with Concept Formulation." *AGARD Conference Proceedings,* No. 424, Toulouse, France, May 1987, pp. 1-1 to 1-7. Discusses ways to reduce risk early in the concept formulation process to avoid costs during the full scale development process. Gives examples of savings that would have occurred if analyses and simulations were done early rather than relying on expensive testing during full scale development.

Culbertson, J. *A Tradeoff Study of Sonar Performance and Powering Requirements for Unconventional Sonar Domes.* U.S. Naval Academy Trident Scholar Project Report. Annapolis, MD: USNA, May 19, 1987. Describes results of a trade study of the resistance characteristics and powering requirements of unconventional sonar domes. Systematically varied the cross-sectional area and the longitudinal length of the domes. Used the changes in powering requirements to fuel costs to do the economic tradeoff analysis.

Defense Acquisition Regulations, Part 15, Effective April 1, 1984. Describes the Federal Acquisition Regulations (FAR) to which government contracts must adhere.

Defense Systems Management College. *Systems Engineering Management Guide.* Contract Number MDA 903-82-C-0339. Fort Belvoir, VA: DSMC, October 3, 1986. Outlines how trade studies are used during the Front End Process and throughout the acquisition life cycle. Gives methodology and applications for trade studies and illustrates with an example from an aft crane configuration on a Navy Logistics ship. Describes techniques such as paired comparisons to achieve weightings, utility curves to translate criteria to a common scale, performance estimates through testing or simulation, and a sensitivity analysis to determine the value of the results. Describes limitations of trade studies. Includes a program managers' checklist for review of trade studies.

Department of Defense. *Total Quality Management Master Plan,* 1986. Describes principles and procedures for quality management and improvement, including quality function deployment.

Hansen, R. C. *Design Tradeoff Study for Reflector Antenna Systems for the Shuttle Imaging Microwave System.* NASA-CR-14C915, Final Report 636-1 on Contract 954026. Encino, CA: R. C. Hansen, Inc., August 30, 1974. Describes the results of a trade study on a crossgrain antenna with regard to meeting a 90% beam efficiency, using numerical integration and estimation of the environmental envelope to take into account tradeoffs between factors that affect beam efficiency, e.g., aperture taper and blockage, feed cross polarization and spillover, reflector cross polarization and backlobes.

Ingels, D. M. *What Every Engineer Should Know about Computer Modeling and Simulation.* New York: Marcel Dekker, 1985. Discusses the purpose and methods of mathematical modeling and simulation, including defining and analyzing the problem, generating mathematical models, solving the models, and performing the simulations.

Institute for Defense Analysis. *IDA/OSD Reliability and Maintainability Study, Volume IV.* Contract Number MDA 903-79-C-0018. Alexandria, VA: IDA, 1983. Summarizes the findings, conclusions, and interrelationships among 15 technology reports. Analyzes future technology impacts and requirements, including the need to mature technology off-line and gain a greater understanding of the causes of failure. Gives examples of mission needs and analyses by functional element groups to evaluate what are necessary and technically feasible reliability levels.

Juran, J. M. and Gryna, F. M. (Eds.). *Juran's Quality Control Handbook.* 4th Ed. New York: McGraw-Hill, 1988. Describes the process of new product development, including procedures for defining design requirements, setting up configuration management systems, and designing experiments for trade studies.

King, R. *Better Designs in Half the Time.* Methuen, MA: GOAL/QPC, 1987. Describes the procedures and tools used in the implementation of quality function deployment. Discusses techniques for listening to the voice of the customer, value engineering, reliability, and failure modes and effects analysis. Discusses how to use these techniques to reduce the design time by focusing on the priorities during the process and reducing the need for redesign, especially on critical items.

Leslie, P. J., and Lilley, E. G. "Development Savings Through Parametric Modeling of Programmer Costs." *AGARD Conference Proceedings*, No. 424, Toulouse, France, May 1987, pp. 2-1 to 2-3. Describes how parametric modeling has been successfully applied to weapons systems in the preliminary design process. The cost models provide the opportunity to base decisions on sound information and avoid costly iterations.

Military Standard 490: *Specification Practices*. Describes how to prepare, interpret, change, and revise specifications prepared by or for the Department of Defense.

Military Standard 1521: *Technical Reviews and Audits for Systems, Equipment, and Computer Software*. Describes reviews typically mandated on government contracts, e.g., System Requirement Review, System Design Review, Preliminary Design Review, Critical Design Review, Functional Configuration Audit, Physical Configuration Audit, Production Readiness Review.

Military Standard 2167: *Defense System Software Development*. Describes the content and scheduling of software reviews that are often mandated on government contracts.

Norman, C. "Quiet Soviet Subs Prompt Concern." *Science*, Vol. 243, March 31, 1989, pp. 1653–1654. Discusses how the latest advances in Soviet submarine design are stretching passive sonar detection technology limits and making active sonar detection an alternative to explore. Describes conclusions of a panel of experts on how to counter the threat.

Packard, D. *A Quest for Excellence. Final Report to the President by the President's Blue Ribbon Commission on Defense Management*. Washington, D.C.: President's Blue Ribbon Commission on Defense Management, June 1986. Describes problems in the Department of Defense's acquisition and procurement procedures. Recommends solutions to solve the problems and prevent problems in the future. Describes the key role of management in bringing about a cultural change in their work on substantial defense contracts.

Pike, R. W. *Optimization for Engineering Systems*. New York: Van Nostrand Reinhold, 1986. Describes optimization methods useful in trade studies, including the theory of maxima and minima, geometric programming, linear programming, single variable and multivariable search techniques, dynamic programming, and calculus of variation.

Priest, J. W. *Engineering Design for Producibility and Reliability*. New York: Marcel Dekker, 1988. Describes a high-level view of environmental and functional profiles, use of trade studies, mathematical models, and simulations to verify that the optimum system approach has been selected, trade studies of design parameters such as cost, schedule, technical risk, reliability, producibility, quality, and supportability, and use of mature vs. leading edge technology. Describes purpose of design requirements, use of trade studies to tailor the requirements for reliability, and producibility. Discusses modeling and simulation and their relation to trade studies.

Rockwell International. *Design Definition Study of NASA / Navy Lift / Cruise Fan V / Stol Aircraft*. Supplemental Report Technology Aircraft Risk Assessment, NASA CR 137727, September 1975. Describes the results of a study to compare and assess risks in three alternative concepts for the technology of the aircraft design. Presents the results of analyses and survey of members of the technical disciplines: aerodynamics, propulsion, advanced systems design, and flight control systems analysis.

Schmid, H., Larimer, S., and Sadeghi, T. "CATS: Computer-aided Trade Study Methodology." *IEEE National Aerospace and Electronics Conference*, 1987. Describes a computerized method of trade studies that improves cost, accuracy, and productivity. The CATS system includes databases of requirements specifications and parameters and spreadsheets for evaluating alternatives. Includes an example that applies CATS to a specific trade study. Outlines and illustrates trade study and design process tasks.

SRI International. *NASA New Technology Identification and Evaluation*. Contract Number NASW-3444. Menlo Park, CA: SRI International, December 1985. Contains a partial list of NASA technology evaluations. Describes the effect the technology is likely to have on performance, cost, and quality standards. Evaluates new technology on novelty, significance, and technical soundness.

Stadler, W. *Multicriteria Optimization in Engineering and in the Sciences*. New York: Plenum Press, 1988. Describes the fundamentals of multicriteria optimization methods. Gives examples of how linear and nonlinear methods are applied in trade studies. Examples include aircraft control systems, focusing systems, and resources planning.

Stefanik, T. *Strategic Antisubmarine Warfare and Naval Strategy.* Lexington, MA: D. C. Heath, 1987. Describes strategic antisubmarine warfare policy and missions to counter potential threats. Discusses submarine design and detection by acoustic and nonacoustic methods.

Sue, M. K., and Park, Y. H. *Second Generation Mobile Satellite System: A Conceptual Design and Tradeoff Study.* Jet Propulsion Laboratory Publication 85-85, June 1, 1985. Discusses the three phases of development for a Mobile Satellite System. Emphasizes the exploration of technologies required to develop a second-generation satellite system that will be deployed in the mid-1990s. Describes a series of trade studies conducted to maximize the number of channels and minimize the overall life cycle cost.

Transition from Development to Production. Department of Defense (DoD) 4245.7-M, September 1985. Describes techniques for avoiding technical risks in 47 key areas or templates, including Design Reference Mission Profile, Trade Studies, Design Requirements, Technical Risk Assessment, and other templates in funding, design, test, production, facilities, logistics, management, and transition plan.

Tyler, P. *Running Critical: The Silent War, Rickover and General Dynamics.* New York: Harper & Row, 1986. Discusses the interactions of the Navy and General Dynamics in the nuclear submarine program, with examples of defining mission profiles and design requirements for the 688 Class submarines.

von Tein, V. "Reasons for Increasing Development Cost of Rotary Wing Aircraft and Ideas to Reverse the Trend." *AGARD Conference Proceedings,* No. 424, Toulouse, France, May 1987, pp. 6-1 to 6-15. Describes the general cost trends of complex systems and the influence of requirements and complexity on cost. Discusses how to reverse the cost trends with good engineering judgment and proper planning and management.

Design Policy, Design Process, and Design Analysis

1

Introduction

To the Reader

System design is a complicated process. There are risks throughout that lead to the development of products that cannot succeed. This part identifies government and industry best practices that can reduce the risks. It involves three critical Transition from Development to Production templates:[1] [2] Design Policy, Design Process, and Design Analysis.

The templates, which reflect engineering fundamentals as well as industry and government expertise, were first proposed in the early 1980s by the Defense Science Board, under the chairmanship of Willis J. Willoughby, Jr. The Board's intent was to encourage everyone who is involved in the acquisition process to become aware of these templates and actually use them on the job.

This part on design policy, process, and analysis is one part in a series of books written to help defense contractor engineers, government program managers, and contract administrators use the templates most effectively. These books are meant to stand-alone as references and textbooks for related courses.

Clustering several templates makes sense when their topics are closely related. For example, the templates in this reference guide interrelate and occur iteratively within system design. Other templates, such as Design Reviews, relate to many other templates and are thus best dealt within individual parts.

---

[1]Department of Defense. *Transition from Development to Production.* DoD 4245.7-M, September 1985.

[2]Department of the Navy. *Best Practices: How to Avoid Surprises in the World's Most Complicated Technical Process.* NAVSO P-6071, March 1986.

*Effective system design and systems
engineering require a keen understanding of
the scientific and technical aspects of a total
system as well as insight and judgment of
how to apply the basic engineering methods
and techniques.*

A Simple View

A simple view of system design in which a system merely transforms inputs into the desired outputs is shown in Figure 1.

With today's complex systems, however, getting from the inputs to the outputs is never simple. Systems draw upon many engineering disciplines with a strong dependence among the hardware, software, firmware, and electrical aspects of a system. An increased focus on reducing costs and shortening schedules also compounds the process. Tradeoffs such as functionality vs. cost or performance vs. schedule are frequently required, yet difficult to integrate with the overall objectives of the system.

Background

The Department of Defense's (DoD) *Transition From Development to Production*[3] manual and the Navy's *Best Practices*[4] manual address

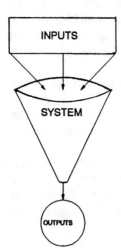

Figure 1 A system transforms inputs to outputs.

---

[3]Department of Defense.

[4]Department of the Navy.

these problems. They identify 48 areas of risk in the system acquisition process and provide templates describing techniques for reducing risk.

Part 2 specifically addresses three design phase templates: Design Policy, Design Process, and Design Analysis. Each of these templates represents a critical element in system acquisition.

Design Policy is a documented corporate policy supported by a set of engineering procedures and guidelines considered the most prudent and effective means for minimizing risk in system design.

Design Process involves the entire set of tasks and activities required to achieve a completed system from its initial requirements, goals, and constraints.

Design Analysis assesses whether a design complies with the requirements, goals, and constraints. It exposes high risk aspects of the design, with respect to meeting performance specifications.

Problems in System Design

System acquisition is an intricate process requiring a fine balance between the quantitative, qualitative, and managerial aspects of system development. Each aspect is essential; yet no aspect alone is sufficient to ensure successful, low risk designs.

Many system problems are a result of errors in the design phase. One study,[5] which examined 850 field failures of relatively simple electronic equipment, found that 43 percent of the problems could be traced to engineering design errors. Another study,[6] which examined seven space programs, concluded that 35.2 percent of component failures could have been avoided with better design or specifications.

Some designs fail because of the lack of formal methods to structure the design process. Others fail because critical design activities are performed improperly, or not at all. When these failures are left uncovered, they cause a rippling effect throughout the entire system. A poorly designed system cannot be adequately tested or produced and the rework necessary to fix design problems can overwhelm system schedules and costs.

Technical Program Risks

In addition to concerns about whether the system itself is designed properly, there are programmatic risks that can mean failure, in-

---

[5]Gryna, Frank M. "Product Development" in *Juran's Quality Control Handbook*. 4th ed. Eds. J. M. Juran and Frank M. Gryna. New York: McGraw-Hill, 1988, p. 13.3.

[6]Gryna, p. 13.3.

creased development costs, longer lead times, and an overall longer transition from development to production. Some of these risks are listed below:

- no provisions for following good design practice
- effective design policies are not recognized or implemented
- inconsistent implementation of design rules and guidelines due to lack of a supported corporate design policy
- poor communication within the project and between contractor and subcontractor
- extensive redesign is required because of inadequate use of design analyses
- cost increases caused by early test failures
- test program is lengthened because of recurring failures
- manufacturing and assembly equipment require costly changes due to lack of early consideration of these product phases
- initial production units require retrofits
- an unproven design is released

Keys to Success

When designers understand and apply fundamental principles of effective system design, they minimize technical risks and improve the overall system acquisition process. The keys to success for applying these principles in the Design Policy, Design Process, and Design Analysis templates are the basis upon which this part was developed.
These keys to success include:

- document and implement a corporate design policy that includes proven engineering methods and guidelines
- ensure adequacy, completeness, and consistency of design requirements
- understand system goals and objectives and adapt acquisition strategies appropriately
- integrate life cycle considerations throughout the design
- ensure producibility is considered and built-in during design

- conduct trade studies to evaluate the impact and risk of using new technologies
- perform pertinent analyses to assess the ability of a design to meet system requirements and use the results to change the design to reduce risk
- incorporate lessons learned to continually improve products and processes

2

Procedures

*Engineering design is the creative process
that is the essential source of all new
products. It involves imagining many
different ways to satisfy a need. Sometimes
multiple, even conflicting requirements and
constraints must be reconciled. There are
various methods by which these different
approaches and requirements can be
synthesized and evaluated.*[7]

Systems engineering is a creative, iterative process by which an idea is transformed into reality. In simplest terms, it is "both a technical process and a management process. To successfully complete the development of a system, both aspects must be applied throughout the system life cycle."[8]

Part 1, Front End Process, details the early part of the systems engineering process—Design Reference Mission Profile, Trade Studies, and Design Requirements. Steps in this part build on that information and detail the design process. Part 6, Design for Testing, discusses testing.

Figure 2 describes the design process in six key steps. While the figure is linear, the steps are more frequently parallel in nature, often iterating several times.

---

[7]Rabins, M. and others. "Design Theory and Methodology—A New Discipline." *Mechanical Engineering,* August 1986, p. 24.

[8]*Systems Engineering Management Guide,* Defense Systems Management College, December 1986. p. 1–2.

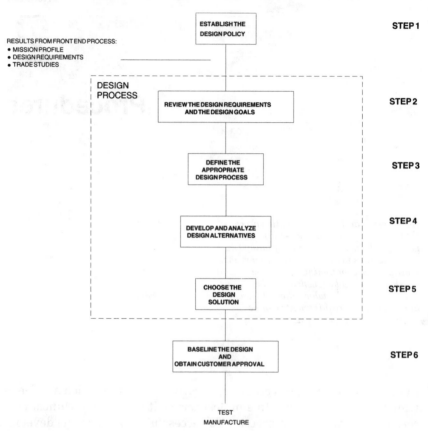

Figure 2 Steps in design policy, process, and analysis.

Design Policy, Process, and Analysis Steps

Step 1 involves establishing a design policy. The objectives of this policy are to identify proven guidelines to reduce risk.

Step 2 signals the beginning of the design process itself. In this step, the requirements and goals generated early in the system acquisition process are reviewed to identify product and process issues, and initiate the change control system.

Step 3 reviews the design goals and refines the design process in order that the best method to meet the goals is identified. Development teams are chosen, cost considerations of designs are examined, methods for reducing downstream risk are identified, and change management is instituted and documented.

Step 4 begins to examine the best way to partition and allocate functions and requirements throughout all levels of the system's hierarchy. Once these values are assigned, analyses are performed to evaluate the levels of risk associated with each decision. For each iteration of analyses, the lessons learned serve to guide the next.

Step 5 involves making a decision regarding the design solution. The criteria for evaluation are carefully established so that the design maturity may be assessed.

Step 6 calls for reviewing the solution with the customer to evaluate the merits of the system and to gain approval to continue development, test, and manufacture.

Step 1: Establish the Design Policy

Establishing and implementing a corporate design policy is critical to successful design and development. A *design policy* identifies tasks and activities that an organization has determined are most likely to produce successful product designs. The policy must include corporate design goals and strategies, along with the associated design practices, procedures, standards, and guidelines for accomplishing them.

Developing a design policy

A corporate design policy covers a wide range of topics. All aspects of the policy must be clearly documented and communicated to everyone involved with design and development. Methods must be established for ensuring that the policy is followed.

Reducing the risks in the design process requires consistent application of a design policy throughout a corporation. A design policy, therefore, must not be developed by only the design organization, but rather should be a corporate policy with support for its implementation from the highest levels of the corporation.

The corporate design policy must be reviewed for each new product, and, in some cases, minor modifications to the policy may be required. When this happens, it is important to communicate changes to the entire organization before product design begins.

Following completion of a system design and development effort, the design team should review and assess their processes and incorporate any lessons learned into the organization's design policy. Doing so will ensure that the policy is kept current with advances in design technology and that future design efforts benefit from the successes and failures of earlier products.

Key aspects of a design policy

The key aspects of product design and development that must be addressed in a corporate design policy are discussed briefly below. They include:

- design process procedures
- design goals
- design practices
- design analyses
- design process control
- assessing design maturity in design reviews

Design process procedures.   The methods and procedures used during the design of a system are critical to its success. Organizations that do not identify, document, and implement proven design methods introduce a high risk of failure into their design processes. To avoid this risk, a design policy must include procedures for process issues such as requirements generation, requirements allocation, design partitioning, trade studies, testing, and design documentation.

Equally important to the success of a design process are the interfaces between the design organization and other groups. A design policy must include guidelines for addressing and representing important interfaces such as manufacturing, assembly, and installation. (This is discussed in more detail in Step 3 of the Procedures chapter and in the Application chapter.)

An additional interface that is crucial during all phases of product design and development is the customer interface. A design policy must include a detailed set of guidelines for establishing and implementing effective and efficient customer interfaces.

One methodology widely used for facilitating customer interaction in the product design process is *Quality Function Deployment (QFD)*. The overall objective of QFD is enhanced customer satisfaction. This objective is accomplished by a methodical translation of customer requirements into technical requirements during each phase of product development.

Reported benefits of QFD include:[9]

- proper interpretation of customer-oriented product objectives throughout product development
- clear understanding and application of production control points
- improved efficiency and shorter product development cycles because of fewer misinterpretations, better coordination, and fewer mid-course changes

---

[9]McEachron, Norman B., Robert J. Lapen, and Ruth A. Tara. *Accelerating Product and Process Development.* Preliminary Draft. Menlo Park, CA: SRI International, September 20, 1989.

Customer satisfaction is difficult but necessary. It is usually done best by working together with the customers to meet their needs. Part 1, Front End Process, describes QFD in more detail.

Design goals. Each design and development organization must determine critical goals for design and structure a policy around those goals. Documenting and communicating the design goals is essential for reducing technical risk in product design.

A design policy must also include techniques and guidelines to help establish realistic and attainable design goals. For instance, being first to develop a particular system may be a desirable goal, but being first with a system that has only marginal performance is probably not.

One technique recommended for use in goal setting is *competitive benchmarking*. With this method, a company conducts periodic benchmarks of itself and competing companies to gauge the advantages and disadvantages of both the products developed and the processes used in their development. From this information, a company can evaluate its strengths and weaknesses and strategically align its individual design goals accordingly.

Setting design goals for an individual product, product line, or an entire design organization cannot be taken lightly. Step 2 provides a more detailed discussion of the types of design goals common in product development and the tradeoffs and risks of these goals.

Design practices. To help reduce the risks in product design, an organization must ensure that the practices used by its design organizations are

- consistent with product development goals and strategies
- competitive
- technologically sound

These practices must be specifically identified in a design policy and their application must be communicated within the organization.

Aspects of design that require documented practices are illustrated in Figure 3. Again, competitive benchmarking can be used to assess the effectiveness of current design practices and to identify areas of improvement.

Design analyses. Design analysis is fundamental in reducing technical risk. To ensure that a complete analysis is achieved, an organization's design policy must include a list of the required design analyses, as well as additional recommended analyses that, when warranted, can further reduce risk. A design policy must also include guidelines specifying who should participate in the design analyses, accepted data

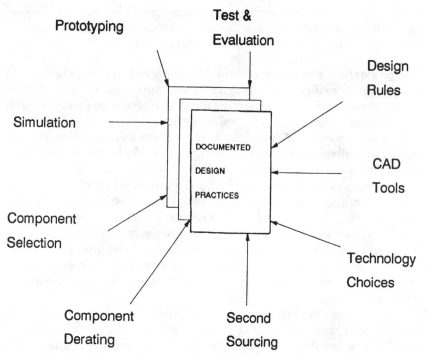

Figure 3 A design policy has documented design practices.

gathering and data generation techniques, practices concerning the use of common databases, and guidelines and standards for reporting the results of the analyses.

Design process control.   Controlling the design process is not simple; seldom does everything work as planned. Without appropriate guidelines, checkpoints, and milestones, critical issues such as scheduling, cost, and quality can get out of control. The results may be systems that are unsafe, unreliable, or do not meet the customer needs.

A key aspect of managing the design process is ensuring that design teams continually focus on improving the product and the processes used to create it. *Total Quality Management (TQM)* is a strategy used both in the Department of Defense and in industry to do that. The objective of TQM is increased user satisfaction. To achieve this objective, a disciplined goal-setting methodology is used to continually improve the product and process development. The key to implementing any effective design process control and improvement strategy is understanding and commitment. Attention to these issues in a design policy is essential to achieve successful design process control and effective product improvement.

Assessing design maturity in design reviews.   Assessing design maturity determines how well a system's design is progressing when measured against its allocated cost, schedule, and quality goals. To ensure accurate and timely assessments of design maturity, an organization's design policy must document key design review procedures and guidelines.

Specific issues covered in the policy include required design reviews, participation, entrance and exit criteria for the reviews, risk assessment techniques, test and evaluation procedures, and design verification practices. Design reviews are covered in detail in Part 8, Design Reviews.

Common design policy shortfalls

Lack of a documented corporate design policy can lead to an undisciplined, inefficient system design process. Common design policy shortfalls include:

- a corporate design policy does not exist or does not reflect government and industry-wide engineering best practices

- systematic implementation of the design policy is not formally supported by its chief executives

- costs, not supported engineering practices, guide the implementation of the policy

- lessons learned are not incorporated into the policy in a timely manner

Design policy in contract awards

Often, portions of system design and development are contracted outside of the initiating organization. With outside contracts, a design policy should be established before a contract is awarded. Also, before selecting a source, the primary contractor should review the design policies of contractors to ensure that they support the overall policy for the system. It is the responsibility of the primary contractor to communicate the design policy to all subcontractors and to certify compliance. For more details, see *Moving a Design into Production.*

Step 2: Review the Design Requirements and the Design Goals

A set of design requirements is generated during the exploration and proposal phase of a system development. Part 1, Front End Process, details the tasks and activities that result in these requirements. Together with the mission profile, this set of requirements provides the basic design parameters from which the system design specifications are developed.

Review design requirements

Before product design begins, the design engineers must review the set of system, functional, and operational requirements to ensure that they are complete and prioritized, and that they contain no contradictions, irrelevancies, or unnecessary restrictions on the system design. Because they form the core from which the entire design is developed, a poorly defined set of requirements can result in an unrealizable design, or one that is much too costly to develop.[10]

Ensuring the completeness of requirements is a very difficult task. As discussed in Step 1, the design policy should include the recommended practices and procedures for developing the system's requirements and the methods used for their review. Frequently, design organizations will develop requirements checklists and use them during reviews to ensure that all critical requirements have been addressed and are stated in sufficient detail. Designers must also review combinations of requirements to certify there are no conflicting elements. Any discontinuities must be resolved with the customer.

The Application chapter gives an example of a checklist that designers use to assess the completeness of the user requirements in a radar system.

Review design goals

The design goals of a system identify important product and process issues that must also be considered during design. Although related to functional and operational requirements, design goals frequently address aspects of a system that are not strictly concerned with product features. In some cases, design goals conflict with each other or with the requirements and thus necessitate tradeoff decisions to be made by the designers and the customer.

Most system design goals fall into one of the following six areas:[11]

- *Product cost:* choose solutions to realize lowest product cost

- *Development cost:* emphasize minimum incurred costs during development

- *Development interval:* base decisions on need to complete system as quickly as possible

---

[10]Petts III, George E. "Radar Systems" in *Handbook of Electronic Systems Design.* Ed. Charles A. Harper. New York: McGraw-Hill, 1980, p. 6–22.

[11]Reinertsen, Donald G. "Whodunit? The Search for the New Product Killers," *Electronic Business,* July 1983.

- *Product performance:* emphasize meeting customer's needs and expectations of quality and reliability

- *Risk taking:* identify specific level of risk that will be accepted

- *Flexibility:* emphasize decisions that enhance system's ability to adapt to program changes

Design goal risks.   Identifying and ranking design goals is not easy. Ideally, a design solution satisfies all goals but, since several are inherently contradictory, no design can possibly satisfy them all. The critical challenge is to identify the right goals for the right product. Table 1 identifies possible advantages and potential risks of the six design goals described above.

Control changes.   After the entire design team reviews the design requirements and the design goals for a product, full scale development can begin. A change control system should be used to baseline the requirements and goals and to control any changes that may be required. When changes are necessary, they must be communicated to and reviewed by all disciplines involved in the system development.

Step 3: Define the Appropriate Design Process

After requirements and design goals are reviewed, the design team must define an appropriate design process for the product. If this is not done with a clear understanding of customer needs, product goals, and design policy and practices, the risks of an inefficient and ineffective design process increase.

Important activities when defining a design process are:

- tailor the design process
- choose the right team
- consider life cycle costs
- consider downstream phases
- manage change
- document the design plan

Although discussed individually below, all these activities are closely related. A poor decision made with respect to one activity will adversely affect all other activities. As a result, valuable time, money, and resources may be wasted.

TABLE 1   Advantages and Risks of Design Goals

Design Goal	Advantages	Risks
Reduce Product Cost	■ creative, low cost solutions	■ lower system quality and performance
	■ low cost (market advantage)	■ higher life cycle costs
		■ fail to meet customer needs
Reduce Development Costs	■ lower system costs	■ lower system quality and performance
	■ low cost (market advantage)	■ higher life cycle costs
		■ fail to meet customer needs
Reduce Development Interval	■ timely delivery to customer	■ increased technology risks
	■ early system introduction (market advantage)	■ higher system and development costs
	■ more efficient development process	■ higher life cycle costs
		■ lower system quality and performance
		■ fail to meet customer needs
Increase Product Performance	■ higher performance (market advantage)	■ higher system and development costs
		■ higher life cycle costs
		■ longer development interval
Increase Risk Taking	■ lower development costs	■ lower system quality and performance
	■ lower system costs	■ higher life cycle costs
	■ advanced technology	■ fail to meet customer needs
Increase System Flexibility	■ lower life cycle costs	■ higher system and development costs
	■ meets customer and user needs	■ longer development interval
		■ poor performance

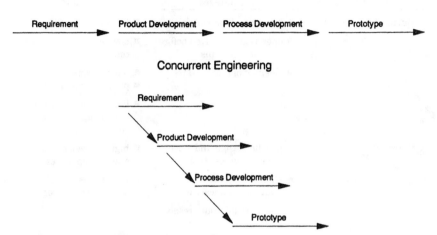

Figure 4 Sequential vs. concurrent engineering.

Tailor the design process

A key factor in the choice of a design process for a particular product is an understanding of both the advantages and disadvantages of different methods. The best design process for a product is one in which all the advantages support the product's goals.

Sequential vs. concurrent engineering. Figure 4 illustrates two different types of product development processes, sequential engineering and concurrent engineering.[12]

Some of the advantages and disadvantages of sequential and concurrent engineering are given in Table 2.

Sequential engineering. Sequential product development efforts have separate organizations with distinct product responsibilities and outputs. This traditional type of development process is used in many organizations because it allows an orderly progression of the product throughout the various phases of the development effort. Lower development risks are often attributed to this method because handoffs between organizations usually do not occur until all open issues and uncertainties are resolved.

---

[12]Winner, Robert A. and others. *The Role of Concurrent Engineering in Weapons System Acquisition*, IDA Report R338. Alexandria, VA: Institute for Defense Analyses, December 1988, p. 12.

TABLE 2   Advantages and Disadvantages

Development Process	Advantages	Disadvantages
Sequential Engineering	■ lower development risk	■ longer intervals
	■ open issues resolved before next phase begins	■ downstream issues not considered early enough
	■ clear set of roles and responsibilities	■ more rework
		■ product waits in queue
		■ little communication among organizations involved
Concurrent Engineering	■ parallel design and development activities	■ higher development risk due to open issues
	■ shorter intervals	■ need for assumptions in design and development
	■ team communication	
	■ downstream issues resolved early	
	■ less rework	

A common problem with sequential product development is that the organizations involved in design and development do not adequately communicate about the design early enough in the product's life cycle. As a result, key issues such as maintainability, manufacturability, or assembly may not be fully addressed until the product has reached the appropriate organization in the flow. Any changes that are then required to support these downstream activities frequently are difficult to make and very costly.

Another problem typical with this method is the increased development intervals because formal handoffs and signoffs between groups take time, thus frequently causing a product to wait in queue at the various group boundaries.

Concurrent engineering.   Many problems associated with the traditional product development process are addressed through the use of a concept called *concurrent engineering.* The DoD defines this as "a systematic approach to the integrated, concurrent design of products and their related processes, including manufacture and support."[13]

Other terms such as *simultaneous engineering* or *concurrent development* refer essentially to the same principles which are based on the con-

---

[13]Winner, p. v.

cept of a parallel development effort. The fundamental advantage of this parallelism or concurrency is that it causes many issues to be addressed simultaneously, resulting in much shorter development intervals. In addition, issues affecting later product phases, such as manufacture or assembly, surface early. These issues are addressed during design, resulting in less rework as the product advances through its life cycle.

Although widely supported, concurrent engineering does have disadvantages to consider when choosing a design process. The most significant is the possibility of increased development risk. The reason this problem occurs is that for parallel activities to exist in the process, all open issues cannot be resolved before proceeding to another phase. Some technical decisions must be based on assumptions that will not be validated or invalidated until later in the process. Thus, an important factor in the success of a concurrent engineering process is managing this level of risk.

Choose the right team

A core ingredient in the success of a product development effort is the development team. Recent emphasis on the integration of the product design with manufacturing and deployment processes has placed new, and often difficult, demands on the team. Three key considerations are membership, size, and location.

Multifunctional teams. When organizing a design and development team, one of the primary goals should be to facilitate communication between people whose work is interrelated. A design team should have representation from other disciplines. This type of team, often referred to as a *multifunctional* or *multidiscipline* team, may typically consist of representatives from systems engineering, electrical design, mechanical design, software development, manufacturing engineering, production engineering, service engineering, materials engineering, quality engineering, reliability engineering, process engineering, marketing, purchasing, production management, installation, maintenance, and suppliers.

Organizational structures. How a company organizes for product development is another aspect that affects the team. There are several alternative ways to do so. One study[14] evaluating team organization and project performance compares two structures: functional organization (i.e., groups organized by their technical specialties) and product organization (i.e., a mix of disciplines). The study's results

---

[14]Allen, Thomas J. *Managing the Flow of Technology.* Cambridge, MA: The Massachusetts Institute of Technology, 1977, pp. 211–219.

indicate the choice of structures should be based on two key factors: project duration and the rate of change in the knowledge base.

Functional organization.　According to the study, a functional organization is better when a project is of long duration (e.g., three to four years or more) or when the technical knowledge base is changing rapidly. This is because state of the art information in technological fields is best transferred when colleagues interact, especially face to face. Removing technical specialists from their areas of expertise for an extended period of time can greatly hamper this communication.

Project organization.　On the other hand, the study indicates that projects of short duration or those based on stable technologies should use a project organization. According to Allen, this type of structure facilitates the coordination of various disciplines and specialties by shortening the communication paths among team members and by identifying a person responsible for a project to whom all team members report.

Matrix organization.　The matrix organization is another organizational alternative. In general, a matrix organization is "a hybrid organization in which the normal vertical hierarchy is 'overlaid' by a lateral project management system."[15]

In a pure matrix structure, a person generally reports to two managers. For example, a person from one of the technical specialties reports to both the technical manager as well as the project manager. The project manager is responsible for defining what needs to be done while the functional managers are concerned with how it will be accomplished. Both parties work closely together and jointly approve workflow decisions.

Proponents of matrix organizations claim that this structure is good in multiproject environments when resources must be shared across different projects. It is also good for providing a formal method of coordinating work across functional boundaries. One disadvantage of a matrix organization is that it requires flexibility due to the dual reporting structure, and the mixture of responsibilities often causes conflicts within the project.

Formal research about which of the three basic organizations (functional, project and matrix) is best is contradictory and inconclusive. The best method for any particular group depends upon the strengths and weaknesses of the group, past successes with different organizational structures, leadership, and the type of product under development.

---

[15]Larson, Erik W. and David H. Gobeli. "Organizing for Product Development Projects," *Journal of Product Innovation Management,* May 1988.

Support organizations.   Also critical to a development team's success are efficient support organizations such as accounting, manufacturing specifications, human resources, management information systems, and purchasing. Accurate and timely information from these groups is essential for many design and development decisions, especially when the speed of the development effort is a high priority.

Team size.   An inherent trap in establishing a team is that its size may grow too large. Determining optimal team size is not easy since it depends on many factors such as project size, product complexity, product innovation, and technology. Manzo[16] points out as team size grows, the number of interrelationships between members also grows. (See Figure 5.) Thus, more time is spent on communication rather than on productive work.

Characteristics of different-sized teams.   Large projects which require many team members should consider team size and communication when subdividing. Doyle and Straus[17] provide the following characterizations for four different-sized teams.

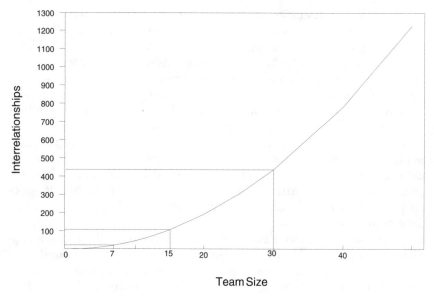

Figure 5 Size and team member interrelationships.

[16]Manzo, John. "Principles for Developing Complex Software and Hardware Systems." Course presented by the National Technological University Satellite Network Professional Development Programs, January 20, 1989, p. 22.

[17]Doyle, Michael and David Straus. *How to Make Meetings Work.* New York: The Berkeley Publishing Group, 1976, pp. 180–185.

2 to 7 members. With two to seven members, team meetings are easy to organize and can be flexible and informal. Detailed technical discussions are easier and logistics problems can be solved quicker. Achieving the best creative problem solving, however, may be difficult because only a few points of view are presented.

7 to 15 members. Seven to 15 members is an ideal size for a problem-solving, decision-making team. Meetings of all types can be held by the team while still allowing for informality and spontaneity. However, the team is complex enough to require some structure to function properly.

15 to 30 members. Fifteen and 30 members makes team communication extremely complex. In addition, spontaneity is lost, and team members tend to form subgroups with hidden agendas.

More than 30 members. With more than 30 members, teams require rigid rules. Even so, chaos is often the result. If team decision-making is critical, it is often best to split the team into two or more subgroups working on subsets of the problem.

Co-location. As discussed above, lack of communication between team members can be a major obstacle to effective teamwork. With only limited exposure to other points of view, people tend to analyze problems from only their own technical reference. Additional communication problems arise because each of the different disciplines involved in system development (e.g., hardware design, software design, and manufacture) use the specialized terminology of their own disciplines which is unfamiliar and confusing to outsiders.

One solution is to co-locate team members. A study supporting this solution suggests that the probability of communication diminishes by 80 percent when the participants are more than 50 meters apart.[18] Some of the benefits attributed to co-location include shorter development cycles and lower costs resulting from fewer design changes and less rework.

Of course, co-location is not feasible in many cases. When it is an option to be considered however, a complete analysis must be conducted to determine if the benefits of geographically locating team members together outweigh the many cost and personnel hurdles of doing so.

Consider life cycle costs

Life cycle costs are another issue to consider when defining a design process. With high technology designs, product and process complexi-

---

[18]Allen, pp. 236–240.

ties provide many opportunities for costs to grow. Using techniques to control costs is a key to success.

For example, estimating the cost of a component over the entire projected life of a system can be based on several key rules:[19]

- estimate costs over the entire production life and base tradeoffs on total costs; do not base costs on today's prices or prices at mid-production

- apply judgment to supplier's estimates based on:

  supplier's conservatism

  competitive environment (forced price reductions)

  maturity of product (new technologies tend to reduce in price faster than old ones)

  total market demand (total volume)

  similarity of components

Design to cost. *Design to cost* is one concept for system design that considers cost just as important as performance or scheduling. It does not imply "designing for the lowest possible cost," but rather implies setting realistic cost goals without sacrificing the system's basic functions and, most importantly, without sacrificing quality.

Design to cost techniques. In design to cost, the philosophy applied to product and process design is to hold cost constant and have the "design converge on the cost," instead of the other way around. To do this, a step-by-step iterative process is used in which analyses and trade studies continually set the limits for acceptable cost, schedule, and performance. These limits are then revisited throughout the program to ensure their validity and the adherence to them.

Figure 6 depicts the steps required for an analysis of system cost-effectiveness.[20]

The feedback loop illustrated in the figure is important, because the feedback ensures that the mission objectives and the intended use of a system will be redefined if they cannot be satisfied within the boundaries of available resources and operational constraints.

Design to cost risks. Although the design to cost concept is used in

---

[19]Harrison Ph.D., Frederick R. "Digital Systems" in *Handbook of Electronic Systems Design.* Ed. Charles A. Harper. New York: McGraw-Hill, 1980, p. 8–19.

[20]Michaels, Jack V. and William P. Wood. *Design To Cost.* New York: John Wiley, 1989, p. 35.

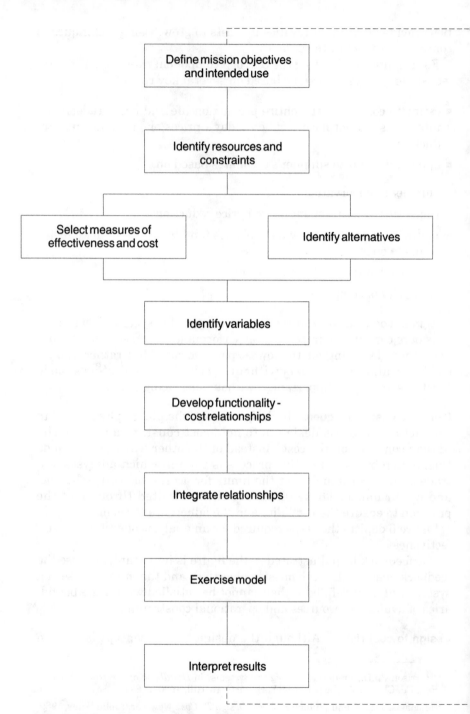

Figure 6 Steps in a cost-effectiveness analysis.

many programs, there are risks associated with its implementation that must be understood. They include:

- as much as a 20 percent increase in the front end costs of design due to the concept's requirement to examine many alternative designs[21]

- program slips into a "lowest cost" mode because of lack of management understanding or commitment to the basic design to cost concepts

- product quality is inferior due to lack of a stringent set of program controls to ensure both organizational and personnel adherence to the concepts

Consider downstream phases

The conventional focus of system design has been product functionality and performance. As the complexities of design continue to grow, the focus is shifting to a more comprehensive view, considering the effects of a design on downstream activities, such as manufacture, installation, or operation.

Off-line quality control.   Several years ago, Dr. Genichi Taguchi developed the framework for this type of comprehensive design strategy with a design method called *off-line quality control*. Its focus is to optimize a design using statistical techniques on a portion of the design, in an attempt to make a design less sensitive to variations in manufacturing, operation, and maintenance. Figure 7 illustrates off-line quality control.[22]

Optimization in off-line quality control.   In off-line quality control, optimization is accomplished through:

- *System design* to find the best technology for a product

- *Parameter design* to find the values which reduce the effect of variability

- *Tolerance design* to determine the tolerances that are required to ensure minimum loss of quality after manufacture, when the product is in use

Robust design.   Taguchi's method provided the foundation for a concept called *robust design* which is a systematic and efficient method of design optimization for performance, quality, and cost.

The key goals in robust design are:[23]

---

[21]Michaels and Wood, p. 8.

[22]Byrne, Diane M. and Shin Taguchi. "The Taguchi Approach to Parameter Design." *1986 ASQC Quality Congress Transaction,* American Society for Quality Control, 1986.

[23]Phadke, Madhav S. *Quality Engineering Using Robust Design.* Englewood Cliffs, NJ: 1989, p. xv.

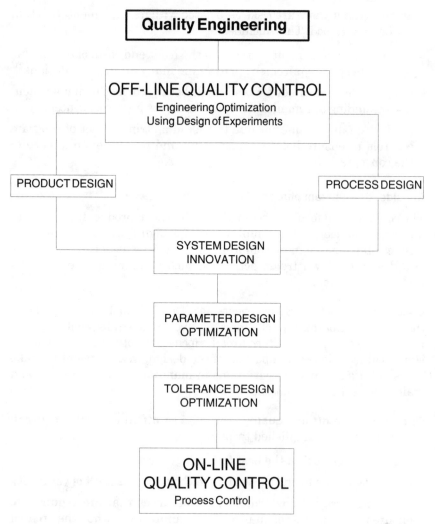

Figure 7   Off-line quality control.

- make product performance insensitive to raw material variations, thus allowing for the use of low grade material and components in manufacturing

- make a design insensitive to the effects of manufacturing variations, thus reducing labor, rework, and scrap

- make the designs insensitive to environmental variation, thus improving reliability and reducing costs

The strategy used to accomplish the goals of robust design is to develop a statistical experiment that will identify specific control parameter values that render the design insensitive to variation. (In robust design, the term "parameter" refers to any aspect of the product or process that can vary.)

Example of robust design.   A simple example of applying robust design is designing the paper tray and paper feed for a printer or copier. It is undesirable to have no paper (misfeed) or too much paper (multifeed) in the machine when the printer or copier is operating. (See Figure 8.) There are several major components that may cause paper problems: the feed roller, the spring, and the paper itself.

The first step in robust design is to identify the parameters of the components that may cause the problems. These are shown in Table 3.

After these parameters are set, the first step determines which characteristics have the most influence on the problem. The next step is another experimental design focusing on the parameter characteristics. A range of values is established for one parameter so that it is insensitive to variation from the other parameters.

Suppose the first step determined that the spring force constant was

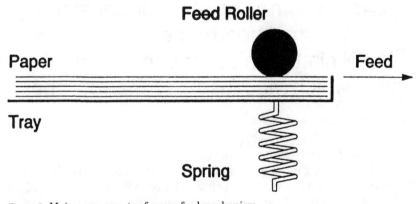

Figure 8 Major components of paper feed mechanism.

TABLE 3   **Potential Causes of Misfeed and Multifeed**

Design Parameters			
Component	Characteristic	Cause of Misfeed	Cause of Multifeed
Feed Roller	Surface	Smooth	Rough
	Size	Small	Large
Paper	Finish	Smooth	Textured
	Weight	Heavy	Light
Spring	Spring Force Constant	Light	Heavy

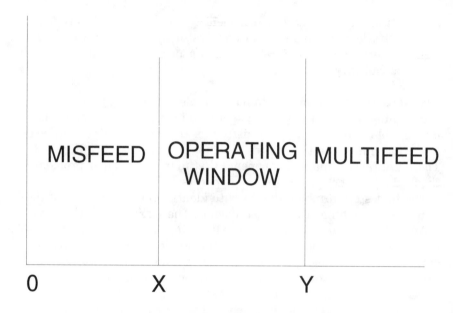

X = Minimum value of spring constant which will feed paper
Y = Minimum value of spring constant which will cause multifeed of paper

Figure 9 Operating window showing optimal range.

the key parameter to focus on. In robust design, an operating window is established (see Figure 9) so that paper will feed properly, regardless of variations in the paper texture or weight, or variations in the roller's surface or size. The robustness of the design is measured by the size of the operating window.

For a more detailed example, refer to the robust design case study in the Application chapter.

Producibility. *Design for producibility* is the ability to design and manufacture products in quantity, with a high degree of manufacturability, quality, reliability, and maintainability.[24]

---

[24]McLeod, Scott. "Producibility Measurement for DoD Contracts." Paper presented at Third Annual Best Manufacturing Practices (BMP) Workshop. San Diego, CA: September 13, 1989.

The three major factors involved in design for producibility are the design of the product, the manufacturing process, and the materials used. Producibility objectives for a system are most likely satisfied when the design approach maximizes the use of:

- simplified design and process
- standard materials and components
- economical manufacturing technology

while at the same time minimizing:

- number and variety of parts and materials
- use of limited availability items and processes
- special manufacturing tests

Planning for producibility early in the product's life cycle is an important part of the design process. The DoD has established a method for *producibility engineering and planning* called PEP. Simply stated, PEP is a systems engineering approach that ensures the end item will be produced efficiently and economically within the specified interval, and that it will meet its performance objectives within its design constraints. The primary purpose of PEP is to identify potential manufacturing problems and suggest a design that facilitates the production process.[25]

Design for "X" (DFX).   *Design for "X"* (DFX), where X stands for any key downstream activity, is a concept based on the strategic approach of taking into consideration the downstream activities related to the design of products and processes. See Figure 10.[26]

Early analysis, planning, and designing for aspects such as installability or operability can significantly reduce the problems that typically surface much later in product development. As the figure illustrates, DFX is comprehensive, covering all phases of a product's life cycle. Table 4 shows terms commonly substituted for the "X" in DFX.

Design for manufacturability and design for assembly.   *Design for Manufacturability (DFM)* and *Design for Assembly (DFA)* are two DFX

[25]Department of the Army. "Systems Acquisition Policy and Procedures." Army Regulation 70-1, Effective November 10, 1988, p. 92.

[26]Gatenby, D. "Design For 'X' (DFX): Assembly, Simplicity, and Beyond." Paper presented at CAD/CIM Alert International Conference on Design For Manufacturability. San Francisco, CA, June 20, 1989.

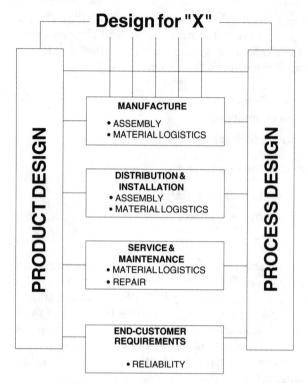

Figure 10 Design for "X."

TABLE 4 Terms Commonly Substituted for the "X" in DFX

Assembly	Reliability
Manufacturability	Quality
Maintainability	Safety
Testability	Operability
Compliance	Simplicity
Material Logistics	Installability

concepts that have been widely publicized and acclaimed for the benefits they achieve. Some of the important benefits from DFM/DFA programs are:

- early problem detection
- reduced costs of fixing problems early in the design process as opposed to fixing the same problems in later phases

- smoother transitions from development to manufacture and assembly

The Application chapter provides DFM and DFA checklists and guidelines as well as examples of what can be accomplished using DFA principles.

Manage change

An important part of the product design process is the scheme used to control changes. Configuration management establishes uniform communication methods among the various organizations involved in product design. It is the system manager's primary tool for controlling and maintaining an integrated and coherent system design and engineering effort.

In most cases, an organization's design policy will provide configuration management guidelines. However, products that have specialized requirements or goals, such as accelerated product development, may require modified change control schemes. For example, informal methods may replace some of the formal procedures normally used in change control.

Informal methods may increase design and development risks. Thus, it is essential to choose modifications carefully and to make sure the entire design and development team understands the modified process. Part 7, Configuration Control, provides a detailed discussion of this topic.

Document the design plan

After the design process for a product is chosen, program managers must ensure it is documented and communicated to the design and development team and, when appropriate, to the customer. Critical elements of the process to document include:

- organizational relationships and responsibilities
- technical approaches
- firm and realistic schedules and cost estimates
- reporting and control methods
- identification of any high risk areas in the design process

Step 4: Develop and Analyze Design Alternatives

After reviewing and baselining the design requirements and goals and defining the design process, the design team's next step is to start generating potential design solutions. The system size and complexity, the constraints on the system, and the technology required are all important aspects in the generation and review of design alternatives. Typically, the team may first choose a design approach appropriate for the type of system under design, and then will begin to partition the system. The combinations of different design approaches and different system partitions provide potential design alternatives that must be analyzed and optimized so that the team can choose the best solution.

Design approaches

Selecting a design approach for a development program is an important step. There are many design approaches documented in engineering literature today, some claiming extraordinary successes. There are also many failures of product design which can, in part, be blamed on the chosen design approach.

Top-down design and bottom-up design.   Traditionally, product design discussions have centered around two basic design philosophies: *top-down design* and *bottom-up design*. Figure 11 represents these two approaches.[27]

As depicted in the figure, top-down design begins with the entire set of required functions for the system and divides them into appropriate subsets until potential technologies are identified for them. Bottom-up

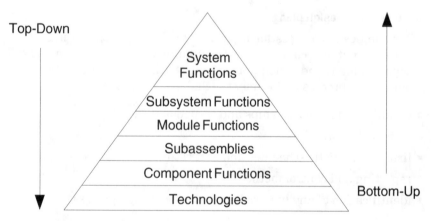

Figure 11 Top-down and bottom-up design.

---

[27]Leis, Charles T. "Dedicated Computer Systems" in *Handbook of Electronic Systems Design.* Ed. Charles A. Harper. New York: McGraw-Hill, 1980, p. 2–31.

TABLE 5   Advantages and Disadvantages

Design Approach	Advantages	Disadvantages
Top-Down Design	■ Parts designed to be optimal for application	■ Long development cycle
	■ Designer has more control during design phase	■ Specialized support required
	■ Relatively shorter system-integration time	
Bottom-Up Design	■ Common parts	■ Relatively long system-integration time
	■ Original manufacturer support	■ System cost possibly higher than necessary
	■ Relatively short development time	

design begins with a set of existing technologies, modules, parts, etc., and attempts to fit them together to satisfy the requirements.

Both design approaches have advantages and disadvantages as shown in Table 5.[28]

Typically, neither approach is used in its pure form. One reason is they both focus solely on functionality which, although important, is not sufficient for high quality system designs. In addition, the increased complexities of today's systems warrant a combination of approaches optimizing both existing and specially-developed technologies.

System partitioning

System partitioning identifies the functional and physical divisions of a system. Although different terms are used for various disciplines, a typical hierarchical ordering may be:

■ system

■ subsystem

■ unit

■ assembly

■ part or device

Partitioning breaks a large, complex, design problem into smaller, more manageable problems. Thus, a solution can be obtained incre-

---

[28]Leis, p. 2–32.

mentally. Partitioning also helps structure the problem, allowing an orderly method for obtaining a solution and the development of many subsections of the system at once.[29]

Although no method is universally accepted, partitioning usually begins with the design team deciding which aspects to base partitioning on. Functionality is the most obvious choice, but reliability, performance, repair and maintenance, testability, assembly, and ergonomics are also considered. It is important to keep the entire system in perspective to ensure interfaces function properly.

Partitioning tradeoffs. During system partitioning, the design team has difficult decisions to make about the tradeoffs that maximize the highest priority design goals, yet still satisfy other important goals. For instance, when partitioning a system to achieve high reliability, the design team may try to minimize the number of interconnections between modules, since these are frequent causes of lower reliability. Fewer connections, however, may result in a need for larger modules, which are generally less flexible, harder to test and maintain, and often more costly. Trade studies are frequently used to help make these design decisions.

Hardware vs. software decision. A common partitioning decision in today's complex, computer-controlled systems is hardware vs. software. Generally, several functions can be allocated to either the hardware or the software. The design team responsible for assessing the tradeoffs considers:[30]

- development costs
- development resources
- costs per system
- speed and accuracy
- reliability and maintainability

Partitioning and system costs. Cost is frequently an overriding factor in partitioning decisions. For example, if partitioning results in a design that includes many, very small modules, manufacturing costs may be

---

[29]Bell Telephone Laboratories, ed. "Systems Organization and Partitioning" in *Physical Design of Electronic Systems, Volume IV: Design Process.* Englewood Cliffs, NJ: Prentice-Hall, 1972, p. 413.

[30]Lemke, Richard L. "Measurement Systems" in *Handbook of Electronic Systems Design.* Ed. Charles A. Harper. New York: McGraw-Hill, 1980, p. 7–82.

disproportionately high as a result of increased material waste, number of connections, assembly, inspection, and test. On the other hand, if the modules are designed so that they are easy to manufacture, the higher yields may actually decrease manufacturing costs.

The costs of maintenance and logistics are additional aspects that are highly affected by system partitioning decisions. Again, module size is one important factor. For example, the complexities of large modules often hinder quick problem isolation, diagnosis and repair. In contrast, use of many smaller modules may mean more frequent module failures requiring complete replacement, thus increasing service costs and spare part costs such as handling, stocking, ordering, and record keeping.

Trade studies help to make these decisions. For more information on trade studies, see Part 1, Front End Process.

The make vs. buy decision.   A common decision a design team must address during system partitioning is whether particular parts (or software programs) should be specially developed for the system, or whether they should be purchased off-the-shelf. Many design factors such as those previously mentioned in this chapter play a major role in the decision. Additional significant factors include:[31]

- contributed value of the end product
- impact on project schedule
- specialization of staff and other resources
- availability of second sources
- future flexibility of the product
- licensing and support

Again, these decisions often require the team to assess design trade-offs. Some decisions may also require substantial involvement from management or from other organizations, especially when the trade-offs affect delivery dates, large capital expenditures, or long-term manufacturing and support arrangements.

Requirements allocation

Allocating a system's operational requirements to the lower levels of the system is part of the partitioning task. Typically, identifying alter-

---

[31]Harrison, "Digital Systems," p. 8-43.

native partitioning and allocation schemes is an iterative process. Designers must continually evaluate alternatives to ensure that no system functions are omitted from a design and that the functions assigned to each subsystem can be handled.

Identify performance budgets. Partitioning and allocation problems can occur when the capabilities of end items are exceeded by the functional demands placed on them. To help alleviate these problems, designers can develop *performance budgets* that apportion system performance values to lower levels of the system. This is often done for system power, optical power, subsystem reliability, and system reliability goals.

Example of budgeting. As an example of a reliability budget, consider a function that consists of two sequentially active elements. Using the "product rule," the reliability of the entire function is determined by multiplying together the reliability of each element. If the function's reliability goal is .94, for example, then each element must be budgeted its own appropriate reliability goal that ensures .94 will be attained by the function.[32] Several budgeting possibilities are shown in Table 6.

Performance budgeting is not an easy task for a design team. Some functions have a constant effect on a system's performance, others vary with environmental conditions or with the level of demand. A critical aspect of system design is accounting for the supply and demand between the requirements and the capabilities of a system.

Successful apportionment of performance requirements is a result of experienced engineering judgment and a solid knowledge of the:

TABLE 6   Example of Reliability Budgeting

Element A Reliability	Element B Reliability	Function Reliability (Reliability of A × Reliability of B)
.97	.97	.94
.96	.98	.94
.99	.95	.94

[32]Chase, Wilton P. *Management of System Engineering.* New York: John Wiley & Sons, 1974, p. 60.

- critical performance requirements for a system
- complexity of the system
- basic objectives for each portion of the design
- interrelationships of various subsystems (including redundancies)[33]

Follow design rules. Over the years, numerous design rules have been developed to ensure manufacturability, testability, reliability, and low cost. *Design rules* are fundamental principles that establish design constraints. During partitioning and requirements allocation, designers must ensure the design rules are not violated.

Design rules cover a wide range of topics. For example, general categories of rules for the design of a printed circuit board may include:

- manufacturing: drilling, plating, soldermask application, component insertion, component placement, soldering, inspection, and repair
- testability: bare-board test, assembled board test, and the number and placement of test points
- electrical characteristics: conductor path width, conductor path spacing, layout dependent crosstalk, net path length, and parallelism[34]

Design rules originate from several sources. Some come from within the design organization as a result of past experiences, and some come from factories that publish their own set of factory-specific rules. Many design rules, however, are from industry wide standards groups or regulatory agencies.

Examples of design rules. Below are illustrations of design rules:

- for the output power density on low-voltage power supplies, do not exceed six watts per cubic inch with switching-mode techniques and .5 watt per cubic inch with linear techniques
- provide positive locking devices to ensure retention of settings on adjustment or alignment devices susceptible to vibration or shock
- for copper thick film technology, use a grid-type or a bus-type plane; do not use solid ground or power planes
- on multilayer interconnect boards, locate vias within the footprint pads if possible

---

[33]Gryna, "Product Development," p. 13.21.

[34]Burling, W. A. and others. "Product Design and Introduction Support Systems." *AT&T Technical Journal,* vol. 66(5), September/October 1987, p. 21-38.

- use alumina or beryllia for ceramic circuit boards to prevent solder joint cracking under extremes of temperature

- use non-activated rosin with 25 to 35 percent solids content; use solder temperature of 245° C within 5° to improve solderability of printed wiring boards using type "R" flux

Design guidelines.   Closely related to design rules are *design guidelines*. These are statements of good design practice, typically derived from years of experience. Design guidelines are directives that should not be violated during the design process without a clear understanding and acceptance of the penalties that may result or of the tradeoffs that must be made.

In order to develop successful designs, a design team must be aware of all the design rules and guidelines that apply to a particular product. Computer-aided design (CAD) systems are used extensively for this purpose. From predefined sets of rules and guidelines, CAD systems assist designers in making choices during design by identifying rule or guideline violations and by verifying that a design meets all standards. Part 5, Computer-Assisted Technology, provides more detail about CAD systems.

Conduct design analyses

Every tentative design for a product must be thoroughly analyzed to determine its compliance with the product's requirements and design goals and to determine the risks of the approach. Effective analyses must be conducted early in the design process by experienced designers. The results of the analyses must be used to refine both the system design and the subsequent test program. When this is done

- fewer system problems are found during development and operational testing

- the technical risk of the design is reduced

- costs savings during test are realized

Figure 12 shows how analysis on the design of a radio at Rockwell International resulted in 333 reliability improvement actions during design.[35]

Numerous CAD systems exist today to assist designers with analy-

---

[35]Willoughby, Willis J. Jr. "Moving R&QA Out of the Red Light District." *The ITEA Journal of Test and Evaluation* vol. 9 (3), 1988, p. 18.

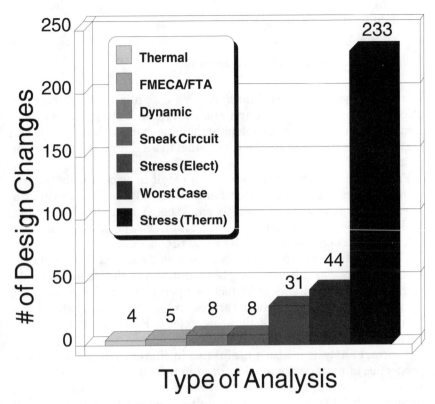

Figure 12 Design analyses improve designs.

sis. Specialized software is available for modeling, prototyping, simulation, and maintaining test data and results. (See Part 5.) To conduct effective design analysis, however, a designer must understand the analyses and how they can be used to reduce technical risks. The analyses listed below are commonly used to evaluate system designs. A brief description of each is provided.

- thermal analysis
- stress analysis
- failure mode, effect, and criticality analysis
- fault tree analysis
- worst case analysis
- sneak circuit analysis
- finite element analysis

- reliability analysis and prediction
- environmental analysis
- safety assessment and hazard analysis

Thermal analysis.   *Thermal analysis* is a discipline that studies the effects of temperature through simulation, modeling, testing, and component qualification. Developing realistic thermal design goals and requirements is critical since each product has a thermal profile unique to its design.

Failures induced by temperature can result in electrical shorts, mechanical failure, electrical failure, and stress failure. By studying the effects of temperature extremes (thermal shock), heat transfer, cooling, and device behavior under operating conditions, designers can make informed decisions about reliability and product life.

Thermal stress is one of the most common sources of failure in electronics. By measuring operating temperatures of electronic parts, determining configurations to reduce temperatures, and incorporating component derating, designers can reduce system failure rates. For a more thorough discussion on derating, refer to Part 3, Parts and Defect Control.

Thermal analysis is also a useful tool in characterizing the physical properties of parts and materials used in a design. For example, analyses can determine the coefficient of expansion, heat capacity, melting temperature, thermal stability, modulus of elasticity, effects of voids, and dielectric constant.[36] Figure 13 illustrates how thermal analysis can be used to understand the effects of temperature on a system's performance. In Figure 13, the projected failure percentages of a TTL device are determined according to the chip's junction temperatures.[37]

Figure 14 is an illustration of a circuit board thermal profile. These profiles are used to determine cooling strategies, component behavior under operating conditions, component placement, layout, and reliability considerations. The Application chapter also presents an example of thermal analysis used effectively in the design process.

Stress analysis.   The primary purpose of stress analysis is either to check a design's integrity against conditional extremes or to categorize

---

[36]Smoluk, George R. "Thermal Analysis: A New Key to Productivity," *Modern Plastics,* February 1989, pp. 67–73.

[37]Jacobson, David W. "Importance of Accurate Thermal Analysis." *Proceedings, 1989 IEEE Annual Reliability and Maintainability Symposium,* p. 466.

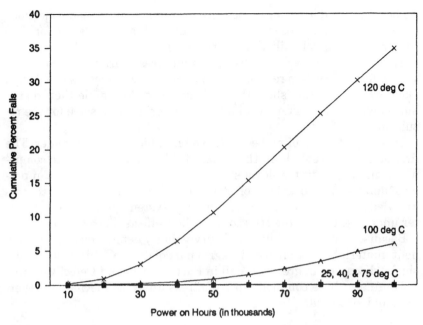

Figure 13 Effect of junction temperature on failure rates.

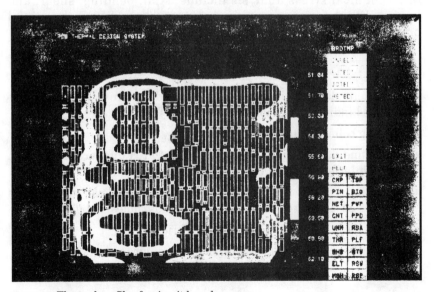

Figure 14 Thermal profile of a circuit board.

its behavior under various loads. By examining these load conditions early in the design process, a designer can evaluate the design for such requirements as reliability and survivability.

Stress analysis involves considering all stress parameters to assess their effects on system reliability and performance. Once a set of design loads has been established, the designer must determine their distribution within the system, evaluate the materials, and solve for the resultant stress levels.

Stress analysis is often best achieved in a hierarchical manner. The first step is to break down the system to the smallest possible components and apply system-level stresses to them. The analyses can provide details on the effects of stress on each component. With this knowledge, it is then possible to build the system back up, with a better understanding of the stresses and their effects on the system.

Because system reliability and survivability depend on each component, material selection is an integral part of design. This aspect of design is discussed in greater detail in Part 3, Parts and Defect Control.

There are three common types of stress analysis: mechanical, thermal, and electrical.

Mechanical stress analysis.   Mechanical stresses are caused by loads induced from the interaction of two physical interfaces. Forces studied in mechanical stress analysis include axial, bending, shear, and torsion. For these analyses, it is important to understand material properties as well as the forces at work.

The most common types of mechanical failures encountered in practice are:[38]

- permanent deformation under static load which causes misalignment or binding of components

- buckling collapse due to elastic instability

- creep (time-dependent deformation) at elevated temperatures which causes interference between components or relaxation of fits

- fatigue failures due to repeated applications of low-level service loads at stress concentrators (e.g., keyways, welds, holes, notches)

- wear in parts caused by inadequate lubrication

- degradation of materials due to corrosive environments

---

[38]Jones, Douglas L. "Mechanical Properties and Fracture Mechanics" in *Materials and Processes.* 3rd ed. Eds. James F. Young and Robert S. Shane. New York: Marcel Dekker, 1985, pp. 115–116.

- unanticipated service overloads such as impact or fatigue loadings on a component designed for static loads

Thermal stress analysis.   Thermal stresses are induced when the environmental temperature of a system (or its components) changes. Again, it is important to understand how the materials will behave under thermal loads. Common problems studied in thermal stress analyses include:

- creep deformation due to elevated temperatures over a period of time
- material instability
- temperature cycling stresses (e.g., failures caused by thermal mismatch of materials)
- plastic deformation caused by loads greater than the limit of elasticity
- elastic deformation, how a solid deforms elastically in response to an applied stress.

Panel (a) of Figure 15 illustrates that a metal rod bent beyond its elastic limit does not recover. Panel (b) illustrates that the glass rod shows perfect recovery.[39]

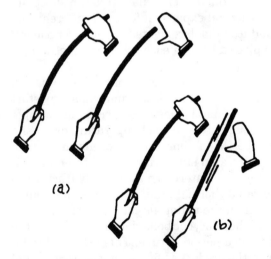

Figure 15 How deformation affects metal and glass rods.

---

[39]Britton, Marvin G., P. Bruce Adams, and J. R. Lonergan. "Ceramics: II. Nature and Properties of Glass" in *Materials and Processes, Part A: Materials*. 3rd ed. Eds. James F. Young and Robert S. Shane. New York: Marcel Dekker, 1985, p. 749.

Electrical stress analysis. Electrical stresses are induced in much the same way mechanical stresses are. The source for the load, however, is often system power, current, or voltage. Much like other stresses, electrical stresses in components can be induced as the system is powered on and off. This power cycling can often have the effect of wearing out devices.

Electrical stresses include:[40]

- field effects and induced currents
- electrochemical reactions in components
- impurity diffusion
- current densities
- degradation
- shorts
- burn-out

It is difficult to examine each of these analyses separately due to their close interactions. As an example, consider an integrated circuit (IC) mounted on a ceramic substrate. This substrate is attached to a printed circuit board with solder joints. As the IC receives power, it begins to generate heat. As this heat is conducted (or radiated) to the substrate and board, the coefficient of thermal expansion for each different material results in different induced displacements for each component. These components begin to place a load on each other and mechanical stresses are induced.

Depending on the environment, a number of scenarios are possible. If the temperature is held at a constant level, the materials may start to relax or creep. If plastic deformation hasn't occurred, the components may return to an unstressed state. If the temperature drops significantly over an extended period of time the materials may creep somewhat before ending in a stressed state. However, if the temperature drops sharply and quickly, the materials may actually fracture as the shear induced from the rapid thermal gradient causes the components to "spring back" faster than the materials can handle.

Finally, the system may see thermocyclic behavior that can cause the assembly to flex over an extended period of time. This may lead to fatigue and possible failure over the product's life. Testing under thermal cycling or thermal shock conditions is explained in greater detail in Part 6, Design for Testing.

---

[40]Hnatek, Eugene R. *Integrated Circuit Quality and Reliability.* New York: Marcel Dekker, 1987, pp. 590–591.

Failure mode, effect, and criticality analysis. A *failure mode, effect, and criticality analysis (FMECA)* is a study of how a proposed system design may fail. The effects of a failure and the criticality of those effects on a system are identified and analyzed. The results help pinpoint areas of improvement needed to reduce the risk of a system failure.

The first step in an FMECA is to identify the failure modes of a system or in the part of a system under study. A failure mode is "the observable or measurable condition or parameter state which accounts for the part [or system] not functioning."[41]   Or more simply, a failure mode is a symptom of a failure, not the cause or mechanism that produced the failure.[42]   For example, a possible failure in a spring-loaded hatch opening may be that one of the metal springs used for tension breaks. The failure mode in this case is a broken spring. The next steps in an FMECA are to determine the cause of the failure, identify effects of the failure on the system or mission, and rate the seriousness of each effect.

An example of a portion of an FMECA report for the spring-loaded hatch opening mentioned above is provided in Table 7. From this in-

TABLE 7   Example from an FMECA Report

Component/ (Part Number)	Failure Mode	Cause of Failure	Effect of Failure	Probability of Occurrence	Degree of Severity (1 is low 10 is high)
spring x (1-XZ7)	broken spring	corrosion of spring	loose hatch	5%	9
		corrosion of spring fasteners		8%	
	tension loose	excessive wear		3%	
spring y (2-RS4)	broken spring	corrosion of spring	loose door handle	5%	5
		corrosion of spring fasteners		8%	

[41]Winans, R. C. "Reliability of Electronic Parts" in *Physical Design of Electronic Systems, Volume IV: Design Process.* Ed. Bell Telephone Laboratories. Englewood Cliffs, N.J.: Prentice-Hall, 1972, p. 334.

[42]Gryna, "Product Development," p. 13.28.

formation, the team can assign design priorities and recommend alternatives for reducing the risk of failure.

Fault tree analysis. A *fault tree analysis (FTA)* is a technique for describing and assessing undesired events in a system. It is similar to the FMECA described above in that it begins with a list of failure modes for the system. It differs, however, in both approach and output. In a fault tree analysis, the sequence of normal and abnormal events, and the combination of events, that can lead up to a failure in the system are hierarchically modeled.

The model used in a fault tree analysis graphically depicts how undesired events can occur in a system. It is developed using a standard set of symbols and when completed resembles an inverted tree-like structure.[43] See Figure 16.

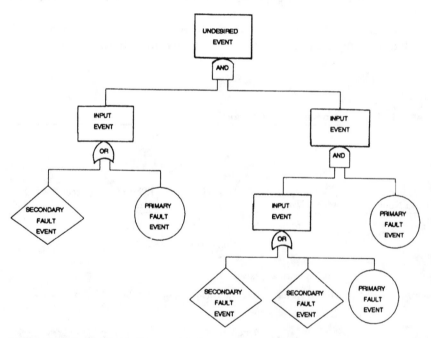

Figure 16 Example of a fault tree analysis.

---

[43]Young, Jonathan. "Using the Fault Tree Analysis Technique" in *Reliability and Fault Tree Analysis*. Ed. Richard E. Barlow, Jerry B. Fussell, Nozer D. Singpurwalla. Philadelphia, PA: Society for Industrial and Applied Mathematics, 1975, p. 828.

As Figure 16 illustrates, the undesired event under study is at the top of the hierarchy. Contributing events branch out at lower levels and then end with a level indicating the origins of the events. Frequently, probabilities are assigned to each of the events in the model. The subsequent analysis of the model calculates the probability of the undesired event under study and thus aids in a prediction of the total system's reliability.

A fault tree analysis is usually an iterative process carried out until the risk of failure falls within acceptable levels. The steps required to conduct a fault tree analysis include:[44]

- become familiar with the system
- define the undesired events of the system with the related contributing and initiating events (e.g., component failure, human error, spontaneous reactions, external conditions)
- develop fault tree(s) for the undesired events
- obtain probabilities for the events on the fault trees
- evaluate the fault trees
- analyze the results and proposals for system improvement
- change the fault trees to reflect proposed improvements and renewed fault tree evaluation

Worst case analysis.   In system design, the purpose of a *worst case analysis* is to verify a proposed design will meet its requirements under the most extreme conditions or combination of conditions. The team must conduct this type of analysis to build reliability into a system.

To perform a worst case analysis, a designer must first assume anything can go wrong, including internal problems such as component degradation due to stress or time as well as external problems due to environmental extremes.

Second, the designer must be aware of what can go wrong in a system and what extremes could possibly be reached. Third, a designer must analyze the effects of these extremes on system operation and performance, either singly or in combination.

For many system designs, real life examples of worst case situations are not available and cannot be created. Advances in computer-aided design (CAD) techniques, however, make it possible to explore many

---

[44]Hauptmanns, Ulrich. "Fault Tree Analysis for Process Plants" in *Engineering Risk and Hazard Assessment, Volume I*. Eds. Abraham Kandel and Eitan Avni. Boca Raton, FL: CRC Press, 1988, p. 25.

different worst case scenarios and, in addition, drastically reduce the required time.

Sneak circuit analysis. A *sneak circuit analysis (SCA)* identifies unexpected paths that may occur in a system. Unlike many of the analyses conducted during system design, sneak circuit analysis does not concentrate on the failures of system components. Instead, SCA attempts to identify latent or hidden conditions in a system that may cause it to fail, even though all of its components are operating according to their specifications. This unique aspect of an SCA is its primary benefit.

Particular attributes of systems have been identified that can cause these latent or "sneak" conditions to be unintentionally introduced into a system. Designers must be aware of these attributes and conduct sneak circuit analyses when appropriate. These attributes include:[45]

- highly complex system designs
- system designs experiencing a high rate of change
- systems with a large number of interfaces to other systems
- systems with complicated operating procedures

Common applications of SCA are seen in electronics, power supply, and control systems. The benefits of an SCA, however, are not limited to these areas. This technique can be successfully conducted on the hardware, the software, the manual procedures used to operate the system, or any combination of these three. For example, the types of problems to examine in the sneak circuit analysis of the operator's procedures include:[46]

- *errors of omission:* the failure of the operator to perform a task or part of a task indicated in the procedures
- *errors of commission:* the operator performs a task or step incorrectly
- *extraneous acts:* the operator introduces some task or step that should not have been performed
- *sequential errors:* the failure of the operator to perform the tasks in the correct order
- *timing errors:* a task or step is performed too early or too late, i.e., not performed within an allotted time interval

---

[45]Department of the Navy. *Sneak Circuit Analysis.* NAVSO P-3634, August 1987, pp. 3–8.

[46]Hauptmanns, p. 35.

Finite element analysis. *Finite element analysis (FEA)* is a mathematical technique for predicting the physical behavior of a system. This is done by first developing a set of simultaneous equations that mathematically describe the physical properties of both the system under study and the problem being analyzed, and then obtaining solutions to the equations to approximate behavior.[47]

The concept underlying the finite element method (FEM) is "divide and conquer." In the analysis, the *domain,* which usually refers to the region of space occupied by the system, is divided into many smaller regions called *elements.* The resulting pattern, as illustrated in Figure 17, is called a *mesh.*[48]

A numerical evaluation of equations that define the elements in the mesh gives a set of equations, often hundreds or thousands of equations, which define the entire system's response.[49]

The types of problems common in an FEA include: stress analysis, heat transfer, fluid flow, vibration, and elasticity. FEA is especially valuable in systems when no other method is available to evaluate the design, such as in a nuclear reactor. It is also an important analysis when a design prototype would prove too costly or too time-consuming, such as an airplane frame. Rodamaker cautions, however, that although FEA can solve many design problems if added intelligently to other design capabilities, some problems are so complex that an FEA cannot produce an accurate model and should be avoided.[50]

When the finite element method was first developed, much of the work was done manually. Today, many finite element computer pro-

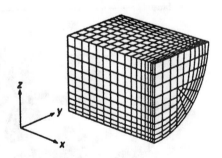

Figure 17 Finite element analysis mesh.

---

[47]Burnett, David S. *Finite Element Analysis.* Reading. MA: Addison-Wesley, 1987, pp. 3–10.

[48]Burnett, p. 8.

[49]Burnett, pp. 5–9.

[50]Rodamaker, Mark. "Optimizing Productivity Through Finite Element Analysis/ Finite Element Meshing." Paper presented at CAD/CIM Alert International Conference on Design For Manufacturability, San Francisco, CA, June 18-20, 1989.

grams allow designers to optimize the development of a system's mesh structure, and thus produce very accurate models. Figure 18 illustrates the impact crushing of a steel nose cone: Panel (a) shows delivery and impact of a bomb; Panel (b) the initial mesh of a nose cone before impact; Panel (c) the deformed mesh after impact; Panel (d) a

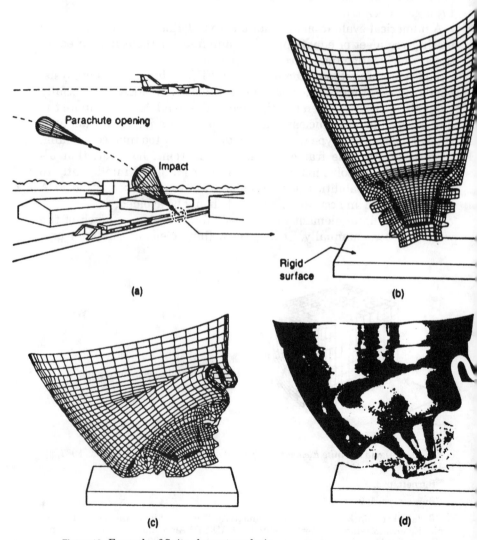

Figure 18　Example of finite element analysis.

photograph of a steel test specimen subjected to impact.[51]

Figure 18 shows an example of how FEA can predict the behavior of a system, and how close the resulting approximation (Panel c) can come to the actual situation (Panel d). The Application chapter also provides a detailed example.

Computer color graphics are also used extensively to illustrate the results of an FEA and are frequently seen in analyses such as heat transfer where a gradation of colors can graphically represent the areas of highest or lowest temperatures.

Reliability analysis and prediction.   Reliability is generally one of the most important concerns in system design, and one that a designer must consider in every aspect of the design. It would be ideal for a designer to base every design decision on the resulting system reliability. This ideal, however, never exists. Time, cost, performance, size, and environment are just a few of the many design constraints that must be balanced with reliability.

Predictions of a system's reliability begin early and continue throughout the stages of the design process. Because these are only *predictions* of reliability, designers must use them cautiously. Designers, however, can use predictions to:[52]

■ determine acceptable system performance levels by considering probable failure rates of its components

■ compare relative hazards within a system to improve performance of the parts most likely to fail

■ analyze and measure actual failures in existing equipment to determine where design changes can improve performance

A description of the factors that reliability predictions are based on and the uses of the predictions in the various phases of system design are provided by Juran and illustrated in Table 8.[53]

Environmental analysis.   During manufacture, shipment, storage, installation, operation, or maintenance, a system may be exposed to an extreme environmental condition or a combination of extreme conditions. A successful system design depends on the system

---

[51]Burnett, p. 21.

[52]Simon, William F. "General-Purpose Computer Systems" in *Handbook of Electronic Systems Design.* Ed. Charles A. Harper. New York: McGraw-Hill, 1980, p. 1-74.

[53]Gryna, "Product Development," p. 13.23.

TABLE 8　Reliability Predictions at Various Design Phases

At the Start of Design	
Prediction Based on:	Primary Uses
Approximate part counts and part failure rates from previous product usage; little knowledge of stress levels, redundancy, etc.	■ Evaluate feasibility of meeting a proposed numerical requirement ■ Help to establish a reliability goal for design
During Detailed Design	
Prediction Based on:	Primary Uses
Quantities and types of parts, redundancies, stress levels, etc.	■ Evaluate overall reliability ■ Define problem areas
At Final Design	
Prediction Based on:	Primary Uses
Types and quantities of parts failure rates for expected stress levels, redundancies, external environments, special maintenance practices, special effects of system complexity, cycling effects, etc.	■ Evaluate reliability ■ Define problem areas

operating during extreme conditions. To achieve such a design, an extensive study of the system's complete environmental envelope is conducted. This study, called an *environmental analysis,* is important in reducing risk in system design. Table 9 provides a list of some environmental factors to consider.

Temperature is one of the most common issues addressed in an environmental analysis. For instance, the physical designer of a system must take into account that the properties of metal and alloys at room temperature differ from those at either very high or very low tempera-

TABLE 9　Factors to Consider in Environmental Analysis

Environmental Factors		
temperature	power disruptions	dust
wind	gravitational fields	electromagnetic interference
chemical pollutants	acceleration	high vacuum
shock	vibration	biological growths
humidity	light	salt spray
noise	radiation	pressure

tures. One example is the drastic reduction of the impact strength of structural steel at subnormal temperatures.[54]

Also important is analysis of the shock and vibration a system may experience during its operation or during any phase of its life cycle. For example, if the design of a system or its packaging has not accounted for the extreme conditions conceivable in shipping and handling, an inoperable system may be delivered to a customer.

Extremes of humidity is another common environmental problem to consider during system design. Corrosion of metal parts in a system, due to a high relative humidity, can be significant in the system's reliability and expected life.

Wind is an important environmental factor in naval operations. Aside from the direct effect of wind on a surface ship, the combination of high winds and low temperatures can cause ice to form on a ship's superstructure. Wind also has an indirect effect on underwater ambient noise and must be considered in naval radar systems.[55]

The list of environmental factors addressed during the development of a design varies substantially depending on the type of system under design. In general, the system's mission profile and the design requirements delineate the extremes to design for.

Statistical information about environmental conditions throughout the world is available. This, coupled with computer-aided design systems, provides designers with the ability to simulate the effects of environmental extremes on a system and, most importantly, to develop a design that can withstand them.

Safety assessment and hazard analysis. *Safety assessment and hazard analysis* are linked very closely to the environmental analyses of a system design since, as Figure 19 illustrates, the environment affects the system and the system affects the environment.[56] The system's effect on the people who manufacture, operate, and maintain it is the primary subject of safety and hazard studies.

A reasonable assumption is that no person or company wants to develop a system that can cause damage to another human being. Unfortunately, unexpected events occur during the life cycle phases of a system that cause human sickness, injury, or death. These events can be classified in one of four ways:

---

[54]Schlabach, T.D. and P.R. White. "Mechanical Properties of Metals" in *Physical Design of Electronic Systems, Volume II: Materials Technology*. Ed. Bell Telephone Laboratories. Englewood Cliffs, NJ: Prentice Hall, 1972, pp. 188–189.

[55]Stefanick, Tom. *Strategic Antisubmarine Warfare and Naval Strategy*. Lexington, MA: Lexington Books, 1987, p. 307.

[56]Leis, "Dedicated Computer Systems," p. 2-34.

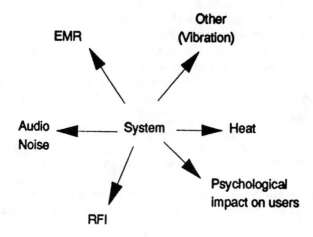

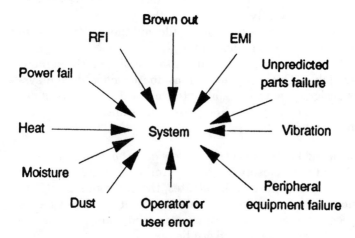

Figure 19 Environmental and system effects.

- unwanted energy transfer (e.g., kinetic, heat, electric)
- unwanted material transfer (e.g., chemicals, microorganisms)
- unwanted energy blockage (e.g., very cold temperatures)
- unwanted material blockage (e.g., air, water)[57]

---

[57]Kumamoto, Hiromitsu, Yoshinobu Sato, and Koichi Inoue. "Hazard Identification and Safety Assessment of Human-Robot Systems" in *Engineering Risk and Hazard Assessment, Volume I.* Eds. Abraham Kandel and Eitan Avni. Boca Raton, FL: CRC Press, 1988, p. 66.

When a potential safety problem or hazard is identified in a system design, the first step is to try to remove the hazard. An obvious way of doing this, for example, is replacing a potentially toxic chemical with a non-toxic chemical.

When hazard removal is not possible, the next step is to identify and analyze causes and provide ways to protect against them. For example, some systems emit an excessive acoustic noise level that becomes a hazard to system operators. Designing a casing for the system that lowers the noise level a human is exposed to is one way of reducing the risk of hearing loss for the operators.

The reality of system design, however, is that preventing every imaginable hazard is impossible. Even if it were feasible, the amount of resources needed are probably not available. Therefore, risk level targets are generally set for a system's hazards. Safety assessment and hazard analysis must show that the upper level of risk for a particular hazard will never be exceeded. In addition, they must show that everything "reasonably practicable" has been done to reduce the risk of a hazard as much as possible.[58]

For some systems, substantial historical data exists about the probability of a particular hazardous event. With data, hazard analysis is relatively straightforward and a system design can be developed using the existing data. Data generally exists when the system under design is not new. For example, much medical data exists about the hazards of human exposure to x-rays.

A new or innovative design lacks historical data. In particular, with long-term hazards, such as human exposure to a new chemical compound, the size of the problem is probably not even known. In these cases, the outcome of the safety assessment or the hazard analysis of a system may result in the rejection of the proposed design.

Step 5: Choose a Design

The most critical decision in the entire design process is choosing the design solution. Although presented here as an individual step, this decision is a culmination of all previous design activities. The various trade studies, benchmarks, and analyses that occur in the design process help to eliminate many of the original choices. Ultimately, the design team, with the customers, must examine the risks and tradeoffs of remaining alternatives and choose a solution for the system that ap-

---

[58]Kletz, Trevor A. "Setting Priorities in Safety" in *Engineering Risk and Hazard Assessment, Volume I.* Eds. Abraham Kandel and Eitan Avni. Boca Raton, FL: CRC Press, 1988, p. 12.

proaches (as nearly as is practical) the best application of requirements, goals, and constraints.

The decision process

When the conditions (e.g., environmental, stress, user) of a system are known with certainty, a decision maker's job is relatively easy. The chosen design solution maximizes (or minimizes) all important payoffs, such as best quality, least noise, shortest development interval, or least cost.

In reality, however, system conditions are generally not known with certainty. Trade studies and analyses usually identify only the probability of some condition and therefore of some payoff. Thus, a decision maker's first job is to determine clear cut decision criteria.

Decision criteria.   Decision criteria provide a way to measure design alternatives. A decision maker must be very certain about the criteria used because failure to consider and examine important aspects of alternatives can greatly increase the risks of success.

Two commonly used decision criteria are:[59]

- the maximin criterion
- the maximax criterion

The *maximin criterion* is based on the view that the chosen alternative should yield the highest value among all the minimum payoffs. Although somewhat pessimistic, this approach is usually low risk.

Example of choosing a design.   As an example,[60] consider a system in which noise reduction at all noise levels is an important goal. For this system, four alternative designs for acoustic treatment are offered. Figure 20 depicts the alternatives, showing predicted percent of noise reduction per unit cost (predicted payoffs) at four different ambient noise levels.

Table 10 is a decision matrix constructed for the same four alternative solutions. The figures in Table 10 are the payoffs (i.e., percent of noise reduction per unit cost) of the four design alternatives at each noise level.

---

[59]Cottingham, W. B. and P. W. McFadden. "Decision Theory" in *Physical Design of Electronic Systems, Volume IV: Design Process.* Ed. Bell Telephone Laboratories. Englewood Cliffs, NJ: Prentice-Hall, 1972, p. 22-30.

[60]Adapted from Cottingham and McFadden, p. 23.

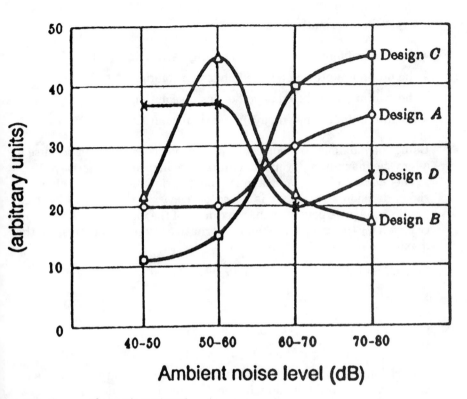

20 Effectiveness of acoustic treatment.

TABLE 10 **Decision Matrix of Design Alternatives**

Design Alternatives	Noise Level (dB)			
	40 to 50	50 to 60	60 to 70	70 to 80
A	21	21	30	35
B	22	44	22	18
C	12	16	40	45
D	37	37	20	25

From the information provided:

- the minimum predicted payoff of design solution $A = 21$
- the minimum predicted payoff of design solution $B = 18$
- the minimum predicted payoff of design solution $C = 12$
- the minimum predicted payoff of design solution $D = 20$

Using the maximin decision criteria, alternative A is chosen because it offers the maximum payoff from all the predicted minimum payoffs.

The *maximax criterion* is a much more optimistic, but possibly more risky, decision strategy. It is based on choosing the alternative that predicts the peak payoff, or the maximum of the maximums. Using this criterion for the sample problem, alternative C will be chosen since the maximum predicted payoffs are:

Design Alternative	Maximum Predicted Payoff
A	35
B	44
C	45
D	37

Decision makers must be careful when using this type of decision strategy. Frequently maximizing one parameter in a design can negatively affect other important factors. In the sample system, design alternative C predicts the best payoff of 45 for the 70-80 dB noise level, but also predicts the worst payoff of only 12 for the 40-50 dB noise level. If noise reduction at both noise levels is equally important, then basing a decision on the best of the best may be the wrong strategy.

Design optimization

Optimization is part of the design decision. This is choosing design parameter values that can produce the best system performance in a specific situation. In complex systems, optimization is a very difficult process, particularly when the values of many parameters are restricted by constraints on the system.

Computer-aided tools are widely available today to assist design engineers with the search for optimum solutions. Designers must be aware, however, that a key risk in optimization is that the benefits of looking for an absolute optimum may reach the point of diminishing returns before the optimum is found. Thus, the most prudent design solution may not be the optimum, but rather an approximation that is a compromise among competing parameters, including cost and time.

Step 6: Baseline the Design and Obtain Customer Approval

Once the design solution for a system is chosen, a formal design review is essential. Primary objectives of this review are to:

- evaluate the design for consistency with the system's operational and performance requirements
- gain client approval or concurrence before significant work is done on the next phase of development

Part 8, Design Reviews, provides detailed information about required design reviews, particularly for government contracts.

Upon approval, a system design should be baselined and placed under configuration control. The baseline is an agreement between the client and the development team that becomes the foundation for all further work on the system. Managing change to the baseline then becomes fundamental for successful program development.

Although changes to a system design baseline can usually be expected, these changes, in most cases, should merely clarify ambiguities or resolve open issues in the original baseline. No changes, however, should be incorporated without a complete evaluation of their impact, particularly on system performance, cost, and schedule. In addition, no changes should be made to a baseline without client knowledge and approval.

More detailed information about baselining and configuration management is available in Part 7, Configuration Control.

3

Application

This chapter provides nine applications of the methods and techniques presented in the Procedures chapter.

- An example of a corporate design policy is presented.

- Development speed as a design goal is difficult to accomplish and introduces many technical risks. A checklist for rapid development projects is provided.

- A complete and consistent set of requirements is essential for effective system design. A checklist to review completeness of user requirements is provided.

- Consideration of downstream phases of system design such as manufacturing is important. A Design for Manufacturability (DFM) checklist is provided.

- The complexities of product assembly significantly impact schedule, cost, and quality. Design for Assembly (DFA) principles are listed.

- The effective use of Design for Assembly principles in the development of a computer printer is illustrated.

- Parameter design is a technique of robust design to make a product insensitive to variations. The steps of the robust design method and a detailed example of parameter design are included.

- An example of thermal analysis is presented to show the importance of lowering junction temperature.

- The use of finite element analysis is illustrated in the design of a fiber optic connector.

Design Policy Example

Below is an example to illustrate sections from a corporate design policy manual.

Purpose

Improve the transition from development to production by setting up design, producibility, supportability, documentation, and manufacturing requirements for programs.

Application

Applies to all Federal Systems programs in design, development, logistics, procurement, and production of hardware, firmware, and software.

Design practices

- Minimize electrical and mechanical adjustments; mention necessary adjustments in the design review package.

- Keep junction temperatures of semiconductor devices as low as possible within technical and economic constraints. Refer to the standardization requirements for the maximum allowable junction temperatures for each class of device.

- Provide test access through the card or unit connector to permit automated testing of circuit boards and assemblies, including fault isolation to the component or circuit level.

- Design circuit boards for automatic insertion and wave soldering, automated placement and reflow soldering, or vapor phase soldering technologies.

- Use standard concepts, configurations and components, whenever possible, as defined by CAD/CAM engineering and manufacturing design rules.

- Use a reliability indicator in all hardware units (e.g., measuring standby time, operating time, laser shot counters).

- Do a system tolerance analysis to support dimensional tolerances on all optical system fabricated parts.

- Have key manufacturing people participate during product and requirements definition, high-level design, and detailed design.

- Include representatives from systems engineering, producibility, material selection, and logistics on the multifunctional team that approves the product design layout.

- Use trade studies to establish product cost objectives.

- Maintain a history file of configuration, technical data, and test results in controlled archives.

Responsibilities

Group President: Approve any exclusions from this policy.

Program Managers: Establish and disseminate program criteria and requirements, ensure the needed funding, and develop the compliance plan.

Systems Engineering: Prepare and control specifications, approve test specifications, retain history of configurations and technical results.

Design Engineering: Implement design requirements and standards, review and approve production and test equipment.

Product Operations: Provide producibility engineering support.

Logistics: Review and approve design for logistics support requirements.

Choosing a Development Goal

Rapid product development is not appropriate in all situations. Speed is vital for short life cycle products like consumer electronics and when a new or upgraded system is needed as soon as possible to counter a military threat. Due to its difficulties and risks, however, speed as an overriding goal is appropriate only when an analysis shows it will have clear advantages over a slower approach that minimizes risk and development costs.

Analysis of data and forecasts

On a commercial project, the project team makes decisions based on an economic analysis of data and forecasts, rather than on philosophical beliefs. An economic analysis has these steps:

- estimate a profit and loss statement and balance sheet
- do several "what if" calculations to see the effect of various scenarios (e.g., being six months late, being on time but exceeding the development budget by 10 percent)
- compare the relative effects of each of these scenarios on profit, customer relations, and accomplishing the mission
- use the results to decide whether to emphasize development speed or a more conservative, lower risk strategy

On a government project, the project team also uses data and forecasts, but not for an economic analysis of profit and loss and market share. Instead, these data and forecasts describe when a new or upgraded system must be ready to meet a need or counter a threat. The team considers tradeoffs between development speed, system and development cost, and minimizing risks. Development speed may be vital, for example, in developing detection devices for antisubmarine warfare. Minimizing costs and safety risks, however, may be the most appropriate goals in space exploration projects. The team must analyze the mission profile and the design requirements to see whether an emphasis on rapid development is the best approach.

Rules of thumb

Table 11 shows useful rules of thumb when development speed is an appropriate goal.[61]

TABLE 11   A Checklist for Rapid Product Development

Choose a clear, limited objective
Don't mix new technology with new applications or customers
Aim for an adequate solution, rather than a perfect one
Meet the needs of leading-edge customers

Select the team
Select experienced, full-time members
Keep size to eight to ten people

Partition the design to allow work in parallel
Define interfaces early
Don't try to optimize each subsystem
Have one person in charge of the overall design

Use informal control systems
Locate all team members together
Limit reporting requirements
Simplify the product specification process
Involve management early to help decide key issues
Ensure all members know the cost of delay

Spend money as needed
Speculate on long-lead time tools, parts, and equipment
Use quick-turnaround outside services as needed

Test key design concepts early
Test technical and marketing concepts separately with simulations and early prototypes

---

[61]Reinertsen, D. G. "Blitzkrieg Product Development: Cut Development Times in Half." *Electronic Business*, January 15, 1985.

User Requirements Checklist

As system design begins, it is important for designers to review the functional, operational, and performance requirements of a system to ensure completeness and consistency. In many cases, designers develop and use checklists as a basis for their review.

A product manager in the Radar Systems Group of Hughes Aircraft Company recommends that radar system designers use a checklist (Table 12) as a guide to ensure completeness of user requirements. If the appropriate system specifics are substituted for the radar specific items, the checklist illustrated in Table 12 can be used in a wide variety of systems.[62]

TABLE 12   User Requirements Checklist

Categories of Requirements	Necessary Elements
Operational scenario	*Target:* Types, number, statistical characteristics, dynamics *Radar:* Structure, platform motion, vehicle type, propagation effects, weather, terrain
Radar modes and functions	Surveillance, detection, tracking, warning, navigation, mapping, weapon control, weapon delivery, terrain following, terrain avoidance, target classification, collision avoidance, approach control, weather detection
Features and capabilities	Built-in test, automatic fault isolation, constant false-alarm rate, electronic scan, mechanical scan, automatic responses, special controls, specific hardware requirements, spectrum and waveform characteristics, programmability
Radar characteristics	Frequency allocation, polarization, time-bandwidth product, electronic counter-countermeasures capability, mutual interference rejection, clutter rejection, up-look, down-look, tail chase, coherent or mono-coherent
Performance	Probability of detection, false-alarm rate, data rate, accuracy, system coverage (range, velocity, acceleration, angle), resolution (angle, range, velocity), clutter rejection, electronic counter-countermeasures capability, interference rejection

---

[62]Petts III, " Radar Systems," p. 6-25.

TABLE 12   User Requirements Checklist (*Continued*)

Categories of Requirements	Necessary Elements
Design for key downstream activities	Quality, safety, maintainability, reliability, manufacturability, electromagnetic compatibility
Cost	Cost of ownership, life cycle cost, acquisition cost
Environment	Temperature, humidity, biological factors, corrosive elements, shock
Mechanical	Size and weight limits, dimensional limits, cooling types and availability, materials restrictions and requirements, human factors
Electrical	Prime power characteristics, quantity, quality, availability, user restrictions
External Interfaces	Electrical, electronic, digital, mechanical, human operator
Instrumentation	Data output, formats, reduction, presentation, and analysis
Optional features	Multiple users, growth, optional modes

Design for Manufacturability (DFM)

Manufacturability has been defined as the "art and science of designing a product that is easy to manufacture."[63] Design for Manufacturability (DFM) has been widely publicized and acclaimed for the benefits achieved when the design and manufacture organizations work together early in the design phase. Manufacturing problems are fewer and potential problems are identified and corrected early in the design phase. As a result, these design changes cost much less than if they occurred later. (Refer to Part 1, Front End Process, for more details.) Also, if the team considers DFM concepts early, a smoother transition from development to production is realized.

A DFM checklist helps assess manufacturability during design and in design reviews. Table 13 has examples of DFM considerations that lead to improved designs, reduced manufacturing costs, and therefore higher product quality.

---

[63]Tanner, John P. "Product Manufacturability." *Automation,* May 1989, pp. 26–28.

TABLE 13   Design for Manufacturability Checklist

Materials

Is material available in standard stock configurations (e.g., bar stock, sheet, standard extrusion)?

Is material compatible with the most desirable manufacturing process (e.g., ease of forming, casting, machining)?

Is the material available from reliable sources?

Do material prices fluctuate widely over time?

Are special alloys and exotic materials used only for environmental or functional demands?

Fabricated parts

Are specified tolerances reasonable for functional requirements?

Are tolerances attainable within the normal capability of the manufacturing process to be used?

Are data points, surfaces, and tooling points clear and accessible?

Does part configuration minimize the need for special processes and special tooling?

Product assembly

Are tolerance dimensions realistic?

Is marking and stenciling defined and visible?

Are assembly notes complete and definitive?

Is internal wiring layout critical? If so, is the location and routing specified?

Are test points and adjustments accessible?

Is harness development required? If so, can the harness be fabricated outside the unit and installed as a subassembly?

Does the design lend itself to automated assembly?

Are component parts accessible for assembly?

Can testing be performed without disassembling the unit?

Are standard connectors and assembly hardware used?

Are circuit cards, if used, designed to plug in?

Has the assembly been analyzed to meet electrical, thermal, vibration, and shock specifications?

Can printed circuit flex cable or molded ribbon be used in place of hard wiring?

Can plastic tie-wraps be used in place of lacing or spot ties?

Design for Assembly

Traditionally, a product's producibility was only measured by manufacturing efficiency and capability. Design for Assembly (DFA) requires that producibility include quality, flexibility, responsiveness, and simplicity. When applied properly, DFA accelerates product introduction, increases quality, and reduces overhead costs. These improvements can reduce the need for design changes later, when the costs are much higher.

Below is a partial list of DFA principles:[64]

- *Design for a minimum number of parts.* Design multifunctional parts—fewer parts mean less of everything.

- *Develop a modular design.* Assign common functional requirements to standard modules. If possible, isolate functions likely to change in one module.

- *Don't fight gravity.* Minimize the number of assembly directions and maximize the use of top-down insertion.

- *Reduce processing surfaces.* Finish all processing on one surface before moving on to the next.

- *Assemble in the open.* Provide a clear view and easy access for part placement. Blind assembly increases quality risks.

- *Eliminate fasteners where possible.* Incorporate the fastening function into another component (e.g., snap-fit parts). If fasteners are needed, use common types and materials.

- *Design for part identity.* Part symmetry yields easier handling, orientation, and reduces quality risks. If symmetry is not possible, overemphasize asymmetry since exaggerated features can also facilitate identification and orientation.

- *Optimize part handling.* Avoid flexible parts that are costly to automate. Maintain part orientation throughout manufacture. Avoid parts that tangle, stick together, or are slippery.

- *Design for easy part mating.* Specify part consistency. Design compliance into the product and the process. Provide adequate guide surfaces to avoid misalignment.

- *Provide nesting features.* Nested parts should not need to be repositioned during final assembly. Parts should not nest, however, during handling or feeding.

---

[64]Shuch, L. K. "DFA Promises and Delivers." *Assembly Engineering,* May 1989, pp. 18–21.

- *Develop a reliable, durable design.* Avoid designs that require unusual or specialized skills. Eliminate clutter from interconnected components. Avoid designs that are sensitive to uncertainties like weld strength or lubrication condition.

Figure 21 illustrates what is possible when DFA principles are applied to a product's redesign.[65] Panel (a) shows an assembly with 4 major components and 20 fasteners. Panel (b) shows a redesign option with separate sheet metal components in which the number of fasteners has been reduced from 20 to 4; the number of parts is 8. Panel (c) shows another redesign in which plastic fasteners hold the spindle in place; the number of parts has been reduced from 24 to 4. Panel (d) shows the ideal redesign; DFA reduced the number of parts from 24 to 2.

Figure 22 shows the total costs of the four designs in the previous figure.[66] For each design, costs were estimated for manual and robotic assembly. As the part count decreased, the materials and manufacturing costs remained relatively stable, but assembly costs were substantially reduced. This example illustrates a key DFA objective—simplify a design so that it is easier and less costly to assemble. An interesting side effect in this example, and in many DFA applications, is that the final design has been so streamlined that the costs of manual assembly are almost the same as the costs of robotic assembly.

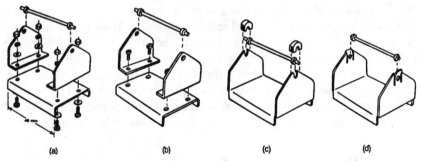

(a)   (b)   (c)   (d)

Figure 21 Product redesign using DFA.

---

[65]Boothroyd, G. "Making It Simple: Design for Assembly." *Mechanical Engineering,* February 1988, pp. 28–31.

[66]Boothroyd, pp. 28–31.

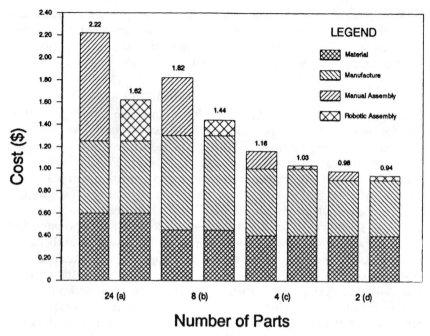

Figure 22   Effect of DFA on product costs.

Design for Assembly Example

The IBM Proprinter, which became available in 1986, was an innovative design because all parts and subassemblies snapped together during final assembly without the use of fasteners or screws. It was an excellent design for manual and robotic assembly. The Proprinter could be assembled ten times faster than the popular and reliable Epson MX80, the Proprinter's primary competitor. This increase in assembly speed was due partly to the significant reduction in the number of assembly operations (32 operations for the Proprinter vs. 185 for the MX80) and partly to the use of effective assembly principles. Table 14 shows how the Proprinter applies design for assembly principles.

Design for key downstream activities

The increased ease of assembly occurred because assembly experts were part of the design team. Teams need representatives from each key downstream activity to minimize the risk of focusing on one aspect and neglecting others. With the Proprinter, for example, ease of maintenance and repair was not given as much attention as ease of assembly. As a result, the Proprinter is hard to service.

TABLE 14   Design for Assembly Principles in the IBM Proprinter

Assembly Principle	Application
Use snap fits instead of screws or fasteners	Molded leaf spring positions the shaft, which eliminates the need for linkages and pins
Use simple downward assembly motions	Steel shafts guide the printhead and paper
Avoid belts and springs (hard to handle)	Printhead has a molded plastic screw, which eliminates the need for belts
Use the printer base as an assembly fixture	Base is a fiber-reinforced molded part that acts as a pallet; two side frames minimize tolerance demand
Design so that no tools or screws are needed for assembly	Locating holes of different sizes guide snap-fit assembly; all parts snap into place including the two motors and the power supply

Proprinter's design for ease of assembly

The Proprinter was designed for simplicity and ease of assembly. Figure 23 shows an exploded view of the Proprinter's 32 parts and sub-assemblies to illustrate its design for assembly.[67] The figure shows that parts can be assembled using simple downward motions, using snap fits rather than screws or fasteners. The Proprinter with no fasteners and 32 parts can be assembled ten times faster than the Epson with 74 fasteners and 49 parts.

Robust Design

As discussed in Step 3 of the Procedures chapter, the principles of Genichi Taguchi and others make designs more robust, i.e., insensitive to variations in materials, components, manufacturing, and users' environments. A useful robust design method is parameter design which improves quality without increasing costs. The example below illustrates the basic principles of parameter design.[68]

Parameter design

The goal of parameter design is to find parameter values or levels that make the product or process function with high quality and minimal

---

[67]Dewhurst, P. and G. Boothroyd. "Design for Assembly in Action." *Assembly Engineering*, January 1987, pp. 64–68.

[68]Adapted from S. Taguchi. *Taguchi Methods: Quality Engineering.* Dearborn, MI: American Supplier Institute Press, 1988.

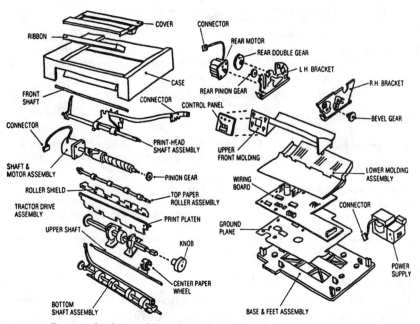

Figure 23 Parts and subassemblies of the IBM Proprinter.

sensitivity to noise or random variation. The basic principle in parameter design is to distinguish between control factors and noise factors. Control factors can be set at a particular level; noise factors cannot be set at specific levels because it is too difficult or expensive. Thus, noise factors will vary. The noise variation can be due to external causes (e.g., temperature, curing time), internal causes (e.g., wear, deterioration), or part-to-part variation (e.g., weight or pressure greater or less than the nominal value). Drawing on knowledge and experience, designers choose control factors that, when combined and used at their optimal settings, are likely to significantly improve the robustness of the product or process under a variety of noise conditions.

Parameter design experiments are valuable because they are an inexpensive way to find optimal settings of control factors that give high quality regardless of the noise conditions. In the experiment described below, the objective was to increase the tear strength of plastic to be coated onto aluminum and make the plastic less likely to tear regardless of the surface roughness of the aluminum, the initial metal temperature, or changes in the plastic supplier.

Control factors

Table 15 shows nine control factors which can increase the tear strength of plastic to be coated onto an aluminum base. For each of these control factors, two settings or levels were examined in a experiment to find the combination of levels that together will produce greater tear strength.

A statistical measure of performance called the signal-to-noise (S/N) ratio was used to evaluate the tear strength. The S/N ratio measures the level of performance and the effect of noise factors on performance. Three types of ratios are available. The choice depends on the objective: improve a desirable characteristic ("larger is better"), reduce an undesirable characteristic ("smaller is better"), or adhere to a standard value ("nominal is best"). This experiment was designed to improve tear strength; thus the "larger is better" S/N ratio was appropriate.

Experimental conditions

Table 16 gives the orthogonal array that shows how the experiment studied the control factors. The array is called orthogonal because the main effects are independent of each other due to the proportionally balanced columns in the array. A carefully selected orthogonal array requires fewer experimental runs than a complete factorial design.[69] In this experiment, for example, 12 runs will give essentially the same information as the 512 runs of a complete factorial design ($2^9 = 512$). With this strategy, however, a follow-up confirmation experiment is necessary to make sure that assumptions made in planning the experiment were not badly violated.

TABLE 15   Control Factors for the Tear Strength of Plastic for Plastic-Coated Aluminum

Control Factors		Level 1	Level 2
A	Molecular Weight	Low	High
B	Type of Aluminum	Existing	New
C	Cleaning Material	Existing	New
D	Percent Catalyst	Low	High
E	Grinding Pressure	Low	High
F	Curing Time	Short	Long
G	Curing Temperature	Low	High
H	Post Curing Time	Short	Long
I	Post Curing Temperature	Low	High

[69]Shoemaker, A. C. and R. N. Kacker. "A Methodology for Planning Experiments in Robust Product and Process Design." *Quality and Reliability Engineering International*, vol. 4, 1988, pp. 95–103.

TABLE 16   Conditions for the Tear Strength Experiment

Run	Level of the Control Factors Used in the Experimental Run								
	A	B	C	D	E	F	G	H	I
1.	1	1	1	1	1	1	1	1	1
2.	1	1	1	1	1	2	2	2	2
3.	1	1	2	2	2	1	1	1	2
4.	1	2	1	2	2	1	2	2	1
5.	1	2	2	1	2	2	1	2	1
6.	1	2	2	2	1	2	2	1	2
7.	2	1	2	2	1	1	2	2	1
8.	2	1	2	1	2	2	2	1	1
9.	2	1	1	2	2	2	1	2	2
10.	2	2	2	1	1	1	1	2	2
11.	2	2	1	2	1	2	1	1	1
12.	2	2	1	1	2	1	2	1	2

In Table 16, a value of 1 indicates the factor is set at level 1; a value of 2 indicates the factor is set at level 2. In the first run of the experiment, for example, level 1 was used for each factor. Tear strength was measured using:

- low values of molecular weight (A), catalyst (D), grinding pressure (E), curing temperature (G), and post curing temperature (I)
- existing types of aluminum (B) and cleaning material (C)
- short curing time (F) and post curing time (H)

In the second run, level 1 was used for A, B, C, D, and E; level 2 was used for F (long curing time), G (high curing temperature), H (long post curing time) and I (high post curing temperature).

Noise factors in parameter design

The experiment also explored three noise factors (Table 17) to find the combination of control factors least susceptible to noise.

A complete factorial design was not needed. The three noise factors, with two levels each, were combined into four representative noise conditions counterbalanced as explained below:

- N1 used the level 1 settings of each noise variable (1, 1, 1)
- N2 used the level 1 settings of surface roughness and level 2 settings of the supplier and metal temperature (1, 2, 2)
- N3 used the level 2 settings of surface roughness, level 1 of the supplier, and level 2 of metal temperature (2, 1, 2)

TABLE 17   Noise Factors for the Tear Strength Experiment

Noise Factors		Level 1	Level 2
M	Surface Roughness	Rough	Smooth
N	Plastic Supplier	Smith	Jones
O	Initial Metal Temperature	Low	High

- N4 used the level 2 settings of surface roughness and supplier and the level 1 settings of metal temperature (2, 2, 1)

Experimental data

The twelve experimental conditions were repeated under the four conditions of noise. The response studied was the tear strength of the plastic, measured using an Instron tester. (The Instron tester measures the stress-strain force needed to tear the plastic held in its two vice grips.) The force data are plotted in the first four columns of Table 18. From data in these four columns, a signal-to-noise ratio is computed for each row, using a formula appropriate for "larger is better." The last column presents these signal-to-noise ratios for each of the 12 conditions.

Finding the optimal settings

These S/N ratios are averaged to estimate the performance at each setting for each of the nine control factors. For example, the average of the first six S/N ratios in Table 18 estimates the performance of level 1 of

TABLE 18   Tear Strength Data Under Four Conditions of Noise

Run	N1	N2	N3	N4	S/N Ratio
1.	33	45	35	50	32
2.	63	68	55	80	36
3.	20	30	30	38	29
4.	20	28	28	50	29
5.	43	55	33	60	33
6.	23	45	48	50	31
7.	60	80	45	63	35
8.	45	48	28	55	32
9.	48	63	75	80	36
10.	55	63	55	48	35
11.	60	45	40	43	33
12.	45	35	20	30	29

TABLE 19   Average Signal to Noise (S/N) Ratios for Levels 1 and 2 of the Control Factors

	A	B	C	D	E	F	G	H	I
Level 1	31.48	33.30	32.45	32.75	33.65	31.36	32.83	30.90	32.25
Level 2	33.33	31.51	32.36	32.06	31.16	33.45	31.98	33.91	32.56
Difference	01.85	01.79	00.09	00.69	02.49	02.09	00.85	03.01	00.31
Rank	4	5	9	7	2	3	6	1	8
Optimal Level	A2	B1	C1	D1	E1	F2	G1	H2	I2

factor A; the average of the last six estimates the performance of level 2 of factor A. Similarly, averages were computed to estimate the performance of level 1 and 2 for each of the other eight control factors. These averages are presented in Table 19.

The ranks of the difference between the level 1 and 2 settings show how important the factor is in improving the tear strength. In this experiment, Factor H (Post Curing Time), with the biggest difference score, contributed the most. Factor E (Grinding Pressure) had the second largest difference score; Factor F (Curing Time) had the third largest. The best tear strength will occur if, with these three control factors, the level is used which had the larger value: longer post curing time (H2), low grinding pressure (E1), and longer curing time (F2). The factors ranked fourth and fifth also contribute to increased tear strength: high molecular weight (A2) and the existing type of aluminum (B1). The settings of the lower-ranked factors are not as important, but may make a marginal contribution.

Confirming the optimal combination

The results indicate the greatest tear strength will come from combining the optimal level of each control factor. To verify this selection, a confirmation experiment was run which compared the baseline combination used before the experiment with the optimal combination from the experiment.

Baseline Combination: A1, B2, C1, D2, E2, F2, G1, H1, I1

vs.

Optimal Combination: A2, B1, C1, D1, E1, F2, G1, H2, I2

Figure 24[70] shows the results of the confirmation experiment. The tear strength data shown in Figure 24 were entered into the appropri-

---

[70]Taguchi, p. I-49.

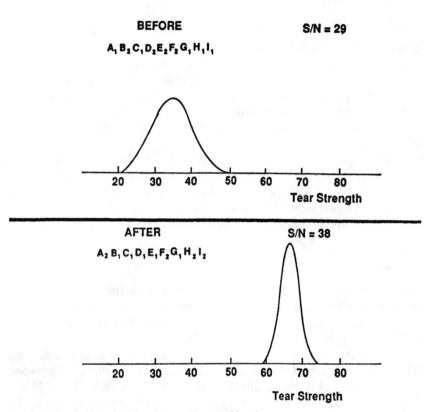

Figure 24 Tear strength at baseline and optimal levels.

ate formulas to obtain the signal-to-noise ratios. The optimal combination was verified: its signal-to-noise ratio is 38 vs. the baseline 29; the variation is much reduced.

Conclusion of robust design example

As can be seen from this example, robust design is an effective method for improving quality (and thus reducing risks) without increasing costs. Two attractive features of robust design are:

- the number of experiments needed to find a better design is manageable
- the final design is less sensitive than the initial design to variations in parts, materials, environments, etc.

Example of Thermal Analysis

The reliability of a computer component is directly related to the junction temperature at which it operates. Junction temperature is the temperature a device sees at a junction (e.g., at the connection to a circuit board or another device). In general, a small increase in junction temperature can result in a decrease in the component's reliability. Thus, it is important to accurately determine the junction temperature of a component.

Underestimating junction temperatures can lead to more field failures than originally projected. However, overestimating junction temperatures can be just as costly. Inaccurate information can lead to unnecessary and expensive design changes or use of components that are more reliable (and probably more expensive) than is necessary.

Background

During the design of a 16-megabyte memory card, the original thermal evaluation of the card indicated a junction temperature of 57°C. The reliability objective of the design was a mean time between failure (MTBF) of 670K power on hours (POH).[71]

Cost constraints required a memory chip whose MTBF was only 500K POH. Clearly, this was unacceptable from a reliability perspective. A memory chip was available which at 57°C had a MTBF of 1100K POH. However, this chip cost approximately $3 more per chip. The memory board design called for 160 chips per board, and the forecast minimum production was 10,000 boards. The added expense was projected to be $4.8 million.

Effect of lowering junction temperature

Since the added costs would have to be absorbed by either the manufacturer or the customer, the designers reviewed the design to try to use the less expensive chip while maintaining the MTBF requirement of 670K POH. By decreasing the junction temperature, the reliability projections for the original, less expensive chip could be increased.

As a result, small design changes lowered the junction temperature to 49° C. At this temperature, the chips now had a MTBF of 700K POH which exceeded the 670K POH requirement. In this case, the costs of

---

[71]Jacobson, pp. 465–469.

modifying the design to lower the junction temperature were minimal when compared to choosing more expensive components. In reducing the junction temperature, both reliability and cost objectives were met.

Finite Element Analysis

Background

The development of a transmission system required the design of a fiber-optic connector. This connector functions by bringing together two pieces of optical fiber. The design called for the ends of the fibers to be machined to a certain radius and held together under a spring-loaded force. See Panel (a) of Figure 25 below.

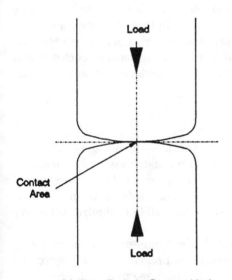

(a) Fiber Ends In Contact Under a Load

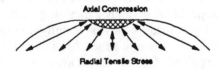

(b) Enlarged View of Contact Area

Figure 25   Contact areas of fiber ends.

Problem description

During the development of the system, it was determined that the fiber ends were fracturing in the connector, thus rendering the connection useless. To understand the problem, it was assumed that the compressive load at the contact area of the fibers was causing a radial tensile force across the surface of the splice. See Panel (b) of Figure 25. These forces were exceeding the fracture strength of the fiber which is 50,000 psi.

The problem was treated as two spheres in contact, because the ends of the fiber were machined to a curve. The radius of curvature determines the contact area where the axial loads are applied. The compression that occurs across this area results in radial tensile forces produced across the rest of the surface.

Spherical contact analysis is a highly complex, non-linear problem requiring intensive and laborious calculation. Often, the problem is one that cannot be realistically solved by hand. In this case, the complexity is compounded because the material (glass fiber) is not a perfectly elastic material, and solving the equations is not practical by conventional methods. Thus, FEA is an excellent tool to understand and solve the problem.

Modeling the problem

By taking advantage of axes of symmetry, a model of the problem can be divided into quadrants. Figure 26 is an illustration of the model divided along the y-axis of symmetry. To generate the finite element mesh, only one quarter of the area in question needed to be modeled, as the symmetry allows for the assumption that all four quadrants behave the same.

The model was developed with 2-D axis symmetric elements, using ANSYS,™ a general purpose finite element program.[72] (See Figure 27.) The detail of the mesh is finer at the top because more information is required there for greater accuracy.

Verifying the validity of an FEA model is a critical step in the analysis. As powerful as the methodology may be, the computers used may simply provide an answer to the problem without explaining it. If the problem is not fully understood, the solutions may be equally invalid.

Using a classical analysis of two spheres in contact,[73] the model was verified by checking the compressive loads for the contact areas of several

---

[72]ANSYS Rel 4.4. Houston, PA: Swanson Analysis Systems, Inc., 1989.

[73]Timoshenko, S. and J. Goodier. *Theory of Elasticity*. 3rd ed. New York: McGraw-Hill, 1970, pp. 409–414.

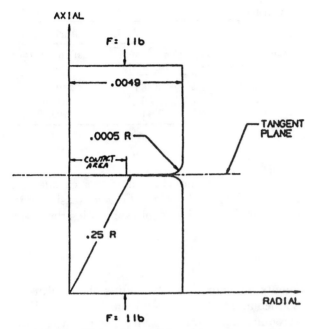

Figure 26 Simplifying the model by using y-axis symmetry.

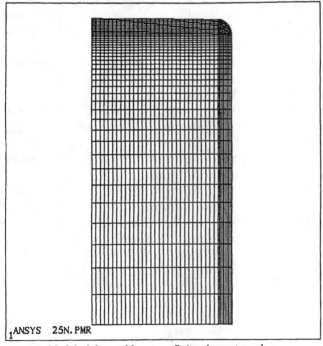

ANSYS **4.3**
JUL 25 1988
16:15:59
POST1 ELEMENTS

ZV=1
DIST=.0055
XF=.00245
YF=.245

ANSYS 25N. PMR

Figure 27 Model of the problem as a finite element mesh.

TABLE 20   Model Verification Analysis

	Classical Analysis		Finite Element Analysis	
Radius	Contact Area (sq. in.)	Axial Compression (psi)	Contact Area (sq. in.)	Axial Compression (psi)
.125	.00265	136,000	.00265	140,000
.25	.0033	94,000	.0032	91,000
.50	.0042	54,000	.0041	60,000

radii. As can be seen in Table 20, differences in contact area calculations between the two analyses are negligible. Also, the compressive axial stresses are within ten percent of each other, again suggesting the model is consistent with assumptions.

Results of analyses

Preliminary measurements indicated that the radius of curvature in the fiber was often .125 inch and less. Figures 28 and 29 on the following pages indicate the troubles that were occurring. The negative numbers indicate a compressive load, and the positive numbers indicate a tensile load. As can be seen in Figure 28, the maximum tensile is 39,927 psi and is highly concentrated in a small area. This is very close to the fracture load of 50,000 psi for the fiber. Since this did not seem to be sufficient to distribute the load on the tip of the fiber, a new radius had to be determined and then set forth as a design requirement. Current manufacturing practices and fiber characteristics limited the new radius to between .25 and 1.00 inch.

Figures 30, 31, 32, and 33 show that as the radius of curvature increases, the maximum tensile load decreases across the surface. It is also important to observe that the region of maximum tension is moving away from the area of contact, toward the body of the fiber, where it can be distributed over a much wider area.

In the final design, the splice's radius was 1.00 inch. As seen in Figure 33, it shows the lowest radial tension, and the load is spread across a large area of the fiber. This figure can be contrasted with Figure 29 which showed a very high tensile load in a very localized area. The design requirement now specifies that the fiber ends are to be machined to a 1.00 inch radius of curvature. The system itself is now in service with no reported failures from the connectors.

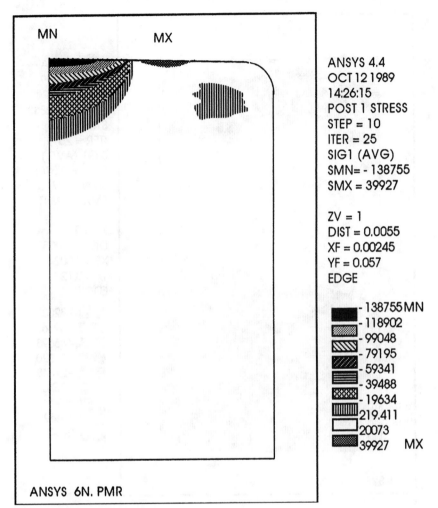

Figure 28 Stress plot: radius = 1/16 in.

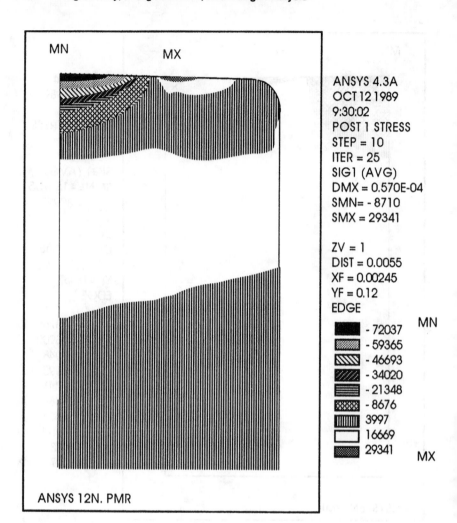

Figure 29   Stress plot: radius = 1/8 in.

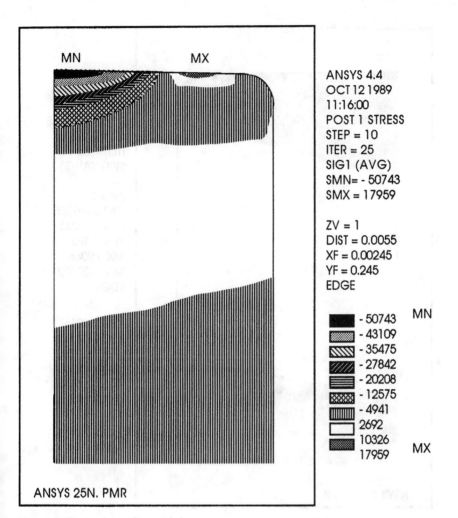

Figure 30 Stress plot: radius = 1/4 in.

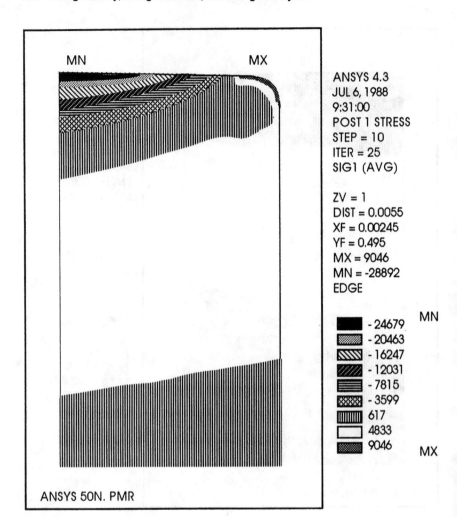

Figure 31 Stress plot: radius = 1/2 in.

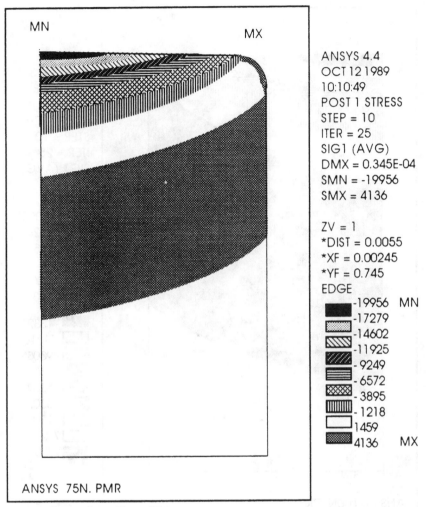

Figure 32 Stress plot: radius = 3/4 in.

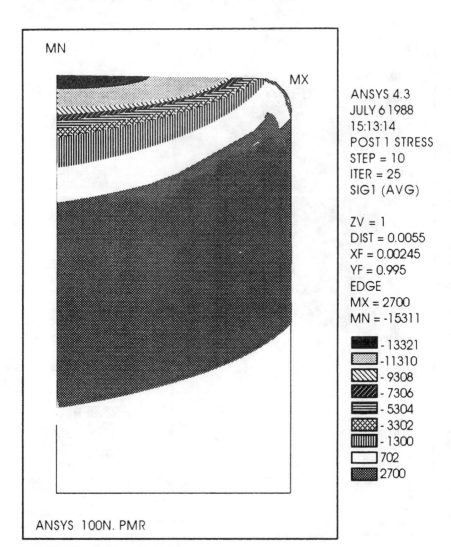

MN

MX

ANSYS 4.3
JULY 6 1988
15:13:14
POST 1 STRESS
STEP = 10
ITER = 25
SIG1 (AVG)

ZV = 1
DIST = 0.0055
XF = 0.00245
YF = 0.995
EDGE
MX = 2700
MN = -15311

- 13321
-11310
- 9308
- 7306
- 5304
- 3302
- 1300
702
2700

ANSYS 100N. PMR

Figure 33 Stress plot: radius = 1 in.

4

Summary

The development effort on any project is a highly complex and iterative process. Risks may appear at every step along the way, increasing the probability of failure. The need for a design policy to guide the participants through the program's life cycle is crucial to its success. Equally, if not more important, is a thorough understanding of the design's goals and the methodologies required of the design process itself. Conducting design analyses are essential for evaluating the technical risks associated with design decisions at all levels of product integration and design maturity.

Summary of the Procedures

A clear awareness of all possible considerations is a fundamental requirement for members of the project. This insight allows the designers and team members to interact and communicate their needs. By knowing what needs to be done, the risks of poor or short-sighted decisions are greatly reduced, and the chances of success are greatly increased. This part has explained the best practices associated with making informed design decisions. Table 21 is a summary of those practices.

TABLE 21   Summary of the Procedures

Step 1: Establish the Design Policy	
Procedures	Supporting Activities
Develop a design policy	■ Document clearly all aspects of policy and communicate to everyone involved
Consider key aspects of a design policy	■ Include procedures for requirements generation
	■ Include guidelines for addressing manufacturing and other key downstream activities
	■ Develop guidelines for setting critical design goals and structure policy around those goals
	■ Ensure design practices are consistent with product development strategy and goals
	■ Develop guidelines for conducting thorough design analyses which are critical for reducing technological risk
	■ Ensure understanding and commitment for an effective design process control and improvement strategy
	■ Include guidelines for conducting design reviews to assess design maturity
Follow the design policy in contract awards	■ Communicate design policy to all subcontractors to obtain support

Step 2: Review the Design Requirements and the Design Goals	
Procedures	Supporting Activities
Review design requirements	■ Ensure requirements are complete, consistent, and contain no contradictions
Review design goals	■ Use design goals of system to identify important product and process issues to consider in design
	■ Examine common system design goals: product cost, development cost, development speed, product performance, risk taking, and flexibility
Control changes	■ Control changes to design requirements or design goals and make sure they are communicated and reviewed by everyone involved in development effort

Step 3: Define the Appropriate Design Process	
Procedures	Supporting Activities
Tailor the design process	■ Understand advantages and disadvantages of different methodologies; the optimum process is one whose advantages support product's goals
	■ Weigh differences carefully between concurrent engineeing and sequential engineering

Step 3: Define the Approriate Design Process	
Procedures	Supporting Activities
Choose the right team	■ Allow for representation from several disciplines to facilitate interaction and communication across development effort ■ Understand project to determine proper organizational structure ■ Consider team size and its effect on communication among team members ■ Co-locate team members when possible to facilitate communication and productivity
Consider life cycle costs	■ Assess how design process will affect product's life cycle costs ■ Consider design to cost to facilitate setting acceptable limits in cost, schedule, and performance
Consider downstream activities	■ Consider robust design for a systematic and efficient method of design optimization for performance, quality, and cost ■ Consider Producibility Engineering and Planning (PEP) to identify potential manufacturing problems and to suggest a design which will facilitate the production process ■ Use Design for X (DFX) to analyze life cycle phases as early as possible to reduce risk of problems late in the development effort
Manage change	■ Use configuration management to control and maintain integrated product and process design
Document the design plan	■ Ensure complete and detailed documentation of the design process and communicate to design team
Step 4: Develop and Analyze Design Alternatives	
Procedures	Supporting Activities
Consider design approaches	■ Choose the best combination of approaches from top-down design and bottom-up design to meet system needs
Do system partitioning	■ Break a large, complex design problem down into small manageable pieces ■ Partition the system carefully to maximize the highest priority design goals while meeting the other goals ■ Keep entire system in perspective while partitioning ■ Track cost implications when making partitioning decisions ■ Consider purchasing rather than producing certain components and subsystems

TABLE 21   Summary of the Procedures (*Continued*)

Step 4: Develop and Analyze Alternatives (*Continued*)	
Procedures	Supporting Activities
Do requirements allocation	■ Develop performance budgets to ensure system capabilities meet performance demands ■ Use design rules to make sure fundamental design principles are followed ■ Use design guidelines to establish good design practices
Conduct design analyses	■ Evaluate system design alternatives ■ Use thermal analysis to study effects of temperature on performance, reliability, and life ■ Use stress analysis to understand effects of load conditions on system reliability and survivability ■ Use failure mode, effect, and criticality analysis to study how the system may fail ■ Use fault tree analysis to describe and assess the effects of normal and abnormal events that may lead to system failure ■ Use worst case analysis to verify system will meet requirements under extreme conditions ■ Use sneak circuit analysis to identify hidden conditions that may cause failures, even if components are operating to specifications ■ Use finite element analysis to study the system's physical behavior when calculation by conventional means is impractical ■ Use reliability analysis and prediction to balance design constraints with system reliability considerations ■ Analyze environmental effects during manufacture, shipment, storage, installation, operation, and service ■ Use safety assessment and hazard analysis to study the system's effects on people

Step 5: Choose a Design	
Procedures	Supporting Activities
Refine the decision process	■ Make sure decision criteria are fully explored ■ Understand intent of decision criteria to recognize impact of decisions ■ Understand that ideal design may be "good enough," not perfect (the benefits of the optimal solution to a design problem may be beyond the point of diminishing returns)

Step 6: Baseline the Design and Obtain Customer Approval	
Procedures	Supporting Activities
Conduct design reviews	■ Use formal design reviews to allow for evaluation of design with respect to requirements and to gain customer approval ■ Document baseline to facilitate change control

Chapter

5

References

Allen, Thomas J. *Managing the Flow Of Technology*. Cambridge, MA: The Massachusetts Institute of Technology, 1977. Reports the findings of an extensive ten-year study about how information is disseminated and exchanged and what are the most effective methods.

Barlow, Richard E., Jerry B. Fussell, and Nozer D. Singpurwalla, eds. *Reliability and Fault Tree Analysis*. Philadelphia, Pennsylvania: Society for Industrial and Applied Mathematics, 1975. Provides a collection of technical papers about reliability and fault tree analyses. Papers range from an introduction of fault tree analysis through in-depth discussions of the mathematical theory.

Bell Telephone Laboratories. *Physical Design of Electronic Systems, Volume II: Materials Technology*. Englewood Cliffs, N.J.: Prentice-Hall, 1972. Emphasizes the fundamental considerations required in materials selection in the physical design of electronic systems.

Bell Telephone Laboratories. *Physical Design of Electronic Systems, Volume IV: Design Process*. Englewood Cliffs, NJ: Prentice-Hall, 1972. Discusses an approach to the design of electronic systems and provides information to improve the efficiency of the system design process. Includes discussions of the decision-making process, reliability allocation and prediction, system partitioning, and optimization.

Bell Telephone Laboratories, ed. "System Organization and Partitioning" in *Physical Design of Electronic Systems, Volume IV: Design Process*. Englewood Cliffs, NJ: Prentice-Hall, 1972, pp. 395–439. Discusses establishing system requirements relating basic parameters and how the requirements influence the physical partitioning of the system. Topics include: establishing system requirements, the nature of the partitioning problem, and partitioning with respect to performance, reliability, and cost.

Boothroyd, G. "Making It Simple: Design for Assembly." *Mechanical Engineering,* February 1988, pp. 28–31. Describes and illustrates how DFA principles can refine and simplify designs. Examples show effect on product cost and assembly.

Boothroyd, G. and P. Dewhurst. "Design for Assembly In Action." *Assembly Engineering,* January 1987, pp. 64–68. Describes the differences between the IBM Proprinter and the Epson MX-80 printer using DFA principles for the analysis.

Bracken, F., Gerard Insola, and Dr. Emory W. Zimmers, Jr. "Considerations in Design: Fundamental Design Methodology." Paper presented at CAD/CIM Alert International Conference on Design for Manufacturability, San Francisco, CA, June 18-20, 1989. Describes a methodology to help guide the design engineer in forming ideas without restricting creativity.

Britton, Marvin G., P. Bruce Adams, and J. R. Lonergan. "Ceramics: II. Nature and Properties of Glass" in *Materials and Processes, Part A: Materials*. 3rd ed. Eds. James F. Young and Robert S. Shane. New York: Marcel Dekker, 1985, pp. 736–783. Discusses the effect of deformation on glass and metal rods.

Burling, W. A., B. J. Bartels Lax, L. A. O'Neill, and T. P. Pennino. "Product Design and Introduction Support Systems." *AT&T Technical Journal,* September/October 1987, pp. 21–38. Describes features of CAD systems which ensure that designs adhere to design rules codifying manufacturability and quality.

Burnett, David S. *Finite Element Analysis*. Reading, MA: Addison-Wesley, 1987. Provides a thorough, graduated discussion of finite element analysis. Includes an introduction of the basic concepts as well as detailed explanations of the analysis methods.

Byrne, Diane M. and Shin Taguchi. "The Taguchi Approach to Parameter Design." *1986 ASQC Quality Congress Transaction*, American Society for Quality Control, 1986. Provides an introduction to Taguchi's philosophy to off-line quality control with emphasis placed on parameter design.

Cottingham, W. B. and P. W. McFadden. "Design Theory" in *Physical Design of Electronic Systems, Volume IV: Design Process*. Ed. Bell Telephone Laboratories. Englewood Cliffs, NJ: Prentice-Hall, 1972, pp. 3–36. Presents an introduction to some formalized procedures for solving design problems. Topics include: decision-making in the design process, utility theory, decision-making under constraints, and decision trees.

Chase, Wilton P. *Management of System Engineering*. New York: John Wiley & Sons, 1974. Describes basic system design and engineering processes. Discusses the creative and decision-making processes necessary to accomplish effective system designs.

Defense Systems Management College. *Acquisition Strategy Guide*, 1984. Provides information for program managers to use to structure and execute system acquisition programs. Includes major strategic alternatives, methods of implementation, and benefits and drawbacks.

Department of Defense. Defense Systems Management College. *Systems Engineering Management Guide*. 2nd ed. December 1986. Provides an educational guide to systems engineering concepts and techniques and identifies relevant directives and references.

Department of Defense. *Total Quality Management Master Plan*, 1986. Describes principles and procedures for quality management and improvement, including quality function deployment.

Department of Defense. *Transition from Development to Production*. DoD 4245.7-M, September 1985. Describes techniques for avoiding technical risks in 48 key areas or templates in funding, design, test, production, facilities, logistics, management, and transition plan.

Department of the Army. *Systems Acquisition Policy and Procedures*, Army Regulation 70-1, Effective November 10, 1988. Covers and consolidates overall policy, procedures, and responsibilities for conducting research, development, and acquisition of materiel and systems to satisfy requirements.

Department of the Navy. *Best Practices: How to Avoid Surprises in the World's Most Complicated Technical Process*. NAVSO P-6071, March 1986. Discusses how to avoid traps and risks by implementing best practices for 48 areas or templates, including topics in funding, design, test, production, facilities, and management. Discusses best practices for design reference mission profile, trade studies, design requirements, and technical risk assessment.

Doyle, Michael and David Straus. *How to Make Meetings Work*. New York: The Berkeley Publishing Group, 1976. Provides information about why meetings are important, what goes wrong in many meetings, what to look out for in meetings, and how to improve meetings.

Gatenby, David. "Design For 'X' (DFX): Assembly, Simplicity, and Beyond." Paper presented at CAD/CIM Alert International Conference on Design For Manufacturability, San Francisco, CA, June 18-20, 1989. Provides an introduction to DFX as a strategic concept for product realization. Also presents a framework and examples for supporting DFX.

Harper, Charles A., ed. *Handbook of Electronic Systems Design*. New York: McGraw-Hill, 1980. Offers an extensive discussion of electronic systems design via seven separate presentations: general purpose computer systems, dedicated computer systems, communications networks, communications systems, communications system analysis, radar systems, measurements systems, and digital systems.

Harrison Ph. D., Frederick. "Digital Systems" in *Handbook of Electronic Systems Design*. Ed. Charles A. Harper. New York: McGraw-Hill, 1980, pp. 8-1 to 8-49. Describes the design process for digital systems. Discusses specifications, program management, the actual design, and production support.

Hnatek, Eugene R. *Integrated Circuit Quality and Reliability*. New York: Marcel Dekker, 1987. Discusses the quality and reliability considerations for integrated circuit design. Includes material on the design and fabrication process, current technologies, sources of manufacturing error, and causes of failure and their remedies.

Jacobson, David W. "Importance of Accurate Thermal Analysis." *1989 Proceedings, IEEE Annual Reliability and Maintainability Symposium*, pp. 465–469. Presents a discussion of how thermal analysis is a valuable design tool. Specifically addresses the importance of understanding the effects of junction temperature on reliability.

Juran, J. M. and Frank M. Gryna, eds. *Juran's Quality Control Handbook*. 4th ed. New York: McGraw-Hill, 1988. Describes the process of new product development, including procedures for defining design requirements, setting up configuration management systems, and designing experiments for trade studies.

Kandel, Abraham and Eitan Avni, eds. *Engineering Risk and Hazard Assessment, Volume I*. Boca Raton, FL: CRC Press, 1988. Discusses the field of risk and hazard assessment. Includes issues such as short- and long-term hazards, modeling, diagnostic systems, and risk analysis of fatigue failure.

Larson, Erik W. and David H. Gobeli. "Organizing for Product Development Projects." *Journal of Product Innovation Management*, May 1988. Presents the results of a study of 540 development projects and how organizational structure affected the projects' cost, schedule, and performance objectives.

Leis, Charles T. "Dedicated Computer Systems" in *Handbook of Electronic Systems Design*. Ed. Charles A. Harper. New York: McGraw-Hill, 1980, pp. 2-1 to 2-63. Describes the design of dedicated computer systems including these topics: functional criteria, system architecture, malfunctions, testing, and installation.

Lemke, Richard L. "Measurement Systems" in *Handbook of Electronic Systems Design*. Ed. Charles A. Harper. New York: McGraw-Hill, 1980, pp. 7-1 to 7-107. Describes the design of measurement systems, emphasizing the basic system building blocks and important design considerations.

Manzo, John. *Principles for Developing Complex Software and Hardware Systems*. Course presented by the National Technological University Satellite Network Professional Development Programs, January 20, 1989. Discusses the nature of complexity and some simple principles and techniques that apply to large-scale, complex projects.

McEachron, Norman B., Robert J. Lapen, and Ruth A. Tara. *Accelerating Product and Process Development*. Preliminary Draft. Menlo Park, CA: SRI International, September 20, 1989. Reports on the challenge to improve responsiveness to the customer. Describes the concurrent development approach, strategies for accelerating development, required support structures, case studies, and implementation fundamentals.

McLeod, Scott. "Producibility Measurement for DoD Contracts." Draft paper presented at Third Annual Best Manufacturing Practices (BMP) Workshop, September 13, 1989. Discusses the principles of producibility measurement and the need for ensuring producibility is properly addressed in all phases of a contract.

Michaels, Jack V. and William P. Wood. *Design To Cost*. New York: John Wiley & Sons, 1989. Describes the design to cost method of system design and development. Identifies fundamental aspects of planning and implementing a design to cost program. Includes examples of design to cost programs.

Military Standard 490: *Specification Practices*. Describes how to prepare, interpret, change, and revise specifications prepared by or for the Department of Defense.

Petts III, George E. "Radar Systems" in *Handbook of Electronic Systems Design*. Ed. Charles A. Harper. New York: McGraw-Hill, 1980, pp. 6-1 to 6-88. Describes the design of radar systems including these topics: frequencies, system architecture, systems engineering, concept definition, performance wave forms, and antennas.

Phadke, Madhav S. *Quality Engineering Using Robust Design*. Englewood Cliffs, NJ:

Prentice Hall, 1989. Describes the theoretical and practical aspects of quality engineering including robust design, matrix experiments using orthogonal arrays, and the design of dynamic systems.

Priest, J. W. *Engineering Design for Producibility and Reliability.* New York: Marcel Dekker, 1988. Describes a high-level view of the contents of environmental and functional profiles, use of trade studies, design processes, design analyses, mathematical models, and simulations to verify that the optimum system approach has been selected, trade studies of design parameters such as cost, schedule, technical risk, reliability, producibility, quality, and supportability, and use of mature vs. leading edge technology. Describes purpose of design policy statements, design requirements, use of trade studies to tailor the requirements for reliability, and producibility. Discusses modeling and simulation and their relation to trade studies.

Rabins, M., D. Ardayfio, S. Fenves, A. Seireg, G. Nadler, H. Richardson, and H. Clark. "Design Theory and Methodology—A New Discipline." *Mechanical Engineering,* August 1986, pp. 23–27. Discusses a National Science Foundation study on design theory and methodology, the need for a formal structure to guide engineers during the development of system designs, and the need for fully integrated design and manufacturing.

Reinertsen, Donald G. "Blitzkrieg Product Development: Cut Development Times in Half." *Electronic Business,* January 15, 1985. Discusses techniques that enable companies to reduce product development intervals. Provides a checklist for rapid product development and discusses the economic analysis required to determine if development speed is needed.

Reinertsen, Donald G. "Whodunit? The Search for the New-Product Killers." *Electronic Business,* July 1983. Discusses key problems in new-product development and the need for economic analyses of a product's market and strategy. Also discusses factors to emphasize in product development.

Rodamaker, Mark. "Optimizing Productivity Through Finite Element Analysis/Finite Element Meshing." Paper presented at CAD/CIM Alert International Conference on Design For Manufacturability, San Francisco, CA, June 18–20, 1989. Provides a discussion and examples of finite element analysis concepts, techniques, and benefits.

Schlabach, T. D. and P. R. White. "Mechanical Properties of Metals" in *Physical Design of Electronic Systems, Volume II: Materials Technology.* Ed. Bell Telephone Laboratories. Englewood Cliffs, N.J.: Prentice-Hall, 1972, pp. 131–196. Discusses basic properties of metals and alloys to understand their various failure modes. Topics include: elastic behavior, plastic behavior, strengthening mechanisms, fracture, fatigue, creep, testing, and metal properties.

Simon, William F. "General Computer Systems" in *Handbook of Electronic Systems Design.* Ed. Charles A. Harper. New York: McGraw-Hill, 1980, pp. 1-1 to 1-78. Describes the design of general-purpose computers including these topics: system topology, partitioning, intersystem connections, operational characteristics, maintenance, control of electrical and acoustic disturbances, and reliability.

Smoluk, George R. "Thermal Analysis: A New Key to Productivity." *Modern Plastics,* February 1989. Summarizes the capabilities of thermal analysis methods. Discusses the shifting of the thermal analysis orientation from laboratory hardware to the production line.

Sneak Circuit Analysis. NAVSO P-3634, Department of the Navy, August 1987. Provides an overview of sneak circuit analysis and its benefits. Includes methods and guidelines for implementation and for estimating the costs of conducting the analyses.

Stefanik, T. *Strategic Antisubmarine Warfare and Naval Strategy.* Lexington, MA: D. C. Heath, 1987. Describes strategic antisubmarine warfare policy and missions to counter potential threats. Discusses submarine design and detection by acoustic and nonacoustic methods.

Taguchi, Shin. *Taguchi Methods: Quality Engineering.* Detroit, MI: American Supplier Institute Press, 1988. Provides an introduction to quality engineering, Taguchi methods, orthogonal arrays, parameter design, case studies, and implementation guidelines.

Willoughby, Willis J. Jr. "Moving R&QA Out of the Red Light District." *The ITEA Journal of Test and Evaluation* vol. 9(3), 1988, p. 18. Discusses the significance of reliability in systems development and the need to focus on the fundamentals of design and manufacturing to achieve reliability.

Winans, R. C. "Reliability of Electronic Parts" in *Physical Design of Electronic Systems, Volume IV: Design Process.* Ed. Bell Telephone Laboratories. Englewood Cliffs, NJ: Prentice-Hall, 1972, pp. 332–371. Discusses the need for understanding the reliability characteristics of electronic parts and their appropriate design considerations. Topics include: the effects of stresses on reliability, failure modes and mechanisms, failure rates, vs. stress levels, electronic parts application considerations, probability distribution functions, reliability control, and prediction for electronic parts.

Winner, Robert I., James P. Pennell, Harold E. Bertrand, and Marko M. G. Slusarczuk. *The Role of Concurrent Engineering in Weapons System Acquisition.* IDA Report R-338, Alexandria, VA: Institute for Defense Analyses, December 1988. Describes the results of a study of concurrent engineering. Describes its use in weapons systems development.

Young, James F. and Robert S. Shane, eds. *Materials and Processes, Part A: Materials.* 3rd ed. New York: Marcel Dekker, 1985. Provides an extensive examination of materials engineering with a focus on the information the designer needs to select, prove, and specify materials for product design.

Young, James F. and Robert S. Shane, eds. *Materials and Processes, Part B: Processes.* 3rd ed. New York: Marcel Dekker, 1985. Provides an extensive examination of the processes involved in materials engineering and management. Focuses on information needed by people in product design, development, and production.

Parts Selection and Defect Control

Introduction

To the Reader

This part, Parts Selection and Control, includes four templates: Parts and Materials Selection, Piece Part Control, Manufacturing Screening, and Defect Control.

The templates, which reflect engineering fundamentals as well as industry and government experience, were first proposed in the early 1980s by a Defense Science Board task force of industry and government leaders, chaired by Willis J. Willoughby, Jr. The task force sought to improve the effectiveness of the transition from development to production. The task force concluded that most program failures were due to a lack of understanding of the engineering and manufacturing disciplines used in the acquisition process. The task force then focused on identifying engineering processes and control methods that minimize technical risks in both government and industry. It defined these critical events in design, test, and production in terms of templates.

The template methodology and documents

A template specifies:

- areas of high technical risk
- fundamental engineering principles and proven procedures to reduce the technical risks

Like a classical mechanical template, these templates identify critical measures and standards. Use of the templates makes it likely that engineering disciplines will be followed.

The task force documented 47 templates and in 1985 the templates were published in the DoD *Transition from Development to Production* (DoD 4245.7-M) manual.[1] The templates primarily cover design, test, production, management, facilities, and logistics. In 1989, the Department of Defense added a 48th template on Total Quality Management (TQM).

In 1986, the Department of the Navy issued the *Best Practices* (NAVSO P-6071) manual,[2] which illuminated DoD practices that compound problems and increase risks. For each template, this manual describes:

- potential traps and practices that increase the technical risks
- consequences of failing to reduce the technical risks
- an overview of best practices to reduce the technical risks

The intent of the *Best Practices* manual is to help practitioners become aware of the traps and pitfalls so they do not repeat them.

The templates are the foundation for current educational efforts

In 1988, the government initiated an educational program, "Templates: Professionalizing the Acquisition Work Force," with courses and books, such as this one, to increase awareness and improve the implementation of the template concept.

The key to improving the DoD's acquisition process is to recognize that it is an industrial process, not an administrative process. This change in perspective implies a change in the skills and technical knowledge of the acquisition work force in government and industry. Many in this work force do not have an engineering background. Those with an engineering background often do not have broad experience in design, test, or production. The work force must understand basic design, test, and production processes and associated technical risks. The basis for this understanding should be the templates since they highlight the critical areas of technical risk.

The template educational program meets these needs. The program consists of a series of courses and technical books. The books provide background information for the templates. Each book covers one or more closely related templates.

How the parts relate to the templates.   Each part describes:

---

[1]Department of Defense. *Transition from Development to Production.* DoD 4245.7-M, September 1985.

[2]Department of the Navy. *Best Practices: How to Avoid Surprises in the World's Most Complicated Technical Process.* NAVSO P-6071, March 1986.

- the templates, within the context of the overall acquisition process
- risks for each included template
- best commercial practices currently used to reduce the risks
- examples of how these best practices are applied

The books do not discuss government regulations, standards, and specifications, because these topics are well-covered in other documents and courses. Instead, the books stress the technical disciplines and processes required for success.

Clustering several templates in one book makes sense when their best practices are closely related. For example, the best practices for the templates in this part interrelate and occur iteratively within design and manufacturing. Designers, suppliers, and manufacturers all have important roles. Other templates, such as Design Reviews, relate to many other templates and thus are best dealt within individual parts.

Courses on the templates. The books are designed to be used either in courses or as stand-alone documents. An introduction to the templates and several technical courses are available. The courses use lectures and other proven instructional techniques such as videotapes, case studies, group exercises, and action plans.

The template educational program will help government and industry program managers understand the templates and their underlying engineering disciplines. They should recognize that adherence to engineering discipline is more critical to reducing technical risk than blind obedience to government standards and administrative rules. They should especially recognize when their actions (or inactions) increase technical risks as well as when their actions reduce technical risks.

The templates are a model

The templates defined in DoD 4245.7-M are not the final word on disciplined engineering practices or reducing technical risks. Instead, the templates are a reference and a model that engineers and managers can apply to their industrial processes. Companies should look for high-risk technical areas by examining their past projects, by consulting their experienced engineers, and by considering industry-wide issues. The result of these efforts should be a list of areas of technical risk which becomes the company's own version of the DoD 4245.7-M and NAVSO P-6071 documents. Companies should tailor the best practices and engineering principles described in the books to suit their particular needs. Several military suppliers have already produced manuals tailored to their processes.

Materials are the heart of engineering design...Selecting the optimum materials and the subsequent processes necessary to turn them into finished parts calls for considerable skill on the part of the designer, a process made more complex by the continuous introduction of new or improved materials.[3]

Parts and materials are selected on the basis of function, performance, reliability, quality, maintainability, methods of manufacture, and cost. Function is the most fundamental reason but the others are also important. Cost refers not only to the cost of the materials but also to the "cost of ownership"—the cost of inventory, special assembly equipment, rework, and delays.

The U.S. Department of Commerce's 1984 survey data on cost distribution show the importance of parts selection and defect control. With commercial systems, the cost of parts and material is about 60% of the cost of the manufactured system. Labor is typically about 20% to 30%.

Figure 1 shows where to find more and more details about risks, best

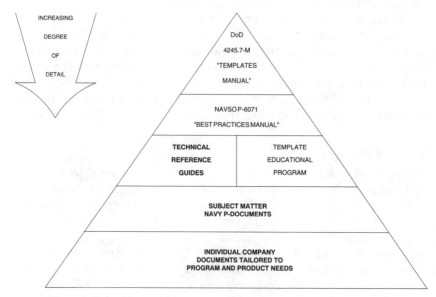

Figure 1 Resources on the acquisition-process templates.

[3]Ball, Graham. "How Well Do Materials Meet Designers' Real Needs?" *Materials: Proceedings of Materials Selection and Design.* London, England, July 1985, p. 208.

practices, and engineering principles. Participants in the acquisition process should have copies of these documents.

Figure 2 shows that the percentages vary from industry to industry.[4] The cost of parts and materials is about 70% of the manufacturing cost in the motor vehicle industry, but about 45% in the ship-building industry. With military systems, the percentages may range from about 10% to about 30%.

Military systems are less sensitive to parts costs, but the total cost of the military systems tends to be higher than commercial systems. Thus, parts selection and defect control are vital in military systems as well as in commercial systems. With both commercial and military systems, false economy at the parts level as well as parts proliferation can cause horrendous costs later in the system's life cycle.

The four templates discussed in this part identify critical areas of risk. They emphasize the importance of carefully selecting parts and materials and controlling defects in piece parts and systems. Screening and prevention are key methods of control. The following pages define each template and give an overview of the risks and their consequences.

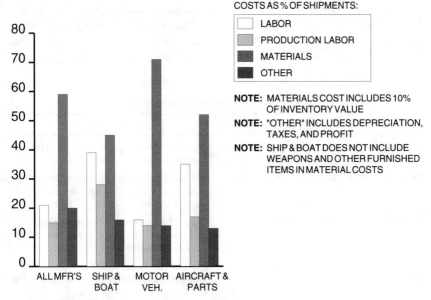

Figure 2 Cost distributions in different industries.

***

[4]Nevins, James L. and Whitney, Daniel E., Eds. *Concurrent Design of Products and Processes*. New York: McGraw-Hill, 1989, p. 39–40.

The Procedures chapter of this part describes engineering fundamentals and best practices to minimize these risks. The Application chapter gives examples of how to use the best practices that reduce these technical risks.

Parts and Materials Selection Template

Designers select parts and materials to meet specified requirements for functionality, performance, reliability, quality, producibility, cost, etc. Designers need to become familiar with the current best practices to avoid poor selections. Table 1 gives risks in parts and materials selection. The Procedures chapter of this part describes best practices to avoid those risks.

Risk: No preferred-parts list

Preferred parts are parts whose quality and reliability are well-known and which the company may already be using for other systems. Without a preferred-parts list, designers choose parts unsystematically, even though the preferred parts would perform as well or better. The result is a proliferation of nonstandard parts, varying in performance and reliability.

Nonstandard parts may have wide-ranging consequences for manufacturing, purchasing, and logistics. Manufacturing engineers must cope with parts that require a variety of assembly methods and tooling. Inventory costs may increase. Ease of automation may decrease.

TABLE 1   Risks and Consequences in Parts and Materials Selection

Risks	Consequences
No preferred-parts list at start of development	A proliferation of nonstandard parts, varying in quality and reliability
Obsolete parts selected	Without a preferred-parts list, designers may choose parts that are obsolete or hard to obtain
New technology parts selected	Without a history of proven reliability and mature technology, parts' reliability may be questionable
Parts unsuitable for particular applications	System fails because design cannot withstand stresses during use, transportation, or storage
Incomplete or inaccurate thermal analysis data on part operating temperature and vibration	System reliability is lowered because parts are not selected and optimally placed using thermal and mechanical analysis data

Purchasing representatives must deal with many different suppliers, making it hard to monitor timely delivery and quality and obtain volume discounts. Logistics specialists must provide spares for many different parts and find storage space for them.

When a preferred-parts list is available at the start of full-scale development, designers can select preferred parts or obtain engineering justification for any nonapproved part. (The justification process should be efficient to avoid unnecessary delays.) Designers can follow design-for-assembly practices in reducing the number of parts and choosing standardized parts that can be used in many assemblies and subassemblies.

Risk: Obsolete, hard-to-obtain, or new technology parts

Without a preferred-parts list, designers may choose obsolete or sole-sourced parts that may be difficult to obtain. Preferred-parts databases help designers identify obsolete parts and how long parts will be available.

Without a preferred-parts list, designers may choose the latest part or an exotic part rather than one with proven reliability and mature technology. Parts should undergo an independent assessment before being put on the preferred-parts list. Information sheets from suppliers may not give complete information.

Risk: Parts unsuitable for particular applications

In selecting parts, designers often fail to consider how the part will be used—what stresses the part will encounter in use and during storage and transportation. Often the stresses during storage and transportation are greater than during use. Likely stresses include shock, vibration, and changes in temperature.

Designers also fail to make their designs able to tolerate wide environmental variations. Similarly, they fail to use *derating* techniques to choose parts manufactured to withstand more than the likely operational, storage, and transportation stresses. With larger and stronger parts, there is a safety margin because parts will be stressed only to a percentage of their capability. Parts can then better withstand hidden stresses in manufacturing, testing, operation, storage, and transportation.

To minimize these risks, companies should establish for all engineers a design policy covering parts and material selection. This policy should set specific limits on allowable stresses.

Risk: Incomplete or inaccurate analyses

Designers' thermal and mechanical analyses are often incomplete and may be based on subjective estimates and probabilities. Without simulations and thermal surveys to measure part operating temperatures, parts may be located too far from heat sinks, thereby lowering the reliability of the system. Mechanical deflection and vibration-related effects are often not simulated during the design process. It is much better to incorporate the results of thermal, mechanical and stress analyses into the design early. If design deficiencies are not found until later during testing, design changes are more difficult and costly. Finite-element analysis and finite-difference analysis are techniques widely used to analyze stresses and temperature profiles. These techniques produce accurate results.[5] Part 2, Design Policy, Design Processes, and Design Analysis, has more details on design analyses.

Piece Part Control Template

Piece part control programs are needed to ensure timely delivery of high-quality, high-reliability parts. Supplier certification programs with timely feedback and exchange of information are a way to ensure incoming parts meet specifications. Table 2 gives risks in piece part control. The Procedures chapter gives best practices to avoid the risks.

Risk: Lack of a formal piece part control program during development

Without a formal parts control program, requirements may not flow down to subcontractors and suppliers. Parts and subassemblies may not meet specifications for quality, reliability, and timeliness. Incoming parts may not meet specifications and thus cause scrap, rework, and higher inventory costs as manufacturing tries to work around the defective parts.

Piece part control depends on parts and material selection. Unless designers have access to a preferred-parts list and a formal parts control program, they are likely to select from suppliers whose parts' quality and reliability are not optimal. They may unknowingly choose suppliers with questionable financial stability. They may unknowingly choose suppliers who have a high reject rate or who fail to meet delivery deadlines. They may choose distributors' parts that are risky because their origin may be uncertain and defects may be introduced during storage and transportation.

---

[5]*Reliability Assessment Using Finite Element Techniques.* RADC Technical Report, TR-89-281, 1989.

TABLE 2   Risks and Consequences in Piece Part Control

Risks	Consequences
Lack of formal piece part control program during development	Requirements may not flow down to subcontractors and suppliers, and thus parts may fail to meet specifications, causing scrap and rework
Failure to qualify and certify suppliers	Parts may vary in quality and reliability, especially if they come from distributors
Poor source management	Piece part quality and reliability may be low if source management techniques are not used (e.g., qualify and certify suppliers, do source inspection, and do receiving inspection if necessary)
Blind dependence on preconceived standards	Even MIL-STD parts may have high defect rates if they are not manufactured with stable process controls
Poor handling during screening or rescreening	Poor handling may damage parts due to mechanical damage, electrostatic discharge, or electrical overstress
Inadequate attention to older-technology parts	Older-technology parts (e.g., relays, transformers, inductors, switches, and connectors) may contribute to failure in many systems

Risk: Poor source management

Piece part quality and reliability may be low if companies fail to use the best approach to eliminate marginal devices. Ideally, companies select parts from qualified and certified suppliers. If this is not possible, companies may send their own inspectors to the supplier's site. This source inspection may fail to ensure high quality and reliability especially if there are strained relations between the supplier and the assembly manufacturer or if source inspectors are assigned to many locations and thus can visit sites only infrequently. When done properly, source inspection may be preferable to rescreening at the assembly manufacturer's site because it saves equipment, labor, and the expense and delays due to return of defective parts.

Risk: Blind dependence on preconceived standards

In the 1980s, electronics parts varied widely in electrical and mechanical quality, even those that were manufactured and screened according to military standards such as MIL-M-38510, General Specification for Microcircuits. To increase system reliability, some assembly manu-

facturers began to rescreen incoming parts, especially integrated circuits.

Screening to find parts with actual and hidden defects requires
resources of time and money. Poorly applied, these resources fail to add
value. Screening is ineffective when it does not remove marginal
devices. Screens should not be chosen arbitrarily without regard to how
they will be used. Screens should be tailored to bring out actual and
hidden defects. The results of screening should be used to find root
causes and corrective actions to prevent the defects.

Risk: Rescreening

Rescreening should be a last resort. Rescreening may reduce a device's
reliability by inducing additional failures. Poor handling may damage
parts due to mechanical damage, electrostatic discharge, or electrical
overstress. Rescreening adds no value when the supplier's process is
stable and producing near-zero defects. Screening at the part level
should be done during original manufacture as part of a process control
program. Rescreening should be performed only on a temporary basis.
An example is using nonqualified vendors until qualified vendors can
be found.

The resources spent on rescreening to ensure usable parts could be
better spent:

- monitoring the supplier's process control, testing, and piece part
  screening

- maintaining records on defect rates from various suppliers

- providing feedback to suppliers and designers on yield and defect
  rates

- setting up alliances with certified suppliers who have a record of
  timely delivery of good parts

Selecting suppliers is critical. Companies must decide what data they
need to select suppliers who have stable processes and near-zero defects.
The Procedures chapter describes what data Texas Instruments is using.
The Application chapter describes the data and procedures AT&T and
Magnavox are using. All the companies use data to make decisions,
although their procedures vary.

Manufacturing Screening Template

Manufacturing screening stresses assemblies such as boards, units,
and subsystems to stimulate assembly, process, and installation de-

fects. Part defects should be detected before they are assembled into boards or units. If part defects are found during manufacturing screening, the part manufacturing and part screening processes should be investigated.

Manufacturing screening is often called environmental stress screening. Temperature cycling and random vibration are the most common screens. The screening is tailored to provide appropriate limits, number of cycles, and rate of changes for temperature cycling, vibration, and shock. Table 3 gives risks that arise from a lack of understanding of environmental stress screening. The Procedures chapter describes best practices to avoid the risks.

Risk: Failure to understand environmental stress screening

To obtain the greatest benefit with least risk and lowest cost, companies must understand how to use screens to stimulate hidden defects. Without this understanding, screens may not be strong enough to stimulate likely flaws. Or screens may be too strong and unnecessarily damage good units. Overly intense screens induce additional defects and consume too much of a unit's operating life.

TABLE 3   Risks and Consequences in Manufacturing Screening

Risks	Consequences
Failure to understand environmental stress screening	Screens may not be strong enough to stimulate likely flaws, or screens may be too strong and unnecessarily damage good units
Screens not tailored to particular process and technology	Workmanship and quality problems may not be detected if screens are chosen inappropriately, and temperature cycling and random vibration screens are not used to find the most defects at the lowest life-cycle costs
Screening is done at inappropriate assembly levels	Screening done at lower assembly levels may not find latent defects in interfaces or interactions, screening done at higher assembly levels may not prevent expensive rework
Lack of corrective action	Assembly and workmanship defects persist even with stress screening if root cause analysis is not used to prevent future defects

Risk: Screens not tailored to process and technology

Ineffective screens may not contain the appropriate level of intensity, duration, range, rate of change, repetitions, etc. Ineffective screening programs may include temperature cycling but not random vibration. Many workmanship and quality problems may escape detection. Temperature cycling and random vibration screens should be tailored to the particular process and technology to find the most defects at the lowest life-cycle costs.

Screening results should be constantly reviewed. The results should be used to increase, decrease, or modify the screens to make them more effective in reducing defects and life-cycle costs. (Life-cycle costs include the cost of repair in the field.)

Risk: Screening is done at an inappropriate assembly level

It is important to consider the advantages and disadvantages of screening at different levels to decide which level is most cost-effective. Ineffective stress screening fails to prevent expensive rework at higher assembly levels. In general, detection efficiency is best at the highest assembly level (the system level), but the cost per failure is least at the lowest assembly level (i.e., the circuit card level). It may be more effective to use stress screening at two or more levels to reduce expensive rework while maintaining detection efficiency.

Risk: Lack of corrective action

Even with effective screening programs, assembly and workmanship defects may persist if root causes and corrective actions are not found. The results of root cause analyses may suggest improvements in the process control systems. These improvements should then be verified.

Defect Control Template

Assembly and workmanship defects result in higher production costs, increased process times, and expensive rework. The best way to prevent these defects is with a disciplined defect control program. Table 4 gives risks in defect control. The Procedures chapter describes best practices to avoid the risks.

Risk: Failure to commit to prevention

Companies should commit to prevention as the prime ingredient of a sound defect control program. Ineffective programs often overlook key

TABLE 4   Risks and Consequences in Defect Control

Risks	Consequences
Failure to commit to prevention	Defects will persist if the emphasis is on short-term fire-fighting rather than on prevention of future defects (e.g., identify critical procedures, equipment, personnel, training)
Poor use of yield and defect data	Defect reports may not be used because they are lengthy and out-of-date when issued
	Factory engineers put too much emphasis on the statistical implications of defect and yield data
	Data and charts are accessible only to factory engineers rather than being visible and understandable by all workers
Lack of root cause and failure analysis programs	Defects continue to occur at a high rate
Lack of timely repair	Delays in identifying, analyzing, and correcting problems may make the results no longer meaningful

elements in preventing and controlling defects—designs, procedures, equipment, personnel, and training. Thus, defects continue to cause delays and disruption. Poor designs may persist when companies fail to make designs less susceptible to variations in materials, processes, testing, or environmental conditions.

Risk: Poor use of yield and defect data

Defect reports are often lengthy and out-of-date when issued. Delays and errors may occur because manual entry of data is cumbersome and time-consuming. Many programs could improve the failure reporting system with an automated tracking system for more timely information. For example, bar codes and scanners can help track and monitor yields and defects.

Factory engineers often overemphasize the statistical implications of defect and yield data instead of using data to trigger corrective actions to improve out-of-control processes. Data and charts are accessible only to factory engineers. Charts on current yields, defect rates, and improvements are not displayed to make them visible and understandable by all workers. Shop-floor workers are not trained to interpret the data.

Successful companies are now training all employees to use statistical process control techniques. These techniques help them collect and use data for continual improvement.

Risk: Lack of root cause and failure analysis

Defects often continue to occur at a high rate if there is too little corrective action to find the root causes of defects and then solve and prevent them. Often, too little feedback goes to the designers and suppliers.

Failure analysis is often "too little, too late." Teams do not include representatives from design, testing, manufacturing, and other areas with key information. Teams are not given the time and resources to investigate failures. They neglect to reproduce the failure. They do not test to find the root cause, especially if the failure occurs intermittently. They also fail to identify and verify all the causes of failure.

The data on the defect's source or causes should be used to correct the process or system to prevent defects.

Risk: Lack of timely repair

Because of time pressures in manufacturing environments, faulty units or assemblies are often set aside for later testing and repair. Often people give higher priority to production testing than troubleshooting defective units. This often results in delays in identifying, analyzing, and correcting problems. Priorities must be constantly reviewed. These reviews ensure the proper balance between production and troubleshooting and help people make trade-offs between the cost of fixing the process vs. the costs of increased overtime, schedule delays, and field failures.

Rationale for an Integrated Parts Control Program

To minimize risks, an integrated parts control program is needed. This program should be multidisciplinary, involving designers, manufacturing engineers, suppliers, purchasing representatives, and subcontractors. Improvements in parts selection and defect control will result in fewer rejects, less rework, less downtime, and fewer failures in the field. These benefits far outweigh the savings in choosing a low-cost bidder with marginal quality.[6] For most large projects, the parts control program can be managed most effectively by a computerized database system.

Below are best practices for design, manufacturing, suppliers, purchasing, and subcontractors to achieve an integrated parts control program.

---

[6]Maass, Richard. *World Class Quality: An Innovative Prescription for Survival.* Milwaukee, WI: ASQC Quality Press, 1988, p. 13.

Best practices for design

- Analyze functionality, reliability, maintainability, testability, and cost requirements.
- Use preferred-parts lists and consult databases and libraries to find standard designs for reuse.
- Make sure part capability matches design needs.
- Analyze failures to find root causes and continuously improve design of products and processes.

Best practices for manufacturing

- Participate in design decisions early.
- Ensure designs reflect producibility and testability requirements.
- Review electrical-test specifications for adequacy.
- Set up a formal piece part control program.
- Decide what stress screening is appropriate and continually assess data to see if screens can be eliminated.
- Establish and maintain an effective failure review and corrective action system.
- Set up supplier and manufacturer quality-improvement teams.
- Give feedback to designers and suppliers.
- Select manufacturing equipment, test equipment, special processes, and resources for high quality and high yields with continual process improvement.

Best practices for suppliers

- Ensure requirements are understood.
- Participate in partnership agreements.
- Use statistical process control and feedback for continual improvements.
- Establish and nurture quality-improvement teams.
- Use failure analysis to improve designs, processes, and part performance.

Best practices for purchasing

- Procure parts from qualified suppliers when possible.
- Develop partnership agreements.
- Help supplier continuously improve quality and cost.

Best practices for subcontractors

- Ensure requirements flow down.
- Participate in design reviews with prime contractor.
- Set up long-term contracts for mutual benefit.

The four templates in this part have been integrated into six steps as Figure 3 illustrates.

Each template is discussed in Step 1: Create an Integrated Parts Management Strategy. Parts and Material Selection is discussed in Step

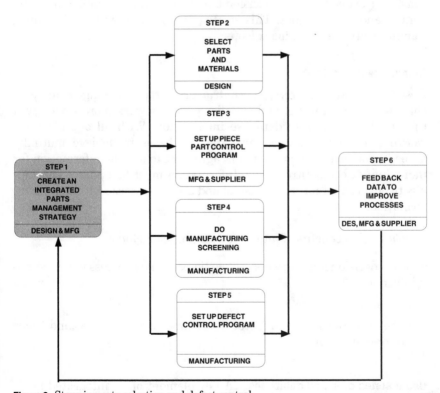

Figure 3 Steps in parts selection and defect control.

2: Select Parts and Materials. Piece Part Control is discussed in Step 3: Set Up a Piece Part Control Program. Manufacturing Screening is discussed in Step 4: Do Manufacturing Screening. Defect Control is discussed in Step 5: Set Up a Defect Control Program. Each of the four templates is also discussed in Step 6: Feed Back Data to Improve Processes.

Figure 3 also shows at the bottom of each box who is primarily responsible for carrying out each step: design, manufacturing, or suppliers.

Step 1: Create an Integrated Parts Management Strategy

To be effective, a parts management strategy should be wide-reaching and truly integrated with designers, manufacturers, quality and reliability engineers, suppliers, subcontractors, and purchasing representatives all working together to meet customer needs. An integrated strategy gives continuity throughout the product life cycle. Systems engineers, designers, and other early decision makers are more likely to consider the needs of later decision makers—testing, manufacturing, logistics, and maintenance. This strategy works only with well-planned communication and feedback loops.

Create integrated databases

Designers, manufacturers, subcontractors, project management, purchasing representatives, and supplier partners work more smoothly if they all have access to database information. With all organizations working together, they can use the databases to achieve mutually agreed-upon goals. Thus, suppliers agree not to use the information for individual or competitive goals. Information must be updated continually to ensure that it remains useful and everyone works with current information.

Benefits.   The benefits of integrated databases include:

- better decision making when all participants have access to up-to-date information
- quicker corrective action
- higher productivity from faster development, less rework, and lower maintenance costs

Use a series of linked databases.   A key element of an integrated parts management strategy is whether information is accessible to each participant. Rather than one large parts database, the most manageable databases are linked databases that have different sets of information for a variety of purposes. For example, one database may include qualification data, another has flammability data, and another has packaging information. Yet all these databases can share data and link with each other.

A few critical index parameters can be used to link these databases to each other. These parameters may include:

- specification number (generic code number rather than specific manufacturer's number)
- supplier by location (locations may have different processes and different quality)
- part description or family (e.g., ceramic capacitor)

Examples of linked databases.   Below are examples of information in specific databases.

A design database should include:

- preferred parts (based on performance, producibility, reliability, power dissipation, availability, cost)
- component selection tools and libraries compatible with design processes and tools
- reliability data (e.g., component failure rates, procedures to calculate failure rates, data from manufacturers' data books)

A supplier database should include:

- supplier's approval rating based on their quality, reliability, and timeliness
- supplier's quality and reliability data
- schedule of audits

A manufacturing database should include:

- forecasts
- producibility and design-for-manufacture requirements
- yields and defect rates for parts, boards, and units
- results of failure analyses of parts that had to be removed from circuits and apparatus

The failure analysis indicates why the failure occurred—defective parts, design-rule violations, missed specifications, wrong applications, inadequate design margins, accidental abuse, or incorrect assembly. These problems are then traced back to their source.

Ideally, feedback is then given to the designer or the supplier. With feedback on troublesome parts, designers understand the specific causes of failure and can avoid them perhaps by adhering more closely to design rules, specification changes, or requalifications. With information on specific defects traceable to the location on the board, designers may be able to avoid these problems in their future designs.

A life-cycle database should include:

- parts that are discontinued but will be needed to support a product over its life cycle

- information collected during the product's life cycle

Feeding all these databases are the quality and reliability databases, which contain information on a supplier's quality and the defects later encountered.

These databases strengthen the interrelationships among design, reliability, quality, purchasing, suppliers, manufacturing, and accounting. Many companies use integrated databases, including AT&T, Texas Instruments, General Dynamics, and GTE. The Application chapter describes how integrated databases are used at McDonnell Douglas Helicopter. For more information on databases and computer-aided tools, see Part 5, Computer-Assisted Technology.

Step 2: Select Parts and Materials

Establish preferred-parts lists

Successful companies have lists of preferred parts and materials that have met criteria for performance, reliability, timely delivery, and reasonable cost. Preferred-parts lists should be continually updated with historical data on delivery schedules, quality, reliability, performance, and costs. Companies that buy large quantities of standardized parts and materials see improvements in inventory costs, quality, ease of automation, and materials handling. Tooling and assembly costs are also less.[7]

---

[7]Priest, John. *Engineering Design for Producibility and Reliability.* New York: Marcel Dekker, 1988, p. 181.

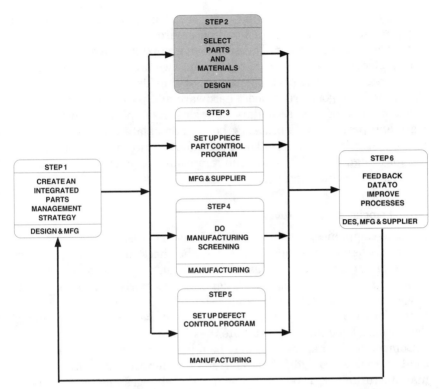

Figure 4 Steps in parts selection and defect control.

In using preferred parts and materials lists, designers are better able to follow good design rules, including design-for-assembly practices. These practices include reducing the number of parts and choosing standardized parts that can be used in many subassemblies and products.[8]

Before developing a new design, designers look in an integrated design database for a part that can be reused or modified to meet new design requirements. To make this possible, a database of coded designed parts must be created. Each part should have its own part number tied to its drawing and its own code number tied to a family of parts with similar attributes.

Designers can improve producibility by selecting parts from a few distinct part families with common design and manufacturing characteris-

---

[8]Nevins, James L. and Whitney, Daniel E., Eds. *Concurrent Design of Products and Processes*. New York: McGraw-Hill, 1989, p. 205.

tics. In *group technology*, similar parts are identified and manufactured together in a common production-line environment. Shorter intervals and improved quality usually result from group technology.[9]

In general, designers select parts from two broad categories. In the first category are the primary parts that provide basic capabilities. In the second category are support hardware and parts used to assemble the item or to provide secondary functions. Examples of support hardware and parts include brackets, hoses, nameplates, passive electrical parts, and indicator meters. These secondary parts offer the most potential for design standardization and improvements in efficiency from the use of group technology.[10]

Submit stocklists for approval

In some companies, designers submit stocklists (lists of the parts their design needs) to component managers who check to see if the parts are in the preferred-parts database. In many companies, component managers provide data to help the designers select parts. The data include power dissipation, reliability, and cost. Component managers get reliability data from factory and corporate databases and manufacturers' data books. They carefully evaluate these data.

Component managers, who may be reliability engineers, occasionally conduct reviews to audit the preferred parts for reliability, availability, manufacturability, and cost. Recommended changes must go through a formal change approval process. Often, the component manager has the responsibility to update the preferred-parts lists.

Select from component equivalents.   Component managers usually work with preliminary stocklists, which are mostly generic codes. A single generic code can correspond to, or be functionally equivalent to, several manufacturers' devices: e.g., parts made by Motorola, Intel, and Texas Instruments may have the same functions. Equivalent device codes have the same generic code but the packaging, testing, and qualification level may vary. Circuit designers use generic device codes to avoid unnecessary restrictions at the early stages of design.

Component managers often approve the designers' selection of parts from among equivalent parts. Circuit designers may select specific device codes for some devices (memory chips and gate arrays) but they typically specify generic codes for most integrated circuits. Circuit designers usu-

---

[9]Priest, John. *Engineering Design for Producibility and Reliability.* New York: Marcel Dekker, 1988, p. 180.

[10]Snead, Charles S. *Group Technology: Foundation for Competitive Manufacturing.* New York: Van Nostrand Reinhold, 1989, p. 121.

ally specify preference among equivalent parts based upon power, reliability, availability, quality, supplier responsiveness, size, weight, cost, etc.

On military contracts, designers often have narrower choices. They may be obligated by contract to use specific parts. For example, on many contracts designers must choose Joint Army Navy (JAN) parts that meet precise specifications. The recent defense initiatives have encouraged efforts to improve and streamline the component selection process. An ad hoc committee with representatives from the semiconductor industry and the defense community is developing an applications guide for contract officers that gives guidelines and trade-offs for using parts in various applications.

Consider quality and reliability

The terms "quality" and "reliability" are often used interchangeably as though they were the same attribute. They are different, however. Quality is the composite of all required attributes, including performance. It must be built in by means of stable design and manufacturing processes. Quality is the number of good parts or products that arrive at the next user.

Reliability is "quality on a time scale."[11] That is, reliability is the ability "to perform a required function under stated conditions for a stated period of time."[12] To produce a reliable device, the designer and manufacturer consider the interactions of the design, how well it is executed during manufacture, and the environmental stresses under which the product will function.

Reliability data should include information on failure rates and stress factors. A failure reporting and corrective action system (FRACAS) should include the failure mode, the failure cause, the corrective action, and the effectiveness of the corrective action. The designer can then use these failure rates to predict a device's reliability when it operates under similar conditions. Simulations are useful in predicting reliability. *Testing to Verify Design and Manufacturing Readiness* has more details on reliability.

Consult MIL-HDBK-217 for component failure rates.   MIL-HDBK-217 was developed to provide a consistent and uniform database for making reliability predictions when there is no reliability experience for a system.

---

[11]Hnatek, Eugene R. *Integrated Circuit Quality and Reliability*. New York: Marcel Dekker, 1987, p. vi.

[12]Klinger, David J., Nakada, Yoshinao, and Menendez, Maria, Eds. *AT&T Reliability Manual*. New York: Van Nostrand Reinhold, 1990, p. 2.

MIL-HDBK-217 gives failure rates for different part families and discusses two procedures for determining reliability: the parts count method and the parts stress method. The parts count method is used early in the conceptual and design phases. In general, the parts count method uses the parts failure rates to estimate the reliability of an assembly. The parts stress method provides input to design trade-off decisions. It gives different formulas for predicting the reliability of microelectronic devices, discrete semiconductors, and other types of components. For most devices, the failure rate is calculated from the circuit complexity, the package complexity, temperature, quality, etc.

Cautions in using MIL-HDBK-217.   This military handbook is sometimes cited as the industry standard for component failure rates because there is no other universally available source of data. The government may require contractors to use models to evaluate the reliability of electronic equipment for government use.

The user should be cautious, however, in using MIL-HDBK-217. Designers can use the failure rates to compare alternative designs but not to predict the field reliability of a device or a system. Field failures, for example, are more often caused by assembly or workmanship failures (e.g., connectors and solder joints) than by electronic-component failures. MIL-HDBK-217 does not address reliability-related problems induced by operators, workmanship, software, or maintenance. The handbook is not meant to predict field reliability and does not do it very well.[13]

Another problem with MIL-HDBK-217 is that the life of solid-state devices and other electronic devices can only be approximated by an exponential distribution. The exponential distribution is a good starting point to estimate times between failures for a system. But additional surveys and analyses are needed to verify the final design.

As the complexity of microcircuit devices increases, the MIL-HDBK-217 models and estimates will become less useful for designers. If possible, designers should substitute known field failure rates of devices into simulations and surveys to prove the design is capable of meeting reliability requirements.

Use other sources of reliability data.   Reliability data are useful for selecting components, for budgeting, appraising, and improving reliability, and for planning environmental stress screening (which will be discussed in Step 4). Many projects calculate reliability with mathematical models or by actual tests. Each has advantages and

---

[13]Morris, Seymour F. *MIL-HDBK-217, Use and Application.* Rome Air Development Center Technical Brief, April 1990.

disadvantages. Simulations provide results quickly and at relatively low cost, but the validity and accuracy of their results depend on the assumptions put into the model. Actual testing would give more valid results, but it is time consuming, expensive, and often virtually impossible.

Consider producibility

Designers make subtle choices about materials, components, fasteners, coatings, and adhesives, often without considering how their choices will affect manufacturing. If manufacturing engineers participate in these decisions, they can help designers consider producibility as one of the key criteria.[14]

To consider producibility in selecting parts, designers often work with component and manufacturing engineers to:

- minimize the number of different parts with the same function
- select parts that are functional, reliable, and conform to the factory's capabilities (e.g., parts that have low defect rates, can be automatically assembled, can be wave-soldered and cleaned)
- ensure enough parts are available at the right time

These considerations help reduce the total manufacturing cost by decreasing assembly errors, tooling and repair costs, labor time, and part shortages.

Designers can also increase producibility by giving preliminary stocklists to the manufacturing engineers early. With early stocklists, manufacturing engineers and purchasing agents can investigate suppliers' backgrounds and capabilities. They can make sure suppliers test components for high reliability and quality. Purchasing agents can negotiate contracts to make sure components are available "just in time" for assembly.

To ensure producibility and reduce component cost, many companies set up component-control committees and preferred-component databases tailored to specific manufacturing needs. The committee evaluates requests to use nonpreferred parts and decides when exceptions are justified. The committee also audits stocklists to check compliance with preferred-parts lists and producibility and reliability requirements.

---

[14]Nevins, James L. and Whitney, Daniel E. Eds. *Concurrent Design of Products and Processes*. New York: McGraw-Hill, 1989, p. 200.

Consider derating

In derating, a designer ensures that the actual stresses will be less than a percentage of the maximum stress according to the manufacturer's rating. In selecting parts and material, designers consider how the part will be used to make sure the applied stress is less than the manufacturer's rating. For example, the stresses for space applications are sometimes greater than ground applications.

Derating techniques increase the circuit's capacity to resist stress or decrease stress variations. Examples include using fans, heatsinks, or packaging. As a rule of thumb, reliability doubles with each 10-degree decrease in junction temperature. If reducing stress is not feasible, designers select larger or stronger parts that can withstand more stress. These parts provide a safety margin because they will be stressed electrically, mechanically, and thermally only to a percentage of their capability.

Use derating curves.   In selecting parts that can withstand the applied stresses, designers use derating curves to find the appropriate derating percentages. At a point in the operating range there is an inflection point where a slight change in stress causes a large increase in the part's failure rate. The appropriate derating percentage should be well below this inflection point.

Figure 5 illustrates a derating curve.[15] On the vertical axis is the derating percentage, which is the actual power rating in watts divided by the manufacturer's rating. On the horizontal axis, the point of inflection is at T_S. After this point, as the ambient temperature (T_A) or case temperature (T_C) increases to the maximum temperature (T_{max}), the derating percentage decreases. The area below the maximum-use rating curve

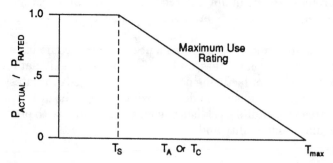

Figure 5 Illustrative derating graph.

---

[15]Anderson, R. T. *Reliability Design Handbook*. Chicago: ITT Research Institute, 1976, pp. 135–143.

provides the optimum margin of safety with no degradation in reliability expected. In the area above the maximum-use rating curve, however, parts would be overstressed.

Derating curves are available for different part types. These curves are sensitive to changes in temperature, electrical transients, vibration, shock, altitude, and acceleration. For example, the junction temperatures of semiconductors and integrated circuits should not exceed 110°C regardless of the nominal power rating. On a transistor, the collector-to-base voltages should be less than 70% of the manufacturer's value.

Figure 6 shows how the acceptable derating ranges for glass capacitors depend on the application. In general, ground applications may be less stressful than airborne or space applications.[16]

Match the derating technique to the part type and application.   The specific derating techniques vary with different types of parts and different applications. For example, capacitors are derated by keeping the applied voltage at a lower value than the voltage for which the capacitor is rated. Semiconductors are derated by keeping the power dissipation below the rated level.

With semiconductors, in addition to reducing part failure, derating reduces the internal operating temperature, thus decreasing the rate of chemical time-temperature reaction that causes part aging. Designers also use functional derating to help them make conservative design decisions with integrated circuits. One decision, for example, is the amount of fanout in a logic circuit, which refers to the number of paths fanning out from one output to various inputs on other devices. If the maximum fanout is to eight input devices, for example, conservative functional derating would restrict the fanout to six input devices. This conservative derating allows the circuits to function even with wide temperature variations and makes the circuits easier to test.

Judge the appropriate derating level.   Projects should collect data to derive derating criteria that meet their needs. Too-stringent criteria unnecessarily boost costs; too-liberal criteria increase failure rates.[17] The Rome Air Development Center has developed a derating "slide ruler" that provides derating guidelines for different part types and design requirements.[18]

---

[16]Anderson, R. T. *Reliability Design Handbook.* Chicago: ITT Research Institute, 1976, p. 166.

[17]Priest, John. *Engineering Design for Producibility and Reliability.* New York: Marcel Dekker, 1988, p. 186.

[18]RADC. *Parts Derating Guideline.* AFSC Pamphlet 800-27, Rome Air Development Center, 1983.

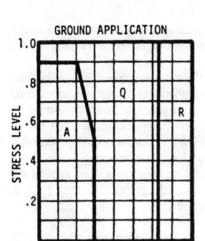

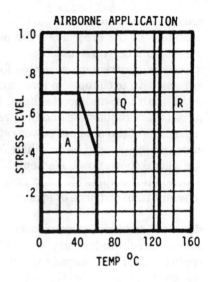

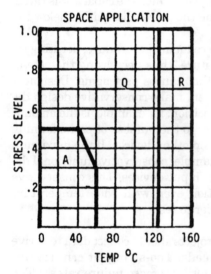

ELECTRICAL STRESS VERSUS
TEMPERATURE DERATING
REGIONS

A. Acceptable

Q. Questionable

R. Restricted

Figure 6 Glass-capacitor derating varies by application.

The advantages of derating often come with some disadvantages: parts having greater capability are often more costly, bigger, and heavier. The designer must decide which trade-offs are most important.[19] The Application chapter gives an example of derating used in the F/A-18 Hornet.

---

[19]Naval Sea Systems Command. *Parts Application and Reliability Information Manual for Navy Electronic Equipment.* TE000-AB-GTP-010, September 1985, p. F-2.

Derating is often used to include safety margins throughout the design as well as the ability to perform additional capabilities in the future. Examples include a power supply that can supply more current than needed, parts that can withstand higher stresses than are applied, a receiver microprocessor that operates at 25% of its rated capacity, and transmitter microcircuits that operate at one-third of their rated voltage.

Using derating principles to add surplus capability or margin to designs contributes more to reliable operation than adding quality improvements to an already high-quality product. The designer judges the appropriate derating level, balancing the need for mission success, cost, schedule, and future enhancements.[20]

Do stress analysis

As stress increases, a component's failure rate increases. Even when a component does not fail, the circuit may fail due to stress-induced reduction in operating margin. One way to prevent these failures is the use of derating criteria. Stress analysis ensures that the derating criteria are met and verifies that the equipment can perform under worst-case conditions.

Stress analysis is difficult with:

- a mix of technologies (e.g., bipolar, CMOS, TTL), which makes interface stress problems more likely and more difficult to detect
- large-scale and very large-scale integrated circuits that make tracing of loads more difficult[21]

Stress analysis is usually done hierarchically. Initially, the stresses on the smallest possible components are analyzed. Then the stresses on higher assembly levels are analyzed. The stress analyses may include mechanical stress (from bending, shear, or torsion), thermal stress (from temperature changes and extreme temperatures), and electrical stress (from power, current, or voltage).

Do thermal analysis

Designers analyze thermal profiles to measure part operating temperatures and compare measured temperatures to derating criteria. Thermal

---

[20]Meinen, Carl. "Reliable Remote Monitoring and Control of Electrical Distribution." *Proceedings of the Annual Reliability and Maintainability Symposium,* 1980, pp. 448–452.

[21]Bannan, M. W. and Banghart, J. M. "Computer-Aided Stress Analysis of Digital Circuits." *Proceedings of the Annual Reliability and Maintainability Symposium,* 1985, pp. 217–223.

analyses are computer simulations that help the designer select parts and materials to avoid thermal coefficient-of-expansion mismatches and other design problems. Thermal analysis is used to select materials for particular applications and to check the quality of incoming raw materials and outgoing products. To make thermal-analysis systems more affordable and flexible, suppliers are using personal computers to analyze and store data.[22] Finite-element analysis is a useful method for analyzing stresses and heat transfer. The results of thermal analyses should be verified with thermal surveys. For more details on stress analysis and thermal analysis, refer to Part 2, Design Policy, Design Process, and Design Analysis.

Thermal mismatches.   One example of the use of thermal analysis is with surface mount devices. As the use of surface mount devices increases, thermal mismatches and failures become likely at elevated temperatures. Thermal analyses often suggest new ways to make the design more functional and reliable.

Figure 7 shows cycles to failure for three multilayer boards (MLB) and one double-sided rigid (DSR) printed wiring board with surface mount devices. The test conditions were 60 cycles per day of temperatures between −20° and +130°C. When the thermal mismatch is large, the solder joints fail in 20 to 40 cycles (see boards J and K).When the thermal mismatch is small, the solder joints survive more than 350 cycles (see board H).[23]

Evaluate the use of composites

Designers must consider many issues in selecting materials such as plastics, wood, ceramics, or composites. Several reference books are useful.[24] [25] Composites are an example for which the technology is rapidly expanding and for which designers must consider trade-offs. A composite is a material that is composed of a high-strength material and a high-toughness material that can be formed, laminated, or molded to replace

---

[22]Smoluk, George M. "Thermal Analysis: A New Key to Productivity." *Modern Plastics,* February 1989, pp. 67–73.

[23]Sherry, W. M. and Hall, P. M. "Materials, Structures, and Mechanics of Solder Joints for Surface-Mount Microelectronics." *Proceedings of the Third International Conference on Interconnection Technology in Electronics.* Fellbach, West Germany, February 18–20, 1986, pp. 47–81.

[24]Young, James F. and Shane, Robert S., Eds. *Materials and Processes, Part A: Materials.* 3rd ed. New York: Marcel Dekker, 1985.

[25]Young, James F. and Shane, Robert S., Eds. *Materials and Processes, Part B: Processes.* 3rd ed. New York: Marcel Dekker, 1985.

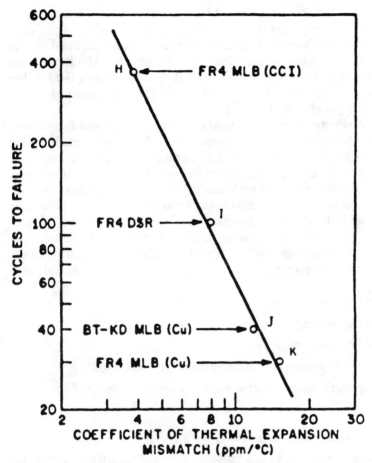

Figure 7 Cycles to failure depend on thermal mismatches.

a metal part. The advantages of composites are reduced weight, better corrosion resistance, better high-temperature resistance, and reduced life-cycle costs.[26]

Composites are used to provide:

- electromagnetic shielding
- electrical and thermal conductivity
- reflective surfaces

---

[26]Daane, John H., Horwath, John A., and Miller, Harold L. "New Materials" in *Manufacturing High Technology Handbook*. Eds. Donatas Tijunelis and Keith E. McKee. New York: Marcel Dekker, 1987, pp. 411–456.

Today, composites comprise about 14% of the structural weight of commercial and military aircraft.[27] For example, McDonnell Douglas used composites for the horizontal stabilizer on the F-14 aircraft and for the tail assembly or empennage on the F-15 aircraft. Bell Helicopter uses composites for its helicopter blades. The skin of Northrop's B-2 bomber is made entirely of composites to provide the stealth characteristics as well as rigidity and weight savings.

Composites may require new methods of processing and fabrication. Automated techniques may be needed to reduce the manufacturing costs. For example, a wing made of aluminum can be manufactured by one worker while a wing made of composites may require 20 workers. There are significant process problems in joining and cutting composites. Key questions are how to handle them, form them, and automate their manufacture.[28] Trade-off studies may show that advantages of composites including fuel economy, temperature resistance, and reduced life-cycle costs may outweigh the higher material and manufacturing costs.

When considering composites or any material, a designer should ask:[29]

- Will the product be safe?
- Can existing tooling be used?
- Are new processes required?
- Are new quality-assurance tools or techniques required?
- Were any weaknesses detected during design simulations?

Devise an overall strategy for selecting parts

Strategies for designers to use in selecting high-quality parts while reducing parts proliferation include:

- select parts that meet requirements for functionality, reliability, producibility, design strategy, and cost
- select preferred parts from approved databases

---

[27]"Materials: Backbone of Aerospace Designs." *Design News,* April 9, 1990, pp. 25-28.

[28]Owen, Jean V. "Assessing New Technologies." *Manufacturing Engineering,* June 1989, pp. 69–73.

[29]Katz, Harry S. and Brandmaier, Harold E. "Concise Fundamentals of Fiber-Reinforced Composites" in *Handbook of Reinforcements for Plastics.* Eds. John V. Milewski and Harry S. Katz. New York: Van Nostrand Reinhold, 1987, pp. 6–13.

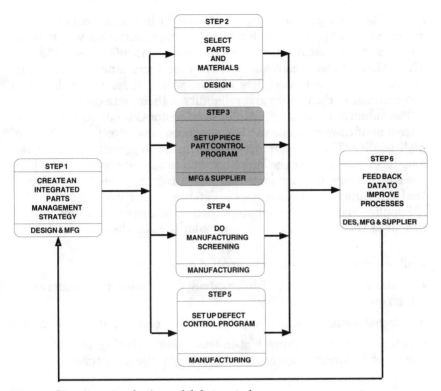

Figure 8 Steps in parts selection and defect control.

- follow procedures for obtaining approval if a needed part is not in the database
- ensure stresses are less than maximum rating
- choose the appropriate derating percentage (e.g., use derating curves)
- ensure derating is cost-effective
- do trade-off studies to decide which material to use (e.g., whether to use composites or metal parts)

Step 3: Set Up a Piece Part Control Program

Qualify and certify suppliers

Suppliers and customers must agree that the quality of the suppliers' parts matches what the customer expects. Customers expect a defect rate of 100 parts per million (ppm) or less for integrated circuits and a defect rate of 10 ppm or less for discrete components. One way to bring about customer confidence is with supplier-manufacturer alliances. These alliances are recent and still evolving. Increasingly, manufacturers want to work with suppliers who are willing to learn and change in order to control the quality and reliability of their piece parts.

The automobile industry, notably Ford Motor Co., has developed supplier-manufacturer alliances that produced cost-effective products of high quality. These programs resulted in an improved supplier base, long-term contracts for the suppliers, and improved operations for both the supplier and the customer. Implementing programs like these will lead to significant improvements in defense acquisition.

Supplier-manufacturer alliances based on cooperation and mutual benefit are becoming increasingly popular because they:

- allow quality standards to be applied consistently
- reduce the rescreening of microcircuits that meet requirements for high quality and reliability
- reduce the number of similar parts and thus costs of parts proliferation
- reduce costs by making just-in-time manufacturing possible (e.g., smaller inventories, lower carrying charges, less downtime)

With supplier-alliances, designers can select parts from proven suppliers chosen for their component quality and reliability, timeliness of delivery, and financial soundness.

Supplier alliances, reduced rescreening, and just-in-time manufacturing are important even with the priority ratings that earmark the first available product for the military contract. The alternative strategy of relying on rescreening adds the risk of delays and slipped schedules due to rejected lots.

Cost of ownership. Many companies select suppliers on a cost-of-ownership basis (e.g., cost impact of rejects, late deliveries, receiving inspections) rather than just the cost of the part.[30] To do this, they consider:

- measurements, including assessment, monitoring, and reassessment

---

[30]Capitano, J. L. and Feinstein, J. H. "Environmental Stress Screening (ESS) Demonstrates Its Value in the Field." *Proceedings of the Annual Reliability and Maintainability Symposium,* 1986, pp. 31–35.

- attributes, including quality, reliability, manufacturing, delivery, service, and cost

- records, including defect rate, mean time between failures, and yield

Suppliers' process improvement.   A key ingredient in fostering understanding between part suppliers and part users occurs when the users share their experience with the suppliers. Many assembly manufacturers also work with their suppliers to improve the suppliers' processes. The assembly manufacturers:

- provide requirements and specifications

- periodically review products and process controls

- help suppliers implement and use statistical process controls (SPC)

Achieving near-zero defects per million parts requires a stable process with all subprocesses in control. The goal is to discover the true sources of variation in the design or the manufacturing process.[31]

SPC must be applied effectively. Control charts should be used to monitor the ongoing process. Any deviations will then be quickly apparent. Deviations should trigger action to correct and eliminate the cause of the deviation. SPC is discussed in more detail in Step 5.

Select parts from qualified manufacturers list (QML).   To reduce the cost of electronic components, semiconductor manufacturers and the government are participating in a joint effort to qualify the manufacturing process, under MIL-I-38535 for monolithic microcircuits and MIL-H-38534/MIL-STD-1772 for hybrids. The QML effort takes advantage of the near-zero defect levels many semiconductor suppliers achieve: incoming component defect levels below 100 parts per million (ppm). The QML program recognizes that incoming inspections are costly and not needed when suppliers implement statistical process control (SPC) programs to ensure quality is built in. The objective of QML is a 10-fold to a 100-fold decrease in the price of silicon microcircuits.[32]

With QML, the manufacturing processes are certified rather than individual parts as in the current Qualified Parts List (QPL) and MIL-M-38510, General Specification for Microcircuits. The traditional approach

---

[31]Maass, Richard. *World Class Quality: An Innovative Prescription for Survival.* Milwaukee, WI: ASQC Quality Press, 1988.

[32]Burgess, Lisa. "Thomas: Pushing the Pentagon Toward QML." *Military and Aerospace Electronics,* February 1990, pp. 39–40.

of certifying parts under the Joint Army Navy (JAN) programs and under the MIL-M-38510 program was costly, lengthy, and inefficient.[33] Parts became obsolete almost as soon as they were qualified. The lengthy audit process had to be repeated with each upgrade. QML will eliminate the need to requalify parts from a certified line.

The challenge for the parts industry is to make QML work. For QML to be successful:

- top management must support the process
- manufacturing must understand the entire manufacturing process
- customers must give feedback

The key element of QML is statistical process control. This in-process monitoring of the manufacturing processes is used to ensure device yield and reliability. Two innovative evaluation features are the standard evaluation circuit and the technology characterization vehicle. The standard evaluation circuit is used to demonstrate the reliability and quality that result from the processes. It is designed to stress the worst-case geometric and electrical design rules and to allow easy diagnosis of failures.

The technology characterization vehicle contains test structures that monitor intrinsic reliability failure mechanisms such as electromigration, time-dependent dielectric breakdown, and hot carrier aging.

The benefits of the QML program are:

- better control of the manufacturing process
- better use of facilities
- fewer government audits and lower qualification costs
- predictable part costs
- improved delivery schedules
- earlier use of advanced commercial integrated circuit technologies in military systems

AT&T's facility in Allentown, PA was the first manufacturing facility to have its processes certified by the government. With QML certification, they can produce their metal-oxide semiconductor integrated circuits for the military on the same line as those for commercial purposes.

---

[33]Gardner, Fred. "Hold Down Ballooning Costs and Boost Quality." *Electronics Purchasing*, June 1988, p. 57.

Commit to continual process improvement.   It is wise to select at least two suppliers for each part who are willing to join in supplier-manufacturer alliances. These suppliers commit to work towards world-class quality of their parts and materials. This commitment means that designers, suppliers, and manufacturing engineers aim for continual process improvement. The designer selects approved and verified materials. The designer also specifies tolerances appropriate to the operating and environmental conditions. The tolerances should not be too tight (difficult to manufacture) nor too broad (cause instabilities and malfunctions). The manufacturing engineer spots abnormal distributions and then designs experiments to find the root causes and appropriate solutions.[34]

Even though the supplier's SPC ensures parts of high quality and reliability, the OEM should have processes in place that provide early warnings of a quality or reliability problem and provide immediate feedback to the supplier. For example, instead of inspecting for an acceptable quality level, the OEM can verify the supplier's test results using standard sampling techniques.[35]

Developing alternative sources may be expensive and difficult. It is difficult, for example, to divide small-quantity orders between two suppliers. Companies should do trade-off studies to balance the benefits with the risks of single sources.

Tailor contracts for reduced rescreening.   Many prime contractors have tailored their government contracts to reduce or eliminate rescreening. Litton Guidance, for example, accepts Texas Instruments parts without rescreening them because of the history of near-zero defects. Texas Instruments certifies that its defect rate is less than 100 ppm and shares its SPC data with Litton.

Standardize military drawings and numbering system.   To improve quality and reduce costs, the Defense Electronics Supply Center (DESC) in Dayton, Ohio set up a program in which suppliers or assembly manufacturers generate standard military drawings (SMD) for standard military parts. This program was set up to stop the expensive proliferation of source control drawings (SCD). In the standard military drawings program, DESC lists suppliers on the SMD whose parts meet

---

[34]Maass, Richard. *World Class Quality: An Innovative Prescription for Survival.* Milwaukee, WI: ASQC Quality Press, 1988, p. 24.

[35]Maass, Richard. *World Class Quality: An Innovative Prescription for Survival.* Milwaukee, WI: ASQC Quality Press, 1988, p. 4.

the specification limits. Assembly manufacturers can generate additional SMDs for parts that are not listed.

DESC has also set up a new numbering system for military semiconductor products to reduce paperwork and errors. Each part will now have one 15-character number. The only variable is a letter that tells users, suppliers, and manufacturers what standards the part is designed to meet—B for JAN Class B devices, S for JAN Class S devices, M for SMDs, and Q or V for devices from the Qualified Manufacturer List (QML).[36]

Manage sources

Companies must decide on the best way to eliminate marginal devices. The ultimate goal is to select parts only from qualified and certified suppliers. To learn more about the supplier's quality and processes, companies may use source inspection or receive inspection temporarily until the data show that the supplier's processes are stable and producing parts with near-zero defects.

In source inspection, companies send their own inspectors to the supplier's site. This practice often has advantages over receiving inspection. Advantages include saving time, equipment and labor, not having to return defective parts, and reducing the failures at higher assembly levels. Also, the inspectors learn about the supplier's practices on-site.

Source inspection may be ineffective if there is a strained relationship between the supplier and the assembly manufacturer or if the source inspector interrupts process flows. Source inspection may be preferable to receiving inspection, but it is less preferable to supplier-manufacturer cooperative alliances, which aim to build in quality through process controls. In these alliances, the manufacturer and supplier jointly share data and improve the processes.

If the supplier fully inspects and screens parts within a stable process-control environment, the assembly manufacturer can:

- eliminate duplicate test equipment
- reduce inventories due to the high percentage of usable parts
- use resources better
- pass along lower costs to the end user

Corporations such as Texas Instruments, General Dynamics, AT&T and many others have instituted supplier programs to reduce incoming in-

---

[36]Keller, John. "One Part, One Number: DESC Simplifies IC Buys." *Military and Aerospace Electronics,* May 1990, p. 1.

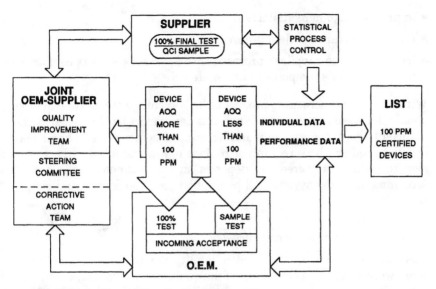

Figure 9 OEM-supplier strategy for defect-free parts.

spection costs. In today's environment of automation, high rework costs, and reduced inventories, companies try to avoid incoming inspections to monitor the supplier's quality. Today, OEMs and suppliers are working together to solve problems. These teams try to correct problems at the source rather than relying on incoming inspection.

Figure 9 shows how Texas Instruments and some of its suppliers are achieving defect-free parts without unnecessary rescreening. (Note that AOQ in the figure means average outgoing quality.)[37]

Use effective screening

Screening ideally occurs as part of the original manufacture of the part. In screening, stress is applied to devices to find marginal devices. Ideally, a failure mode analysis team uses the results of the screening to find root causes of the marginal devices and how to prevent them.

Screening helps the supplier, manufacturer, and customer. The benefits include:

■ fewer failures at higher assembly levels and in the field
■ less rework and repair on assemblies using screened parts

---

[37]Bindhammer, Carl and Krog, John. "An Electrical Test Correlation Experience." *Integrated Circuit Screening Report,* Institute of Environmental Science, November 1988, pp. 4-5.

- improved customer satisfaction
- reduced warranty or field failure costs
- feedback on the supplier's process-control system and more confidence in the supplier's components and material

With an effective screening program, the percentage of marginal devices should decrease to near zero. When this zero-defect level is reached and sufficient process control in manufacturing is achieved, companies may decide to reduce the screening or replace it with sampling.[38] This decision to reduce the screening depends on the technology, the cost of screening, how the system will be applied, and the contractual obligations.

Evaluate piece part quality and reliability

Ensuring piece part quality and reliability is vital. During the 1980s, there were divergent perceptions of the quality and reliability of military parts. The defect rate, which was actually quite high in the early 1980s, improved in the middle 1980s when many suppliers upgraded and standardized their screening and worked toward continuous improvement of their processes. By the mid 1980s, suppliers were reporting an average defect rate of about 1,000 parts per million (ppm). Assembly manufacturers and the military, however, perceived the average defect rate to be an order of magnitude higher, about 10,000 ppm.[39]

Even with the supplier improvements, the assembly manufacturers' and the military's perceptions of poor quality persisted. Strained relationships were common. Suppliers blamed assembly manufacturers for inaccurate testing, and assembly manufacturers blamed suppliers for poor quality. Many DoD contracts required 100% rescreening of incoming components in spite of the improvements in quality.[40]

Root causes for the defects.   There were many reasons for these different perceptions. The data did include parts with inadequate quality and limited performance and environmental capabilities that were improperly used in military systems. And, in many cases, the assembly

---

[38]Klinger, David J., Nakada, Yoshinao, and Menendez, Maria, Eds. *AT&T Reliability Manual.* New York: Van Nostrand Reinhold, 1990, p. 52.

[39]Golshan, Shahin and Oxford, David B. "ESSEH Parts Committee Overview." *Integrated Circuit Screening Report,* Institute of Environmental Sciences, November 1988, p. 3-1.

[40]Oxford, David B. "Total Quality Management: Business Aspects and Implementation." *Integrated Circuit Screening Report,* Institute of Environmental Science, November 1988, p. 2-1.

manufacturers and the suppliers used different equipment and screening methods. But perhaps the main cause of the different perceptions was that the average defect rate was misleading.

During 1986, for example, Texas Instruments' incoming-inspection data showed an average defect level of about 4,000 ppm. This average was computed by calculating the total confirmed rejected parts divided by the total parts screened and then normalizing to one million. This method of calculating the average, however, obscured key information. The 4,000 ppm average could result if each part type had a defect rate of about 4,000 ppm or if a few part types had a very high defect rate and most part types had a lower defect rate. The average would be misleading in that case.

Preliminary analyses at Texas Instruments showed that the average was indeed misleading: 31% of the part types contributed to the high defect rate, but 69% of the part types were defect-free when they left the supplier's site.

Figure 10 shows these data.[41]

Of the 31% part types that were responsible for the high defect rate, analog devices (e.g., linear interface, control, and converter circuits) showed the most correlation failures. That is, the assembly manufacturer and the supplier obtained different results. Correlation problems are likely with analog devices because it is harder for the supplier and the assembly manufacturer to obtain the same results screening analog

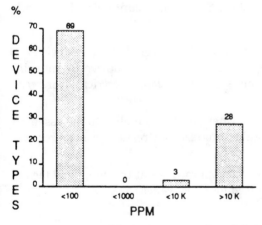

Figure 10 Percent of device types at various defect rates.

[41]Bindhammer, Carl and Krog, John. "An Electrical Test Correlation Experience." *Integrated Circuit Screening Report,* Institute of Environmental Science, November 1988, p. 4-1.

devices than digital devices, even with the same equipment and screening methods. The Texas Instrument analyses also showed that some parts were damaged during rescreening due to mishandling, insertion errors, and electrical overstress.[42]

In 1986, to investigate this correlation problem, the Institute of Environmental Sciences formed an ESSEH Semiconductor Parts subcommittee with representatives from suppliers, assembly manufacturers, test labs, and the military. This subcommittee operated within a larger committee called the Environmental Stress Screening of Electronic Hardware (ESSEH) committee.

The subcommittee set up a pilot program of three pairs of suppliers and original-equipment manufacturers (OEMs) to analyze the integrated-circuit screening data and resolve the divergent perceptions. In early 1986, one pair, Texas Instruments Defense Electronics Equipment Group (the OEM) and Texas Instruments Semiconductor Group (the supplier), formed joint OEM-supplier quality improvement teams to focus corrective action on the part types with high defect rates.

As a result, Texas Instruments increased the percentage of near-zero defect part types from 69% in 1986 to 95% by 1987.[43] This improvement, shown in Figure 11, resulted from failure analyses done by the OEM-supplier quality improvement teams. Their failure analyses helped prevent defects due to overstress and poor handling.

Conclusions.   After studying the failure-analysis data from the three pairs in the pilot program, the ESSEH Semiconductor Parts subcommittee concluded:

- rescreening does not improve the quality or reliability of a device and in many cases may reduce the reliability of the device by inducing additional failures (e.g., from wrong insertions, electrical overstress, poor handling)

- rescreening should be avoided when the data supplied to the assembly manufacturer indicate a stable process that results in near-zero defects

- test procedures and manufacturing processes at the suppliers' facility should be periodically reviewed for effectiveness

---

[42]Golshan, Shahin and Oxford, David B. "ESSEH Parts Committee Overview." *Integrated Circuit Screening Report,* Institute of Environmental Sciences, November 1988, p. 3-1.

[43]Bindhammer, Carl and Krog, John. "An Electrical Test Correlation Experience." *Integrated Circuit Screening Report,* Institute of Environmental Science, November 1988, p. 4-4.

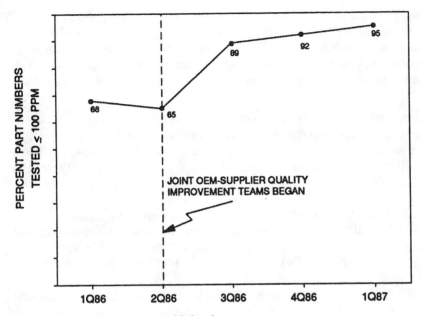

Figure 11   Improved percentages of defect-free part types.

- handling procedures should be reviewed for possible damage from electrostatic discharge and other problems
- customer-supplier teams are effective in solving problems

Screen at the piece part level

Screening at the piece part level is usually cost-effective for the supplier. Finding and removing defective parts early before they are shipped to the assembly manufacturer pays off. The supplier usually gets more business with a reputation for quality and reliability. The assembly manufacturer is spared the expense and delays of rescreening or failures at higher assembly levels.

Examples of screening at the piece part level include particle impact noise detection, high-temperature burn-in, highly accelerated stress testing, thermal shock, and hermeticity. Step 4 describes environmental stress screening that is applied to boards, subassemblies, and finished systems.

By definition, screening is done on 100% of the devices. Whether to use sampling or screening depends on the type of defects expected or the customer's requirements. For random defects (e.g., bonding defects), 100% screening is needed. For defects related to the particular manufacturing batch (e.g., defective metallization due to masking defects), sampling may be effective. Sampling tests the manufacturing process, 100% screening tests the components.[44]

Particle impact noise detection (PIND).   For some device types (e.g., open-cavity devices, metal cans), a test of particulate contamination is particle impact noise detection (PIND), which can detect microscopic particles. To find loose particles in a sealed package using the PIND test, the packaged device is shaken with highly sensitive acoustic monitors attached to it.

Screening is more difficult for particles that are not actually loose but have the potential to come loose during mechanical or other operational stresses. These failures are often unexpected, unrepeatable, and hard to screen out. They usually come from defects and flaws that have been aggravated during manufacture.[45]

PIND has been controversial for the last ten years. Some government agencies, NASA for example, are strong advocates. Others question its usefulness, saying it is better to correct the process that is causing the loose particles rather than trying to screen for them. When improvements in the process result in near-zero defects, PIND may be no longer cost-effective.

High-temperature burn-in.   Before suppliers ship the devices to assembly manufacturers, they expose devices to elevated temperatures while all electrical connections are exercised. The purpose of this burn-in is to find marginal devices that would otherwise fail in the assembled equipment. Suppliers analyze devices that fail burn-in and use the results to improve their design or manufacturing processes.

In the typical burn-in process, devices experience temperatures of 125°C for at least 48 hours.[46] High-temperature tests effectively screen microchemical flaws (e.g., contamination) in semiconductors and also weak bonds due to heel cracking. Some failure mechanisms (e.g., oxide defects) may not be sensitive to temperature but may be accelerated by overvoltage, overcurrent, or other stress. Tests that combine elevated

[44]Amerasekera, E. A. and Campbell, D. S. *Failure Mechanisms in Semiconductor Devices.* New York: John Wiley and Sons, 1987, p. 114.

[45]Amerasekera, E. A. and Campbell, D. S. *Failure Mechanisms in Semiconductor Devices.* New York: John Wiley and Sons, 1987, p. 112.

[46]Buck, Carl N. "Improving Reliability." *Quality,* February 1990, pp. 58–60.

temperature with mechanical stress are usually the most effective screens.[47] This combined thermal and mechanical screening is not yet frequently used.

Highly accelerated stress testing. Many semiconductor manufacturers are using highly accelerated stress testing (HAST) to stimulate corrosion failures in plastic-encapsulated devices. The devices are exposed to high temperature and humidity under pressure, which greatly accelerates wear and forces failures due to corrosion. This new accelerated approach takes less time than the traditional approach of exposing devices to 85°C and 85% relative humidity at ambient pressure for at least 1,000 hours. The traditional approach, called 85/85 testing, requires several weeks of testing. With HAST testing, however, the devices are exposed to 120°C and 85% relative humidity for only 100 hours.

Many studies have shown correlations between 85/85 testing at 1,000 hours and 120/85 testing (HAST) at 100 hours. Intel, for example, found that the failure mechanisms that HAST finds at the wafer level are similar to those found in actual packaged devices. Intel stressed the need to keep the test environment free from contamination to make sure the failure data reflect fabrication problems only. Intel also pointed out the need for careful handling during HAST and traditional testing to avoid contamination and corrosion from chlorine and salts.

Even though the HAST test equipment is more costly, the shorter testing times make it cost-effective. HAST has also pointed out ways to improve reliability and reduce defects from contamination, corrosion, and handling.[48]

Thermal shock. Thermal shock exposes a component or system to a rapid change in temperatures over a specified range at a determined rate to stimulate latent defects into failure. Thermal shock uses a liquid-to-liquid medium to provide the severe temperature shock environment when transferring between the temperature extremes. For example, thermal shock may subject devices to temperatures of −65°C and +125°C for about 10 seconds at each level. Thermal shock detects crystal defects and other packaging defects. Thermal shock is a cost-effective method of screening components and assembled circuit boards. It may be used at higher levels of assembly but the cost of the

---

[47]Amerasekera, E. A. and Campbell, D. S. *Failure Mechanisms in Semiconductor Devices*. New York: John Wiley and Sons, 1987, p. 104.

[48]Comerford, Richard. "Turning up the Heat on Stress Testing." *Electronics Test*, January 1990, pp. 20–23.

monitoring equipment may be high. It may be destructive and should be used cautiously.[49]

Mechanical shock.   Shock tests determine the device's relative resistance to damage when dropped. The shock level is measured in multiples of gravity (G's).[50]

Temperature cycling.   Temperature cycling screens for manufacturing defects (e.g., wire-bond defects, poor package seals, cracked dies). In a typical test, devices are put in hot/cold air-to-air cycling chambers for 10 cycles of −65°C to +150°C with 10-minute cycles.[51]

In temperature cycling, the temperature gradually increases or gradually decreases. In thermal shock tests, the temperature changes rapidly from one extreme to another.

Destructive physical analysis.   Destructive physical analysis (DPA) may be used on prototypes, models, and samples to look for defects and identify corrective action. Because it is destructive, DPA is done on a small, representative sample. DPA might include x-ray tests, hermeticity residual gas analysis, particle impact noise detection, and internal disassembling to check for wire bonding, die-attach quality, and internal wire construction. On complex or submicron processes, a scanning electron microscope might be used.

Hermeticity.   Hermeticity screens check for broken or cracked package seals. These hermetic seals control the gas leakage over the device's life cycle. Fine and gross leak tests may be used to stimulate these types of latent failures. Fine and gross leak tests are both necessary to detect different magnitudes of leak rates. Two fine leak tests are often used: helium leak testing and krypton gas testing. The most commonly used gross leak test is the fluorocarbon bubble test. For more details, see the Institute of Environmental Sciences' guidelines.[52]

---

[49]Amerasekera, E. A. and Campbell, D. S. *Failure Mechanisms in Semiconductor Devices.* New York: John Wiley and Sons, 1987, p. 105.

[50]Sanders, Robert T. and Green, Kent C. "Proper Packaging Enhances Productivity and Quality." *Material Handling,* August 1989, pp. 51–55.

[51]Amerasekera, E. A. and Campbell, D. S. *Failure Mechanisms in Semiconductor Devices.* New York: John Wiley and Sons, 1987, p. 106.

[52]Institute of Environmental Sciences. *Environmental Stress Screening Guidelines for Parts.* Mount Prospect, IL: Institute of Environmental Sciences, September 1985, p. 2-9.

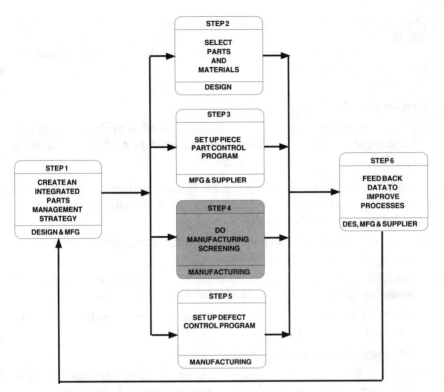

Figure 12 Steps in parts selection and defect control.

Step 4: Do Manufacturing Screening

Environmental stress screening (ESS) consists of exposing a circuit card, unit, or system to an environment that will induce latent defects to fail. These failures are *latent defects* because they are not detected through standard test methods, but they will emerge over time.

Screens to consider are thermal shock, temperature cycling, and random vibration. Screens may be used in combination to stimulate failures, e.g., temperature cycling and random vibration. The key factor with combination screens is whether the added stresses will expose significantly more failures and justify the cost.

Failures found by ESS fall into two categories:

- workmanship errors
- process defects

Define ESS objectives

The objective of environmental stress screening is to establish that the manufacturing process is in control and capable of manufacturing the product. Even a qualified process will produce a few products that are outside the acceptable limits, because of random variations within the process itself. It is better to find and eliminate these latent defects in the factory than in the field.

Develop plans for ESS

Because screening is expensive, planning is needed to obtain the greatest benefit at the lowest cost and at the least risk to good units. Planning helps:[53]

- select screens that stimulate the likely flaw without damaging good units (e.g., temperature cycling, random vibration, thermal shock, high-temperature burn-in)

- select the appropriate level of intensity, duration, temperature range or rate, vibration, repetitions, etc.

- use ESS at the lowest manufacturing level to find workmanship and assembly defects when rework is least expensive

- select the appropriate screen for technology being used (e.g., through-hole vs. surface-mount)

- demonstrate the screens' cost-effectiveness

- decrease ESS when the manufacturing processes are in control and the defect rate is too low to make ESS cost-effective

Select screen levels.   Engineers must choose levels carefully to find as many flaws as possible without false positives (non-defects) or false negatives (missed defects) and with minimal damage to the product. Choosing correct levels is an iterative process. First, a low-level screen is run. If that does not stimulate latent defects, a higher level is run. If the higher level damages the unit, the engineers may decide to lower the level or modify the design.[54]

Five methods are in use, for example, for selecting the appropriate level for random vibration screening. Table 5 presents advantages and disadvantages of the five methods. For more details, see the Institute of Environmental Sciences guidelines.[55]

---

[53]Hobbs, G. K. "Development of Stress Screens." *Proceedings of the Annual Reliability and Maintainability Symposium,* 1987, pp. 115–119.

[54]Hobbs, G. K. "Development of Stress Screens." *Proceedings of the Annual Reliability and Maintainability Symposium,* 1987, pp. 115–119.

[55]Institute of Environmental Sciences. *Environmental Stress Screening Guidelines for Assemblies.* Mount Prospect, IL: Institute of Environmental Sciences, March 1990, pp. 37–41.

TABLE 5   Advantages and Disadvantages of Methods to Determine Screening Level for Random Vibration

Methods	Advantages	Disadvantages
Tailored Spectral Response: Uses flaw precipitation threshold to develop input spectrum and tailor screening level	■ Only method able to develop spectral characteristics ■ Least likely to damage good hardware ■ Shortest vibration exposure during development	■ Needs spectral analysis equipment and skilled operators ■ More expensive and time-consuming than tailored overall response method ■ May not be effective for new technology
Tailored Overall Response: Uses overall internal response levels to develop screening level	■ Similar to tailored spectral response method but less expensive and complex	■ Unable to adjust spectral characteristics
Step-Stress Tests: Sets the screening level between the operating level and one-half the design or tolerance limits	■ Straightforward empirical method useful for existing and developing technology ■ Defines item design limits and makes equipment stronger	■ Some risk of overstress if design limits are unknown ■ Design may have to be changed to make it stronger
Fault-Replication Tests: Increases screening level until seeded (i.e., known) faults are replicated	■ Supplements step-stress tests	■ Hardware may not have replicable failure modes and it may be hard to seed hardware faults realistically
Heritage Screen: Derives screening level from past experience	■ Minimum development resources required and thus easier to obtain resources	■ Transparent dissimilarities may make screen inadequate or damaging

ESS can be used successfully if the engineers carefully calculate stress levels to avoid degrading the equipment for normal use.[56] However, the long-term effects of ESS may limit a product's life. If this is likely, simulation may be used to find potential failures.

Choose screens.   An effective screen stimulates likely latent flaws without damaging good products. No one screen will find every type of flaw. Some flaws are stimulated by thermal cycling, some by vibration, and some by voltage cycling. Some flaws can be stimulated by several screens. Ignoring these flaw-stimulus relationships would lessen the effectiveness of environmental stress screening.

---

[56]RADC. *Stress Screening of Electronic Hardware.* RADC Technical Report TR-82-27. Rome Air Development Center, 1982.

It is important to distinguish between latent or hidden defects and actual defects. Latent defects, which arise from irregularities in manufacturing processes or materials, become actual defects when exposed to environmental stimuli.[57] Effective screens find these latent defects without causing additional harm. Table 6 illustrates characteristics of effective and ineffective screens.[58]

Verify screens.   One way to verify a screen is to seed the hardware, that is, to put known defects into the hardware. Effective screens will find most of these defects; ineffective screens will not.[59] It is not always possible to seed realistically.

Decide when to use ESS

ESS can be useful at the circuit-card, unit, or system level. In general, the cost per failure is lowest if ESS is used at the lowest possible level, but the detection efficiency is best at the highest levels because interface errors as well as other types of errors can be detected. Each project must decide which level is most cost-effective. Table 7 presents some advantages and disadvantages of ESS at the circuit-card, unit, and system levels.[60]

TABLE 6   Characteristics of Useful and Poor Screens

Useful Screens	Poor Screens
Precipitate flaws quickly	Fail to stimulate latent flaws
Stimulate adequate proportion of latent defects	Induce additional defects
Apply accelerated stress but not overstress beyond the equipment's design limits	Consume too much of equipment's operating life

---

[57]Hobbs, G. K. "Development of Stress Screens." *Proceedings of the Annual Reliability and Maintainability Symposium,* 1987, pp. 115–119.

[58]Department of the Navy. *Navy Manufacturing Screening Program.* NAVMAT P-9492, May 1979.

[59]Hobbs, G. K. "Development of Stress Screens." *Proceedings of the Annual Reliability and Maintainability Symposium,* 1987, pp. 115–119.

[60]Fuqua, Norman B. "Environmental Stress Screening." Paper presented at the Joint Government-Industry Conference on Test and Reliability, AT&T Bell Laboratories, May 4, 1990.

TABLE 7   Advantages and Disadvantages of ESS at Different Levels

Level	Advantages	Disadvantages
Circuit-Card	Cost per failure is lowest Small size and mass permit batch screening and fast rates of temperature change	Detection efficiency is relatively low Interface errors are not detected
Unit	Can power and monitor performance during screen Detection efficiency is higher than at circuit-card level Interconnections (e.g., backplanes) are screened	Large mass precludes fast rates of change without costly equipment Cost per failure is higher than at circuit-card level Temperature range is lower than at circuit-card level
System	All potential sources of failures are screened Unit interconnections are screened Detection efficiency is highest	Screening at temperature extremes is difficult and costly Large mass precludes vibration screens without costly equipment Cost per failure is highest

Implement environmental stress screening

Examples of environmental stress screening include temperature cycling and random vibration.

Temperature cycling.   Temperature cycling is widely used to find defects in electronic equipment including circuit cards, assemblies, and systems. It is used during development to find and eliminate design problems and during production to find and eliminate defective units, processes, and workmanship. With temperature cycling, the temperature gradually increases and then gradually decreases for a set of predetermined times. The rate of change of temperatures depends on the specific heat of the unit or system being screened.

How many temperature cycles are needed depends on the complexity. Test and failure rate data show that six cycles are adequate for equipment with about 2,000 parts; 10 cycles are needed for equipment with about 4,000 or more parts. When unscreened parts are used, more than 10 cycles may be needed.

Temperature ranges of –65°F to +131°F are common. The maximum safe range of component temperature and the fastest rate of change of hardware temperatures will give the best screening. The optimal rate of change depends on the size and mass of the hardware.[61]

Examples of defects screened stimulated by temperature cycling include poor solder joints, welds, and seals; shorted wire turns and cabling due to damage or improper assembly; fractures, cracks, and nicks in materials due to unsatisfactory processing.[62]

Temperature cycling is an accepted procedure for testing the reliability of surface-mount attachments under accelerated conditions. Fatigue and relaxation-type mechanisms commonly cause solder-joint failures in the field. Many fatigue failures are due to cyclic thermal variations that cause the component and the substrate to expand differently.

Random vibration.   Random vibration is appropriate when the design is sufficiently mature, i.e., no major unsolved design problems exist and no major design changes are expected. Random vibration, which excites many modes at once, has been found to be more effective than the single-frequency sine test. Using an apparatus with 100 simulated defects, Grumman compared random vibration with sine fixed-frequency and sine sweep tests at different levels for varying amounts of time. Random vibration was the most effective. It found faults the other tests missed. Eliminating these faults can prevent degradation in reliability.

Damaging vibration may occur during product use, handling, or transporting. To determine a product's susceptibility to vibration, a product is subjected to randomly changing frequencies (usually from 2 to 200 cycles per second). A critical component may resonate, fatigue, and then fail. Failure is likely if the vibrating range that damages the unit is the same as the vibrating range of the truck, train, or airplane that transports the unit. If the ranges are the same, cushioning or dampening is needed.[63] Examples of latent failures stimulated by random vibration include poor solder joints, poor connections, improperly seated connectors, and improperly mounted components.

---

[61]Department of the Navy. *Navy Manufacturing Screening Program.* NAVMAT P-9492, May 1979, pp. 5–7.

[62]Department of the Navy. *Navy Manufacturing Screening Program.* NAVMAT P-9492, May 1979, pp. 5–7.

[63]Sanders, Robert T. and Green, Kent C. "Proper Packaging Enhances Productivity and Quality." *Material Handling,* August 1989, pp. 51–55.

Use ESS results for prevention

The effectiveness of a screen must be evaluated with factory and field failure rates and the screen parameters adjusted accordingly. ESS results can help find root causes and corrective actions to prevent defects.[64] The purpose is to remove the sources of the defects. Many companies use feedback from ESS results to improve their design and manufacturing process. Fault rates are lower, as well as scrap and rework. Figure 13 illustrates the dynamic, closed-loop feature of an effective ESS program.[65]

With corrective action in place, the defects should fall to near zero. If the near-zero trend continues and the team is confident that the sources of defects have been removed, the team may decide to decrease or eliminate that screening and perhaps concentrate on other areas with greater sources of defects.

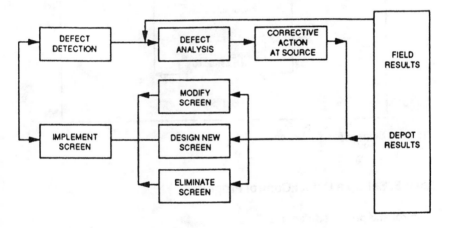

CLOSED-LOOP SYSTEM

Figure 13 ESS as a closed-loop system.

---

[64]Department of the Navy. *Best Practices: How to Avoid Surprises in the World's Most Complicated Technical Process.* NAVSO P-6071, March 1986, pp. 6-51 to 6-54.

[65]Institute of Environmental Sciences. *Environmental Stress Screening Guidelines for Assemblies.* Mount Prospect, IL: Institute of Environmental Sciences, March 1990, p. 6.

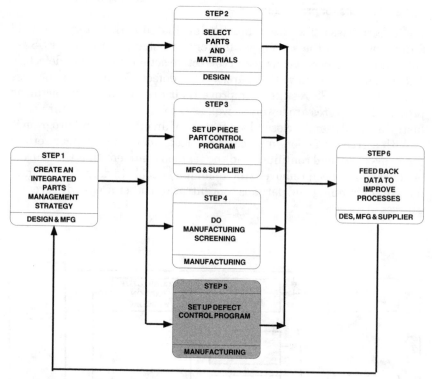

Figure 14 Steps in parts selection and defect control.

Step 5: Set Up a Defect Control Program

Control and prevent defects

Screening components at the suppliers' site will reduce component defects in assembled circuit boards. But defects originate during the assembly process through bending and cutting of component leads, overheating during soldering, and electrostatic discharge. Thus, the assembly process should be analyzed to find the root causes of defects.[66]

Below are strategies for defect control:

■ use robust design to optimize the design, manufacturing, and assembly processes

---

[66]Amerasekera, E. A. and Campbell, D. S. *Failure Mechanisms in Semiconductor Devices*. New York: John Wiley and Sons, 1987, p. 113.

- use statistical process control to reduce variation in all processes and prevent defects
- use a closed-loop defect control system
- use automated tracking and automated reporting for yield and defects
- use failure mode analysis (FMA) to find root causes and prevent recurring defects in components and in the manufacturing process

Use robust design

The goal of robust design is to make designs less susceptible to variations in materials, processes, or environmental conditions. An example of a robust-design technique is *design of experiments,* which engineers use to improve design and manufacturing processes.[67] For example, the techniques of Genichi Taguchi, an internationally known quality expert, combine statistical methods with engineering.

Advocates of robust design point out that the effort to reduce product failures in the field will also reduce the number of defective products in the factory. Systems that are designed to withstand wide variations in actual use can better withstand variations in factory processes and conditions. Part 2, Design Policy, Design Process, and Design Analysis, has more details on the principles and application of robust design.

Use statistical process control

Statistical process control techniques are relatively simple tools for process monitoring, problem solving, and communication. When managers and employees use the statistical tools properly, they generate the continual improvements that prevent defects.

The logic of using statistical process control techniques is:

- Variability in processes causes many defects.
- Variability can be analyzed using statistical methods.

Teach basic SPC techniques.   Use of the statistical methods promotes better communication. Graphs, for example, clarify data and stimulate discussion on which processes are working well and which need to be improved. People can more readily see patterns and identify problems. A

---

[67]Phadke, Madhav S. *Quality Engineering Using Robust Design.* Englewood Cliffs, NJ: Prentice Hall, 1989.

small amount of data displayed graphically gives a lot of information about the process and the potential sources of defects. People can discuss them objectively and decide the best strategies for improvement. More information on SPC concepts and techniques is available in a number of reference books.[68][69][70][71]

Effective companies teach basic SPC techniques to all employees, including machine operators and other shop-floor people. These techniques require familiarity with simple mathematical calculations, not statistical theory. The techniques are best taught with as many everyday terms as possible, with job aids for the step-by-step procedures, and with examples based on realistic situations.

Below are some useful SPC techniques that many companies teach to all employees:

- use check sheets to collect data
- use histograms to describe and analyze a process
- use Pareto analysis to separate the "significant few" from the "trivial many" and thus identify the key improvement opportunities
- do cause-and-effect analysis to find root causes of problems (e.g., people, methods, materials, equipment, and environment)
- construct control charts, including mean and range (X-bar and R) charts
- use control charts to identify which processes are acceptable and which processes are producing a high rate of defects (e.g., P charts, which show percentages of defective parts, and C charts, which show counts of the defects)

Figure 15 shows how SPC techniques can be used together to collect data, identify defect causes, and assess whether the improvements are stabilizing the process and preventing defects.[72]

---

[68]Aft, Lawrence S. *Quality Improvement Using Statistical Process Control.* New York: Harcourt Brace Jovanovich, 1988.

[69]Coppola, Anthony. *Basic Training in TQM Analysis Techniques.* Griffiss Air Force Base, NY: Rome Air Development Center, 1989.

[70]Juran, J. M. and Gryna, Frank, Eds. *Juran's Quality Control Handbook.* 4th ed. New York: McGraw-Hill, 1988.

[71]Kane, Victor E. *Defect Prevention: Use of Simple Statistical Tools.* New York: Marcel Dekker, 1989.

[72]Kane, Victor E. *Defect Prevention: Use of Simple Statistical Tools.* New York: Marcel Dekker, 1989, p. 354.

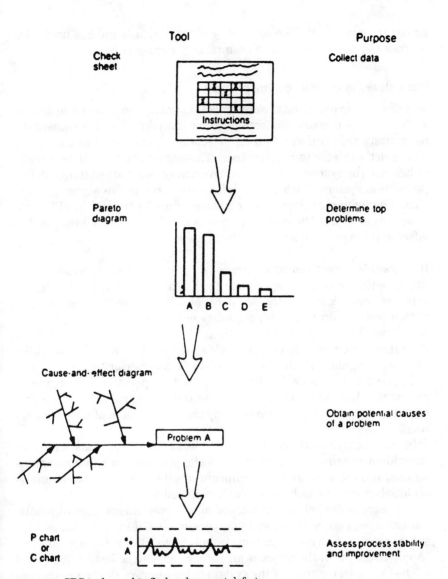

Figure 15 SPC tools used to find and prevent defects.

Teach advanced SPC techniques. Advanced SPC techniques help engineers and other technical staff reduce variation, and thus defects, in products and processes. These techniques include use of multivariate charts to identify the major sources of variation, design of experiments to verify solutions, and capability charts to find opportunities for continual improvement. The capability charts include indices useful for process engineers, machine operators, and managers.

Use a closed-loop defect control program

In a closed-loop program, the objective is to prevent mistakes and defects with real-time process monitoring rather than detect and document defects afterwards. Information on defects and yields goes to the people whose activity affects performance. This information must be timely. Otherwise the system is a defect correction system rather than a defect prevention system. With timely information, corrective action can and should occur in real time. Each employee must be trained in SPC techniques to interpret the real-time results and immediately measure the effect of the corrective action on the process.

Honeywell's defect-reduction process. Honeywell, for example, is implementing a closed-loop system to eliminate the "hidden factory" of extensive rework in its torpedo operations. They establish teams, each with a production engineer, a quality engineer, and a factory representative. The teams meet weekly to discuss yield and defect data. Managers from the three areas also attend. The weekly meetings promote a regular and disciplined look at factory performance.

To prepare for the weekly formal review, the operators and engineers review the data daily. They use process control charts to measure process stability and to show immediately the effectiveness of corrective action.

In the weekly meetings, the teams select targets, set goals, interpret data, identify causes of significant defects, prioritize and implement solutions, and measure and communicate results. They use Pareto analysis to select targets with consistently high defects.

The teams establish a time-phased plan of continuous improvements. For example, Figure 16 shows improvements in defects per unit as a result of a series of corrective actions from 1980 to 1987 at Honeywell. (The "Apple Orchard" in the figure is an automated data collection system.)[73]

The closed-loop nature of the system helps improve the processes continually. With timely and accurate information, the teams focus on the areas with the lowest yields and the highest number of defects. As these areas improve, others receive attention.

Figure 17 compares actual defects per unit with the goals.[74]

---

[73]Raasch, Daniel. *The Defect Reduction Program.* Hopkins, MN: Honeywell, 1985.

[74]Raasch, Daniel. *The Defect Reduction Program.* Hopkins, MN: Honeywell, 1985.

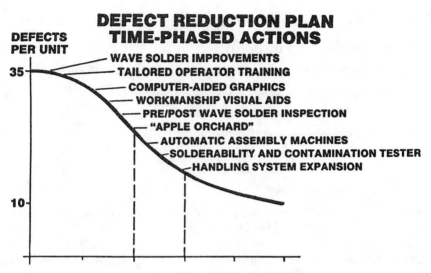

Figure 16   Corrective actions reduce defects at Honeywell.

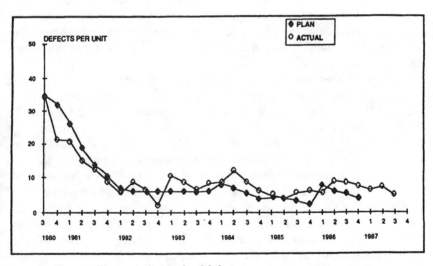

Figure 17   Honeywell's quarterly trends of defects.

Among the keys to success are:

- management actively participates
- shop-floor employees help recognize and solve problems as they occur
- a designated leader organizes, directs, and monitors
- data are collected accurately and timely
- process improvement occurs continually

Track yields and defect rates

Many companies have instituted automated systems to help track yield, which is the percentage of usable boards, units, or systems. The systems can also track removal rate by part, lot, supplier, and date. The yield and defect data are entered into databases that can provide daily, weekly, or monthly reports and trends for different products, manufacturing lines, and suppliers. With these data, engineers can work on low-yield, high-defect-rate areas.

The key feature of automated systems is automatic identification and tracking of components, units, and systems. Companies are using advanced technologies to enter data directly into computer systems, thus avoiding the time and errors of manual entry.[75] Table 8 describes some of these technologies.

For automated systems to facilitate data reduction, correlation, and trend analysis, it is important to make sure they have inherent controls and safeguards. Information such as data, part number, serial number, and test location is entered into the database system automatically. Often, however, trouble symptoms, troubleshooting results, and rework is entered manually. This manual entry can cause the database to be incomplete or inaccurate. Database entries must be reviewed and corrected often to ensure the failure data are accurate and complete.

Analysis of the failure data often suggests corrective actions to improve the yield and decrease the defects. These actions may include increasing the frequency of process checks, notifying suppliers of high defect rates, changing the test programs, and making reports available sooner. Table 9 gives examples of problems and solutions that may emerge after a failure analysis team investigates defect data.

Effective companies graph the yield and defect-rate data and post them prominently to make the data visible to each employee. Employees

---

[75]Beckert, Beverly, Knill, Bernie, and Rohan, Thomas. "Integrated Manufacturing: New Wizards of Management." *Automation,* April 1990, pp. 1–26.

TABLE 8   Technologies for Automatic Identification of Parts, Units, and Systems

Technology	Definition	Comments
Bar code scanning	Machine-readable code of bars and spaces	Fast, accurate, and inexpensive data entry
Radio frequency	Transponder located on the object being tracked operates on a unique frequency	Requires antenna to pick up and deliver signal to host and reader to interpret the distinctive data stream; electrical noise may cause problems
Optical character recognition	Stylized numbers and letters recognizable by scanner that transmits data to the computer	Characters can be read by people and machines
Voice data entry	Operator speaks into a microphone using a preprogrammed vocabulary	Frees operator's hands and avoids typing errors
Machine vision	Optical scanning of identification labels, objects, or documents	Imaging process is more complex and more expensive than bar codes and optical character recognition

TABLE 9   Examples of Problems and Solutions Found with Defect Analysis

Problem	Solutions
High failure rate for a particular capacitor	Examine the design and the component, tell supplier about component failures
False indications from test sets	Clean test probes, clean connectors, check calibration, check power supply
Test points hard to probe	Make test points accessible, use low solids flux to prevent flux build-up on component side, clean test points, increase pressure
Bent pins	Investigate handling and assembly processes

at all levels learn how to interpret graphs and control charts in classes on statistical process control techniques. Continual improvements in processes and products come about when employees use data to classify the type and cause of defects. Failure mode analysis is used to find the root causes of problems and likely solutions.

Use failure mode analysis

Failure data often indicate which components failed on which boards, the location of the removed components, and the quantity removed. These data on removal rates and defect rates, however, may not be meaningful without failure mode analysis (FMA). FMA can determine the root causes of the high removal rate. Is the removed component actually defective? Is the diagnostic method effective and efficient? Is the test procedure properly applied? Is the test set miscalibrated? Or is the component being damaged by poor handling?

When designers and engineers have access to FMA data, they can monitor trends and investigate problems early to prevent scrap, rework, repairs, and line shutdowns.

FMA helps answer these questions:

- How can a product or process fail? Potential failure modes in a manufacturing process include machines out of alignment, inconsistent temperatures, etc.

- What will happen if a product or process does fail? Effects range from minor delays to major failures to catastrophic safety issues.

- How can design and manufacturing prevent these failures? Teams set priorities based on how severe the effect is, how likely it is, and whether they can find a way to prevent it.

Use effective FMA techniques.   Typically, FMA begins at the early design stages and continues throughout the production process. Effective companies form FMA teams of designers, manufacturing engineers, quality engineers, reliability engineers, and others who can evaluate data on the design and the processes and combine their different skills and experience to analyze and solve problems quickly. FMA teams identify types of failures and find ways to prevent them, using the following techniques:

Reproduce the failure.   Reproducing the failure is essential in order to vary the stimulus conditions and obtain additional data on the sources of the problem.

Determine the root cause.   The root cause is the original event that triggered the problem. Extensive testing is often needed, especially if the failure occurs intermittently at a low rate. To find root causes, teams ask "why" repeatedly until they arrive at the underlying cause of the failure. Teams begin by looking for common problems. In manufacturing, for example, common problems include bending and cutting of component leads, overheating during soldering, electrostatic discharge, etc.

Identify the failure mechanism.   The team must determine the entire failure mechanism to find cost-effective solutions and ways to prevent future problems. It is important to find the entire failure mechanism because there may be several failure mechanisms activated at different stress levels.

Verify the failure mode analysis.   After the team identifies the failure mechanisms, the team chooses the most cost-effective and feasible solution. The next step is to test the solution under stress, including the worst-case configurations. The team verifies the root cause with tests that show the failure occurs when the failure mechanism is activated and does not occur when the failure mechanism is not activated. When the solution is verified, the corrective action can be implemented.

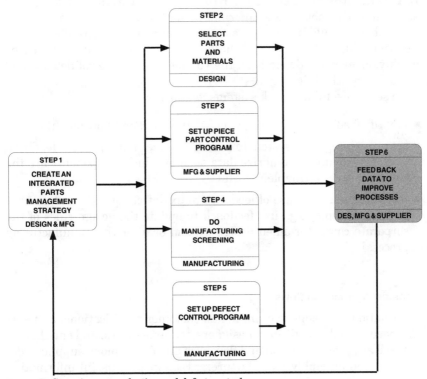

Figure 18   Steps in parts selection and defect control.

Step 6: Feed Back Data to Improve Processes

Obtain useful feedback information

Use of feedback leads to improved processes and systems. With feedback, designers, suppliers, and manufacturers can see what is going well and what needs to be improved. Effective feedback gives information to those who can correct or change the product or process. Ineffective feedback results when information goes only to people who have little power to correct and prevent the problem.

Feedback is most helpful when it improves processes and prevents problems in the future. It is less helpful when it is just used in short-term fire-fighting. Prevention is the key to successful feedback.

Feedback should be presented in a usable form. Key information should not be buried in lengthy, hard-to-read reports that require readers to extract and summarize the useful information in separate reports. In general, it is more efficient for the report preparers to abstract and summarize the metrics and information for the users of the data.

Feedback should be timely, rather than arriving months after the action that caused the problem. Whether the appropriate interval is daily, weekly, or monthly depends on the volume, the number of defects, and the delivery schedules.

To set up useful feedback systems, teams must:

- identify feedback flows (e.g., who, what, when, and how much)
- set up an integrated system to collect, analyze, and report the feedback (e.g., extract and summarize data such as which components are the source of several trouble reports)
- institutionalize the use of feedback to fix defects quickly and prevent future problems (e.g., use feedback to update the design rules of the corporate computer-aided design system or alter the manufacturing process)

Consider information flows

Information on design requirements, customer specifications, and special needs should be given to designers, manufacturers, and suppliers.

Feedback information should flow back and forth among suppliers, designers, and assembly manufacturers. Examples of useful information include:

- defect rates and yield categorized by product, component, and supplier
- results of FMA analyses

- periodic reports on supplier performance (e.g., quality, reliability, delivery, service, and cost)

- changes in manufacturing processes or assembly equipment

Designers need appropriate feedback from manufacturing. If feedback is available in a database or in a few timely reports, it can be used to improve designs and the design process.

With feedback from the assembly manufacturer, the supplier can do root cause analysis to see what is causing the defects. For example, some defects may originate at the assembly location due to mishandling, electrical overstress, or reverse insertion during retesting. Sharing the results of the analysis may lead to improved processes at the supplier and the assembly location.

Consider many sources of feedback information

Factories have data from test results, failure mode analyses, and all stages of manufacturing. For example, circuit boards that fail in-circuit or functional tests are analyzed to find the cause. Trends of the defect rates for the past month, quarter, and year can provide useful feedback to suppliers and can be used to upgrade the component and supplier databases.

Useful feedback also comes from field operations. Below are sources of field feedback and possible results of analyzing that feedback. (For more information, see AT&T's *Performance Limit Reference Guide* which includes the Field Feedback template.)

- Warranty and Repair Feedback—What are the root causes and corrective actions for damaged and non-operative systems? Which units are often returned for repair?

- Installation and Maintenance Feedback—What parts are easily broken or damaged during shipping and installation? Are the cables and connectors the right length and configuration? Which units have more maintenance than expected?

- Customer Trouble Reports—What are the most frequent sources of problems? What percentage of the problems originate in design, manufacturing, or supply?

- Customer Surveys—Which customer needs are being met? Which needs are not being met?

- DoD Field Failure Return Program—What are reliability trends for parts in use?

Government information data exchange program (GIDEP). The Government Information Data Exchange Program (GIDEP) is a cooperative activity between government and industry to make maximum use of existing knowledge. GIDEP coordinates and distributes information that participants send to them. Reports are available on mechanical, electrical, electronic, hydraulic, and pneumatic devices. Information on failure experiences is continually collected on parts, materials, safety and health hazards, and test instrumentation. Information is indexed by subject, supplier, and part number. For critical items, the GIDEP Operations Centers issues an alert or a safe-alert to inform all participants immediately.

The four GIDEP information exchanges include:

- failure experience data interchange, which has information on failure analysis, failure experiences, problems, and diminishing manufacturing sources

- engineering data interchange, which distributes reports on the testing and qualification of parts, materials, and systems

- reliability-maintainability data interchange, which has information on operational field performance, accelerated life testing, and reliability and maintainability tests on systems and equipment

- metrology data interchange, which has calibration procedures, maintenance manuals, and measurement techniques

An organization can use the information to improve the reliability, maintainability, safety or cost of systems being designed, being manufactured, or in field use. Participants are alerted about potential failures and hazards so they can avoid costly errors, prevent malfunctions, and save resources.

For example, the June 1986 GIDEP Alert traced catastrophic electrical failures in circuit card assemblies to cracks in the glass frit used to seal microcircuit packages. Contamination entered through the cracks and caused corrosion. The cracks were caused by thermal shock as the microcircuits were dipped in hot solder and then into a cleaning solvent without cooling. For more details, contact the GIDEP Operations Center, Corona, CA.

Use feedback data to improve processes

Successful companies use feedback data to improve processes continually. They make sure the feedback information clarifies problems, rather than obscures them. For example, averages may be misleading and obscure the root causes of problems. For example, an average defect rate of

2,000 ppm could result even if 80% of the part families were defect-free and 20% of the part families had defect rates much higher than 2,000 ppm. It is important to have additional data to identify the problem and find root causes and solutions.

Here are some likely questions that feedback data will help answer:

- How can the designers' component selection process be improved?
- How can the supplier management process be improved?
- How can the suppliers' piece part control process be improved?
- How can the manufacturers' screening and defect control programs be improved?

The answers will help improve processes and update information in databases accessible to each participant in part selection and defect control.

Application

In this chapter are examples of the use of the principles and processes discussed in the Procedures chapter. The examples include:

- McDonnell Douglas's integrated database
- Improving reliability in the F/A-18 Hornet
- AT&T's supplier-management program
- AT&T-supplier alliance on a widely used component
- Magnavox's supplier program
- Stress screening at Hughes Aircraft
- Monsanto's defect control program
- Motorola's six-sigma program

McDonnell Douglas's Integrated Database

To record, structure, and communicate information and to evaluate the downstream impact of design decisions, a diverse life-cycle team at McDonnell Douglas Helicopter uses an integrated database called the Integrated Design Environment—Aircraft (IDEA).[76] This common database collects and translates information from all the computer-aided tools the team uses. Figure 19 shows the interrelationships among engineering, producibility, supportability, cost, schedule, and accounting.

---

[76]Meyer, Stephen A. "Integrated Design Environment-Aircraft (IDEA): An Approach to Concurrent Engineering." Paper presented at the American Helicopter Society 46th Annual Forum and Technology Display, Washington, DC, May 21-23, 1990.

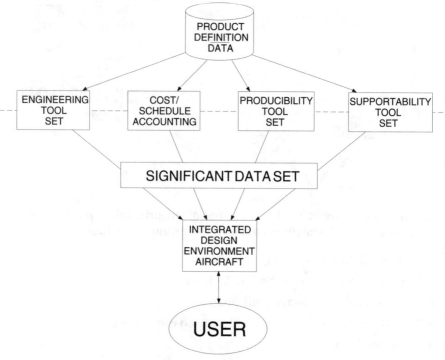

Figure 19 Schematic view of the IDEA integrated database.

Database characteristics

McDonnell Douglas Helicopter developed IDEA using an object-based, commercial database called the Integrated Development Environment. IDEA is object-based: its data structures are objects that make up complex networks. A landing gear picture, for example, can be used to get its definition and data.

IDEA is hierarchical—lower-level drawings relate to assemblies, assemblies relate to installations, installations relate to subsystems. An IDEA data object has many attributes and network associations. Figure 20 shows the data clusters for the LH aircraft system are subsystems or assemblies of parts that can be broken down into smaller parts and attributes.

Managing the data

Product definition data include system and subsystem specifications, drawings, process planning, technical documentation, supplier data, make-buy decisions, computer-aided design two-dimensional draw-

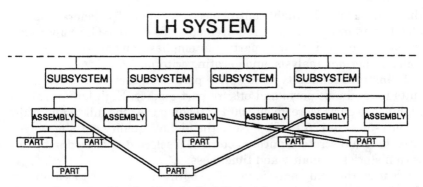

Figure 20 Hierarchical levels of data for the LH aircraft.

ings, weight and cost estimates, reliability and maintainability estimates, and lessons learned. Organizational tools manage design information and enhance the transition from development to production. These tools associate product information and stocklists with the drawings to facilitate communication. The database has information on design criteria, manufacturing processes, and support methods.

Using the data

Different work groups can use the data for their individual needs. To ensure the integrity of the data, everyone has access to the data in a view-only mode. The ability to edit, move, or delete data from the database is restricted by work group. All users can build flexible views of the data for their personal use without compromising the data. IDEA ensures that changes in one subsystem, for example, are communicated to all other affected groups.

This integrated database helps facilitate teamwork among people from design, component engineering, reliability, maintainability, human factors engineering, and producibility. It allows product changes to be communicated rapidly to other members of the development team. It provides efficient, paperless communication of design decisions.

Improving Reliability in the F/A-18 Hornet

Improving reliability with part selection, thermal analysis, and derating

The Navy fighter attack plane F/A-18 Hornet built in the late 1970s by McDonnell Douglas was designed to meet performance and reliability requirements three times more stringent than the fleet capability at that time. Thus, the flight control electronics built by General Electric also had to meet stringent requirements. To do this, GE emphasized "reliability by design," with particular emphasis on parts selection and control, thermal analysis, and derating criteria.[77]

To improve reliability at the piece part level, they reviewed the circuits to minimize parts and ensure part quality. High-failure rate or high-usage parts were given 100% screening over the full temperature range (−55°C to 125°C). To ensure the highest quality parts for microcircuits and discrete semiconductors, GE selected approved suppliers with histories of quality and timeliness.

GE used thermal analysis to define optimum piece part locations on the circuit boards and optimum circuit board locations within the unit for effective cooling. Heat sink rails on the circuit card were held in pressure contact with the unit by means of cam-action extractors. This design gave a positive path to remove heat from the circuit boards to the unit's cooling system.

Figure 21 shows how derating characteristics were imposed on the

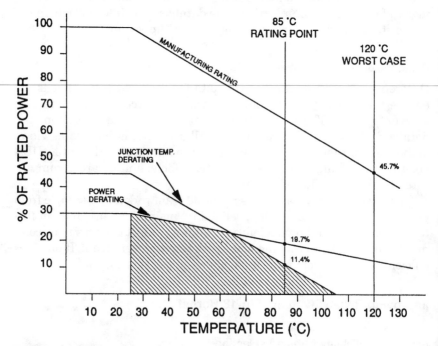

Figure 21 Application of three derating conditions.

[77]McGrath, J. D. and Freedman, R. J. "The New Look in Reliability—It Works." *Proceedings of the Annual Reliability and Maintainability Symposium,* 1981, pp. 304-309.

devices. The figure uses a 2N6193 transistor operating curve. Each device had to meet three stringent derating criteria:

- operate at a maximum case temperature of 85°C
- maintain derated maximum junction temperatures on semiconductors
- remain within the manufacturer's rating at 120°C case temperature

The shaded area is the derated operating temperature range. The maximum percentage of rated power for this device is the intersection of the 85°C case temperature and the maximum junction temperature derating. To meet the three conditions, this transistor should not be operated at more than 11.4% of its rated power.

As the example shows, the operating power limit was often well below the manufacturer's rating. Exceptions were made for about 2% of the parts due to board area limitations and to avoid adding a significant number of integrated circuits. But overall, the flight control electronics met the stringent performance and reliability requirements.

AT&T's Supplier-Management Program

To improve quality and reduce costs, AT&T created a four-phase supplier-management program with selected suppliers. The four phases are: identify suppliers, evaluate their capabilities, select the suppliers who commit to near-zero ppm defect rate, manage the process to achieve continual improvements.

Phase 1: Identify needs

In this first phase, the goal is to identify needs of the customer as well as design and manufacturing. These needs are then translated into product requirements.

Phase 2: Evaluate suppliers

The quality, reliability, and manufacturing systems of potential suppliers are audited.

Phase 3: Select suppliers

Suppliers are selected after the evaluation data are reviewed by representatives from design, purchasing, quality, reliability, and manufacturing.

Phase 4: Manage and improve suppliers' process

AT&T collects quality data from its suppliers and analyzes and maintains the data in a central on-line database. Designers, customers, and suppliers have access to its quality data (defect rate measured in parts per million) and reliability data (measured in failures in time).

When a supplier has a record of near-zero defects for some time, AT&T no longer inspects that supplier's incoming parts. Instead, AT&T monitors the supplier's quality control. The monitoring becomes less and less intrusive as the quality becomes proven and more stable. The progression typically follows this sequence:

- 100% reinspection
- lot-by-lot acceptance sampling
- periodic review of products and product controls
- audit of process controls
- quality leadership programs and awards

The ultimate level is set using a total cost perspective that includes the cost of mission failure, repair, replacement, and lost opportunities. An undersea cable that costs $1,000,000 to $10,000,000 to repair will undergo more inspection and stress testing than a cable used on land.

The essence of the AT&T's supplier-management program is assess, monitor, and reassess quality, reliability, manufacturing processes, service, delivery, and price. Table 10 illustrates AT&T's supplier-management form. In the cells of the table would be the dates and results of the system audits during the three phases.

The initial assessment provides the baseline from which subsequent progress is measured. Joint supplier-manufacturer quality improvement plans result from monitoring. Reassessment and periodic updates offer assurance of continued improvement. Monitoring with

TABLE 10　AT&T's Supplier-Management Form

Element	Assess	Monitor	Reassess
Quality			
Reliability			
Manufacturing Processes			
Service			
Delivery			
Price			

statistical process control is encouraged. Data are mutually shared. Joint failure mode analysis teams are formed as needed to identify problems and find corrective action.

System audits occur during assessment of quality, reliability, manufacturing, service, delivery, and price. The quality system audits, for example, examine the control of measurement and test equipment, quality in specification and design, production control, and corrective action procedures. The reliability system audits examine the collection, computation, and use of reliability data, the failure mode analysis procedures, and the past reliability history. The manufacturing system audits examine the facilities, process controls, and problem resolution procedures.

During the monitoring phase, AT&T uses data from the system audits on the six elements (quality, reliability, manufacturing, service, delivery, and price) to classify the suppliers as Preferred, Reliability-Monitored, Acceptable, Restricted, or Unclassified. The goal is two preferred suppliers for each product family (e.g., capacitors, resistors, potentiometers, inductors). Preferred suppliers have a proven history of excellent quality and reliability, superior performance in service, plus competitive pricing.

Measure quantifiable results

With the database, engineers can continuously monitor results to make sure the program is meeting its goals. The program is designed to improve quality, provide conforming material, reduce variability, and reduce the number of suppliers. These goals are being met. Quality engineers can spend more time training and consulting, rather than inspecting.

AT&T-Supplier Alliance on a Widely Used Component

AT&T set up a supplier alliance on a widely used ceramic capacitor that reduced parts proliferation, inventory, and a 12-week ordering interval to one week. After a year, AT&T had zero defects in 24 million pieces. An integrated team with representatives from quality management and engineering, design, manufacturing, and purchasing achieved these results.

Work with the supplier

The team focused on a bypass capacitor that was used at least once in every circuit board. Previously, 30 different parts from five suppliers were used to accomplish the same function, with the same form and fit, making it difficult to manage quality and stock levels. The team used feedback from designers and manufacturing people to select the best two of the 30 parts: one through-hole device and one surface-mount device. The team then worked with the supplier to enhance and optimize the parts.

To reduce the 16-week lead time between the order and the delivery of the parts, AT&T set up an alliance with the supplier. AT&T shared yield data and forecasts, with actual equipment orders translated to part levels. The supplier agreed to ship full reels only with 5,000 parts per reel in increments of five reels per box to prevent errors and bad counts. AT&T adjusted its production to use box units of 25,000 parts. The sharing of data and supplier agreements allowed just-in-time manufacturing and eliminated counting errors. The supplier receives the forecasts each Monday and ships on Tuesday. The parts arrive at the storeroom on Thursday when the new production week begins.

Measure quantifiable results

To measure the effectiveness of the program, AT&T measured the number of parts with quality defects and the number of parts not on hand when needed. In the first year of the program, there were no missed deliveries and no quality defects in 24 million parts. (The process has been proven to be sensitive to seeded defects.) AT&T is now planning to apply these techniques to other mature parts and mature suppliers who can commit to high levels of quality, reliability, manufacturing, service, and delivery at competitive prices.

Magnavox's Supplier Program

Magnavox Electronic Systems worked with suppliers to cut rework costs by 80% and reduce lamination, interconnect fractures, and copper fractures on multilayer boards.[78]

To correct recurring problems in laminate separation and plated copper fractures, Magnavox began a program to learn more about their suppliers' capabilities. They first asked suppliers to complete a questionnaire about their measurement units, quality standards, and

---

[78]Gardner, Fred. "Magnavox's Message: Don't Settle for PC Board Garbage." *Electronics Purchasing,* March 1989, pp. 84-87.

procedures. They reviewed the answers to find the most knowledgeable and capable suppliers whom they then visited to inspect their facilities.

Visit supplier facilities

On their plant visits to the multilayer-board producers, Magnavox evaluated the suppliers' capabilities. One of the items Magnavox looked for was whether the supplier had automatic optical inspection equipment. Optical inspection reduces the amount of scrap by helping pinpoint flawed laminates before the board is sandwiched together. Magnavox also looked for a functioning statistical process control (SPC) program, particularly in the areas of wet chemistry and plating. Daily charts of temperature, acidity, and impurity rates are keys to monitoring the process. Any discrepancies from specifications trigger corrective action. For example, too much tin may cause solderability problems. Another item was whether the printed circuit board templates are stored under correct temperature and humidity control. Most important, however, was whether quality is built in rather than bad material screened out.

Communicate expectations

Magnavox then gave a seminar for designated suppliers and Magnavox field inspectors. The goal of the seminar was to clarify specifications, describe expectations for supplier process audits, and establish clear accept-and-reject criteria for the circuit boards.

For example, a supplier must meet the following standards to qualify as a designated supplier with an effective SPC program:

- All process variables must be controlled unless specifically excluded. An example is charting the specific temperature range for an operation.

- Each variable must be charted for at least 25 measurements before evaluating its importance.

- After 25 measurements, a variable may be adjusted or dropped if SPC shows the variable does not affect the process.

- Each facility must have an SPC policy available for review at all times. The charted variables must be prominently displayed in the workplace.

- The SPC charts must demonstrate yields, problems, and impact. The data must be able to predict results and show whether the process is in control.

Measure quantifiable results

Magnavox's efforts paid off. They learned how to manage their suppliers. Dollars spent on scrap and rework of finished boards were reduced by $400,000 between 1987 and 1988.

Stress Screening at Hughes Aircraft

This case study illustrates stress screening of the AIM-54C Phoenix missile built by Hughes Aircraft.[79]

Analysis, strategy, and implementation

The Phoenix missiles use complex state-of-the-art hardware with demanding performance requirements. The missiles track six different targets as far away as 125 miles and varying from sea level to high altitudes. Two of its electronic units have 1,300 and 1,500 parts, respectively. The missiles must have a mean time between failure of 500 hours of flight conditions, even after being stored for extended periods of time. To achieve this, Hughes used a stress screening program to prevent operational failures and find defects at the lowest level of assembly.

The components undergo screening at their manufacturers' site. Circuit cards undergo functional and temperature tests. Units, which are collections of cards, undergo random vibration tests. Missile sections, which are collections of units, undergo temperature, voltage, shock, and random vibration testing. The Phoenix missile, which has four sections, undergoes two operational tests. Table 11 shows the stress screen tests from the lowest level (semiconductor) to the highest level (missile).

Measure quantifiable results

Evaluation of 3,000 tests showed that increasing the number of temperature cycles from 6 cycles to 20 cycles at the circuit-card or chassis level increased the percentage of defects uncovered from 30% to 60% of

---

[79]Wong, C. L. and Zimmerman, R. L. "Stress Screening Can Benefit a Pipeline Requirement." *Proceedings of the Annual Reliability and Maintainability Symposium,* 1987, pp. 125-129.

TABLE 11   Stress Screens and Tests for the Phoenix Missile

Level	Stress Screens and Tests
Semiconductor	High and low temperature Particle impact noise detection
Hybrid	Bond pull Particle impact noise detection Functional tests
Chassis (Circuit Card)	Temperature cycling Functional tests
Unit	Temperature cycling Random vibration Functional tests before and after vibration tests
Guidance Section	Harmonization Temperature and voltage Shock Random vibration Functional tests before and after vibration tests
Control Section	Shock Random vibration Functional tests before and after vibration tests
Missile	System functional tests

the possible temperature-related defects. Figure 22 shows that super-imposing the temperature and voltage extreme tests with 20 cycles increases the percentage to 100%. Thus, the screen uncovers all of the temperature-related defects, which are about 90% of the total defects.

Hughes also analyzed the data to see if the percentage of defects the screens actually found varied by assembly technology (e.g., wave-soldering, hand wiring, point-to-point wiring). The screens were most effective with the point-to-point wiring technology and least effective with the hand-wiring technology. Most of the defects were workmanship defects in manufacturing rather than component or engineering defects.

Hughes then used random vibration to trigger nontemperature-related defects. They decided to double the unit vibration test time and start section-level testing in the factory because they were finding many vibration-triggered defects at the missile level. They also used two cycles of temperature and voltage extreme tests at the section level. Functional tests are done before and after the stress screening tests.

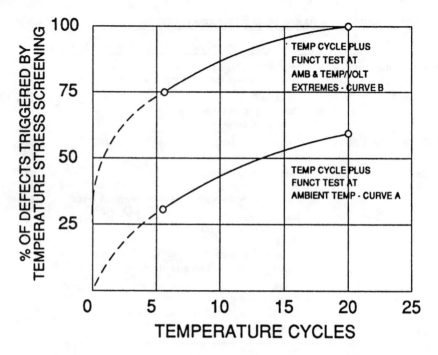

NOTES:

(1) CYCLES FROM -50 °C to +85 °C, 15 °C/MIN

(2) SAMPLE SIZE: OVER 600 CHASSIS

(3) TEMP/VOLT EXTREME: +71 °C, -55 °C

Figure 22   Adding temperature cycles and extreme tests.

Report the benefits

These tests plus the two system tests of the missile were expected to reduce storage and flight problems to one third of the previous level. For
example, the percentage of missiles out for repair should decrease from
22% to 7%. The number of missiles required to be in the inventory
dropped from about 1,176 to 938 with substantial cost savings.

Monsanto's Defect Control Program

Monsanto's defect control program boosted yields 35%, cut rejects 75%,
and cut inventory 25%.[80] In the mid 1980s Monsanto's customers com-

[80]Elliot, Marc. "Monsanto's Quality Turnaround." *Electronics Purchasing,* March 1989,
pp. 95-98.

plained about poor service and inconsistent quality in the wafers Monsanto produced. In the past five years, however, Monsanto has established a model defect control program.

Reduce defects

To reduce the defects, they formed quality improvement teams with members from different disciplines. Some of the improvements reflect the input from several groups working together. For example, because the wafer-slicing department and the test and measurement department were far apart, it took several hours to find out if the slicing was accurate. The team proposed reconfiguring the equipment and layout. The new arrangement cut the cycle time from 50 hours to 6 hours.

They also improved the defect rate by closely monitoring their process. The statistical process control charts allow them to stay more closely within the optimum ranges. For example, one type of wafer used to require resistivity between 5 and 10 ohms per square even though the optimum resistivity is 7.5 ohms per square. They are narrowing the range closer and closer to the optimum.

Measure quantifiable results

Monsanto's defect rate went from 20,000 ppm to 1,500 ppm and is still improving. All 3,000 of its employees have been trained in statistical process improvement techniques and how to manage quality improvement as a daily part of their job.

Motorola's Six-Sigma Program

By 1992, Motorola wants to achieve a rigorous goal of "six-sigma quality" which is a defect rate of 3.4 parts per million units or 99.9997% accuracy.[81] Overall, defects have been cut from nearly 3,000 ppm in 1983 to less than 200 in 1989. Motorola was one of three companies to receive the first Malcolm Baldrige Award in 1988 on the strength of these improvements.

The term "six sigma" comes from the statistical distribution of variability around an average, as Figure 23 illustrates. In a normal distribution, 68% of the values fall within plus or minus one sigma or one standard deviation away from the mean, 99.7% fall within plus or minus three sigma or three standard deviations away from the mean.

---

[81]Therrien, Lois. "The Rival Japan Respects." *Business Week,* November 13, 1989, pp. 108-118.

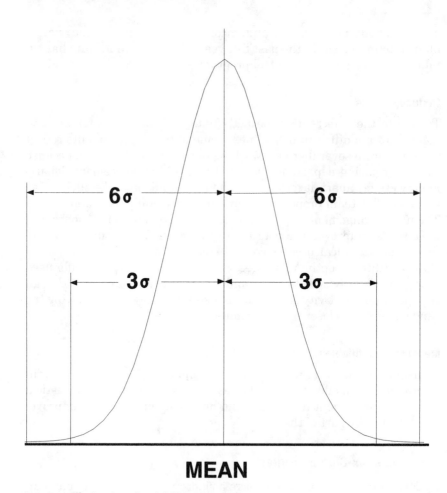

MEAN

Figure 23 Illustration of variability around an average.

Plus or minus three sigma is 2,700 defects per million. 99.9997% of the values fall within plus or minus six sigma. To achieve six sigma or 3.4 defects per million, engineers must create more-robust designs (i.e., those that tolerate more than the normal variation) or use tight control over the process to minimize variations from the optimal value.

Example

For example, a circuit board with a 10,000-ohm resistor can achieve a 3.4-ppm defect rate with robust design or with close control over the variation. A robust design would allow the board to operate within specifications even if the resistor varied about 5,000 ohms above or below the 10,000 average value. If such a robust design is not feasible, the variation of the resistors themselves must be kept within a narrow range.

Measure quantifiable results

To reach the six-sigma goal, Motorola needs a much higher level of quality in its suppliers and from their own workers. They are helping suppliers implement statistical process control and cross-disciplinary teams. Design, manufacturing, and marketing participate early in new-product development to ensure quality, producibility, and shorter cycle times.

The six-sigma goal of 3.4 ppm has already been reached for simple products like calculators and is the goal for all products for 1992. Measurements are used to check progress toward these tough goals. Each employee receives 40 hours of training on the principles of statistical process control linked directly to applications on the job. Cross-disciplinary teams often go through training together.[82]

---

[82]Haavind, Robert. "Motorola's Unique Problem: What to Do for an Encore." *Electronic Business,* October 16, 1989, pp. 60-66.

4

Summary

This part has explained the best practices associated with informed decisions on parts selection and defect control.

The best practice that is key to minimizing the technical risks is setting up an integrated parts control program. This program should be far-reaching. Designers, manufacturing engineers, quality and reliability engineers, suppliers, purchasing representatives, and subcontractors should work together to meet customer needs. These groups are able to work more smoothly if they have common access to up-to-date information in integrated databases that share data, provide feedback, and foster continual improvements.

Best Practices for Each Template

The following are key best practices for the templates included in this part. When implemented, they will make an integrated parts selection and defect control program possible.

Parts and material selection template

A best practice in parts and material selection is a preferred-parts list to avoid a proliferation of nonstandard parts that vary in performance and reliability. In many companies, component managers and component-control committees help designers choose preferred parts suitable for the particular application.

Derating is useful to ensure the parts can withstand the stresses during use as well as during storage or transportation. Effective designers also use the results of stress analyses and thermal analyses in choosing parts and in locating them optimally in the circuits.

Piece part control template

A best practice in piece part control is a formal parts control program in which requirements flow down to subcontractors and suppliers. Another best practice is effective screening. Screening is most cost-effective when it is done at the supplier's site within a statistical process control program.

Successful companies look for root causes and corrective actions to prevent defects in the future. They actively manage their sources of supply by moving from receiving inspection, to source inspection, to supplier-manufacturer cooperative alliances that aim to build in quality through process controls.

In these alliances, the manufacturer and supplier jointly share data and work to improve the processes. When the data show the supplier's process is stable and producing near-zero defects, rescreening may not add value and may actually cause damage due to electrostatic discharge or electrical overstress.

Manufacturing screening template

A best practice in manufacturing screening is to understand environmental stress screening. Engineers need this understanding to choose screens that are strong enough to stimulate latent flaws in circuit cards, units, and systems and yet not too strong to cause unnecessary damage.

Another best practice is to tailor the screens to the particular process and technology. Temperature cycling and random vibration are useful in stimulating hidden assembly and workmanship defects.

Engineers also need to consider the appropriate assembly level. Screening done at lower assembly levels may not find latent defects, especially in interfaces and interactions among circuit cards and units. But screening done at higher assembly levels may fail to prevent expensive rework. Trade-off studies are needed to find the most appropriate assembly levels. Root cause analysis and corrective action should be incorporated into the screening program to prevent future defects.

Defect control template

Best practices in defect control focus on prevention, rather than on firefighting. Robust design techniques are used to optimize the design, manufacturing, and assembly processes. Statistical process control techniques are used to reduce variation in all processes and prevent defects. Automated systems are used to track and report yield and defects, thus avoiding the time and errors of manual entry. Failure mode analysis is used to find root causes and prevent recurring defects in components and in the manufacturing process.

Summary of the Procedures

Table 12 is a summary of the key practices in parts selection and defect control.

TABLE 12   Summary of the Procedures

Step 1: Create an Integrated Parts Management Strategy	
Procedures	Supporting Activities
Create integrated databases of preferred parts and forecasts, yields, and defects	■ Give database access to designers, manufacturers, subcontractors, project management, purchasing representatives, and supplier partners ■ Update information continuously to keep it current ■ Use data only to achieve mutual goals
Use series of linked databases	■ Include in each database information specific to it as well as index parameters to link databases to each other (e.g., specification number, supplier by location, part description or family)

Step 2: Select Parts and Materials	
Procedures	Supporting Activities
Establish preferred-parts lists	■ Include parts that meet criteria for performance, reliability, timely delivery, reasonable cost
Submit stocklists for approval	■ Make sure parts are approved ■ Audit parts for reliability, availability, manufacturability, and cost ■ Help designer select parts from equivalent components that meet precise specifications
Consider quality and reliability	■ Distinguish between quality (number of good parts that arrive at the next user) and reliability (ability to function for an expected time at an expected level) ■ Use reliability data in selecting components, for improving reliability, and for planning stress screening ■ Use MIL-HDBK-217 and other sources for component failure rates
Consider producibility during the design phase	■ Work with manufacturing engineers in selecting parts ■ Minimize number of parts ■ Select parts that are functional, reliable, available, and meet factory needs

TABLE 12   Summary of the Procedures (Continued)

Step 2: Select Parts and Materials	
Procedures	Supporting Activities
Consider derating	■ Include safety margins throughout the design ■ Use derating curves to show points of inflection where small increases in stress produce large increases in failure rates
Do stress analysis and thermal analysis	■ Use stress analysis and thermal analysis to measure stresses and temperatures and compare them to derating criteria
Evaluate use of composites	■ Consider advantages and disadvantages of composites in trade-off studies
Devise overall strategy for selecting parts	■ Select parts that meet requirements for functionality, reliability, producibility, design strategy, and cost ■ Use preferred parts that meet derating criteria

Step 3: Set Up a Piece Part Control Program	
Procedures	Supporting Activities
Qualify and certify suppliers	■ Ensure near-zero defect rates while reducing screening and rescreening that damages parts ■ Choose suppliers based on their record of quality, reliability, timeliness, and financial soundness ■ Work with suppliers to improve and stabilize their processes with statistical process control techniques ■ Select parts from suppliers on qualified manufacturers list (QML) ■ Use suppliers who commit to continual process improvement ■ Tailor contracts for reduced retesting ■ Standardize military drawings and numbering system
Manage sources	■ Decide on best approach to eliminate marginal devices (e.g., qualify and certify suppliers, source inspection, receiving inspection)
Use effective screening	■ Apply stress screening to devices to remove marginal devices ■ Find root causes of marginal devices ■ Reduce screening as marginal devices decrease to near zero

Step 3: Set Up a Piece Part Control Program	
Procedures	Supporting Activities
Evaluate piece part quality and reliability	■ Analyze defect rates to see which parts are defective ■ Find root causes for defects (e.g., mishandling, insertion errors, overstress, poor process controls) ■ Focus on corrective action for part types with high defect rates
Screen at piece part level	■ Consider advantages of particle impact noise detection to find loose particles vs. corrective action to eliminate sources of loose particles ■ Consider high-temperature burn-in to find contamination flaws ■ Consider highly accelerated stress testing to stimulate corrosion failures ■ Consider thermal shock to detect crystal defects and other packaging defects ■ Consider mechanical shock to determine device's resistance to damage when dropped ■ Consider temperature cycling to find manufacturing defects ■ Consider destructive physical analysis on prototypes, models, and samples to find defects and identify corrective action ■ Consider hermeticity screens to find broken or cracked package seals

Step 4: Do Manufacturing Stress Screening	
Procedures	Supporting Activities
Define environmental stress screening (ESS) objectives	■ Improve reliability of circuit cards, units, and systems by stimulating factory failures rather than field failures
Develop plans for ESS	■ Select screens that stimulate flaws without damaging good units ■ Select appropriate level of intensity, duration, temperature range, vibration, repetitions, etc. ■ Choose effective screens ■ Demonstrate screens' effectiveness
Decide when to use ESS	■ Consider advantages and disadvantages of ESS at circuit-card, unit, and system level ■ Use ESS at lowest effective manufacturing level

TABLE 12   Summary of the Procedures (Continued)

Step 5: Set Up a Defect Control Program	
Procedures	Supporting Activities
Implement ESS	■ Consider temperature cycling to find defective units, poor solder joints, shorted wires due to improper assembly, and cracks due to unsatisfactory processing ■ Consider random vibration to find poor solder joints, poor connections, improperly seated and mounted components
Use ESS results for prevention	■ Use ESS results to find root causes and corrective actions to prevent defects ■ Decrease ESS as defects fall to near zero
Control and prevent defects	■ Analyze assembly process to find root cause of defects and strategies to prevent them
Use robust design techniques	■ Use robust design techniques to make design less susceptible to variations in materials, processes, or environmental conditions
Use statistical process control (SPC) techniques	■ Use SPC for process monitoring, problem solving, and communication ■ Teach basic SPC techniques to all: check sheets, histograms, Pareto analysis, cause-and-effect analysis, and control charts ■ Teach advanced SPC techniques to engineers and operators: multivariate charts, design of experiments, and capability charts
Use closed-loop defect control program	■ Prevent mistakes and defects with real-time process monitoring ■ Give information to people whose activity affects performance
Track yields and defect rates	■ Use bar code systems and computer databases to track yield and defects automatically ■ Analyze data to find frequent defects ■ Use results to find root causes and corrective actions for low-yield and high-defect rates
Use failure mode analysis (FMA)	■ Reproduce failure if possible ■ Determine root causes ■ Identify failure mechanisms ■ Verify failure mode analysis

Step 6: Feed Back Data to Improve Processes	
Procedures	Supporting Activities
Obtain useful feedback information	■ Feed back information to designers, manufacturers, and suppliers on part, circuit, and equipment failure rates
Consider information flows	■ Arrange for information to flow back and forth among designers, manufacturers, and suppliers ■ Feed back information to designers on circuits with too-tight tolerances ■ Feed back information to manufacturers on design requirements, customer specifications, changes in suppliers' processes, suppliers' failure rates ■ Feed back information to suppliers on results of FMA analyses and report card on supplier performance (quality, reliability, manufacturing, delivery, service, and cost)
Consider many sources of feedback information	■ Use factory FMA data and field data from installation, maintenance, warranties, trouble reports, and customer surveys ■ Use information from the Government Information Exchange Program (GIDEP)
Use feedback data to improve processes	■ Use data to improve component selection, supplier processes, stress screening, and defect control program

Chapter
5
References

Aft, Lawrence S. *Quality Improvement Using Statistical Process Control.* New York: Harcourt Brace Jovanovich, 1988. Describes concepts and procedures for statistical process control, including Pareto analysis, cause-and-effect analysis, control charts, and capability charts. Includes case studies and examples.

Amerasekera, E. A. and Campbell, D. S. *Failure Mechanisms in Semiconductor Devices.* New York: John Wiley and Sons, 1987. Describes screening techniques for semiconductor defects including particle impact noise detection (PIND) for particulate contamination, high-temperature burn-in to eliminate device failures due to manufacturing defects, thermal shock to detect crystal defects and other packaging defects, temperature cycling to find structural defects (e.g., wire-bond defects, poor package seals, cracked dies). Discusses analysis of the entire manufacturing process to identify the root causes of defects. Discusses when to use sampling or 100% screening.

Anderson, R. T. *Reliability Design Handbook.* Chicago: ITT Research Institute, 1976. Provides design information, data, and guidelines for engineers to use to ensure a reliable end product. Describes methods for part control, derating, environmental resistance, redundancy, and design evaluation.

AT&T. *Performance Limit Testing Reference Guide.* Describes government and industry best practices for design limit, life, and field feedback.

Ball, Graham. "How Well Do Materials Meet Designers' Real Needs?" *Materials: Proceedings of Materials Selection and Design.* London, England, July 1985, p. 208. Discusses the selection of materials on the basis of function, reliability, appearance, methods of manufacture, and cost.

Bannan, M. W. and Banghart, J. M. "Computer-Aided Stress Analysis of Digital Circuits." *Proceedings of the Annual Reliability and Maintainability Symposium,* 1985, pp. 217-223. Discusses stress analysis and derating to prevent stress-induced failures. Stress analysis ensures that the derating criteria are being met and verifies that the equipment can perform under worst-case conditions. Discusses manual and computer-aided methods of stress analysis.

Beckert, Beverly, Knill, Bernie, and Rohan, Thomas. "Integrated Manufacturing: New Wizards of Management." *Automation,* April 1990, pp. 1-26. Describes technologies to identify parts, units, and systems automatically and to track and analyze yields and defects.

Bindhammer, Carl and Krog, John. "An Electrical Test Correlation Experience." *Integrated Circuit Screening Report,* Institute of Environmental Science, November 1988, pp. 4-1 to 4-14. Discusses the reasons for the different perceptions of the quality and reliability of integrated circuits in the 1980s. Gives data that define the problem and the root causes. Describes improvements that resulted from joint supplier-manufacturer corrective action teams.

Bronikowski, Raymond J. *Managing the Engineering Design Function.* New York: Van Nostrand Reinhold, 1986. Discusses the design function and the design process, including parts and material selection.

Buck, Carl N. "Improving Reliability." *Quality,* February 1990, pp. 58-60. Describes the theory and practice of burn-in to screen for hidden failures in piece parts.

Burgess, Lisa. "Thomas: Pushing the Pentagon Toward QML." *Military and Aerospace Electronics.* February 1990, pp. 39-40. Describes the qualified manufacturers list (QML) which certifies manufacturing processes rather than individual parts.

Capitano, J. L. and Feinstein, J. H. "Environmental Stress Screening (ESS) Demonstrates Its Value in the Field." *Proceedings of the Annual Reliability and Maintainability Symposium,* 1986, pp. 31-35. States major cause of failures is due to defective component parts, not quality or design errors. Components are often defective because component manufacturers focus on decreasing unit cost rather than removing latent defects. Also, environmental test programs for components are often less stringent than the actual field environments.

Caruso, Henry. "Environmental Stress Screening: An Integration of Disciplines." *Proceedings of the Annual Reliability and Maintainability Symposium,* 1989, pp. 479-486. Discusses the distinct but complementary technical disciplines and backgrounds that are essential ingredients for developing and implementing an environmental stress screening program. Provides guidelines for developing consistent and effective environmental stress screening programs for electronic assemblies and systems.

Clech, J-P. and Augis, J. A. "Engineering Analysis of Thermal Cycling Accelerated Tests for Surface Mount Attachment Reliability Evaluation." Paper presented at the Seventh Annual International Electronics Packaging Society Conference, Boston, MA, November 8-11, 1987. Describes thermal cycling, which is an accepted procedure for testing the reliability of surface-mount attachments under accelerated life conditions. Discusses fatigue and relaxation-type mechanisms commonly responsible for solder-joint failures.

Comerford, Richard. "Turning up the Heat on Stress Testing." *Electronics Test.* January 1990, pp. 20-23. Describes a procedure called highly accelerated stress testing (HAST) to stimulate corrosion failures that takes less time than the traditional approach of exposing devices to 85°C and 85% relative humidity at ambient pressure for at least 1,000 hours. Describes the new approach in which devices are exposed to 120°C and 85% relative humidity for only 100 hours. States that the HAST failure-rate data are comparable to data found after 1,000 hours of traditional testing.

Coppola, Anthony. *Basic Training in TQM Analysis Techniques.* Griffiss Air Force Base, NY: Rome Air Development Center, 1989. Describes techniques for statistical process control which are useful in total quality management (TQM).

Daane, John H., Horwath, John A., and Miller, Harold L. "New Materials" in *Manufacturing High Technology Handbook.* Eds. Donatas Tijunelis and Keith E. McKee. New York: Marcel Dekker, 1987, pp. 411-456. Discusses physical characteristics, advantages and disadvantages, and applications for metal and fiber composites.

Department of Defense. *Total Quality Management Guide. Volume 1: A Guide to Implementation.* DoD 5000.51-G, January 1990. Provides a basic understanding of total quality management (TQM). Describes principles, tools, and techniques for continuous process improvement.

Department of Defense. *Transition from Development to Production.* DoD 4245.7-M, September 1985. Describes techniques for avoiding technical risks in 48 key areas or templates in funding, design, test, production, facilities, logistics, management, and transition plan.

Department of the Navy. *Best Practices: How to Avoid Surprises in the World's Most Complicated Technical Process.* NAVSO P-6071. March 1986. Discusses how to avoid traps and risks by implementing best practices for 48 areas or templates including parts and materials selection, piece part control, defect control, and manufacturing screening.

Department of the Navy. *Navy Manufacturing Screening Program.* NAVMAT P-9492, May 1979. Discusses characteristics of effective environmental screening programs including selecting the number of temperature cycles, the temperature range and rate of change, and the random vibration parameters.

Elliot, Marc. "Monsanto's Quality Turnaround." *Electronics Purchasing.* March 1989, pp. 95-98. Describes Monsanto's defect control program that increased yields, cut rejects, and cut inventory.

Finney, John W. "Pentagon, in Effort to Replace Army's Tanks, Finds Industry is Unwilling or Unable to Expand Production." *New York Times*, September 30, 1974, p. 12. Describes the pitfalls of relying on a single supplier who could not provide castings for tanks to replace those given away much faster than predicted.

Fuqua, Norman B. "Environmental Stress Screening." Paper presented at the Joint Government-Industry Conference on Test and Reliability, AT&T Bell Laboratories, May 4, 1990. Describes the advantages and disadvantages of environmental stress screening at different assembly levels.

Gardner, Fred. "Don't Settle for PC Board Garbage." *Electronics Purchasing*. March 1989, pp. 84-87. Describes how Magnavox worked with suppliers to cut rework costs by 80% and reduce lamination and interconnect separation and copper fractures.

Gardner, Fred. "Hold Down Ballooning Costs and Boost Quality." *Electronics Purchasing*, June 1988, p. 57. Compares the advantages of the qualified manufacturers list (QML), which certifies manufacturing process with the traditional parts certification method, which was costly, lengthy, and inefficient.

Garfield, Jerry and Bazovsky, Igor. "Economical Fault Isolation Analysis." *Proceedings of the Annual Reliability and Maintainability Symposium*, 1985, pp. 480-484. Discusses failure mode analysis at a radio-control development unit at Gould NAV-COM Systems Division.

Golshan, Shahin and Oxford, David B. "ESSEH Parts Committee Overview." *Integrated Circuit Screening Report*, Institute of Environmental Science, November 1988, pp. 3-1 to 3-6. Discusses the events that led to the formation of an ESSEH parts committee and joint supplier-manufacturer teams to isolate and solve quality problems in integrated circuits.

Guitard, Roger. "Reliability Data: A Practical View." *Microelectronics Reliability*, vol 29(3), 1989, pp. 405-413. Describes effective reliability data systems that include the failure mode, the failure cause, the corrective action, and the effectiveness of the corrective action. Describes how designers can then use the data to predict a device's reliability when it operates under similar conditions.

Haavind, Robert. "Motorola's Unique Problem: What to Do for an Encore." *Electronic Business*, October 16, 1989, pp. 60-66. Discusses Motorola's efforts at continued quality improvement after having won the prestigious Malcolm Baldrige award.

Heindenreich, Paul. "Supplier SPC Training: A Model Case." *Quality Progress*, July 1989, pp. 41-43. Describes how a supplier used Motorola's training materials to implement statistical process control (SPC) techniques at every level of the company.

Hnatek, Eugene R. *Integrated Circuit Quality and Reliability*. New York: Marcel Dekker, 1987. Discusses the quality and reliability considerations for integrated-circuit design. Includes material on the design and fabrication process, current technologies, sources of manufacturing error, and causes of failure and their remedies.

Hobbs, G. K. "Development of Stress Screens." *Proceedings of the Annual Reliability and Maintainability Symposium*, 1987, pp. 115-119. Discusses key aspects of environmental stress screening, including eliminating future defects by corrective action rather than defect detection and repair, selecting a screen that can stimulate likely flaws, selecting what level of stimulus to use with regard to intensity, duration, temperature range, vibration spectrum, and number of repetitions. Discusses how to prove a screen is effective and nondamaging to good products. Gives examples of defects screened out by temperature cycling.

Hyland, Charles and Shea, Joseph. "Environmental Stress Screening at Raytheon." *International Test and Evaluation Association Journal*, vol 9(3), 1988, pp. 32-37. Describes screening programs for two Raytheon programs including the screening environments and the results from factory testing and actual use.

"Incentive Payments to Concerns May End Tank Production Snag." *New York Times*, December 7, 1974, p. 58. Describes how the Pentagon found a second supplier and gave incentive payments for the two suppliers to increase their production of the castings needed for the Army tank.

Institute of Environmental Sciences. *Environmental Stress Screening Guidelines for Assemblies*. Mount Prospect, IL: Institute of Environmental Sciences, March, 1990. Describes concepts and procedures for random vibration and temperature cycling stress screening.

Institute of Environmental Sciences. *Environmental Stress Screening Guidelines for Parts.* Mount Prospect, IL: Institute of Environmental Sciences, September 1985. Describes concepts and procedures for stress screening for parts.

Juran, J. M. and Gryna, Frank, Eds. *Juran's Quality Control Handbook.* 4th ed. New York: McGraw-Hill, 1988. Describes the theory and procedures for basic statistical methods, for statistical process control, and for the design and analysis of experiments.

Kane, Victor E. *Defect Prevention: Use of Simple Statistical Tools.* New York: Marcel Dekker, 1989. Describes the concepts and tools needed to establish a defect prevention system for any work activity. Discusses statistical process control concepts and simple statistical tools useful in solving a variety of manufacturing or administrative problems.

Katz, Harry S. and Brandmaier, Harold E. "Concise Fundamentals of Fiber-Reinforced Composites" in *Handbook of Reinforcements for Plastics.* Eds. John V. Milewski and Harry S. Katz. New York: Van Nostrand Reinhold, 1987, pp. 6-13. Discusses the elements, properties, design, and testing of fiber-reinforced composites.

Keller, John. "One Part, One Number: DESC Simplifies IC Buys." *Military and Aerospace Electronics,* May 1990, p. 1. Describes the Defense Electronics Supply Center's new numbering system for semiconductors to reduce paperwork and errors.

Kidwell, George. "Aircraft Design—Performance Optimization." in *Engineering Design: Better Results Through Operations Research Methods.* Ed. Reuven Levary. New York: North-Holland, 1988, pp. 276-293. Discusses procedures to optimize aircraft design, including wind design, parts and material selection, and engine constraints.

Klinger, David J., Navada, Yoshinao, and Menendez, Maria, Eds. *AT&T Reliability Manual.* New York: Van Nostrand Reinhold, 1990. Discusses reliability concepts, device and system reliability, device hazard rates, and techniques to monitor reliability.

Maass, Richard. *World Class Quality: An Innovative Prescription for Survival.* Milwaukee, WI: ASQC Quality Press, 1988. Discusses the importance and benefits of improving component quality and reliability. Discusses supplier alliances and when incoming inspection is needed. Discusses techniques for achieving near-zero defects per million parts from a stable process with narrow tolerances with all subprocesses in control.

Malcolm, John G. "R&M 2000 Action Plan for Tactical Missiles." *Proceedings of the Annual Reliability and Maintainability Symposium,* 1988, pp. 86-92. Describes the results of an extended storage study on tactical missiles and the implications for stress screening.

Martini-Vvedensky, J. E. "Computer-aided Selection of Materials." *Materials: Proceedings of Materials Selection and Design,* London, July 1985, pp. 269-273. Discusses sources of information including structured handbooks that help the designer choose parts and materials. Also lists computerized data bases for plastics, metal alloys, ceramics, and other nonmetallics. These data bases include information from suppliers and the technical literature.

"Materials: Backbone of Aerospace Designs." *Design News,* April 9, 1990, pp. 25-28. Discusses benefits and applications of advanced composites which are now 14% of the structural weight of commercial and military aircraft.

McGrath, J. D. and Freedman, R. J. "The New Look in Reliability—It Works." *Proceedings of the Annual Reliability and Maintainability Symposium,* 1981, pp. 304-309. Discusses how General Electric improved the reliability of the F/A-18 Hornet's flight-control electronics. GE emphasized "reliability by design," with particular emphasis on parts selection and control, thermal analysis, and derating criteria. GE selected approved vendors with histories of quality and timeliness. Each part had to meet stringent derating criteria. GE used thermal analysis to define optimum piece part locations on circuit boards and optimum circuit-board locations within the unit for effective cooling.

Meinen, Carl. "Reliable Remote Monitoring and Control of Electrical Distribution." *Proceedings of the Annual Reliability and Maintainability Symposium,* 1980, pp. 448-452. Describes a broad view of derating including safety margins throughout the design as well as the capability to perform additional capabilities in the future. Includes examples.

Meyer, Stephen A. "Integrated Design Environment—Aircraft (IDEA): An Approach to Concurrent Engineering." Paper presented at the American Helicopter Society 46th Annual Forum and Technology Display, Washington, DC, May 21-23, 1990. Describes the integrated database used by diverse life-cycle team members who are participating early in the helicopter design.

Military Handbook 217. *Reliability Prediction of Electronic Equipment.* Describes techniques for obtaining base failure rates for various types of electronic components including the parts count method and the parts stress method.

Morris, Seymour F. *MIL-HDBK-217 Use and Application.* Rome Air Development Center Technical Brief, April 1990. Describes frequent misconceptions about the use and application of MIL-HDBK-217.

Naval Sea Systems Command. *Parts Application and Reliability Information Manual for Navy Electronic Equipment.* TE000-AB-GTP-010, September 1985. Describes derating techniques in which designers select parts and material to ensure the applied stress is less than rated for a specific application. Discusses benefits of derating, uses of derating curves, and possible trade-offs that may be necessary.

Nevins, James L. and Whitney, Daniel E., Eds. *Concurrent Design of Products and Processes.* New York: McGraw-Hill, 1989. Discusses multidisciplinary teams. Discusses why designers should consider producibility when choosing parts and materials. Discusses design for assembly practices and the benefits of manufacturing engineers participating in design decisions.

Owen, Jean V. "Assessing New Technologies." *Manufacturing Engineering,* June 1989, pp. 69-73. Discusses process problems with composites including how to handle them, form them, and automate their manufacture.

Oxford, David B. "Total Quality Management: Business Aspects and Implementation." *Integrated Circuit Screening Report,* Institute of Environmental Science, November 1988, pp. 2-1 to 2-8. Discusses the problems arising from antagonistic relations between suppliers and manufacturers and the advantages arising from cooperative alliances.

Phadke, Madhav S. *Quality Engineering Using Robust Design.* Englewood Cliffs, NJ: Prentice Hall, 1989. Describes the theoretical and practical aspects of quality engineering including robust design, matrix experiments using orthogonal arrays, and the design of dynamic systems.

Priest, John. *Engineering Design for Producibility and Reliability.* New York: Marcel Dekker, 1988. Discusses techniques for component management and control including environmental stress screening, group technology, and thermal analysis. Discusses principles and application of derating to ensure the stresses on the part are no greater than some percentage of the maximum rating. Discusses use of lists of preferred parts and materials that have met criteria for performance, reliability, timely delivery, and reasonable cost.

Raasch, Daniel. *The Defect Reduction Program.* Hopkins, MN: Honeywell, 1985. Describes the closed-loop defect reduction program at Honeywell.

RADC. *Parts Derating Guideline.* AFSC Pamphlet 800-27. Rome Air Development Center, 1983. Describes how to select derating levels for different parts families and applications using the part derating guide.

RADC. *Stress Screening of Electronic Hardware.* RADC Technical Report TR-82-27. Rome Air Development Center, 1982. Describes how to select levels for environmental stress screening of hardware.

Sanders, Robert T. and Green, Kent C. "Proper Packaging Enhances Productivity and Quality." *Material Handling,* August 1989, pp. 51-55. Discusses how damaging vibration may occur during product use, handling, or transporting. Discusses procedures for testing a product's susceptibility to vibration in which a product is subjected to gradually changing frequencies.

Sherry, W. M. and Hall, P. M. "Materials, Structures, and Mechanics of Solder Joints for Surface Mount Microelectronics." *Proceedings of the Third International Conference on Interconnection Technology in Electronics.* Fellbach, West Germany, February 18-20, 1986, pp. 47-81. Describes experiments that analyze how cycles to failure depend on thermal mismatches. Points out that as the use of surface mount devices increases, thermal mismatches and failures become likely at elevated temperatures.

Smoluk, George M. "Thermal Analysis: A New Key to Productivity." *Modern Plastics,* February 1989, pp. 67-73. Describes how thermal analysis is used to select materials for particular applications. Describes efforts to make thermal analysis systems more affordable and flexible and use personal computers for data analysis and storage.

Snead, Charles S. *Group Technology: Foundation for Competitive Manufacturing.* New York: Van Nostrand Reinhold, 1989. Describes group technology that codes and classifies parts into families. Discusses advantages for design and manufacturing engineers. Gives examples of group technology applications.

Testing to Verify Design and Manufacturing Readiness. Describes industry and government best practices for integrated test; test, analyze, and fix; failure reporting system; and uniform test report.

Therrien, Lois. "The Rival Japan Respects." *Business Week,* November 13, 1989, pp. 108-118. Discusses Motorola's six-sigma program to reduce its defect rate to 3.4 parts per million by 1992.

Wong, C. L. and Zimmerman, R. L. "Stress Screening Can Benefit a Pipeline Requirement." *Proceedings of the Annual Reliability and Maintainability Symposium,* 1987, pp. 125-129. Describes improvements in stress screening to reduce operational failures and repair time of the Phoenix missiles built by Hughes Aircraft.

Young, James F. and Shane, Robert S. Eds. *Materials and Processes, Part A: Materials.* 3rd ed. New York: Marcel Dekker, 1985. Discusses the information a designer needs to select, prove, and specify materials for product design.

Young, James F. and Shane, Robert S. Eds. *Materials and Processes, Part B: Processes.* 3rd ed. New York: Marcel Dekker, 1985. Discusses the processes involved in materials engineering and management. Focuses on information needed by people in product design, development, and production.

Software Design and Software Test

1

Introduction

To the Reader

Software design and test is a process that covers all the steps in software engineering. It involves two key Transition from Development to Production templates:[1] [2] Software Design and Software Test. These templates cover all activities in software development.

The templates, which reflect engineering fundamentals as well as industry and government expertise, were first proposed in the early 1980s by the Defense Science Board, under the chairmanship of Willis J. Willoughby, Jr. Their intent was to encourage everyone who is involved in the acquisition process to become aware of these templates and actually use them on the job.

This part on Software Design and Test is one in a series of books written to help defense contractor engineers, government program managers, and contract administrators use the templates most effectively. These books are meant to be stand-alone references and textbooks for related courses.

Clustering several templates makes sense when their topics are closely related. For example, the templates in this part interrelate and occur iteratively within the software development process. Other templates, such as Design Reviews, relate to many other templates and are thus best dealt within individual parts.

Over the past 10 or 15 years, software use has grown exponentially in every industry segment and government department in this country.

---

[1]*Department of Defense. September 1985. Transition from Development to Production.* DoD 4245.7-M.

[2]*Department of the Navy. March 1986. Best Practices: How to Avoid Surprises in the World's Most Complicated Technical Process.* NAVSO P-6071.

Our money systems, medical systems, communications systems, defense systems, and even our educational systems depend highly on computers. The success of these critical systems requires closely integrated hardware, firmware, and software; problems with any of these three can cause problems for the entire system. Increasingly, however, software has come under the heaviest fire from critics because of high cost, schedule slippage, and poor quality.

This part covers two templates: Software Design and Software Test. In particular, this part covers all the steps in designing and testing software. These steps include related issues such as planning, maintenance, operations, and quality assurance. For more information about the Transition from Development to Production templates, see *Transition from Development to Production*[3] and *Best Practices: How to Avoid Surprises in the World's Most Complicated Technical Process.*[4]

Software System Cost

A generation ago, when computers first became part of business and government operations, the purchase price and operating cost of hardware were high; software expenditures were low.

One study (Figure 1) estimated that the amount of money spent for computerized systems in 1960 was 80% hardware and only 20% software. By 1980 the figures had inverted: 20% hardware and 80% software. The study further estimated that by 1990, the amount of money spent for software would likely be more than 90% of total systems' expenditures.[5]

As Figure 1 shows, money spent today on software in relation to hardware is increasing rapidly. In large, complex software systems, most of this increase can be attributed to the increasing cost of labor. Some systems, such as military control systems, air traffic control systems, and mainframe computer operating systems, require several thousand programmers working for years to produce more than one million lines of code. International Business Machines (IBM) recently reported that it takes 30 times more software code to support a shuttle mission for NASA than was required by the Apollo mission that took astronauts on their first trip to the moon.[6]

---

[3]Department of Defense.

[4]Department of the Navy.

[5]Fairley, Richard E. 1985. *Software Engineering Concepts.* New York: McGraw-Hill. 8.

[6]Joyce, Edward J. 1989. "Is Error-Free Software Achievable?" *Datamation* (February 15): 53.

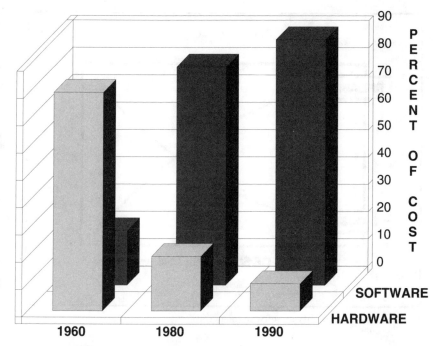

Figure 1 Trends in hardware and software costs.

With systems of such size and complexity, however, the cost of fail-ure is extremely high; unfortunately, so is the cost of success. During the intense scrutiny of the space shuttle program after the 1986 disas-ter, the shuttle's software system was applauded for the high quality that was achieved, especially for realizing such a low rate of software defects. It is reported that for each thousand lines of code (KLOC) in the shuttle software, only 0.1 errors were detected after release. Industry defect rates average 80–100 times worse, with 8 to 10 errors per KLOC common. The achievement of such high quality, however, did not come cheaply. IBM reported that it cost NASA a total of $500 million to build the shuttle software system or about $1,000 per line of code.[7]

The cost of quality is a common criticism of software systems. Many observers complain that software systems are much too error-prone. It is claimed that the increased role of software in systems amplifies the costs, while it decreases the reliability.[8] (See Figure 2.)

---

[7]Joyce.

[8]DeMillo, Richard A. 1987. *Software Testing and Evaluation*. Menlo Park, CA: Benjamin/Cummings. vi.

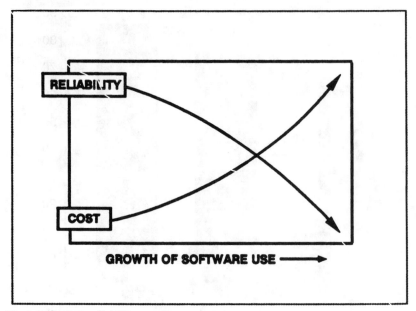

Figure 2 Cost vs. reliability as software use grows.

Cost and reliability concerns are further compounded by the belief that software systems are almost never delivered on time. Those that do meet schedules are then often criticized for not satisfying the system's specifications or the users' requirements. Some experts assert that accountability for software systems is so dispersed that development groups cannot learn from their own errors and thus cannot improve future efforts.

Software Risk Items

Cost and reliability are not the only technical risks in a software project. A leading software expert, Barry Boehm, has identified ten top software risk items (Table 1).[9] The items include not only budgets and performance shortfalls, but other concerns such as user needs and poor requirements.

---

[9]Boehm, Barry. 1989. *Risk Management Techniques.* Presentation at AT&T Bell Laboratories, Holmdel, NJ, May 3.

TABLE 1   Top Ten Software Risk Items

Risk item
Personnel shortfalls
Unrealistic schedules and budgets
Developing the wrong software functions
Developing the wrong user interface
Gold plating
Continuing stream of requirements changes
Shortfalls in externally-furnished components
Shortfalls in externally-performed tasks
Real-time performance shortfalls
Straining computer science capabilities

Keys to Success

NASA did achieve excellent software quality in the space shuttle project, but at a very high cost. Clearly, few other government agencies or companies can afford that cost, yet all organizations that produce software require good quality. There are five keys to success (Figure 3) that not only ensure good quality software, but also lower software costs and reduce the mentioned risks. These five keys are: Plan, Organize, Communicate, Control, and Measure. These five principles are illustrated throughout the Procedures and Application chapters of this part.

Plan

Plan for a project's entire life cycle, not just coding. Production and maintenance are critical stages of a project often completely overlooked. Sometimes, those two phases alone ultimately account for more than half of the total effort devoted to a project.

Plan the approach to be used in development. Various methodologies exist—which is best for the given project? What tools and techniques make sense?

Plan for software verification and validation throughout the entire development life cycle. Consideration of verification and validation methods, tools, techniques, and staff is important for building quality into the system from the start.

Organize

Organize the project based upon the architecture and the development method. Ensure that project accountability is clear.

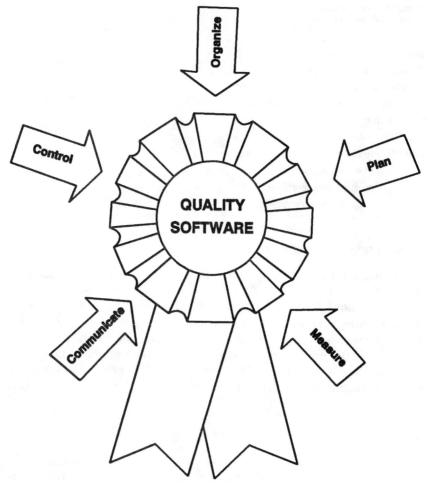

Figure 3 Key principles of quality software development.

Organize an appropriate team. Consider technology, skill levels, diversity of knowledge, and leadership.

Communicate

Communicate project goals, objectives, guidelines, milestones, and status with all team members. For large teams, a consistent method of timely communication is critical.

Communicate with users as appropriate for the project. Many experts believe that too much communication is better than too little.

Control

Control the changes to all aspects of the system, document the changes, and communicate them. Formal methods of configuration control are available and should be used.

Measure

Measure the progress of the system and communicate the status—both good and bad—with the project team regularly. Accurate measurements are essential for good management of software development projects.

Measure to improve the process, not to penalize individuals. Statistics about productivity, for example, may be collected from individuals, but results should be published for the group as a whole. Thus, you remove the fear of incrimination (being identified as a contributor to poor productivity) which can produce often unreliable measurements.

2

Procedures

The importance of high-quality, reliable software cannot be overemphasized. A disciplined approach, with consideration given to best practices, helps ensure success.

Software design and test encompasses a broad range of activities, often referred to collectively as *software engineering*. This term is used frequently throughout this part.

This Procedures chapter describes an eight-step process for developing a software system. For ease of discussion, the steps in this part are presented in sequence. In software development, however, these steps are highly interrelated. They frequently overlap and occur iteratively, depending upon the software development model chosen. See Figure 4.

Chronological Steps

Below is a sequence of steps that often occurs in software engineering.

Step 1 involves establishing a clear definition of the software system. The objectives are to assess user needs and determine the feasibility of the system.

Step 2 describes critical elements of software planning. It addresses issues relating to team, cost, life cycle, documentation, design reviews, change control, metrics, and tools.

Step 3 examines important factors involved in software requirements definition. Key aspects for determining *what* the system will accomplish are identified.

Step 4 involves software system design and *how* the software will meet the requirements. It describes preliminary design, system architecture, hardware architecture, software architecture, and detailed software design.

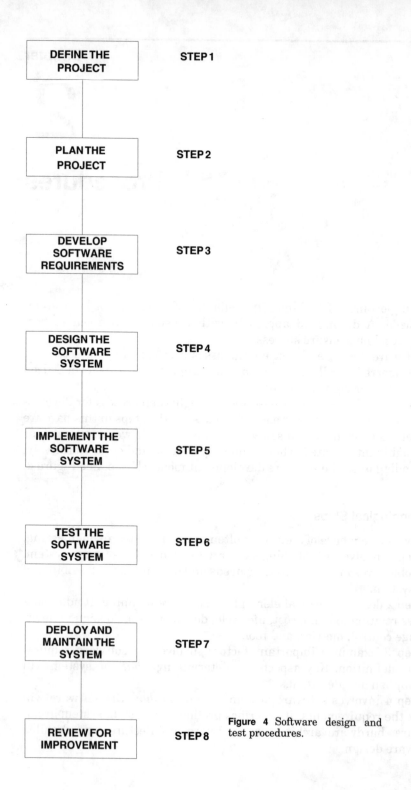

DEFINE THE PROJECT	STEP 1
PLAN THE PROJECT	STEP 2
DEVELOP SOFTWARE REQUIREMENTS	STEP 3
DESIGN THE SOFTWARE SYSTEM	STEP 4
IMPLEMENT THE SOFTWARE SYSTEM	STEP 5
TEST THE SOFTWARE SYSTEM	STEP 6
DEPLOY AND MAINTAIN THE SYSTEM	STEP 7
REVIEW FOR IMPROVEMENT	STEP 8

Figure 4 Software design and test procedures.

Step 5 discusses the software implementation. Use of structured programming techniques, coding standards, and code walkthroughs and inspections are highlighted.

Step 6 examines the critical issues involved in testing a software system. Module testing, integration testing, and system testing methods and techniques are discussed.

Step 7 details important aspects involved with installing a software system, operating it, and maintaining it.

Step 8 involves a review of both the software product and the development processes with a goal of continual improvement. Specific metrics and techniques are discussed.

Step 1: Define the Concept

The first step necessary in software development is a clear definition of the concept, opportunity, or problem, together with its proposed solution. For embedded software systems, look for the definition in the design requirements or in the design reference mission profile; with systems that are software only, that may not be the case.

Writing a clear definition of the major functions of a software system at the outset is critical. This information may have been provided by the customer, or it may be developed in-house. Part 1 provides details for embedded systems. See Figure 5 for an overview of the tasks involved in concept definition.

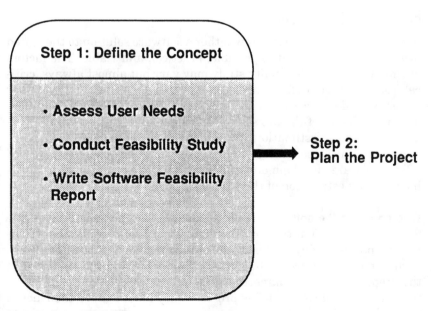

Figure 5 Concept definition.

Assess user needs

Before undertaking a software development project, examine whether the software will address user needs. User needs encompass not only what the user *says* he or she needs, but also what is not stated. For example, a user may complain about how unreadable a document is when in actuality if the system were designed more logically, complex user documentation would not be required.

Either the customer or the software development team undertakes a *software needs analysis*. This is a formal process of studying the intended users of the software and how they presently do their jobs, and then attempting to project how the job will be performed with the new system. Techniques recommended for gathering data about the users include:

- user surveys
- user interviews
- observation of the users and the tasks associated with the current situation or the new system
- actual execution of the user's job as it relates to the new system

It is important to realize that a new system may significantly change users' jobs. The analysis of user needs must include new user tasks and activities.

Conduct a feasibility study

A *feasibility study* determines whether a software solution to the stated problem is feasible, that is possible, practical, and viable. The factors examined during a feasibility study may vary but almost always, critical issues include performance, reliability, and maintainability. Other factors include: an assessment of the complexity of a software solution; a review of state-of-the-art software technology pertaining to the proposed system; an estimation of resources required to develop, operate, and maintain the software; the regulatory or policy controls that will affect software development and operation; and economic analysis including an estimation of the payback period.

Who conducts the software feasibility study?   The persons responsible for a software feasibility study should be skilled software systems engineers who are highly proficient in conducting the required analyses. Communication between the systems engineers and the customer of the proposed system is important. Customers must be completely satisfied with the outcome of this process. They must believe their needs are understood and addressed. A wrong turn can be corrected quickly and at very little expense compared to the consequences of changes in subsequent stages of software development.

Software prototyping and system simulation may be helpful tools in determining feasibility. These techniques are discussed later in this part.

Write a software feasibility report

The outcome of the concept definition period should be a formal Software Feasibility Report. This must be a clear, concise statement of why a software development effort is proposed. The quality depends upon the amount of time available as well as the size of the project. The report should include the high-level expectations of the software, an analysis of the risks associated with the software development, unresolved software development issues, and all assumptions made during the feasibility study. The report should be written in terms easily understood by the customer, and a formal design review of the report should be conducted. (Chapter 3: Application includes a detailed description of the contents of a Software Feasibility Report.)

The end of the concept definition period marks a major milestone. Based on the Software Feasibility Report, management must make a "go" or "no-go" decision. If a decision is made to continue development efforts, funding is made available to begin detailed analysis activities and the development of the software requirements.

Step 2: Plan the Project

Effective planning for software development involves much more than planning for coding. In fact, coding is one of the least critical concerns in the early stages of a software project. The important early development issues are:

- choose and train a competent team
- define a software cost estimation plan
- choose a software life cycle model
- define a software documentation methodology
- define requirements for reviews, inspections, and walkthroughs
- define a software change control scheme
- define appropriate software metrics
- choose appropriate software tools

Address and define these tasks during the early stages of a software project. (See Figure 6.)

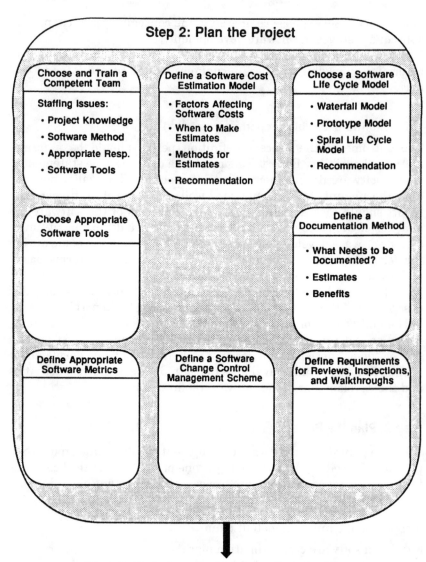

Figure 6 Project planning.

Choose and train a competent team

The skill levels and training needs of a software development team are frequently overlooked during the early phases of a project. Lack of attention to these important issues often results in more time and money being required for a project than originally anticipated.

A number of studies show wide differences between the abilities of programmers. In one study, a particular programming problem was presented to each of several experienced programmers. They were assigned to design, develop, and test their solutions; programmer activities were then measured and compared.[10] Eliminating the best and worst cases, the study found that a variation of 5-to-1 in programmer productivity is normal. The results showed, however, that depending on the activity, variations much higher than 5-to-1 also could be found. For example, comparisons of the times required to debug the programs exhibited a 28-to-1 variation, and the execution times of the finished programs varied by as much as 8-to-1.

The skill levels of a project's programming staff can have a significant effect on productivity and schedule. Since software development is labor intensive, it is important that management address the skill levels and training needs of all project team members: analysts, planners, systems engineers, designers, programmers, testers, documenters, and trainers.

Staffing issues to consider. Several issues that management needs to consider when choosing and training the software development team are: project knowledge, software methodology, and appropriate responsibilities.

- *Project Knowledge.* Is there project-specific and application-specific expertise on the project team? Is this project similar to others in the organization, or is this project so different that an overall project training course needs to be developed for team members? What level of detail is needed in the training? Who will develop it? Who should attend?

- *Software Methodology.* What is the level of understanding on the team about software methodologies in general, and about how to choose the methodology appropriate for this project? Are there particular aspects of the project that require one methodology over another? Will any specialized training related to software methodology be required?

- *Appropriate Responsibilities.* Are the right people trained in the right skills? Are the systems analysts knowledgeable about how to work with users to gather data and information necessary to develop an accurate set of system requirements? Do the systems analysts

---

[10]Fairley 65.

know what a good system requirement is? Do the programmers know effective techniques for writing efficient code or for debugging a software program? Is the system test group trained in the art of software testing? If some members of the team will be asked to do multiple tasks, is cross-training planned?

Define a software cost estimation plan

The amount of money associated with software costs in this country today staggers the imagination. A report estimated that in 1980, software costs were $40 billion annually—or approximately two percent of the U.S. Gross National Product! A more recent study by the Electronic Industries Association showed that in 1985 the U.S. Defense Department alone spent approximately $11.4 billion on mission critical software costs and predicted that costs would grow to $36 billion by 1995.[11]

In an industry with such high costs and such rapid growth rates, one would think that sound, sophisticated tools for producing accurate and verifiable cost estimates would be available. In reality, cost estimation for software projects is far from an exact science. Cost estimation frequently emerges as management's most difficult and error-prone responsibility. According to research by Capers Jones, a software industry expert, manual calculation of software cost estimates can be off by 50 to 150 percent.[12]

Factors affecting software costs.   There are numerous factors affecting software costs. Management must be aware of these factors and must consider their effects when planning and estimating costs for a software development project.

Listed below are six of the most important factors affecting software costs.

- *Project Size.* The most obvious software cost factor is the size of the project; that is, the ultimate number of lines of code that need to be written to meet the requirements of a system. This factor, however, cannot be used by itself to estimate resources since required lines of code is variable depending upon the programming language used, the use of code generators, and the skill of the programmer.

---

[11]Boehm, Barry W. and Philip N. Papaccio. 1988. "Understanding and Controlling Software Costs." *IEEE Transactions on Software Engineering.* vol. 14, no. 10 (October). 1462.

[12]Contese, Amy. 1988. "Estimating Tools Reap 85% Accuracy, Some Say." *Computerworld* (November 14): p. 33.

- *Product Complexity.* The type and complexity of the software is a sinificant cost factor. Software projects are often categorized into three basic types: application programs (for example, business data processing); utility programs (for example, language compilers or debuggers); and system-level programs (for example, operating systems or real-time systems). Several years ago, Brooks stated that it is three times more difficult to write utility programs than it is to write application programs and nine times more difficult to write system-level programs.[13]

- *Required Reliability.* Software reliability should be a significant consideration in project plans. A study conducted on more than 60 software development projects and 25 software maintenance projects showed that, if all other factors were equal, the cost of developing software in which reliability was critical would be almost twice as much as developing software with only minimal reliability requirements. The study further showed, however, that the costs of maintenance were reversed, with the minimally reliable systems more costly to maintain.[14]

- *Level of Technology.* Items that increase or reduce the productivity of team members on a software project ultimately cause proportional cost increases or reductions. The level of technology used on a project is one factor that can substantially affect productivity. Level of technology includes analysis and design tools and techniques, the programming language, debugging aids, testing tools, simulators, text editors, and database management systems. Management must be aware of available software technologies and plan for their use.

- *Product Novelty.* Whenever a new project is introduced into a software organization, there are various unknowns that can produce inaccurate software cost estimates. For example, the programming group has experience developing spread sheet applications, while the new project involves real-time transactions. The introduction of different programming languages, operating systems, or computer processors can also substantially effect cost estimates.

In the planning stage, management should determine if there are significant differences between previous software projects and the one under development. If such differences exist, previous cost estimation techniques should be adjusted to compensate for all new project factors.

---

[13]Brooks, Jr., Frederick P. 1979. *The Mythical Man-Month: Essays on Software Engineering* Reading, MA: Addison-Wesley. 4–6.

[14]Boehm and Papaccio, 1463.

More recently, Fairley noted the differences in programmer productivity for these three types of projects. He states, as a general rule of thumb only, the typical figures for lines of code that can be produced per programmer day (based on project complexity) are: 25 to 100 lines of code for an application program, 5 to 10 lines of code for a utility program, and less than one line of code per day for a system-level program.[15]

- *New vs. Reuse.* One method of controlling, and often reducing, the costs of software projects is to reuse software components rather than to build new components. Experts claim that costs for reusable software generally average only 30 percent of the cost to build a new component.[16] Software reuse, however, must be considered early in a project's life cycle and it must be planned. Force-fitting "old" software for the sake of reuse can end up costing a project much more than originally estimated.

Other factors also affect software costs and must be carefully reviewed when estimating the people, time, and computer resources needed for each phase of a software project. Table 2 lists 18 other important software cost factors.

When to make software cost estimates. The first cost estimates for a software development project are generally required during a product's feasibility stage. Estimates are very rough at this point since the understanding of the problem is still usually high-level. It is important that initial cost estimates are not cast in concrete. Instead, they should be reviewed and updated, if necessary, when each of the following mile-

TABLE 2   Other Software Cost Factors

Problem understanding	Appropriate goals
Stability of the requirements	Design stability
Available development time	Number of lines of code
Use of notations	Management skill
Change control	Team ability
Required skills	Team communication
Training	Staff build-up
Personnel turnover	Morale
Concurrent hardware development	Equipment delays

[15]Fairley, 17.

[16]"Managing Software Development: Solving the Productivity Puzzle." 1985. *Course Moderator's Guide.* New York: John Wiley.

stones is completed: the System Requirements Review, the Software Requirements Review, the Software Preliminary Design Review, and the Software Detailed Design Review. Additionally, maintenance cost estimates should be reviewed upon system delivery and revised if necessary.

Methods for estimating software costs. There is no single prescribed method or technique for estimating costs for every size and type of software project being developed today. Some of the common techniques, with their advantages and disadvantages, are:

- *Expert Estimation.* Expert estimation is a technique in which one or more highly experienced people use expert judgment, past experiences, and an understanding of the current problem to develop a cost estimate.

  Advantages: Projections can be developed quickly; better for determining interactions and exceptional circumstances.

  Disadvantages: Results are only as valid as the expert's experiences and biases; data is rarely verifiable.

- *Project Similarity.* This technique involves estimating costs by comparing similarities and differences for the proposed new project and past projects. (Expert judgment is not usually expected.) Cost estimates are then extrapolated.

  Advantages: This method can be very accurate, quick, and quite reliable.

  Disadvantages: No two projects or systems are exactly alike.

- *Lines of Code Estimation.* With this method, the total number of lines of code to be developed for a project is first estimated, and then a cost figure is approximated based on the resources (human and machine) needed to produce the estimated amount of code.

  Advantages: It is an easy method to use for quick estimates.

  Disadvantages: No clear relationship exists between lines of code and what the customer wants to buy and only a gross relationship exists between lines of code and the time it takes to produce them. Accurate estimates of lines of code is difficult and lines of code vary by programmer and project complexity.

- *Cost Estimation Models.* In recent years the software industry has seen the emergence of a number of cost estimation models. These models first require the identification (or estimation) of specific information about a project, mostly related to the software cost factors listed earlier. The information is then fed into the model (many are automated software tools) and estimates of development time and personnel resources are calculated. Some of the models even provide recommended staffing levels for various phases of the project.

Several of the most frequently used cost estimation models for software are: COCOMO, (the COnstructive COst MOdel developed by B. Boehm of TRW); SPQR/20 (from Software Productivity Research, Inc.); Function Points (developed by A. Albrecht at IBM); and SLIM (developed by L. Putnam of Quantitative Software Management). Chapter 3: Application includes a description of COCOMO cost estimating.

Advantages: Estimates are consistent internally; results are repeatable.

Disadvantages: Accuracy based on broad industry data is suspect; estimates are calibrated to data from the past that may not be accurate in the future.

Cost estimation recommendation.   Since each of the available cost estimation methods has advantages and disadvantages, many experts recommend using a combination of techniques to develop several estimates that can be compared and consolidated into one.

Each software development organization should also collect and maintain actual cost data about the projects it develops. This historical data can then be used as a basis for determining cost for new projects and for calibrating cost models to the particular environment in which a project will be developed.

Choose a software life cycle model

Critical activities in the planning stage of software development are choosing and defining a life cycle model for the project. In software development, the term life cycle covers all aspects of development—from the earliest phases of the project's definition through installation, operation, and maintenance.

Three commonly used software development life cycle models are: the waterfall model, the prototype model, and the spiral model. Each of these methods is discussed briefly below.

Waterfall model.   The waterfall model for software development is the basis for many software development projects in industry and government today. This model was first acknowledged in the late 1960s and early 1970s as an excellent tool for identifying and classifying the various phases in the life cycle of a software project. The phases identified include: feasibility, requirements, design, implementation, test, operations, and maintenance. The term "waterfall" implies that phases flow neatly and logically from beginning to end, from one phase into the next logical phase.

Given the nature of software projects, however, it is almost impossible for a project to finish one phase completely before moving on to the next as the term "waterfall" might suggest. The basic model has therefore evolved to look like Figure 7, with both forward and backward flows. The model includes verification and validation activities for each phase, with feedback to previous phases. The primary emphasis of the waterfall model is a systematic, disciplined, phased approach to software development.

Prototype model. A prototype reduces overall program risk by early modeling of an unclear aspect of the project such as user needs or available technology. In software development, the prototype life cycle model is based upon the use of various system prototypes throughout the development process.

Prototypes are commonly used to demonstrate a critical aspect of the system to users or to refine the users' needs. User interface requirements (screens, interface languages, and reports) are frequently uncovered using a prototype.

Technical risk reduction. Another reason for using prototypes is to reduce technical risk. A prototype is used to investigate or prove a particular technical aspect of a system. Some examples:

- Can performance and response time be improved? Can key algorithms be optimized?

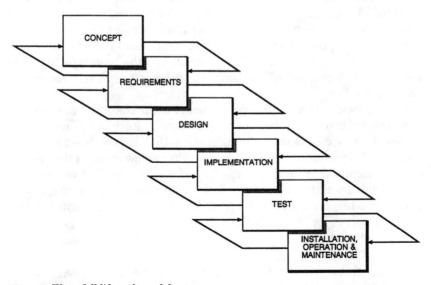

Figure 7 Waterfall life cycle model.

- Can the required response time be achieved using a new hardware or software component? Will the chosen component be compatible with the system?

- Can a new database management system adequately handle the load of the targeted system?

Implementation of the basic prototype life cycle model can follow two very different approaches, concentrating on: *system exploration* or *system evolution.* Project management and the project team must clearly define the preferred approach.

The first approach, *system exploration,* is one in which the prototype development is purely for exploration and data gathering. Once that job is done, the prototype is "thrown away." Experts who encourage this method claim that lessons learned from the prototype activity allow developers to "do it right" the next time and therefore produce a much higher quality product.

The second approach, *system evolution,* is often also called the *evolutionary development model* or the *successive versions model.* In this approach, prototypes are continually enhanced to provide more and more detail and refinement of the users' requirements for the software. Each succeeding version of a prototype is an improvement of a previous one. This technique provides quality checkpoints throughout the product development cycle.

Spiral model.   The spiral life cycle model for software development, as depicted in Figure 8, was developed by Barry Boehm.[17] The model uses a risk-driven assessment to determine if and when to move from one stage of development to the next. The spiral model provides and supports various aspects of both the waterfall and the prototype life cycle models and, depending on the amount of risk, often closely resembles one or the other.

One important facet of the spiral model is the continual assessment of whether the software system meets its goals. Each cycle ends with a review by the development organization that covers previous work cycles and plans for future work cycles. The review ensures that everyone is committed to the approach for the next phase. The review may range from a walkthrough of the design of a single programmer's component to a major requirements review for all interested organizations.

---

[17]Boehm, Barry W. 1988. "A Spiral Model of Software Development and Enhancement." *Computer* (May).

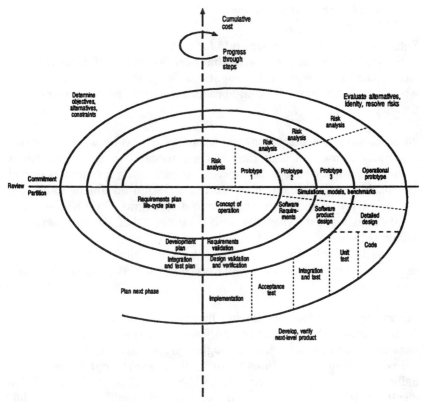

Figure 8 Spiral life cycle model.

If the software development fails to pass a review, the cycle or steps are repeated, or the process is terminated.

Life cycle model recommendation. A software life cycle model brings order and control to the development process. The model identifies specific activities that must occur and establishes unambiguous definitions for each stage of the process. No particular life cycle model is best for all software projects and, for some projects, a combination of several must be used. Most software experts believe that software development success does not necessarily depend on which specific model is chosen. Rather, what is important is that a model is chosen and defined and then clearly communicated to all project team members. If this is done well, and the model is understood, accepted, and followed by the team, then the flow of information on the project can be enhanced; project control becomes easier because it is well-defined. Ultimately, the product quality is improved.

Define a software documentation methodology

Software documentation accounts for a large part of a project's efforts and costs. Boehm has estimated that over 50 percent of a software project's activities result in documents as the immediate end products, while code is the immediate end product of only 34 percent of a project's activity.[18]

What needs to be documented? Management must be careful to choose a document methodology that is compatible with the development model of the project. For instance, a project using a typical waterfall approach to development may require a design document to be completely reviewed and accepted before coding begins. On the other hand, a project using prototypes is usually documented after a successful prototype is built.

Items that need to be documented include project plans, requirements, system architecture, software design, test plans and results, and the actual code itself.

Documentation estimates. Boehm has compared documentation rates from several software studies and found that the amount of software documentation for a project is roughly proportional to the number of delivered source instructions.[19] His analysis shows that for each 1,000 delivered source instructions (KDSI), the range of printed pages (PP) of documentation varied from 12 to 162 pages. Boehm also extracted data from the studies about the number of team-member hours (MH) required per page to write, review, and revise the documentation (excluding documentation design and planning). He estimates that:

for a small project and for the manuals of large projects	2 MH/PP
for the specifications and plans of a large project	4 MH/PP

Using 50 PP/KDSI as an average estimate of the number of pages of documentation and 3 MH/PP as an average estimate of the member hours per page of documentation, the calculation:

$$(3 \text{ MH/PP})(50 \text{ PP/KDSI}) = 150 \text{ MH/KDSI}$$
$$\text{(approximately 1 month)}$$

shows that a software project requiring 1,000 lines of code needs one team-member a month for the documentation.

---

[18]Boehm, Barry W. 1981. *Software Engineering Economics.* Englewood Cliffs, NJ: Prentice-Hall. 488–489.

[19]Boehm, *Software Engineering Economics*, 571–575.

Benefits of documentation.   Frequently, those involved with a software project question why so much documentation is required. Vincent, Waters, and Sinclair offer the following five reasons (the first three of which are from Frederick P. Brooks) that documentation is essential to the success of a software project:[20]

1. "Writing down decisions is essential: only when documented will gaps and inconsistencies appear, and all the hundreds of mini-decisions which must be made come into clear focus."

2. "Documentation communicates decisions to others and considerably lightens the manager's job of keeping everyone going in the same direction."

3. "Documentation offers the manager a data base and checklist—and periodic reviews will show the manager where he is and what changes in emphasis or shifts in direction are necessary."

4. "Documentation offers a clearly definable position—essential in demonstrating to the customer how requirements and specifications are being met, and how the project is evolving."

5. "Documentation provides a resource that may be used again in the development of future projects."

Another important consideration is that clearly written software documentation can greatly reduce the effort required to produce good user documents and training materials.

Define requirements for reviews, inspections, and walkthroughs

The goal of reviews, inspections, and walkthroughs within the software development project is to assess the design maturity and improve the quality of the workproducts (code and related documents). Effective design reviews and inspections are oriented toward error detection and the assessment of maturity. In contrast, management reviews concentrate on costs, schedules, and resources. Walkthroughs provide an overview of functions to be provided. They are especially useful when the design or code is complex.

The review and inspection team should have expertise in the key technical areas of the project, including systems analysis, software design, programming, and testing. Effective teams follow formal pro-

---

[20]Vincent, James and others. 1988. *Software Quality Assurance: Volume I, Practice and Implementation.* Englewood Cliffs, NJ: Prentice Hall. 90–92.

cedures, with well-defined roles and entry and exit criteria. Specific guidelines for individual reviewing time and meeting pace increase the chance of finding errors when the errors are easy to fix.

Although reviews and inspections take time and resources, they prove cost-effective—with savings of 10-to-1 if errors are found in requirements rather than in coding and with savings of 100-to-1 if errors are found in requirements rather than in the field.

Management support and commitment is vital. Time and resources for reviews and inspections must not be sacrificed when deadlines become tight. Managers should praise people for finding errors and ways to improve the process. (See Part 8 for more details about these principles.)

Define a software change control management scheme

During the planning phase, a change control management system should be set up to capture changes made during the product life cycle. An effective change control system can identify, maintain, store, and retrieve changes and information on a product. Change management should be applied to all code and written outputs.

The change control system may be manual, automated, or a mix of manual and automated elements. The choice depends on the complexity of the software system, the availability of resources to develop the system, the frequency of required reporting, and the number of locations that must have access to the system. Features included in a configuration control system are:

- *Configuration Control.* This process establishes a frozen baseline and coordinates changes to the baseline. A baseline is an intermediate product, designated at a milestone, and used as a basis for comparison to a future product. Changes must be approved before altering software and updating files. Changes should be visible to ensure developers work with the latest information. (Refer to Part 7 for more details.)

- *Configuration Identification.* This process defines and identifies versions of software components and systems (release or version numbers). Versions may be labeled to indicate development phases and customer releases.

- *Configuration Accounting.* This process enters potential solutions for defects. The configuration accounting system must provide feedback on the solutions to modification requests. When the defects are corrected, the files are automatically updated, but previous versions are kept. Developers may then review the reasons for the changes. Prior versions of the product may be restored, if necessary.

- *Configuration Auditing.* This process monitors the administrative and technical integrity throughout software development to improve the process continually. Experienced auditors, independent of the development efforts, establish baselines and implement metrics. They may also oversee the use of military specifications, development of standards, test planning, design reviews, error tracking, and tools certification.

Define appropriate software metrics

Software metrics measure various aspects of the software, with an eye toward eventual improvement of the final product (*outcome metrics*) or the development process (*process metrics*). The term "metric" means "measure." An ideal collection of metrics for software development has the following characteristics:

- measures the process from the requirements stage through delivery and use by the customer
- can be customized to suit small, medium, and large projects
- provides measures of software quality, accuracy, completeness, timeliness, and productivity that allow action to be taken early to correct any deficiencies
- provide guidelines on how to correct deficiencies as they arise
- pays off in results for the effort spent collecting, plotting, and interpreting the metrics
- includes both outcome and process metrics

Outcome metrics.   Outcome metrics measure whether the final product has met customer needs. They are implemented after the product has been delivered. Customer needs often include quality, performance, availability, and cost. If quality is defined as the degree to which a product or service meets evolving customer expectations, then quality software meets customer expectations for performance, availability, and cost. Quality software is not only free from defects, it is also available on time, easy to use, and easy to maintain.

Outcome measures for software development usually measure software faults found internally and by customers. A fault occurs when the software fails to conform to requirements or fails to meet customer needs. The level of severity indicates the seriousness of the fault from the customer's point of view: interruption of basic service, degradation of basic service, inconvenience, and minor deficiencies of little consequence.

Examples of outcome metrics for software development. Examples of outcome metrics that measure quality, performance, timely availability, and customer satisfaction are:

- results of customer satisfaction surveys
- cumulative density of faults found internally and by customers from the start of system test to one year after release
- results of acceptance tests conducted in the customer environment
- timeliness of the documentation measured by the percentage of volumes shipped on the release date

Process metrics. Process metrics are related to outcome metrics, but give immediate feedback on the quality of the development process. Process metrics fine-tune the development process to achieve outcomes and are usually implemented *during* the process. Process metrics can be used to:

- track the progress of software development against milestones and estimates of completion
- measure the quality of the software
- evaluate the effectiveness of the reviews and inspections
- evaluate the effectiveness of the development team and management
- identify areas for improvement

Examples of process metrics for software development. Examples of process metrics are:

- *Requirements availability.* These metrics compare the rate of completion of the system requirements to the amount of completed development effort.
- *Change management.* These metrics measure the impact of major feature changes on baselined requirements. The goal is to identify the staff months associated with deleting features, adding features, or making major changes to features.
- *Fault detection profile.* This metric shows fault densities found in each development phase. Compare fault densities found early with those found later; compare projects of similar size and complexity. Remember that improvements to prevent design and coding errors

reduce errors there and in later testing phases. Earlier improvements (requirements, for example) increase error density early but reduce it in later phases (testing, for example). This metric also helps in defect analysis and root cause analysis.

- *Fault cause profile.* This metric shows the kinds of faults that occur often in each development phase. Steps can then be taken to avoid introducing these faults or to find them earlier. The data can also be used to see whether process improvements are actually working to reduce and prevent faults.

Refining the metrics. Projects get the most benefit from metrics when the metrics are refined based on actual use. As the project team collects historical data on outcome and process, they can set objectives and guidelines based on their own situation and past performance. The payoffs will be seen quickly.

Choose appropriate software tools

In recent years, there has been considerable research in the area of software engineering. Much of that research has produced new tools and techniques whose proponents claim improved programmer productivity and decreased overall system development time frames. Unfortunately, many of the tools available today (for example, code generators, application generators, report writers, library managers, test case generators) focus on later stages of system development. Software analysis and software design tend to be performed much as they have been for the past 10 to 25 years, although some of the standard techniques have now been enhanced with automated "companion" tools.

One example often cited is Computer-Aided Software Engineering (CASE) which is discussed below. There have also been extensions to many of the classical analysis and design techniques to accommodate real-time systems development such as state-transition modeling and control flow analysis. Advances have been made, however, using expert systems technology to improve systems development, but many of these systems remain only experimental or are limited to a very narrow selection of software systems.

CASE. CASE is the general name for a wide range of automated products used in the development of software. Each CASE product contains a set of tools which are utilized at different stages throughout the development activity. In the best products, all the tools are closely integrated and provide a complete development environment. The base set of tools for a CASE product usually contains

- a tool for specifying the requirements of a system
- a tool for determining system design
- a data dictionary for storage of the system's data information
- a code generator
- a test generator

Selecting and using a CASE system requires careful planning. Although most CASE products are user-friendly, their effective use still requires substantial training and experience. This is especially true when the CASE product supports a new development methodology.

Step 3: Develop the Software Requirements

Following concept definition, a software project enters the software requirements or requirements analysis phase. The objective is a more specific understanding of *what* the completed software system will do, instead of *how* the system will do it (which is discussed in Step 4). The tasks involved in developing software requirements are shown in Figure 9.

Requirements vs. feasibility

The two major differences between the requirements phase and the feasibility phase are:

- By the requirements phase, a decision has been made to continue with development; the feasibility phase lacks that assurance.
- The requirements phase is concerned with detailed specifications; the feasibility phase is appropriately high-level.

Who conducts the software requirements analysis?

Software systems analysts develop the software requirements with participation from any of the following: hardware specialists, communications specialists, lead programmers, database specialists, and documentation specialists. Additional key participants should be the customer and users of the system. Their subject-matter expertise is critical to the success of the future system. Lack of either group's involvement at this stage often results in substantial system rework.

The Software Feasibility Report (from Step 1) is a primary input into the analysis and development of the software requirements. In fact,

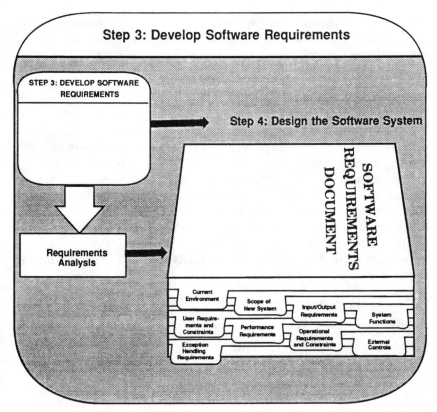

Figure 9 Software requirements development.

many of the aspects of the software system that were analyzed during concept definition will be revisited during requirements. The analysts refine the high-level system objectives into a prioritized, detailed, and quantifiable set of conditions. Note that the system requirements developed and documented during this period become the basis for the development of the overall software system design, the software system test plan, and the software user-acceptance test plan (see Steps 4–7). It is critical, therefore, that requirements be complete, explicit, and contain measurable acceptance criteria.

Software requirements document

The software requirements document should be developed to describe both what the system does and what it does not do. Too often, it is tacitly assumed that anything left out of the document will not be added to the system.[21] Functional requirements can be documented in text or graphics of varying degrees of formality and automation suiting the complexity of the system.

Key aspects of the software requirements specification

The final prioritized set of requirements for a software system is often developed through a series of negotiations between customers and software systems analysts with special consideration given to user needs, the technical feasibility of the desired system, system controls, system scheduling, and ultimately, the system cost. Various requirements may be stated in terms of alternatives that are purposely left open-ended at this point in the development process. Later, during software system design, all alternatives will be resolved.

Ten key activities that systems analysts must address during requirements analysis are:

- describe the current environment
- define the scope of the new system
- develop input and output requirements
- define the system's functions
- catalog user requirements
- list performance requirements
- itemize operational requirements
- identify maintenance requirements
- specify external controls
- develop exception handling requirements

Describe the current environment.   A new system often replaces an existing system or automates tasks currently performed manually. In either case a detailed study of the current environment should be completed. This study should give special attention to user activities and to the flow of data in the environment.

Define the scope of the new system.   An important activity in requirements analysis is to define clearly the scope and limits of the new sys-

---

[21]Quirk, W. J., ed. 1985. *Verification and Validation of Real-Time Software*. New York: Springer-Verlag. 23.

tem. Kirk urges analysts to "stretch the limits of the system to their natural boundaries."[22]

The requirements document should detail the scope of the new system and identify related activities that will not be part of the new system. This approach highlights these items and clarifies the system's limitations.

Define the system's functions. A system's functions are those processes that occur to transform the identified inputs into the required outputs. The system is designed and implemented based on these functions.

System functions should be described logically rather than physically; by *what* should be done rather than *how*.[23] This is due to the volatility of the requirements. New aspects of the system are discovered frequently, impelling frequent changes to the requirements. It is better to tie functionality to implementation in the design step when volatility has decreased.

Develop input and output requirements. Each input and each output for the new system must be identified and described, including input source(s), and output destination(s). References to external interfaces must also detail their requirements. Feedback—where input is dependent on output—must be considered.[24] Although many input and output details are specified during requirements analysis, methods for implementing them are not, since that is a function of system design.

The following list of input and output attributes specifies aspects to explore:

- content
- format
- accuracy
- security
- known or predicted volumes
- peak periods

---

[22]Kirk, Frank. 1973. *Total System Development for Information Systems.* New York: John Wiley and Sons. 85.

[23]Davis, Alan. 1988. "A Comparison of Techniques for the Specification of External System Behavior." Communications ACM (*September*).

[24]Wymore, A. Wayne. 1976. *Systems Engineering Methodology for Interdisciplinary Teams.* New York: John Wiley and Sons. 111–165.

- retention duration
- geographical considerations (example: time zones)

Catalog user requirements.   Identifying the users of the new system and the associated requirements is a large part of the requirements effort. Necessary details include: who the users are, how many users, the users' skill levels, the training required for the new system, the stability of the user population, the type and amount of documentation needed, and geographical considerations. Special consideration may be required for external users, as opposed to internal users.

It is vital that users perceive the system as meeting their needs. This perception is separate from whether the system actually does meet their needs. Too often, users do not base appraisals either on how well-designed the system is or on how well-managed the development effort is, but rather on the adequacy of their interface to the system. This interface includes not only user-machine communications, but also ease of using the documents.[25]

List performance requirements

All system requirements must be measurable and must have associated acceptance criteria, especially when detailing the system performance expectations. Inadequate or unacceptable system performance is a primary ingredient of user dissatisfaction.

Performance requirements in particular should be bounded rather than open-ended and so prevent over- or under-design. Here is an example of an open-ended requirement:

Response time for task X must not exceed 5 seconds.

Given this requirement, several questions remain unanswered:

- What may the user expect as the normal or average response time? Is it 5 seconds or something less than 5 seconds?
- If average response time is less than 5 seconds, how much less?
- Should the system be designed for an average response time of 1 second with a response time not to exceed 5 seconds under stressful conditions?

To remove the ambiguity, a better statement of the response time requirement might be:

---

[25]Page-Jones, Meilir. 1988. *The Practical Guide to Structured Systems Design.* 2nd ed. Englewood Cliffs, NJ: Yourdon Press. 286.

Response time for task X under average load conditions for the system should be in the 2- to 3-second range.

During peak load periods, the response time must not exceed 5 seconds.

Optimizing the design for the 2- to 3-second response range can be very different from trying to design for the lowest possible response time. With the revised requirement, a designer has a specific design target and users and system testers have specific acceptance criteria. Note: Average load and peak load must be clearly defined in the requirements for this to be an effective response time requirement.

Any system, particularly real-time systems, may have many performance requirements. The success of the system hinges on the specificity of the requirements. Several aspects of a system for which performance attributes should be examined and described include:

- response time
- start-up time
- availability
- fault detection and recovery
- maintenance
- communications
- reliability

Itemize operational requirements. The requirements for a system's operations cover a broad range. In most development efforts, the hardware and software components are not chosen until the design phase. Specifications may cite a particular processor size (that is, mainframe, minicomputer, or microcomputer) or require compatibility with a particular system. The physical location of the new system may also pose constraints, and these must be documented. Additional requirements may include resource parameters: use of memory, disk storage, file size, conversion requirements (for replacement systems), data redundancy, processor redundancy, off-site storage, backup capability, and recovery capability.

Identify maintenance requirements. The requirements for a system's maintenance cover every aspect of the system. The basis for good itemized maintenance requirements is a well thought out maintenance concept that describes how and who will do the maintenance. The maintenance concept should cover both software and hardware maintenance. It should also separate maintenance tasks performed by the user from maintenance tasks performed by trained maintenance personnel. The maintenance requirements should cover items such as level and types of tests, backup capability and schemes, and routine and emergency maintenance procedures. The overall maintenance concept and requirements may significantly impact the system architecture and the actual software.

Specify external controls. Many systems being developed today must adhere to external controls placed on the system. These may include business or government regulations, environmental controls, safety controls, or even political and social controls. All such requirements must be clearly specified during system requirements analysis.

Develop exception handling requirements. The exception handling requirements of a system involve a definition of the actions to be taken when undesired events or conditions occur. In a critical system involving life-threatening situations, exception handling receives top priority. In such a case, requirements for damage control or confinement may need to be developed. In less critical systems, requirements are concerned with system or user error messages, or with exception reports. Analysts often create categories of exceptions and define error handling in terms of the severity of the error category.

Software requirements review

The final requirements specification documents for a software system must be formally reviewed and accepted in a Software Requirements Review. Those involved in the review include: customers, users, software analysts, hardware designers, software designers, hardware testers, software testers, and management.

Software requirements specifications, although accepted, are not valid until it is determined that the resulting software system can be built for a reasonable cost. This requires the development of one or more software designs, which is generally the next step in software development.

After the Software Requirements Review, the Software Requirements Document should be baselined and placed under change control. Additionally, system resource estimates and schedules for the software should be updated to reflect modifications resulting from the review.

Most large, complex projects can expect changes to the system's requirements as software development progresses. Proposed changes to the requirements must be closely monitored, however, since they have the potential for large rippling effects as the software system development matures. Accepted revisions to the requirements must be reflected in the requirements specifications documents and they must be quickly and clearly communicated to the entire development team so that resource adjustments, if necessary, can be managed and coordinated.

Step 4: Design the Software System

Once the requirements have been detailed and accepted, determine how the software system will accomplish its specified tasks. Software system design is the process of allocating and arranging the functions of the system so that it meets all specified requirements. See Figure 10.

Since several different designs may meet the requirements, alternatives must be assessed based on technical risks, costs, schedule, and other considerations. A design developed before there is a clear and concise analysis of the system's objectives can result in a system that does not satisfy the requirements of its customers and users. In addition, an inferior design can make it very difficult for those who must

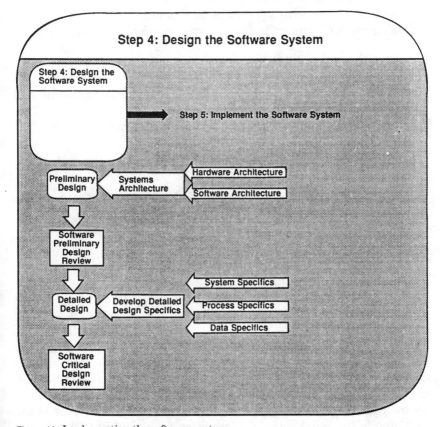

Figure 10 Implementing the software system.

later code, test, or maintain the software. During the course of a software development effort, analysts may offer and explore many possible design alternatives before choosing the best design.

Frequently, the design of a software system is developed as a gradual progression from a high-level or logical system design to a very specific modular or physical design. Many development teams, however, choose to distinguish separate design stages with specific deliverables and reviews upon completion of each stage. Two common design stages are *preliminary design* and *detailed design*.

Preliminary design

Preliminary or high-level design is the phase of a software project in which the major software system alternatives, functions, and requirements are analyzed. From the alternatives, the software system architecture is chosen and all primary functions of the system are allocated to the computer hardware, to the software, or to the portions of the system that will continue to be accomplished manually.

During the preliminary design of a system, team members

- develop the architecture

  system architecture—an overall view of system components

  hardware architecture—the system's hardware components and their interrelations

  software architecture—the system's software components and their interrelations

- investigate and analyze the physical alternatives for the system and choose solutions

- define the external characteristics of the system

- refine the internal structure of the system by decomposing the high-level software architecture

- develop a logical view or model of the system's data

This list is not all-inclusive or in any special order.

Develop the architecture

The architecture of a system describes its parts and the ways they interrelate. Peters notes that the lack of a clear, concise architectural description for a system makes communication between customer and designer difficult. An effective system architecture, however, acts "as an information pipeline at the highest conceptual level." This is the first step in the refinement of information from concept to implementation.[26] Like blueprints for a building, there may be various software architectural descriptions, each detailing a different aspect. Each architecture document usually includes a graphic and narrative about the aspect it is describing.

The *system architecture* shows the system hardware, software, and human aspects and the interfaces among them. A high-level system architecture is vital in a large, complex system, especially when functions are performed on different computers at different locations. It presents the overall system concept, enables understanding of the large issues, and helps uncover key areas of risk.

A system functional model is usually included as part of a system's architecture. It identifies the system's high-level functions and their interrelationships, external interfaces, input data sources, system outputs, and output destinations. The model also depicts the system's functional allocation—the allocation of functions to either the users or the system itself.

Primary considerations during system architecture development are the system's performance, cost, and reliability requirements as well as any technological and developmental constraints. Specific goals for the system design may also be identified. For instance, one goal may be to design a system so that the human interface is minimized. Another goal may be to create a system that will be easy to maintain in the future. Finally, a design goal frequently considered today is that of software reusability—either reusing previously developed software, or developing a new system whose software can be easily reused in the future.

A *hardware architecture* for a software system identifies the number, type, and location of items such as:

- processors (for example, personal computers, minicomputers, mainframes)
- storage, back-up, and recovery facilities
- end-user terminals (for example, display terminals, hand-held portable terminals)
- printers and output devices
- administrative and operator terminals
- protocol converters
- communications links (among processors and from terminals to processors)

---

[26]Peters, Lawrence J. 1981. *Software Design: Methods & Techniques.* New York: Yourdon Press. 44.

The development of a hardware architecture can help to highlight possible communications bottlenecks, capacity problems, and performance problems. It can also be very helpful in verifying the fault-tolerance of a system.

The *software architecture* for a system describes the internal structure of the software system. It breaks high-level functions into subfunctions and processes and establishes relationships and interconnections among them. It also identifies controlling modules, the scope of control, hierarchies, and the precedence of some processes over others. Areas of concern that are often highlighted during the establishment of the software architecture include: system security, system administration, maintenance, and future extensions for the system.

Another aspect of the software architecture may be the allocation of resource budgets for CPU cycles, memory, I/O, and file size. This activity often leads to the identification of constraints on the design solution such as the number of customer transactions that can be handled within a given period, the amount of inter-machine communication that can occur, or the amount of data that must be stored.

The first software architecture model for a system is usually presented at a very high level with only primary system functions represented. An example of a high-level software architecture is presented in Figure 11. As design progresses through detailed design, the architecture is continually refined.

Choose physical solutions. Unless a software system has been given a pre-defined physical solution, an activity called *environmental selection* occurs during the preliminary design of a system. This is the process of investigating and analyzing various technological alternatives to the system and choosing a solution based upon the system's requirements, the users' needs, and the results of the feasibility studies.

Aspects of a system that are generally selected at this time are: the hardware processing unit; computer storage devices; the operating system; user terminals, scanners, printers and other input and output devices; and the computer programming language.

In some cases, hardware and software items such as communications hardware and software, report writers, screen management systems, or database management systems are available "off-the-shelf." This is especially true for aspects of a system that have no specialized functionality or performance requirements. In other cases, unique requirements of the system may dictate the development of specific hardware and software items, specially designed for the system. The additional resources required to customize the system must be estimated and reviewed.

Define external characteristics. Following the software system's functional allocation and physical environment selection, the details of the

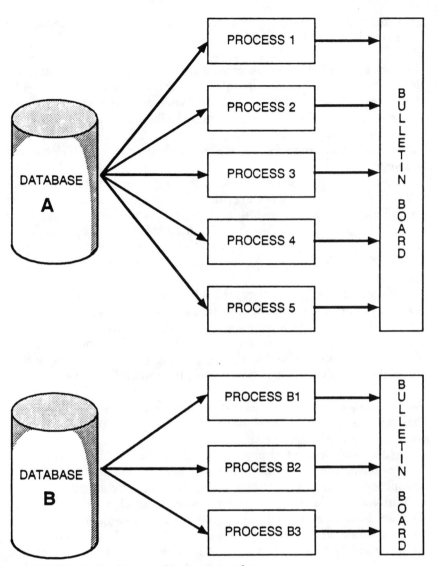

Figure 11 High-level software architecture example.

external or observable characteristics of a system can be developed. Included here would be terminal screen displays, report formats, error message formats, and interfaces to other systems.

A human factors engineer may be part of the design team concerned with the observable characteristics of a software system. This person specializes in the analysis of the human-machine interface. When a system's targeted users are novice computer users or when a system requires extensive manual data entry, human factors engineering can be a very important aspect of the design.

Perform system functional decomposition. The activity of breaking a high-level system architecture into distinct functional modules or entities is called *functional decomposition*. When preparing to decompose a software system, the design team must decide what strategy they will use. Many decomposition strategies have been written about and are advocated; most are variations of the widely used *top-down* or *bottom-up* approaches.[27]

Top-down design is the process of moving from a global functional view of a system to a more specific view. Stepwise refinement is one technique used in top-down design. With this method, design begins with the statement of a few specific functions that together solve the entire problem. Successive steps for refining the problem are used, each adding more detail to the functions until the system has been completely decomposed.

A bottom-up design strategy for a software system is often used when system performance is critical. In this method, the design team starts by identifying and optimizing the most fundamental or primitive parts of the system, and then combining those portions into the more global functions. Object-oriented design is a newer bottom-up strategy that promotes reusability and enhances the maintainability of a system.[28][29]

Module relationships. The design team also determines how the decomposed set of modules will be interrelated. Data flow, control flow, and concurrency are common concepts.

An excellent design technique is called *information hiding*. This particular technique is used often for systems with a small number of very complex portions or for systems subject to frequent change.

With information hiding, the complex or changeable sections of a system are isolated in a small number of modules and hidden from other components of the system. When a lot of system change is expected, this method confines that change to a narrow subset of the components, thus aiding future maintenance activities. For a system with a few highly complex components, information hiding also aids system testing and debugging.

---

[27]Fairley, 161–163.

[28]Booch, Grady. 1986. "Object-Oriented Development," *IEEE Transactions of Software Engineering* (February).

[29]Meyer, Bertrand. 1988. *Object-Oriented Software Construction*. Englewood Cliffs, NJ: Prentice Hall.

Regardless of strategy, functional decomposition begins when the software architecture has been determined and usually continues throughout design. A difficult aspect of functional decomposition is knowing when it is done. Experts prefer erring on the side of decomposing too much and then combining low-level modules as needed to aid performance or enhance high-usage areas of the system.[30] They also suggest that a system should be decomposed until no subset of elements in the modules or entities can be used alone.

Data modeling. In almost every software system, quick access to data has a high priority. Therefore, how the system receives its data, how the system stores its data, and how data can be retrieved from the system are key factors to consider during system design. This is called *data modeling*. Data modeling and database design are usually the responsibility of highly skilled database experts who are part of the design team.

Requirements for data and data handling should have been developed during the requirements analysis activities (Step 3). When system design begins, the design team must have clear goals and objectives concerning data access performance, data volumes, data storage, data integrity, data reliability, and data security.

A *data flow diagram* is a common tool used during preliminary design of a software system to represent system processes and their data interconnections. With this tool and others, a logical view of all the data stores for the system is developed as well as a view of how data flows to various processes.

Data abstraction is a concept often used in data design. Much like information hiding, data abstraction hides the actual structure of data stores from the system's processes. With this method, a common set of database access routines is designed so that when the system is implemented, each process needs to know only these routines to access data.

Poor design of a system's data-handling activities can cause inefficient data access and poor system performance. Common problems caused by ineffective data design include:

- data corruption resulting from ineffective data file locking
- system or database deadlock resulting from the mishandled contention for access and release of common data files
- high usage of processing resources due to inefficient data access paths

---

[30]Fairley, 152.

Software preliminary design review

At the end of the preliminary design phase of a software system, a formal design review must be conducted. Customers of the system often mandate the Software Preliminary Design Review. The review examines design rationale and design assumptions, along with all the other preliminary design aspects mentioned above, to ensure that the resulting software system will meet the stated requirements. Since user manuals and software test plans develop in parallel with the software design activities, these items may also be considered for examination during the preliminary design review, or they may be reviewed separately. Particular attention should be given to the high-priority aspects of the system such as performance, security, maintainability, and system recovery.

The results of the Software Preliminary Design Review should provide management with enough information to decide whether to continue the software development effort. Acceptance or rejection of particular design alternatives may also mean that cost estimates need to be reviewed and revised before further work on the system is approved. See Part 8 for details.

Detailed design

Detailed design or low-level design determines the specific steps required for each component or process of a software system. Responsibility for detailed design may belong to either the system designers (as a continuation of preliminary design activities) or to the system programmers.

Information needed to begin detailed design includes: the software system requirements, the system models, the data models, and previously determined functional decompositions. The specific design details developed during the detailed design period are divided into three categories: for the system as a whole (*system specifics*), for individual processes within the system (*process specifics*), and for the data within the system (*data specifics*). Examples of the type of detailed design specifics that are developed for each of these categories are given below.

Detailed design examples.   System specifics:

- physical file system structure
- interconnection records or protocols between software and hardware components
- packaging of units as functions, modules or subroutines
- interconnections among software functions and processes

- control processing
- memory addressing and allocation
- structure of compilation units and load modules

Process specifics:

- required algorithmic details
- procedural process logic
- function and subroutine calls
- error and exception handling logic

Data specifics:

- global data handling and access
- physical database structure
- internal record layouts
- data translation tables
- data edit rules
- data storage needs

Detailed design tools

Various tools such as flowcharts, decision tables, and decision trees are common in detailed software design. Frequently, a structured English notation for the logic flow of the system's components is also used. Both formal and informal notations are often lumped under the term *pseudocode*. This is a tool generally used for the detailed design of individual software components. The terminology used in pseudocode is a mix of English and a formal programming language. Pseudocode usually has constructs such as "IF ..., THEN ...," or "DO ... UNTIL ...," which can often be directly translated into the actual code for that component. When using pseudocode, more attention is paid to the logic of the procedures than to the syntax of the notation. When pseudocode is later translated into a programming language, the syntactical representation becomes critical.

Level of detailed design. A question frequently asked about detailed design is, "What level of design should be achieved?" As with other aspects of software development, there are no exact measures. Suggestions have been made that each statement from the detailed design notation should translate into some given number or fewer statements in the coding language. One expert recommends that the translation be ten or fewer lines of source code.[31]

Software critical design review

Once the detailed design of a software system has been completed, the system can be coded. Before coding begins, however, it is extremely important that the total system design down to each low-level component be reviewed. This final software design review is frequently called the Software Critical Design Review. Participants should include key representatives from the analysts, the system designers, the database designers, the programmers, and the system testers. Customers are also often represented at this review. See Part 8 for details.

The completion of the system design establishes a major milestone in a software system development process. When the design is accepted, resource estimates and allocations are updated, code and test commitments are made, schedules are confirmed, and funding sign-offs occur.

Step 5: Implement the Software System

Software implementation transforms the detailed design specifications for a system into the source code of an actual program. Development of the code is the responsibility of the programming group. A critical goal of this activity should be to develop code that is easy to read and understand, and is therefore easy to debug, test, and maintain. See Figure 12.

Information required before the start of the software implementation phase includes: the system's software requirements, the architecture documents, the functional design specifications, and the detailed design specifications. Coding objectives, such as minimizing the number of lines of code or minimizing the use of primary memory, may also be provided.

Structured programming

Structured programming or structured coding is an established technique for helping to resolve the potential problem of complex, unreadable code. A structured computer program is written so that the result can be read like a book from start to finish without flipping pages back and forth. The general rule for a programmer attempting to write structured code is to develop each routine so there is only one entry into

---

[31]Fairley, 182.

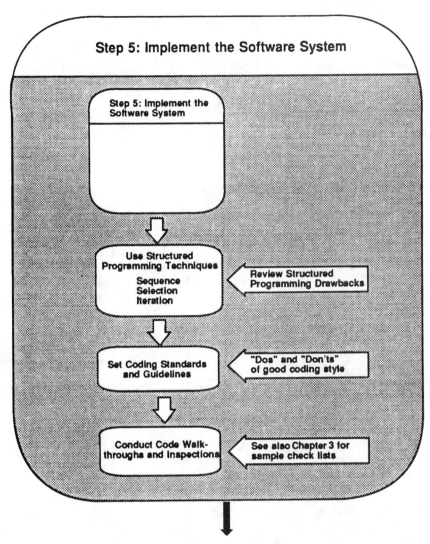

Figure 12 Implementing the software system.

it and one exit out. Excessive and undisciplined use of the "GOTO" command, if it exists, is usually prohibited.

Experts have demonstrated that only three different programming constructs are necessary to write any single-entry/single-exit routine. The three constructs, illustrated in Figure 13, are: *sequence, selection,* and *iteration*. In the figure, a box represents a procedure, function, or statement; a diamond-shaped symbol represents a decision point; and an arrow indicates control flow.

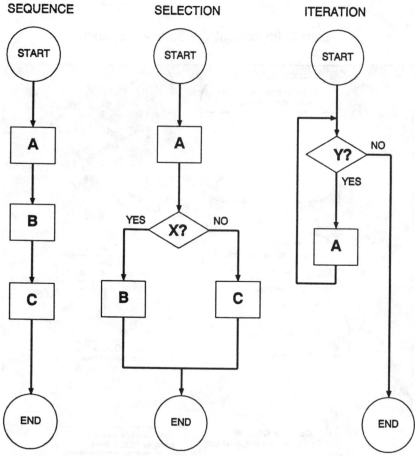

Figure 13 Sequence, selection, and iteration constructs.

Sequence.   Sequence is a programming construct in which procedures follow directly after each other in consecutive order. In Figure 13, procedure A is executed first, then B, and finally C. Each pass through that construct always follows the exact same set of procedures.

Selection.   Selection is a programming construct in which some decision must be made to determine the path to follow. The example in Figure 13 shows procedure A being executed first, then a question represented by X. If the answer to the question is "yes," procedure B is executed; if the answer to the question is "no," procedure C is executed.

Iteration.   The third programming construct, iteration, is one in which, based upon some decision point, a procedure may be repeatedly executed.

Figure 13 shows a question, X, which if answered "yes" causes procedure A to be executed and then the question is asked again. Procedure A is executed again and again until the answer to question X is "no."

Structured programming drawbacks.   There is some criticism of the inefficiencies caused by strict adherence to structured programming techniques. Most analysts believe, however, that if the general principles are followed, with exceptions allowed for critical portions of a system, then the readability and maintenance of source code can be improved.

Coding standards and guidelines

Most development organizations have their own set of implementation standards and guidelines to help ensure similarity of coding and documentation formats. These standards usually pertain to a particular programming language but may also contain general rules for good programming. Fairley offers a set of eight "Dos" and eight "Don'ts" of good coding style.[32] They are:

- DO
  Use a few standard, agreed-upon control constructs
  Use GOTOs in a disciplined way
  Introduce user-defined data types to model entities in the problem domain
  Hide data structures behind access functions
  Isolate machine dependencies in a few routines
  Provide standard documentation prologues for each subprogram and/or compilation unit
  Carefully examine routines having fewer than 5 or more than 25 executable statements
  Use indentation, parentheses, blank spaces, blank lines, and borders around comment blocks to enhance readability

- DON'T
  Don't be too clever
  Don't use null THEN statements
  Don't use THEN_IF statements
  Don't nest too deeply
  Don't rely on obscure side effects
  Don't sub-optimize
  Don't pass more than five formal parameters to subroutines
  Don't use one identifier for multiple purposes

---

[32]Fairley, 210, 215.

Code walkthroughs and inspections

Code walkthroughs and inspections should be a routine practice in every software development organization. Consider the project being developed to determine which reviews are required. In some cases, every piece of code may need to be examined; in others, only critical components require inspection. During a walkthrough, the programmer leads other team members through a module of code or part of the design while they attempt to discover faults by questioning. Sample code review and code inspection checklists are provided in Chapter 3: Application. As a best practice, thorough code inspections are strongly recommended. See Part 8 for details.

Step 6: Test the Software

The testing phase of software development is when individual components of the software are tested and combined to verify that the software system meets the documented requirements and adheres to the predefined design. See Figure 14.

Independent testing organization

Most software experts recommend that an independent organization test a software system. One option is to contract with an outside organization for the testing. If this is not possible, the testing organization should be managerially separate from the design and development groups assigned to the project.

This recommendation is based more on observations of human nature than on substantiated fact. Effective testing groups need to have somewhat of a "destructive" view of a system, so that they can flush out errors and "break" the system. The design and development groups who have built the software system have a "constructive" view, and may therefore find it too difficult to develop the frame of mind required for testing.

Plan for software testing

A critical concept that experts strongly stress is the importance of planning and documenting software test activities *before they begin.* Whereas software testing begins after the coding phase, the plans for what to test and how to test must begin in the early stages of software development.

The Institute of Electrical and Electronics Engineers (IEEE) standards on software development require that *Software Verification and Validation Plans* be developed and documented for all software pro-

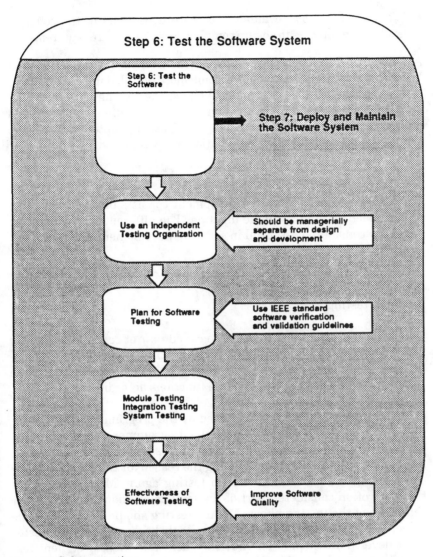

Figure 14 Software testing.

jects.[33] In the standard, the terms verification and validation are defined as follows:

- *Software verification*—determines whether or not the products of a given phase of the software development cycle fulfill the requirements established during the previous phase.

---

[33]IEEE Computer Society. 1986. *IEEE Standard for Software Verification and Validation Plans Std. 1012-1986.* IEEE, Inc. New York. 9–10.

■ *Software validation*—evaluates software at the end of the software development process to ensure compliance with software requirements.

As software development progresses, the *Software Verification and Validation Plan* (SVVP) should become more detailed until a final version is published. The plan must include: the testing objectives, the scope of the testing, the testing environment, acceptance criteria, and any constraints to the testing. The Application chapter includes the IEEE standard for software verification and validation.

Types of software testing

The testing step of software development consists of several distinct phases, each serving a unique function. The different types of testing generally considered essential in software development are

■ module (or unit) testing

■ integration testing

■ system testing

Module testing.   Module testing (also called *unit* or *component testing*) is the testing of one individual component (that is, one program module, one functional unit, or one subroutine). The objective of module testing is to determine if the module functions according to its specifications.

Module testing is usually conducted by the programmer of the module being tested. It is closely tied to the programmer's development of the code and often becomes an iterative process of testing a component, finding a problem, debugging (finding the reason for the problem in the code), fixing the problem, and then testing again. Module testing is therefore often considered part of the implementation (Step 5) rather than part of the testing phase. Module testing should nevertheless be recognized as a separate function, and should be disciplined. The tester must develop a test plan for the component and must document test cases and procedures. Too often, this discipline is overlooked and testing of individual components becomes "ad hoc" testing with no records about the actual cases, the procedures, or the results.

White box testing is frequently used during module testing. White box testing means that the tester is familiar with the internal logic of the component and develops test cases accordingly.

Code coverage (how much of the code is covered by the testing) and logic path coverage (how many of the logical paths in the code are tested) are two primary considerations when developing test cases for module testing.

Special testing procedures and tools often need to be developed to isolate modules for testing. Many programming organizations keep a library of tools (such as stub and driver routines) which can be borrowed. Other groups develop their own modules as needed. In any case, before module testing begins special procedures must be designed and documented as part of the test plan for each component.

Integration testing

After module testing, the next step in the software testing phase is integration testing. This activity involves combining components in an orderly progression until the entire system has been built. The emphasis of integration testing is on the interaction of the different components and the interfaces between them.

Most often, the programming group performs software integration testing. As with module testing, integration testing is very closely linked to the programming activity since the tester needs to know details of the function of each component to develop a good integration test plan.

Integration test techniques.   An important decision when planning for integration testing is determining the procedure to be used for combining all the individual modules. There are two basic approaches for doing this: non-incremental testing and incremental testing.

In *non-incremental integration testing,* all the software components (assuming they have each been individually module tested) are combined at once and then testing begins. Myers calls this "big bang" testing.[34] Most experts, Myers included, do not recommend this approach to integration testing because problem isolation is very difficult. Since all modules are combined at once, a failure could be in any one of the numerous interfaces that have been introduced.

The recommended approach for the integration of system components is planned *incremental testing.* With this method, one component is completely module tested and debugged. Another component is then added to the first and the combination is tested and debugged. This pattern of adding one new component at a time is repeated until all components have been added to the test and the system is completely integrated.

Incremental testing requires another decision about the order in which the components will be added to the test. There are no clear-cut rules for doing this. Testers must base a decision on their knowledge of what makes the most sense for their system, considering logic and use of resources. There are, however, two basic strategies: top-down or bottom-up.

---

[34]Myers, Glenford J. 1979. *The Art of Software Testing.* New York: John Wiley and Sons. 89.

The structure of many software systems can be defined as a hierarchy of modules with a single starting point, and with several logical paths available to traverse to the bottom of the hierarchy. Figure 15 shows a structural representation of a small system that contains eight modules: A, B, C, D, E, F, G, and H.

A tester using *top-down integration testing* on this system begins by module testing and debugging the A component. The next step is to add a new component to the test. In this case, either B or C is added. If B was chosen and tested, either C or D could be the next choice. Some testers prefer to follow one path to completion, while others prefer to complete all the modules on the same level before proceeding to a lower level of the hierarchy.

Bottom-up integration testing reverses top-down testing. With this approach, a tester simply starts at the bottom-most level of the hierarchy and works up. As shown in Figure 15, a tester might start by module testing component G. With bottom-up testing, all the components at the bottom of the hierarchy are usually module tested first and then testing proceeds in turn to each of their calling components. The primary rule in bottom-up testing is that a component should not be chosen to be the next one added to the test unless all of the components that it calls have already been tested.

System testing.   The end of integration testing is usually a major milestone in the development process. It is at this point that most develop-

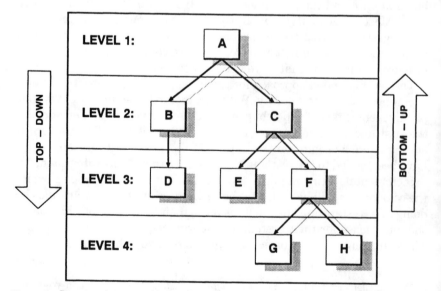

Figure 15 Structural representation of a software system.

ment organizations relinquish the code as an entire system to the appropriate organization for system testing.

System test is considered the most difficult type of testing. With the move to system test, there is also a move from structural testing to functional testing. Now, rather than testing the internal logic of the software, the tester concentrates on what the program is supposed to do and how well it does it.

The definition for system test used in the IEEE software testing standards is "the process of testing an integrated hardware and software system to verify that the system meets its specified requirements."[35] The hardware aspect of this definition is important. Frequently, module and integration tests are conducted within a development environment. Ideally, in system test, the software is tested on the actual target hardware for a system. In many cases, however, the target hardware is being built in parallel with the software and system testing may need to be conducted on prototype hardware or with a simulator.

The responsibility for software system test belongs to the group designated as the test organization, ideally independent from the software designers and programmers. System testers should be involved with a project from the beginning and should have access to the most up-to-date documentation of the system objectives, the software requirements and specifications, the architecture, the development and design methodologies used during implementation, and a complete set of the user documentation.

When the system is delivered to the testers from the development organization, a system test plan, already reviewed and accepted, should be ready for execution. Similar to the plans in the earlier testing activities, the software system test plan should be complete with scheduling, resource estimates, test cases, procedures, and acceptance criteria. Since this is generally the last step in the development life cycle before the software is passed to the customers, it is extremely important that a system test plan cover all aspects of the software and its operation.

System testing techniques. System testing is often referred to as "testing the whole system." Translated literally, that could mean that every input or output condition in the software needs to be tested for every possible logical path through the code. Even in a small system this task could become quite lengthy. In a large, complex system, it would be prohibitively time-consuming and expensive.

The system test organization must develop a strategy for testing a particular system and determine the amount of test coverage required.

---

[35]*IEEE Standard for Software Verification and Validation Plans Std. 1012-1986.* 11.

There is no cookbook for doing so. In a small noncritical system, a very low degree of test coverage may be acceptable. High coverage is needed in a critical software system involving human life. The testers must decide the best plan based on system characteristics, the environment in which the software system will operate, and the testers' experience.

In general, software system testing is done using *black box testing*. The tester, viewing the system as a black box, is not concerned with the internals, but rather is interested in finding if and when the system does not behave according to its requirements.

One technique often used for identifying specific test cases is called *equivalence partitioning*. In this method, an equivalence class is identified so that one test case covers a number of other possible test cases.

Boundary analysis is another technique used in which testing is performed on all the boundary conditions. This method tests the upper and lower boundaries of the program. In addition, it is usually wise to test around the boundaries.

A third technique that should always be applied to the testing of a program is called *error guessing*. With this method, testers use their intuition and experience to develop test cases. A good system tester is usually very effective at doing this.

The following example shows how test cases might be developed using these three techniques.

The program

For this example, consider a program into which the input can only be a whole number (X) between 10 and 100 inclusive. To test all valid inputs to this program, a system tester must test every single whole number between 10 and 100 inclusive or 91 separate test cases. Testing all invalid inputs would be impossible since the number of whole numbers less than 10 and greater than 100 is infinite.

Equivalence partitioning

To solve this testing problem, a system tester could first use equivalence partitioning to create several equivalence classes of test cases. Each equivalence class in this instance is developed to represent many possible inputs, either valid or invalid. By testing a few of the options in the equivalence class, the whole equivalence class is considered tested. Three possible equivalence classes are listed below:

- All whole numbers greater than 100 (Try $X = 254$, $X = 1000$)
- All whole numbers less than 10 (Try $X = 1$, $X = 5$)
- All whole numbers between 10 and 100 (Try $X = 47$, $X = 85$)

Boundary analysis

The three equivalence classes listed above, however, do not completely test the input conditions of the given program. Test cases need to be developed to test what happens when a user enters either the number 10 or the number 100. Since these are the boundaries of the valid input group, this technique is called boundary analysis.

Testing around the boundaries

Several more test cases that might then be added to test around the boundaries are: $X = 9, X = 11, X = 99, X = 101$.

Error guessing

Lastly, a system tester should apply the error guessing technique to develop additional test cases. For the sample program, an experienced tester may choose the following test cases:

- a negative number
- zero
- a decimal number ($X = 15.3$)
- a non-numeric
- a blank

Other test case development techniques exist besides the ones mentioned above and there are numerous automated testing tools available today that address specific testing problems. System testers should explore the tools available for their particular testing environment.

Regression testing. Regression Testing is retesting software that has been modified in order to ensure that new errors have not been introduced into the system. Software maintenance involves changing programs as a result of errors found or a modification in user requirements. Such changes are liable to introduce new errors into the software. There are a variety of static analysis and dynamic testing techniques for performing regression tests as well as several regression testing tools. Regression testing should be an integral part of the test plan and software test activities.

System test categories. Reiterating that system testing is the process of ensuring that the system does what it is supposed to do, Myers has identified 15 different categories for which test cases need to be developed for effective system testing.[36]

Facility Testing. Ensuring that the program performs each function specified in the objectives. This type of testing is at a very high level and does not measure how well or how effectively the program does what it is expected to do.

Volume Testing. Testing with a heavy volume of data. For example, if a database program is expected to accept an input file of data values, volume testing might mean creating an extremely large number of data values to be entered at one time. Often this type of testing finds failures with declared buffer sizes or input arrays.

---

[36]Myers, *The Art of Software Testing,* 110–118.

Stress Testing.   Testing a system under heavy load conditions. If an objective of the system is to support 20 simultaneous users with a given response rate, stress testing may begin with 20 users and continue to add additional users until response rates noticeably decrease. For this type of testing, a system test group often develops test simulators. Simulators are generally software programs or hardware devices that can be tuned to simulate various levels of stress conditions. In this case, the simulator appears as if 20 or more users were on the system when there may be only one system tester.

Usability Testing.   Testing the human or user interface of the system. Many development organizations today use human factors engineers to analyze the user aspects of a system, including making sure error messages are meaningful, the system is easy to use, and the number of keystrokes required is not excessive.

Security Testing.   Testing a system to determine if it inadvertently allows unauthorized entry into the system or unauthorized access to sensitive data. System security is extremely important in most government projects and for many systems in the business world.

Performance Testing.   Testing a system against the objectives that have been specified regarding response time, throughput rates, start-up times, and capacity. Performance standards in operating systems, database management systems, communications software, and real-time applications are critical to success.

Storage Testing.   Testing the system to reveal problems related to the requirements for amounts of storage used by the program, the size of the system, or the size of the files the program uses.

Configuration Testing.   Testing the system to determine if it supports the various hardware and communications configurations for which it was designed. In particular, many systems have a base set of required components and a set of optional components. Each combination should be tested.

Compatibility Testing.   Testing to ascertain whether the system meets objectives concerning compatibility with other systems.

Conversion Testing.   Testing objectives for converting from a former system to the new system. This is an important concern when the old system must remain operational during the conversion or when large amounts of data need to be converted.

Installability Testing.   Testing to uncover problems in the installation procedures for the system.

Reliability Testing.   Testing against specific objectives that detail how reliable a system is expected to be. Examples of reliability objectives are: Mean Time To Failure (MTTF), Mean Time Between Failures (MTBF), maximum number of errors detected after deployment, and percent of downtime.

Recovery Testing. Testing to ensure that a system can recover from its own failures or failures by related hardware, software, and communications failures. Recovery requirements are very common in operating systems and in critical applications systems.

Serviceability Testing. Testing the objectives that have been specified for maintaining and operating the system. Auxiliary programs such as audit routines and diagnostic routines are examples of programs that need to be tested for serviceability.

Documentation Testing. Testing the documentation that will be delivered to the user with the system. This includes end-user manuals as well as operations and maintenance documentation.

Procedure Testing. Testing the human procedures associated with the system. Examples include database administration procedures and system back-up procedures.

Quality, cost, and time are the three customer concerns when contracting for a software system. If system testing is well-planned and executed, errors are uncovered that would eventually have been found by customers and would conceivably have caused severe impact on their operations. System testing, therefore, is a major contributor to the delivery of a more reliable system within the scheduled timeframe and within budget.

Effectiveness of software testing

Musa, Iannino, and Okumoto collected data about the number of faults found in software during the various stages of development and estimated the mean fault density remaining at the beginning of each development phase. Their data strongly supports software testing as an effective means for significantly improving the quality of a software system.

Figure 16 shows the data from the Musa, Iannino and Okumoto study. According to their estimates, a software product in the coding stage, after compilation and assembly, has a mean fault density of 99.5 faults per 1,000 source lines. Their next estimates show by the beginning of module test, the number of faults per 1,000 source lines has been reduced to 19.7. This dramatic reduction is attributed to desk checking and code inspections. Module testing and integration testing cause another big drop in the number of faults so that as the software enters system test, the mean fault density is estimated to be 6.01 faults per 1,000 source lines. Finally, by the beginning of system operation, the estimated fault density is only 1.48 faults per 1,000 source lines.[37]

---

[37]Musa, John D., Anthony Iannino, and Kazuhiro Okumoto. 1987. *Software Reliability: Measurement, Prediction, Application.* New York: McGraw-Hill. 118.

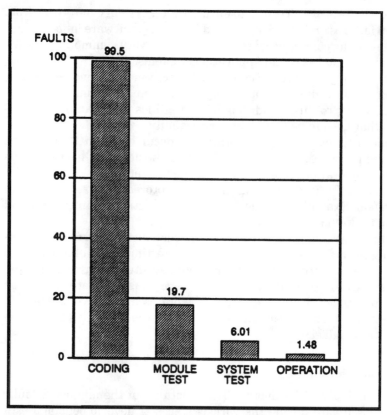

Figure 16 Faults per 1,000 source lines (found on entering the indicated phase).

Step 7: Deploy and Maintain the Software System

When a software organization has completed all the development and testing of a system, it continues to the next phase, deployment. This may include installation, acceptance test, operation, and maintenance of the software system.

Software deployment activities vary substantially depending on the type of system being installed and the particular circumstances related to that system. For instance, the system may be

- a replacement for an older system, thus requiring system conversion to phase out the old
- a completely new system being installed for the first time
- an upgrade of a real-time system requiring uninterrupted service during the deployment

- part of a complex hardware and software system requiring considerable coordination activities

An overview of the process is shown in Figure 17.

Software installation and conversion

The initial installation of any new or replacement software system is usually done via a beta test. If the completed system will eventually be installed at numerous user sites, one of these sites is chosen as the beta test site. During the beta test period, the new system is installed on the target hardware, and system conversion activities, if required, are completed. Users then begin to use the new system. Before any software system is installed, however, the development organization should prepare a complete system installation plan and review it with the customers, the user organization, and the system operations group.

During the initial installation or the early use of a software system at a user's site, errors often surface due to unexpected aspects of the user's environment. A system failure during this time could cause serious setbacks for users, especially on critical real-time systems. Because of this, the system installation plan should cover the installation, conversion, and beta test periods, and should include prearranged agreements about

- installation and conversion schedules
- resources required
- error repair timeframes
- system contingencies and plans
- system recovery
- training

Acceptance test

The customers, the users, and the development organization should also agree about a time during the beta test period when the software will be baselined and a formal customer test of the software, called an *acceptance test,* will begin. As with all other tests of the system, the acceptance test must be a planned activity that includes a test plan, test procedures, test cases, and the acceptance criteria.

During acceptance test, the system's users should attempt to perform every system activity. User training before the acceptance test is a critical success factor for a software system, as are the quality and the accuracy of the user documentation.

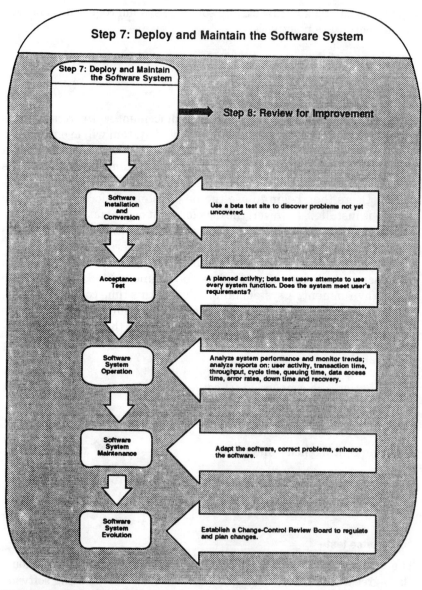

Figure 17 Deploy and maintain the software system.

When the acceptance test is complete, there should be documented data about whether the system meets the customer's requirements. If the system is accepted, it officially enters the operation and maintenance phase of its life cycle. If there are multiple target user sites for this system, a schedule should be developed for deployment at those sites.

If the software system is not acceptable, the customer and the developers should agree on what changes are necessary for the system to meet the objectives. They should also agree on the proposed schedule and procedures for delivering the revised system. In some cases, an unacceptable system remains in operation at a customer's site and revisions are added to the system and tested as they become available. If the new system is a replacement system, a customer may request that the old system be reinstalled until the new system can be revised to meet stated objectives.

Software system operation

The operation of a software system is the ongoing process of running the system. Some software systems are required to run 24 hours a day, 7 days a week; some have Mean Time Between Failure (MTBF) requirements that cannot be exceeded; and others may have specific performance tolerances within which the system must conform. As with every other phase of a software system, the operation phase should be planned and documented. Procedures for managing the operation, for evaluating the operation, and for providing information about operational problems are required.

System performance reports should be periodically scheduled and analyzed to ensure proper system operation and to monitor operational trends in the system's performance. Common software operational reports include statistics about the following aspects of the system:

- user activity
- transaction time
- throughput rate
- cycle time
- queuing time
- system access time
- data access time
- error rates

Software system maintenance

The maintenance period of a software system covers customer acceptance of the initial system to system retirement. The primary activities included in the maintenance period are: correcting software problems, developing system enhancements, and adapting the system to a new environment. Figure 18 illustrates the typical distribution of these

MAINTENANCE EFFORT DISTRIBUTION

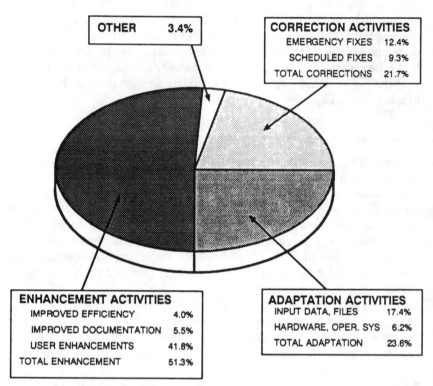

OTHER   3.4%

CORRECTION ACTIVITIES
EMERGENCY FIXES   12.4%
SCHEDULED FIXES   9.3%
TOTAL CORRECTIONS   21.7%

ENHANCEMENT ACTIVITIES
IMPROVED EFFICIENCY   4.0%
IMPROVED DOCUMENTATION   5.5%
USER ENHANCEMENTS   41.8%
TOTAL ENHANCEMENT   51.3%

ADAPTATION ACTIVITIES
INPUT DATA, FILES   17.4%
HARDWARE, OPER. SYS   6.2%
TOTAL ADAPTATION   23.6%

Figure 18 Distribution of maintenance efforts.

maintenance activities, determined from a survey of 487 business data processing installations.[38]

In some cases, the maintenance period of a software system may be as long as 10 to 15 years and may account for as much as 70 percent of the total system costs. This means software maintenance should be planned early in the development life cycle, and maintainability should be a design and implementation consideration.

Enhancing software maintainability. Key development activities that can enhance the maintainability of a software system include:

- standardization of design notations for data flows, structures, algorithms, interface specifications, pseudocode, and data dictionaries

---

[38]Fairley, 83.

- emphasis on clarity of code and modularization through the use of information hiding, data abstraction, and single-entry and single-exit constructs
- coding style standards including program prologues, comment lines, indentation, parameters, and error handling
- guidelines and requirements for supporting documentation such as a user manual, a maintenance guide, an error message manual, test documentation, and cross-reference directories
- keeping all documents up-to-date

Configuration management during maintenance. During the maintenance phase of a software system, it is very important to use a configuration management system to track and control both the software and the documentation of the system. A description of change control and configuration management is described in Step 2 of this chapter. Important features of a configuration system during software maintenance include a history of component revisions, version control by site, an accounting of the number of errors reported, fixed and released to the field, and documentation updates.

Software system evolution

Over time, enhancements and adaptations may cause substantial cumulative changes in a software system. These changes often increase the size and complexity of a system while decreasing its flexibility. When this happens, the maintenance of the software system can become difficult and even small fixes or enhancements become costly. Often, a minor problem results in a major system failure.

Change Control Review Board. To help avoid these problems, system changes should be carefully regulated. Many software organizations use a Change Control Review Board to consider and approve all changes. The membership of this board (and the rules by which it operates) vary depending on the nature of the software. The two primary objectives of the board are to review all change requests and establish priorities and constraints for making changes. In some cases, however, a software system simply outgrows its originally planned and designed bounds. It then becomes more cost-effective to replace the entire system than to continue to patch the old one.

Step 8: Review for Improvement

The goal for software projects, like other development projects, is continual improvement in quality or productivity. This goal is typically achieved little by little rather than with major bursts. These incremental improvements, however, can add up to large differences in the software system over the course of time. In quality, the improvements come from reducing errors, finding errors early, and meeting requirements for function and usage. In productivity, improvements come from meeting customer needs while avoiding waste from redesign, repair, or maintenance.

Sustained high quality and productivity require a disciplined application of process improvement. To do this, software development organizations must allocate resources to review the outcome of each of their projects and the processes under which each project was developed. (See Step 2 for suggested outcome and process metrics.)

Figure 19 shows an overview of the improvement process.

Software development postmortems

A *postmortem* provides an excellent way for a software organization to evaluate its development process. A postmortem is an assessment done after the completion of a project or major development phase, when team members can look at the processes objectively.

A postmortem identifies successes so that they can be maintained. Data gathered from the postmortem can also be used for planning future projects and for identifying specific areas for quality improvement within the organization. Since the key focus of a postmortem is generally the *process* under which the software was developed, it can almost be considered a debriefing of the project members.

Customers and users may also be interviewed during a project's postmortem. They frequently provide valuable information and insight into the project's development processes.

During a postmortem, risk areas are studied. Items of technical risk are listed and prioritized; attendees offer ways of managing the risk. In this way, the cost of doing postmortems is justified since improving the process will reduce costs in the long run.

Examples of postmortem topics. Examples of topics explored during postmortems include:

- the management techniques used on the project
- the assignment of project responsibilities
- the development methodology chosen
- the project's standards (for example, documentation and change control) and how well standards were followed
- the amount of project-specific and technological experience on the team

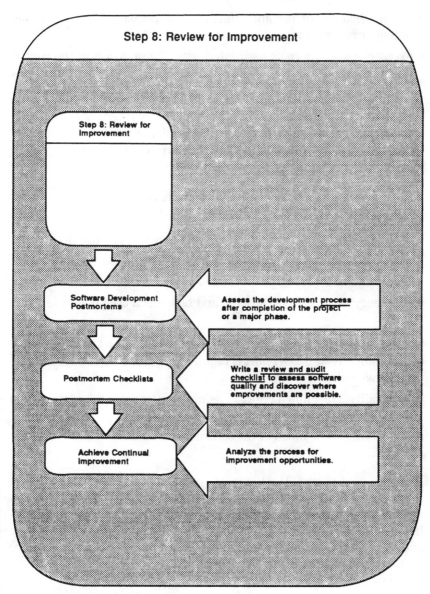

Figure 19 Review for improvement.

- the level of training provided to team members
- the tools and technologies used
- the degree of communication between management and the team, between the various organizations involved with the project, among the team members, and between the development team and the customer

- the methods used for and timeliness of transferring deliverables among organizations
- assessment of the risks associated with the project and the contingency plans

Postmortem checklists

Postmortem checklists help assess the quality of the software deliverables and discover where improvements are necessary. These review and audit checklists often include these factors:

- Correctness: Does the software perform as expected?
- Reliability: How accurate is it?
- Usability: How much effort does it take to learn and to use the system?
- Flexibility: How much effort is required to modify the system?
- Maintainability: How much effort is required to find and fix errors?
- Efficiency: How many resources and how much code does a function take?
- Integrity: Is unauthorized access controlled?
- Testability: How much effort is required to make sure the system performs as expected?
- Portability: How much effort is required to transfer the system from one machine or environment to another?
- Reusability: How much of the software can be used in another application?

It is important to consider the correlations between factors. Correlations may be positive in that improving one factor also improves another factor; for example, improving reliability also improves correctness. Correlations may be negative in that improvements in one factor negatively affect another, for example, the added code and processing to improve usability may make the program less efficient. Thus, consider the tradeoffs in deciding which improvements to pursue.

Achieving continual improvements

Continuous process improvement is iterative. Process analysis involves deciding which improvements to pursue, implementing the improvements, checking and assessing the results, and then planning the next improvements. The efforts pay off, however, in improved customer satisfaction, increased productivity, and lower costs.

3

Application

This chapter provides five specific examples of applications for the procedures in Chapter 2.

- Concept definition is illustrated by a sample Software Feasibility Report. It is meant to be a suggested framework for guiding feasibility studies.

- Project planning is often considered to be the most difficult segment of software development, and its greatest challenge lies in correct estimation of cost, staff, and time. The COCOMO model for effort and cost estimation is one way to meet this challenge, and its methodology is shown as an example.

- The IEEE Standard for a Software Verification and Validation Plan is presented as a framework, to be customized for every project.

- The heart of any software project is, of course, its code. The code inspection example describes items of concern during a code review. The list of items, while extensive, should be considered only as a jumping-off point for actual reviews.

- Two Software System Test Plans are presented, an example of a good plan and an example of one that needs more work.

Software Feasibility Report

The first step in a software engineering project is to define the concept. This chapter presents a practical application of the concept definition stage: a sample of a Software Feasibility Report.

The flow of ideas

Ideas for new products can come from several sources: the research lab, marketing department, or the proposal development organization, to name a few. From there, a formal selection process should be encouraged, with research into alternatives. Those selected should be subject to a formal feasibility study, with the result being a feasibility report. Figure 20[39] shows the four major analyses: customer needs, competitive, functional, and economic.

Target of the software feasibility report

The outcome of the concept definition stage should be a formal software feasibility report targeted for management, users, and developers.[40]

This high-level document must be a clear, concise statement of user needs and technical risks, and a description of how the proposed system will or will not meet those needs and handle the risks. Indeed, the feasibility study will be reviewed with the users, and it should be written in user terminology that is easily understood.

In addition to a general description of the proposed system in its "finished" state, the report should contain preliminary estimates for development cost, staff, and time. Management's go/no go decision should be based on the contents of this report; the importance of their approval is signified by making this a "sign-off document."

Framework for the report

There are many factors involved in the consideration of feasibility, and the report must address the ones that are important to the company and the project. Software development organizations should develop their own standard feasibility report framework based on items identified as significant. A sample framework begins on page 342. Remember, it is just a sample; you must customize it to your needs.

---

[39]Starr, Martin K. 1988. *Operations Management.* Englewood Cliffs, NJ: Prentice-Hall. 88.

[40]Frank Kirk. *Total System Development for Information Systems,* 57–63.

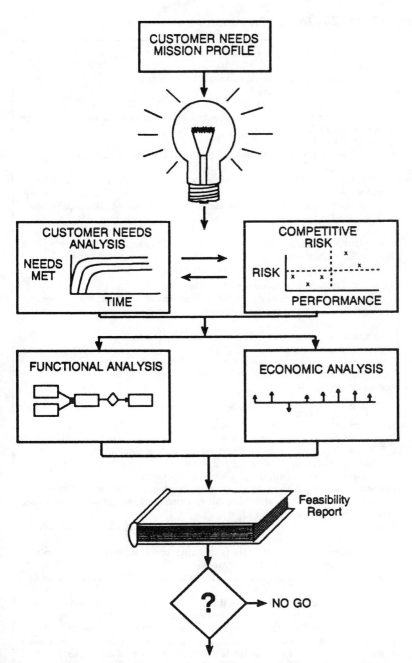

Figure 20 Evaluating a new product for feasibility.

Sample Feasibility Report Framework

1. Abstract—executive summary
2. Introduction
   A. scope—limitations on the information in the document
   B. organization of the document
   C. intended audience—assumptions about reader
3. Statement of Need—expanded statement of the opportunity to be addressed by the system, including:
   A. the circumstances
   B. description of how the proposed system would correct or eliminate the problem or fulfill the need
   C. competitive situations, strategic advantages, and benefits to the organization
4. Overview—high-level narrative summary of the proposed system that clearly identifies the system and defines boundaries that limit the set of its requirements
5. External and Internal Interfaces—display of key elements that interact with or in the system
6. System Goals—unambiguous description of objectives with a level of detail that allows a full understanding of the system's capability but without the level of detail typically found in a requirements document

   The types of information discussed should be
   A. output—content, frequency, physical form, use of symbols; with particular attention to data and processing reports
   B. required user skills
   C. estimated operating costs
   D. estimated production volume
   E. geographic locations for operation
   F. communication requirements within and between systems
   G. data storage
   H. data security
   I. available operating modes
   J. backup systems (or reference to alternative operating modes)
7. Capabilities Analysis—clarification of three key factors for each capability:
   A. function—the procedure by which people and equipment produce results, without description of method of achievement. The six primary functions are: receive, transfer, record, store, process, and present.
   B. inputs and outputs—medium (display, hard copy), volume/quantity, frequency (per day, week, month), quality (accuracy, dependability), and source (input) or destination (output)
   C. personnel interface—number of people and types of skills required, human factors engineering, amount and type of training
8. Expected Users—description of users and user implications (i.e., how the new system will change the way the user works)
   A. organizational—number of people affected, type of effect, recruitment, selection, training
   B. business—financial, managerial, time, analytical, forecasting, measurement
   C. customer—social, service, error, environmental, product performance
9. Preliminary Reliability and Performance Requirements—strictures to be stated as quantifiably as possible, with ranges of acceptability; like the system goals (described above), these requirements should have a level of detail that allows a full understanding of the system's capability but without the level of detail typically found in a requirements document
   A. response time—definition (expressed in terms of access time) and projections of duration from input to output

    B accuracy—definition and frequency of significant errors

    C. availability of total system—including frequency and duration of degraded performance modes

    D. flexibility—type and number of exceptional conditions handled by the system

    E. security—including legal safety and degree of vulnerability

    F. capacity—average and peak loads

    G. customer acceptance—by management and by user

    H. efficiency or productivity—performance ratio

    I. quality—tolerance and appearance

10. Assumptions, Items for Resolution—any items discovered while doing the study, stated with best- and worst-case scenarios

11. Resources and Constraints—factors already in existence and/or for which there is planned expansion

    A. physical plant facilities, supplies

    B. hardware, software, and firmware technology

    C. development staff, time, and budget

    D. current and potential system interfaces

    E. environmental conditions such as company organization, union contracts, laws, regulations

    F consonance with short- and long-range company goals

    Each resource should be examined for development and for the finished product and should be listed together with its

    A. best- and worst-case analyses

    B. degree of confidence of the analyses

    C. identification of the criticality of resources and constraints; if very critical, the ease of changing the restriction

12. Possible System Configurations—these should describe functionality only (without reference to particular hardware); e.g., several terminal communications media, a central processing unit, some off-line storage

13. Impact and Cost Analysis—estimates for the cost of development (personnel, facilities), operation, maintenance, and support; forecast of demand or usage and expected revenues; other impacts on the state of the art in weapons systems (or other government activities); impact on the overall DoD procurement effort. Describe the estimates, forecasts, and impacts

    A. for all the possible system configurations

    B. with identification of possible savings and/or additional costs

    C. with numbers expressed as high-low ranges, stated together with a probable figure

14. Estimate of Development Schedule—high-level display of time vs. staff, coordinated by life cycle phase, including interactions with users and other involved groups, and taking into account the following items:

    A. communications

    B. budget

    C work space

    D. job aids

    E. special equipment

    F. staff requirements (number of, skills, knowledge)

    G. development objectives

    H. scheduling network

    I. activity time status reports

    J. function bar charts

    K. cost milestone reports

(Continued)

Sample Feasibility Report Framework (Continued)

> L. staff loading
> M. financial plans
> N. status reports
> 15. Ranked Control Objectives—preliminary proposed orders based on importance to the user and on economic, technological, and operational feasibility. Although they will be stated here in high-level terms, these goals may be expected to remain constant and not be significantly altered in any forthcoming requirements document.
>   A. system administration
>   B. data storage and security
>   C. error detection
>   D. regulatory and company policy requirements
>   E. audit trail
>   F. system reliability
>   G. system maintainability
> 16. Customer Acceptance Criteria—the features and timing that prospective customers are willing and committed to accept and on which they will base their "go" or "no-go" decisions
> 17. Management Acceptance Criteria—high-level discussion of the basis for "go" or "no-go" on the project
> 18. Risk Assessment—high-level discussion of "what-ifs" and key issues of concern as well as the likelihood of the realization of these concerns
> 19. Recommended Action—"go" or "no-go" and the elements that are paramount to the recommendation
>   Appendix I   Terminology—project-specific terms, acronyms, conventions, and language standards
>   Appendix II   Methodology of the Study—ongoing research; interviews with users, developers, subject matter experts; literature search; type of cost analysis; analysis of current and projected staffing levels, funding levels, physical plant; market research and analysis
>   Appendix III   References—a brief list of references used during the study; not an exhaustive historical survey

Effort and Cost Estimation

An important activity in planning a software project is predicting the resources that are needed for the project and the costs that will be incurred during its life cycle. Several different types of software costing methods were discussed in Chapter 2: Procedures: expert estimation, project similarity, lines of code estimation, and cost estimation modeling.

This chapter gives an example of a software cost estimation method. The COCOMO (COnstructive COst MOdel) cost estimation method originated with Barry Boehm and is widely used, particularly in government projects. This discussion is neither an endorsement of the model nor a complete tutorial for using the model. Instead, the model is used in this chapter to highlight the issues and factors that must be considered when estimating software resources. In fact, Boehm believes that the model's biggest asset is not the methodology itself, but rather COCOMO's provision of a "common universe of discourse" for understanding software tradeoffs, for management negotiation and control, and for improving software productivity.[41]

The remainder of this chapter

- defines the COCOMO model
- describes the steps of the COCOMO Intermediate model
- applies that model to a software communication project

Although detailed, the example does not cover every aspect of the COCOMO model.

What is COCOMO?

COCOMO is a method of estimating the effort, schedule, and cost of a planned software project by analyzing the attributes of the project and applying that information to a set of standard equations. The basis for the model, its equations, and its charts is a study conducted in the late 1970s at TRW in which the estimated and actual project team member-months of 63 projects were analyzed along with data about the type of projects, the complexity, and the skill level of analysts and programmers on the projects.

As shown in Figure 21, the COCOMO model requires a cost estimator. The cost estimator is an analyst or manager familiar with cost estimation methods who determines estimates of software size, and product, computer, personnel, and project attributes. These estimates and attributes are used with various look-up tables to determine factors. The factors are used with various look-up equations to estimate costs, staffing estimates, and rough schedules. There are three different levels of COCOMO models—basic, intermediate, and detailed—each respectively building on the lower level model and providing more accurate and more detailed estimates. There are also COCOMO extensions available for various programming language development environments.

Steps in effort and cost estimation

The steps required to estimate the effort, schedule, resources, and costs of a software project differ depending on the model being used. This dis-

---

[41]Boehm, Barry W. 1981. *Software Engineering Economics,* 42; and Presentation, Washington, D.C., May 1989.

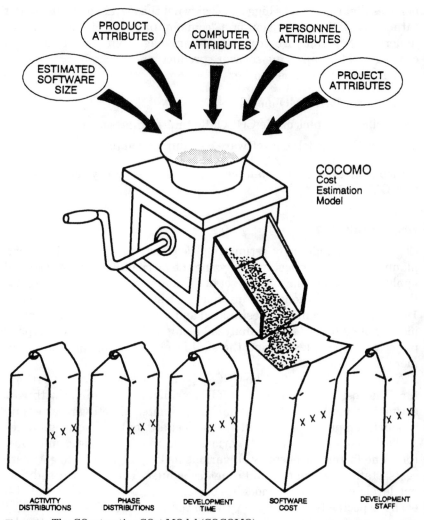

Figure 21 The COnstructive COst MOdel (COCOMO).

cussion focuses on the Intermediate COCOMO model. The steps of the Intermediate model are listed below and each is then briefly discussed:

- classify the project mode
- size the project
- determine the nominal effort
- rate the cost factors

- assign effort multipliers to the cost factor
- calculate the effort and schedule estimates
- calculate the phase distributions
- estimate the cost

Classify the project mode. First, the cost estimator must pick one of three development modes: the organic, the semidetached, or the embedded mode. This classification is important because the effort and schedule equations in the model are based on the mode chosen for the project.

The organic mode classification refers to projects that are relatively small (less than 50,000 lines of code), have little concurrent development of new hardware, and have a minimal need for innovative data processing. Examples of organic mode projects are batch data reduction, scientific models, business models, and simple inventory or production control systems.

The semidetached mode is usually chosen for medium to large projects (up to 300,000 lines of code) that have requirements for complex interfaces with hardware or other software systems. Typical semidetached systems include transaction processing systems, new operating systems, new database management systems, and simple command-control systems.

Any size project can be classified as embedded mode if it requires complex interfaces with hardware, firmware, and other software systems, extensive or innovative data processing architecture and algorithms, and if there are tight constraints on reliability and performance. Large and complex transaction processing systems, very large operating systems, ambitious command-control systems, and avionics systems are all classified as embedded mode.

Size the project. Next, the cost estimator must determine a size estimate for the project. Many of the software cost estimation models available today (including COCOMO) are based on an estimate of the number of lines of code required for a project. Sizing is not easy since there are many variables and dependencies that must be considered in the estimate. One way of guessing this number is to use PERT sizing: this requires estimation of the lowest possible size, the most likely size, and the highest possible size. These estimates are then put into various PERT formulae to give estimated total size.

Determine nominal effort. The COCOMO model next requires the cost estimator to determine the nominal effort for an average project of this size (as determined by lines of code).

Rate the cost factors.   Boehm has chosen 15 different attributes to be considered when you estimate software development efforts and costs. The attributes are:

- product attributes
  required reliability (RELY)
  database size (DATA)
  software product complexity (CPLX)
- computer attributes
  execution time constraint (TIME)
  main storage constraint (STOR)
  virtual machine volatility (VIRT)
  turnaround time constraint (TURN)
- personnel attributes
  analyst capability (ACAP)
  applications experience (AEXP)
  programmer capability (PCAP)
  virtual machine experience (VEXP)
  programming language experience (LEXP)
- project attributes
  use of modern programming practices (MODP)
  use of software tools (TOOL)
  required development schedule (SCED)

The cost estimator must examine each attribute as it relates to the given project and rate its applicability as: Very Low, Low, Nominal, High, Very High, or Extra High.

For instance, the execution time constraints (TIME) of a business data processing project may be considered average, so a cost estimator would rate the execution time constraints factor as nominal. However, a real-time software system would have tight requirements on response time and processing time and would therefore require the execution time constraints factor to be rated Very High or Extra High.

Boehm has provided some criteria that a cost estimator may use for assigning these ratings. (See Table 3.)

Assign effort multipliers to the cost factors.   In the Intermediate COCOMO model, the rating of each cost factor has associated effort multipliers. These multipliers are used to determine how the effort for this project will vary from the average project. The values for these multipliers (included in Table 4) are from the TRW study mentioned earlier.

As an example of how these effort multipliers work in the COCOMO model, consider the ACAP factor—analyst capability. If a cost estima-

TABLE 3 Software Cost Factors

Cost factor	Very low	Low	Ratings nominal	High	Very high	Extra high
			Product Attributes			
Required Reliability (RELY)	Effect: slight inconvenience	Low, easily recoverable losses	Moderate, recoverable losses	High financial loss	Risk to human life	
Database Size (DATA)		$\dfrac{\text{DB bytes}}{\text{Prog.DSI}}$ <10	$10 \leq \dfrac{D}{P}$ <100	$10 \leq \dfrac{D}{P}$ <1000	$\dfrac{D}{P} \geq 1000$	
Software Product Complexity (CPLX)		See Boehm, *Software Engineering Economics*, p. 122.				
			Computer Attributes			
Execution Time Constraint (TIME)			≤50% use of available execution time	70%	85%	95%
Main Storage Constraint (STOR)			≤50% use of available storage	70%	85%	95%
Virtual Machine Volatility (VIRT)		Major change every 12 months Minor: 1 month	Major: 6 months Minor: 2 weeks	Major: 2 months Minor: 1 week	Major: 2 weeks Minor: 2 days	
Turnaround Time Constraint (TURN)		Interactive	Average turnaround <4 hours	4–12 hours	>12 hours	
			Personnel Attributes			
Analyst Capability (ACAP)	15th percentile	35th percentile	55th percentile	75th percentile	90th percentile	
Applications Experience (AEXP)	≤4 months experience	1 year	3 years	6 years	12 years	
Programmer Capability (PCAP)	15th percentile	35th percentile	55th percentile	75th percentile	90th percentile	

TABLE 3 Software Cost Factors (Continued)

Cost factor	Very low	Low	Ratings nominal	High	Very high	Extra high
			Personnel Attributes (Continued)			
Virtual Machine Experience (VEXP)	≤1 month experience	4 months	1 year	3 years		
Prog. Language Experience (LEXP)	≤1 month experience	4 months	1 year	3 years		
			Project Attributes			
Modern Prog. Practices (MODP)	No use	Beginning use	Some use	General use	Routine use	
Use of Software Tools (TOOL)	Basic microprocessor tools	Basic mini tools	Basic mini/maxi tools	Strong maxi programming	Add requirements, design, management documentation tools	
Required Development Schedule (SCED)	75%	85%	100%	130%	160%	

TABLE 4 Software Development Effort Multipliers

Cost factor	Very low	Low	Ratings nominal	High	Very high	Extra high
		Product Attributes				
Required Reliability (RELY)	0.75	0.88	1.00	1.15	1.40	
Database Size (DATA)		0.94	1.00	1.08	1.16	
Software Product Complexity (CPLX)	0.70	0.85	1.00	1.15	1.30	1.65
		Computer Attributes				
Execution Time Constraint (TIME)			1.00	1.11	1.30	1.66
Main Storage Constraint (STOR)			1.00	1.06	1.21	1.56
Virtual Machine Volatility (VIRT)		0.87	1.00	1.15	1.30	
Turnaround Time Constraint (TURN)		0.87	1.00	1.07	1.15	
		Personal Attributes				
Analyst Capability (ACAP)	1.46	1.19	1.00	0.86	0.71	
Applications Experience (AEXP)	1.29	1.13	1.00	0.91	0.82	
Programmer Capability (PCAP)	1.42	1.17	1.00	0.86	0.70	
Virtual Machine Experience (VEXP)	1.21	1.10	1.00	0.90		
Prog. Language Experience (LEXP)	1.14	1.07	1.00	0.95		
		Project Attributes				
Modern Prog. Practices (MODP)	1.24	1.10	1.00	0.91	0.82	
Use of Software Tools (TOOL)	1.24	1.10	1.00	0.91	0.83	
Required Development Schedule (SCED)	1.23	1.08	1.00	1.04	1.10	

tor perceives that the analysts available for a project have average capabilities, the ACAP cost factor should be given a nominal rating. The effort multiplier associated with a nominal rating for any of the cost factors is 1.00 (see Table 4), thus ACAP is assigned a 1.00 value.

If, however, the analysts are perceived as having better than average capabilities, ACAP may be assigned a high rating. The high rating has an associated multiplier of .86. Using .86 as a multiplier for the average effort estimate for the project decreases the estimate. This is understandable since management can usually expect reduced effort and shorter schedules when there are experienced and talented resources on a project.

The reverse is also true. That is, if analysts' capabilities are perceived as less than average, management might expect longer estimates for work and possibly more rework. The problem of less than average capabilities is handled by assigning the ACAP factor a low rating and a multiplier of 1.19, with a resulting higher-than-average estimate of resources for the project.

Calculate the effort and schedule estimates. Once the size, the development mode, and the appropriate cost factor ratings and values have been identified, the next step in the Intermediate COCOMO model is to apply that information to a given set of effort and schedule equations.

The effort equations provided by Boehm give an estimate of the total development effort in team member-months (MM) required by the project. This effort estimate is then used within the scheduling equation to calculate an estimate of time, in months, required for a project. The cost estimator may then determine the number of people needed for the development phase of the project by calculations using the effort and time estimates.

The attributes of different types of software projects affect development efforts and schedules. Because of this, Boehm provides specific effort and schedule equations for each of the development modes (organic, semidetached, and embedded) in the model. The example that follows at the end of this chapter gives the equations used for an embedded mode project.

Calculate the phase distributions. For some projects, especially smaller ones, an estimate of the resources required and the duration of the development effort is enough information for further project planning. For large projects, further refinement of effort and schedule estimates may be necessary. The Intermediate COCOMO model provides a method of distributing portions of the effort and schedule estimates to each of the different phases of a software project.

TABLE 5   Phase Distribution (Effort)

Effort distribution	Project Size				
Phase	Small 2 KDSI	Intermediate 8 KDSI	Medium 32 KDSI	Large 128 KDSI	Very large 512 KDSI
Product Design	18%	18%	18%	18%	18%
Programming	60%	57%	54%	51%	48%
Integration and Test	22%	25%	28%	31%	34%

TABLE 6   **Phase Distribution (Schedule)**

Schedule distribution	Project size				
Phase	Small 2 KDSI	Intermediate 8 KDSI	Medium 32 KDSI	Large 128 KDSI	Very large 512 KDSI
Product Design	30%	32%	34%	36%	38%
Programming	48%	44%	40%	36%	32%
Integration and Test	22%	24%	26%	28%	30%

The COCOMO model provides tables that can be used to estimate the distribution of effort and schedule across the phases of the project. The distribution of the estimates is shown as a percentage. (See Tables 5 and 6.)[42]

Estimate the cost.   A common method for estimating software development cost is first to identify the number of months of effort required for a project, then multiply by the average monthly cost.

Using the COCOMO model, a cost estimator can provide a high-level cost breakdown of a total project effort as well as costs broken down by schedule or phase.

A COCOMO example

The following is an example (adapted from Barry Boehm) of how to use the steps of the COCOMO Intermediate model to determine the effort, schedule, resources, and cost estimates for a project.[43]

---

[42]Boehm, *Software Engineering Economics,* 90.

[43]Boehm, *Software Engineering Economics,* 524–529 and Presentation.

Sample Application of COCOMO Model

Project Description

The project is a software product that will be used to process communications on a new commercial computer.

Classify the Project Mode

The development of communications software is complex because of performance constraints and because of requirements to interface with hardware, firmware, and other software in the system. Thus, the project is classified as an embedded mode development.

Size the Project

The size estimate for this project is 30 KDSI (that is, 30,000 delivered source instructions, or 30,000 lines of code).

Determine Nominal Effort

The cost estimator uses the following equation determined by Boehm from the TRW study to calculate the nominal development effort, in team member-months (MM), for embedded mode projects:

$$MM_{nom} = 2.8 \, (KDSI)^{1.20}$$

where: MM_{nom} = the team member-months of effort required for a nominal project, and

KDSI = the estimated number of delivered source instructions in thousands.

The result is a nominal effort estimate of about 165 MM for the communications software project. This is calculated by substituting 30 KDSI into the equation:

$$MM_{nom} = 2.8(30)^{1.20} = 165 \, MM$$

Note: For the COCOMO model, as well as for other cost estimation models, on-line programs for doing the necessary calculations are generally available.

Rate the Cost Factors

The next step is to examine the 15 attributes and apply a rating of Very Low, Low, Nominal, High, Very High, or Extra High.

Assign Effort Multipliers to the Cost Factors

For many projects, not all 15 cost factors used in the COCOMO model will be assigned the Nominal rating. The cost estimator for the communications project in our example decided that the factors given in the following table apply to the project. The table includes explanations for the ratings.

Cost Factors for Example Project

Cost Factor	Situation	Rating	Effort Multiplier
Required Reliability (RELY)	Local use of system. No serious recovery problems	Nominal	1.00
Database Size (DATA)	20,000 bytes	Low	0.94
Software Product Complexity (CPLX)	Communications processing	Very High	1.30
Execution Time Constraint (TIME)	Will use 70% of available time	High	1.11
Main Storage Constraint (STOR)	45K or 64K store (70%)	High	1.06
Virtual Machine Volatility (VIRT)	Based on commercial microprocessor hardware	Nominal	1.00
Turnaround Time Constraint (TURN)	Two-hour average turnaround time	Nominal	1.00
Analyst Capability (ACAP)	Good senior analysts	High	0.86
Applications Experience (AEXP)	Three years	Nominal	1.00
Programmer Capability (PCAP)	Good senior programmers	High	0.86
Virtual Machine Experience (VEXP)	Six months	Low	1.10
Prog. Language Experience (LEXP)	Twelve months	Nominal	1.00
Modern Prog. Practices (MODP)	Most techniques are in use more than one year	High	0.91
Use of Software Tools (TOOL)	At basic minicomputer level	Low	1.10
Required Development Schedule (SCED)	Nine months	Nominal	1.00

Calculate the Effort and Schedule Estimates

When a non-nominal rating is assigned to any of a project's cost factors, it implies that the nominal effort estimate for the project must be modified.

The Intermediate COCOMO model has a method to calculate an effort adjustment factor based on the cost factor ratings: multiply all effort multipliers together to yield an effort adjustment factor. Then this adjustment factor is used to estimate how the effort for this project is different from the average project.

(Continued)

Sample Application of COCOMO Model (Continued)

In the communications project example, nine factors (refer to Table 5) were given non-nominal ratings. To find the effort adjustment factor (EAF) for the project, multiply together all effort multipliers (shown in the table on page 355):

EAF = (1.0)(.94)(1.3)(1.11)(1.06)(1.0)(1.0)(.86)(.86)(1.0)(1.1)(1.0)(.91)(1.1)(1.0) = 1.17

The value, 1.17, means that the effort for this communications project will require 17% more effort than the average project.

The new effort estimate is 193 MM. It is found by multiplying the nominal effort of the project (165 MM) by the effort adjustment factor (1.17):

$$MM_{adj} = MM_{nom} \times EAF = (165) \times (1.17) = 193 \text{ MM}$$

The next step for the cost estimator is to use this new effort estimate of 193 MM to determine an estimate of the schedule for the project. The COCOMO model provides specific equations, again based on development mode, for this translation. The embedded mode equation for estimating the schedule is:

$$TDEV_{emb} = 2.5(MM)^{0.32}$$

where: $TDEV_{emb}$ = the number of months estimated for the software development.

For this example:

$$TDEV_{emb} = 2.5(193)^{0.32} = 13 \text{ months}$$

Thus, the communications project will require 13 months to develop.

Calculate the Duration

Once the effort and schedule estimates for a project have been calculated, an estimate of the number of people required on the project is calculated by simply dividing the effort estimate (193 MM) by the schedule estimate (13 months). The effort of 193 team member-months over a 13 month period equals approximately 15 people:

Resource estimate: (193 MM + 13 months) = 14.8 non-management
team members (about 15 people)

Note: The average project would have required 12.7 people.

Calculate the Phase Distribution

The next step in the COCOMO model is to determine the distribution of people across the phases of the project's life cycle.

Refer to Tables 5 and 6 for the percentage bases for the calculations to follow.

Table 5 is used to calculate the phase distribution of the effort estimate. Table 6 is used to calculate the phase distribution of the schedule estimate. The tables are further broken down by the approximate size of a project and then by its development mode.

The size of the communications software project was estimated to be about 30 KDSI. Therefore, when calculating phase distributions using Table 5, the cost estimator for this project chooses to use the percentage values for medium size projects (32 KDSI).

The approximate distributions of effort across phases is calculated by multiplying the total effort estimate of 193 MM by the appropriate percentage from Table 5:

Product Design Phase Effort
Estimate: (193 MM) × (.18) = 35 MM

Programming Phase Effort
Estimate: (193 MM) × (.54) = 104 MM

Integration and Test Phase Effort
Estimate: (193 MM) × (.28) = 54 MM

Similarly, using Table 6, the schedule estimates for each phase are calculated by multiplying the total schedule estimate (13 months) by each of the percent values provided in the table. These approximations are:

Product Design Phase Schedule
Estimate: (13 months) × (.34) = 4.5 months

Programming Phase Schedule
Estimate: (13 months) × (.40) = 5 months

Integration and Test Phase Schedule
Estimate: (13 months) × (.26) = 3.5 months

The graph in Figure 22 shows the relationship of the effort estimate to the schedule estimate.

Estimate the Cost

The last step illustrated here is to calculate the cost estimates for the communications software project. Although presented last, it is not necessarily the last step in software estimation.

For example, once the cost estimator had determined the nominal effort estimate (165 MM) for this project, a high-level cost estimate could be calculated by multiplying that number by the average cost per person per month.

For our example, assume that management uses an average cost of $7,000 per person per month (including overhead costs) when estimating project costs. The high-level cost estimate for this project would be as follows:

$$(165 \text{ MM}) \times (\$7,000) = \$1,115,500$$

As the cost estimator proceeds through the COCOMO model, estimates of increasingly greater accuracy and detail may be calculated. For instance, when the cost estimator changed the cost factor ratings, the effort adjustment factor of 1.17 caused the effort estimate to change from 165 MM to 193 MM. This in turn causes a 17% increase in estimated costs:

$$(193 \text{ MM}) \times (\$7,000) = \$1,351,000$$

Using the schedule estimates and the phase distribution values calculated in the COCOMO model, management can further refine these cost estimates as needed.

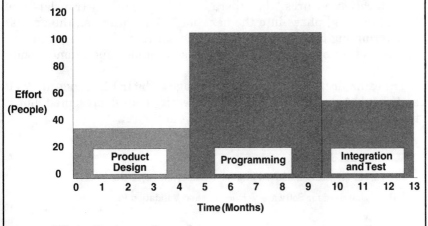

Figure 22 Effort estimate on a time scale.

Effort and cost estimation cautions

The COCOMO model considers many factors that affect the effort, schedule, resource, and cost estimates of a project. Its considerations are by no means exhaustive, however, and there are some factors not specifically addressed in COCOMO. They include:

- language level
- requirements volatility
- amount of documentation
- customer interface quality
- management quality
- security/privacy restrictions
- personnel continuity
- hardware configuration

Each of the many available cost estimation models use some subset of the cost factors discussed throughout this part. Before choosing a model, cost estimators should look at the factors considered significant in each model and determine the ones that provide the best fit for their projects. Although sometimes none will fit exactly, a reasonable effort and cost estimate of effort and cost may be determined by a comparison of the results from different models.

Software Verification and Validation Plan

The Software Verification and Validation Plan (SVVP) is the strategic planning document that describes the standard for quality assurance over the life cycle of the product.

Verification ensures "that there has been a faithful translation of [each life cycle] phase into the next one." "Validation is...the process of determining the level of conformance between the system requirements and an operational software system under operational conditions."[44]

The example beginning on the next page is the IEEE standard SVVP and should be used as a model.[45] The entire plan is presented, but it may need to be customized to apply to your project.

---

[44]Quirk, ed. *Verification and Validation of Real-Time Software,* 32–34.

[45]"IEEE Standard for Software Verification and Validation Plans," 18.

Sample Software Verification and Validation Plan

IEEE Standard SVVP

1. Purpose—brief definition of project; scope and goals of SVVP
2. Referenced Documents—list of binding compliance documents, historical references, supporting documents
3. Definitions—interpretation of terms, acronyms, notations
4. Verification and Validation (V&V) Overview
   A. Organization—definition of the V&V effort for developers, management, quality assurance, operations, users; definition of lines of communication, authority for problem resolution, authority for product approval
   B. Master Schedule—phases of project life cycle (with dated milestones)
   C. Resources Summary—staffing, facilities, tools, finances, special procedures (e.g., security, access rights, documentation control)
   D. Responsibilities—"who does what," tied to the "when" of the master schedule
   E. Tools, Techniques, Methodologies—summary of use and purpose of ancillary processes and products, together with a plan for acquisition, training, and support (or reference to a V&V Tool Plan)
5. Life Cycle V&V—the linkage between specific tasks and V&V goals. Each phase (plus the management "phase") should be examined as follows:

   - identify methods and procedures for each task and define the evaluation criteria
   - identify inputs (often written) for each task, together with source and format
   - identify outputs (often written) for each task, together with source and format
   - provide task schedule, with dated milestones for initiation and completion, for (input) receipt, and for (output) delivery
   - describe source, use, and availability of resources (e.g., staff, equipment, facilities, schedule, travel, training, tools)
   - identify the risks and assumptions for each task and risk contingencies
   - assign specific task responsibilities to individuals or small groups

Note: The identified phases may be overlapping and iterative.
   The tasks are listed below.
   A. Management of V&V—tasks include SVVP generation, baseline change assessment, management review of V&V, review support.
   B. Concept Phase V&V—tasks include creation and evaluation of concept documentation.
   C. Requirements Phase V&V—tasks include software requirements traceability analysis, software requirements evaluation, software requirements interface analysis, test plan generation (system, acceptance).
   D. Design Phase V&V—tasks include software design traceability analysis, software design evaluation, software design interface analysis, test plan generation (component, integration), test design generation (component, integration, system, acceptance).
   E. Implementation Phase V&V—tasks include source code traceability analysis, source code evaluation, source code interface analysis, source code documentation evaluation, test case generation (component, integration, system, acceptance), test procedure generation (component, integration, system), component test execution.

(Continued)

Sample Software Verification and Validation Plan (Continued)

> F. Test Phase V&V—tasks include acceptance test procedure generation, test execution (integration, system, acceptance).
> G. Installation and Checkout Phase V&V—tasks include installation and configuration audit, final V&V report generation.
> H. Operation and Maintenance Phase V&V—tasks include SVVP revision, anomaly evaluation, proposed change assessment, phase task iteration.
> 6. Software V&V Reporting—tasks include: identification of the content, format, and timing of all V&V reports; task reports, phase summary reports, anomaly reports, final report, special studies report.
> 7. V&V Administrative Procedures
>   A. Anomaly Reporting and Resolution—method for reporting and resolving anomalies: definitions of anomaly, criticality criteria, distribution list, timing of resolutions
>   B. Task Iteration Policy—the extent to which a task will be performed again when its input is changed; assessment of change magnitude, criticality, effect on cost, schedule, and quality.
>   C. Deviation Policy—proper deviation as defined by rationale; effect on quality and authority
>   D. Control Procedures—plan for configuration, protection, storage of product and data
>   E. Standards, Practices, Conventions—internal organizational standards that govern task performance

Code Inspection

Conducting formal code inspections is a part of the overall effort to develop high quality software systems. With proper planning in place and a verification and validation scheme in use, technical risks are reduced and overall quality is improved.

Code inspections are rigorous and formal peer reviews of the software. They are conducted to make sure the code agrees with the requirements and the design and to reduce the number of faults transmitted from one development stage to the next (and thereby reduce the cost of removing faults in later phases or after the product has been released). Another goal is to make sure the software can be enhanced and modified later, if needed. Inspections should *not* be an evaluation of the coder's performance.

During a code inspection, software errors are classified by severity (major or minor) and by class (missing, wrong, or extra). Refer to Part 8 for more details about the mechanics of conducting inspections.

Inspection checklist

The following Sample Code Inspection Checklist identifies some concerns during a code inspection. Though long, the list is not meant to be all-inclusive. This is an example; your own list will need to be customized.[46][47][48]

---

[46]Page-Jones, *The Practical Guide to Structured Systems Design,* 299–301.

[47]Myers, 30–32.

[48]Fagan, Michael. 1976. "Design and Code Inspections to Reduce Errors in Program Development." *IBM Systems Journal* vol. 15, No 3, (July): 182–211.

Sample Code Inspection Checklist

Data Reference and Usage

1. Make sure there are no off-by-one faults in array subscript or indexing.
2. Check input parameters to ensure they are within bounds.
3. Check value of variables to ensure they are within bounds.
4. Define pointers appropriately for the objects they point to.
5. Do type-casting correctly.
6. Use macros properly with the right parameters.
7. Don't use absolute (literal) constants in place of symbolics.
8. Don't use uninitialized variables or unset variables.
9. Don't use uninitialized pointers or unset pointers.
10. Don't use non-integer array subscripts.
11. Eliminate dangling references.
12. Match record and structure attributes.
13. Make sure there are no computation of bit-string addresses or passing of bit-string arguments.
14. Make sure based storage attributes are correct.
15. Make sure string size limits have not been exceeded.

Data Declaration/Definition and Initialization

1. Declare all variables.
2. Match structure definitions across procedures.
3. Make sure variables initialized by declaration are fault-free on restarts.
4. Make global variable definitions consistent across modules.
5. Declare variables in data structures (to conserve memory).
6. Declare indexing variables as short integers (to conserve memory).
7. Make sure sign extension is correct.
8. Make sure lengths, types, and storage classes are declared correctly.
9. Define all constants.
10. Make sure variables have dissimilar names.
11. Make sure arrays and strings are properly initialized. Indexing variables should be properly initialized (0 or 1).
12. Make sure initialization is consistent with storage class.

(Continued)

Sample Code Inspection Checklist (Continued)

General Computation

1. No arithmetic faults.
2. No indeterminate expressions.
3. No computations on non-arithmetic variables.
4. No mixed-mode computations.
5. No mixed computations on variables of different sizes.
6. Target size the same as the size of assigned value.
7. No intermediate result overflow or underflow.
8. No division by zero.
9. Operator precedence understood correctly and properly parenthesized.
10. Correct integer divisions.

Comparison and Booleans

1. No mixed comparisons of variables of different types.
2. Correct and properly parenthesized comparison relationships.
3. Correct and properly parenthesized Boolean expressions.
4. Compiler evaluation for Boolean expressions understood.
5. Correct condition tested ("if x = on" vs. "if x = off").
6. Correct variables used for test ("if x = on" vs. "if y = on").

Iteration

1. Each loop should terminate.
2. Program should terminate.
3. No loop bypasses because of entry conditions.
4. Possible loop fall-throughs should be correct.
5. No off-by-one iteration errors.

Control Flow

1. Every case statement should have a default case.
2. Each case in a case statement should have a break, a return, or a comment that the code is falling through.
3. There should be no unnecessary GOTO statements.
4. IF/THEN/ELSE statements should handle all cases and be exhaustive.
5. The branch most frequently exercised should be the THEN clause.
6. Null THEN or ELSE statements should be included as appropriate.
7. DO/END statements should match. IF/THEN/ELSE statements should match.

Intra- and Inter-Process Interfaces

1. All modifiable parameters should be "call by reference."
2. Global variables should contain the correct data.
3. Function declaration and return value should match.
4. Operating system calls, sends, and receives should be correct.
5. All libraries should be shared.
6. Registers should be saved on entry and restored on exit.
7. The number, attributes, and order of arguments transmitted to called modules should match number, attributes, and order of parameters.
8. The number, attributes, and order of arguments to built-in functions should be correct.
9. There should be no alteration of input-only arguments.
10. No constants passed as arguments.
11. Feature interaction(s) should not be missing.
12. Future feature interactions should not be precluded.

Input/Output

1. File attributes should be correct.
2. OPEN statements should be correct.
3. Format specification should match I/O statement.
4. Buffer size should match record size.
5. Files should be opened before use and closed after use.
6. End-of-file conditions should be handled.
7. I/O errors should be handled.
8. Output information should contain no textual errors.

Project Specification/Design

1. The code should achieve what is specified in the specification document and what is specified in the design document.
2. There should be no missing or incorrect specification(s) in the documentation against which the code is being verified.
3. There should be no architecture violations.

Achievement of Design Goals

1. Code should meet all design goals (e.g., should be functional, accurate, reliable, portable, maintainable, efficient, cost-effective, reusable).

Commentary

1. Comments should be meaningful to someone other than the author.
2. Comments should comply with the standards for commenting code.

Coding Standards

1. Code should comply with coding standards.

System Resource

1. Memory usage should not be excessive.
2. Resources should be allocated and freed correctly.

Post-Compilation Checks

1. There should be no unreferenced variables in the cross-reference listing.
2. Attribute list should be as expected.
3. There should be no post-compilation warning or informational messages.
4. There should be no missing functions.
5. There should be no misplaced punctuation (e.g., semicolons, parentheses, brackets, braces).

Inspecting rewritten code

If the code to be inspected has been rewritten to fix a problem, several other questions must be answered. See example below.

Checklist for Rewritten Code

1. If a global variable has been changed or deleted, all functions using that variable should be modified.
2. Additions, changes, or deletions of parameters should be reflected in calling modules.
3. If modifications to an inter-process message structure increase the size of the structure, the size should be within the message size limit.
4. The solution should not break another feature.
5. The solution should not preclude adding feature interactions if needed.
6. The solution should not adversely affect system performance.

Software System Test Plans

An important part of the SVVP is the collection of system test plans that address individual software features. These test plans are the tactical implementation of the SVVP strategy. Two examples are given on the next few pages.

- Sample 1 shows the concerns that should be addressed in a system test plan and a convenient format for their discussion.
- Sample 2 demonstrates some ways in which a system test plan may fall short.

An analysis of the strong and weak points of each test plan follows the samples.

Sample 1: A Good System Test Plan

I. Name and Description of Feature to be Tested

BigSys 'date' command

Given a date as input from the user (valid format: mm/dd/yy), display the day of the week on the user's terminal.

II. Responsible Tester

Connie Conrad of the BigSys System Test Group (QQQ Department), Maintown.

III. Test Objective(s)

Primary Objectives

Basic Functionality—meets facility specifications according to BigSys Detailed Design Document, p. 16.

Performance—meets performance requirements according to BigSys Requirements Document, pp. 20–21.

Fault Tolerance—meets error recovery specifications according to BigSys Requirements Document, p. 11, and BigSys Preliminary Design Document, p. 19.

Secondary Objectives—to be done in parallel with tests to meet primary objectives.

Documentation—actual system responses should match user document, BigSys User Guide.

IV. Referenced Materials

BigSys Requirements Document
BigSys Preliminary Design Document
BigSys Detailed Design Document
BigSys User Guide

V. Test Environment

Hardware: DEC VAX 8530, AT&T 615 CRT

Software: BigSys Version III, Release 2.0

Memory: 16Mb

Storage media: ten RPO-5 drives

VI. Method of Test Invocation and Evaluation, Result Documentation

Testing is partially automated: command invocation is automated via the T_SCRIPT tool (see Test Tools), input/output capture is automated via T_SCRIPT, evaluation is manual, and documentation is semi-automated via the DOC_IT tool (see Test Tools).

VII. Test Tools

The T_SCRIPT program will be used to invoke the command under test and to capture the input and output. On completion (pass or fail), on-line test results will be manually moved to files that are named according to the project guidelines and kept there until no longer needed.

The DOC_IT program will be used to generate end-of-the-day reports about the progress of testing, including the success rate, the current status of bug fixes, and the number and results of the day's tested cases.

VIII. Test Cases

Test cases, script files, and expected output files are:

Test Case	T_SCRIPT Script	Test Output
1	scr.date.1	out.date.1
2	scr.date.2	out.date.2
3	scr.date.3	out.date.3
4	scr.date.4	out.date.4
5	scr.date.5	out.date.5
6	scr.date.6	out.date.6
7	scr.date.7	out.date.7
8	scr.date.8	out.date.8

All scripts and test outputs are located on the alfa3 machine. The precondition for all test cases is that BigSys is operating in a non-degraded mode.

Test Case	Test Objective	Input	Expected Output	Comments
1	function	date 02/29/88	Monday	leap year
2	function	date 02/01/89	Wednesday	
3	fault	date 2/1/89	invalid input format	
4	fault	date 02/29/89	invalid date	not a leap year
5	fault	date 13/13/89	invalid date	
6	fault	date 02/89	invalid input format	
7	perform	date 02/01/89	Wednesday	max. time: 4 sec.
8	perform	date 02/89	invalid input format	max. time: 4 sec.

(Continued)

Sample 1: A Good System Test Plan (Continued)

IX. Test Case Failure Procedure

If a test case fails, the procedure is as stated in the SVVP; it is included here for completeness.

Test case failure will be summarized to first-line management on a daily basis. Each failed test case will be documented via the opening of an on-line Modification Request (MR) within one day of failure. The MR system will automatically print a copy of the MR for the developer, the developer's supervisor, and the tester. When the problem is resolved, the test case will be retested; if the tester deems the problem fixed, then the MR will be closed.

Sample 2: A Poor Test Plan

I. Name and Description of Feature to be Tested

BigSys family of report-generation commands that create reports automatically.

II. Responsible Tester

Harry Harrier.

III. Test Objective

Stress at designed peak; system should be easy to use.

IV. Referenced Materials

All BigSys Design documents.

V. Test Environment

BigSys on a DEC VAX.

VI. Method of Test Invocation and Evaluation, Result Documentation

Testers will log into system and pretend to be users, invoking the report generating programs.

VII. Test Tools

None used.

VIII. Test Cases

1. List the available reports.
2. Request reports for which the user is not authorized.
3. Generate sales report—New York, USA, International.
4. Generate sales report and spool for later printing.
5. Generate sales report and spool for later printing, then cancel.

IX. Test Case Failure Procedure

For every detected error, tester will write a memo to the developer.

Analysis of sample test plans

Sample 1 and Sample 2 are distinguishable primarily by the degree of preciseness: the good test plan is very explicit, stating assumptions and goals with clarity; the poor test plan lacks detail and is vague. That imprecision is apparent in every section of Sample 2 as discussed in the comparison below:

  I. **Name and Description of Feature to be Tested.**   The specific ("date") is preferable to the vague "family of commands" because it is exact. All readers need a definitive list of commands to be tested, be they document reviewers, the system test manager, other members of the test group, or the person performing the test.

   A brief description should provide a simple overview, mentioning the general nature of input, output, and the method of invocation (manual or automated).

 II. **Responsible Tester.**   The tester should be named; the organization and location should be identified.

III. **Test Objective.**   Every test must have one or more goals. Sample 1 has three primary goals; its secondary goal is ancillary and is accomplished in parallel with the primary ones. Sample 2 lacks clear goals.

   A test objective should identify a test category and include a statement about the quantifiable goal (or a reference to the specifying document).

   The following are some examples of test categories.

Examples of Test Categories:

- basic functionality (i.e., facility)
- data overload tolerance/performance (i.e., volume)
- user overload tolerance/performance (i.e., stress)
- judgment of human interface (i.e., usability)
- security
- performance
- data and program storage
- configuration
- compatibility with other systems
- conversion from a previous release
- coupling (i.e., installability)
- reliability
- fault tolerance (i.e., error recovery)
- maintainability (i.e., serviceability)
- documentation
- ancillary (support) procedures

 IV. **Referenced Materials.** All sources should be specifically cited.

  V. **Test Environment.** All hardware and software should be specifically listed.

VI. Method of Test Invocation and Evaluation, Result Documentation. A choice of automated, semi-automated (that is, with tester intervention), or manual should be stated for each: invocation, evaluation, and documentation.

VII. Test Tools. Any automation (identified in Method of Test Invocation and Evaluation, Result Documentation) should be specifically listed.

This section should contain:

- list of auxiliary software and hardware required for testing
- description of test equipment and "capture" method for input and output
- notes on procedures

VIII. Test Cases. Generality in this section guarantees an unsuccessful test program. Sample 2 suffers from extreme imprecision, raising many questions:

- Test Case 1: Does it mean "execute the 'list' command to view the menu of available reports" or "review the documentation to familiarize yourself with the options" or something else? What are the expected responses?
- Test Case 2: How is authorization set up? Verified?
- Test Case 3: Does it mean "generate three sales reports, one of each?" What are the expected responses? How much time will be spent on the test? Will the testers be concurrently logged on? Might automated or manual scripts be used? What features of the report generators will be exercised? What are the expected results?
- Test Case 4: To which sales report does this refer? How does the user "spool"?
- Test Case 5: How does the user "cancel"?

IX. Test Case Failure Procedure. This section should contain:

- reference to management information and escalation
- time frame for evaluation of failure and reporting to management
- tracking process for bug fixes

Summary

Software is becoming an ever-increasing part of our military systems. This growth in software use has resulted in both increased system costs and increased system complexity. The two keys to successful software development are: careful planning for each step in the process; and verifying and validating the results of each step. (See Figure 23.) The actual writing of code is only one small aspect of software development. Thus, software design becomes the process of planning for quality before, during, and after the writing of code. Software test becomes the process of ensuring that a quality product is achieved.

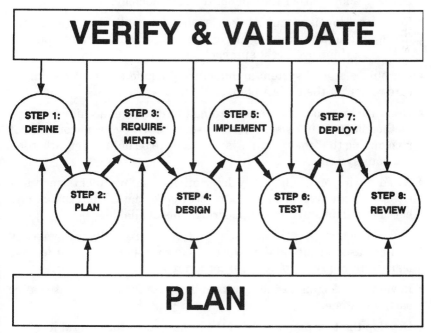

Figure 23 Software engineering process.

Over the past decades, software engineers have shifted away from fixing software failures after the system has been deployed to developing proven methodologies and techniques that ensure software quality from the start. These efforts have also yielded increased productivity and better adherence to schedules.

The Procedures chapter describes an eight-step process for developing a software system. Two concepts—Verification and Validation, and Planning—are of critical importance to system development. They influence every step, and together they determine the success or failure of the project.

Things to remember

In following best practices, experienced software engineers highly recommend the following:

- Plan carefully and completely with attention to details. There is always a temptation to start development of the software before the project is really understood.

- Don't be short-sighted when planning the feasibility report. A feasibility report for a large project requires time and effort.

- Choose appropriate and well-proven methodologies carefully. The key is to adapt rather than adopt the methodology.

- Consider "downstream" phases such as installation, operations, maintenance, and training in your plans, schedules, and budgets. Take a total life cycle cost approach.

- Formulate a good system architecture. Include a detailed architecture review in the project plan.

- Organize human resources according to the system architecture and methodology. Otherwise, you risk using the wrong people or risk not recognizing that the current system needs to be built differently than previous systems.

- Avoid relying on English-only descriptions of requirements and functions. Use graphs and diagrams where appropriate. There are a number of excellent software engineering tools available.

- Make sure that the "throw-away" prototypes are really thrown away and not used in the final product. Make sure, however, that lessons learned from prototyping are incorporated.

- Review and document the results and update the project status after each milestone.

- Develop the test plans early in the project. Plan for several levels of testing and avoid a "big bang" approach.

- Use good documentation.
- Use good configuration control.

Summary of procedures

Good quality software is the result of careful attention to the quality of the planning and the standards for that software, as well as attention to the quality of the code itself. This part has presented the step-by-step procedures for implementing the principles and templates for good quality software. Table 7 is a summary of those steps.

TABLE 7   Summary of the Software Procedures

Step 1—Define the Concept

Procedure	Supporting activities
Identify a Need for Product	■ Review need for product ■ Refer to Part 1 for methods
Assess User Needs	■ Define user base via user surveys and interviews ■ Observe users and their tasks ■ Analyze user needs data
Analyze Feasibility	■ Assess risks ■ Assess complexity ■ Estimate required resources ■ Determine constraints ■ Define and use feasibility criteria to analyze "go" or "no-go" ■ Produce Feasibility Report

Step 2—Plan the Project

Procedure	Supporting activities
Form the Project Team	■ Assess personnel needs for project design, development, review, inspection, test, deployment, maintenance ■ Assess skill levels and training needs of potential team members ■ Choose and train competent teams
Define the Cost Estimation Plan	■ Identify those cost-affecting factors that are appropriate to the project ■ Select one or more costing techniques that emphasize those factors ■ Document the selection and the criteria

TABLE 7   Summary of the Software Procedures (Continued)

Step 2—Plan the Project (*continued*)

Procedure	Supporting activities
Define the Life Cycle Model	■ Compare available life cycle models to identify structures and concerns that are appropriate to the project ■ Select a model ■ Define entry and exit criteria for the stages of the model ■ Document the selection and the criteria
Define the Documentation Requirements	■ Analyze project and life cycle model for information needs ■ Identify type and content of documents ■ Define criteria for the acceptability of the documents ■ Document the selection and the criteria (part of draft SVVP)
Define Coding Standards	■ Use corporate coding guidelines ■ Define criteria for the acceptability of the code ■ Document the standards (part of draft SVVP)
Define Testing Standards	■ Define entry and exit criteria for the acceptability of the modules ■ Document the standards (part of draft SVVP)
Define the Method for Reviews, Inspections, Walkthroughs	■ Analyze project for quality requirements ■ Select a method ■ Define entry and exit criteria for reviews ■ Document the selection
Define the Change Control Scheme	■ Analyze alternatives and select one to use ■ Document the selection (part of draft SVVP)
Define the Project Metrics	■ Assess project and corporate measurement needs ■ Use corporate measurement guidelines ■ Select appropriate metrics ■ Document the selection
Conduct SVVP Review	■ Review draft SVVP Plan, which contains documentation requirements, coding and testing standards, and change control scheme ■ Produce SVVP Plan
Choose Software Tools	■ Analyze alternatives available, for example, CASE development products

Step 3—Develop the Software Requirements	
Procedure	Supporting activities
Develop System Requirements	■ Describe the current environment ■ Define the scope of the new system ■ Develop I/O requirements ■ Define the system's functions ■ Catalog user requirements and constraints ■ List requirements for performance, operations, testing, and maintenance ■ Specify external controls ■ Produce draft Requirements Document
Conduct Requirements Review	■ Review draft Requirements Document ■ Produce final Requirements Document ■ Follow change control guidelines for revisions
Step 4—Design the Software System	
Perform Preliminary Design	■ Define the system architectures—overall, hardware, software ■ Define the user interface ■ Choose the physical implementation ■ Perform ongoing functional decomposition ■ Perform data modeling ■ Identify test tactics and strategies ■ Produce draft Preliminary Design Document
Conduct Preliminary Design Review	■ Refer to Part 8 for methods ■ Review draft Preliminary Design Document for fidelity to requirements, rationale, assumptions, and goal attainment ■ Produce final Preliminary Design Document ■ Follow change control guidelines for revisions
Perform Detailed Design	■ Choose design tool ■ Develop implementation details: specifics of system, process, and data ■ Produce draft Detailed Design Document
Conduct Critical Design Review	■ Refer to Part 8 for methods ■ Review draft Detailed Design Document for fidelity to requirements, rationale, assumptions, and goal attainment ■ Produce final Preliminary Design Document ■ Follow change control guidelines for revisions

TABLE 7   Summary of the Software Procedures (Continued)

Procedure	Supporting activities
Step 4—Design the Software System (*continued*)	
Review and Revise Cost and Scheduling Estimates and Metrics	▪ Collect and analyze data ▪ Publish any revisions
Step 5—Implement the Software System	
Produce Code	▪ Follow design ▪ Attain requirements goals ▪ Adhere to coding standards ▪ Follow change control guidelines for revisions
Conduct Code Inspections	▪ Inspect, as per review guidelines ▪ Document the results
Develop Unit Test Plans	▪ Document overall test strategy, test cases, entry and exit criteria, scheduling
Review and Revise Cost and Scheduling Estimates and Metrics	▪ Collect and analyze data ▪ Publish any revisions
Step 6—Test the Software	
Perform Unit Tests	▪ Test, as per Unit Test Plans (as per SVVP) ▪ Publish test results
Perform Integration Test	▪ Combine components incrementally until the system is built ▪ Test the system and its interfaces, as per Integration Test Plan ▪ Publish test results
Perform System Test	▪ Test, as per SVVP, for adherence to requirements, functionality, and other factors ▪ Conduct regression testing ▪ Publish test results
Review and Revise Cost and Scheduling Estimates and Metrics	▪ Collect and analyze data ▪ Publish any revisions
Step 7—Deploy and Maintain the Software System	
Train the Users	▪ Provide user documentation ▪ Provide instruction in system use ▪ Obtain feedback on training from users

Procedure	Supporting activities

| Step 7—Deploy and Maintain the Software System (*continued*) | |

Deploy the System	■ Replace older system
	■ Use change control and configuration management to track revisions, version control by site, error reports, fixes, releases, documentation updates
Perform Acceptance Test	■ Perform user test, as per SVVP
	■ Negotiate any needed redevelopment
Generate Periodic System Performance Reports	■ Monitor operational trends: user activity, transaction time, throughput rate, cycle time, queuing time, system access time, data access time, error rates
	■ Publish ongoing reports
Perform Ongoing Maintenance	■ Correct problems
	■ Develop enhancements
	■ Adapt software to evolving environment
Revise Cost and Scheduling Estimates and Metrics	■ Collect and analyze data
	■ Publish any revisions

| Step 8—Review for Improvement | |

Perform Product Postmortem	■ Compare the deployed product to design objectives
	■ Analyze the deployed system for correctness, reliability, usability, flexibility, maintainability, efficiency, integrity, testability, portability, reusability
	■ Publish the results of the analysis
Perform Process Postmortem	■ Compare the development process to planning objectives
	■ Analyze the development process for management techniques, responsibilities, methodology, adherence to standards, experience, training, tool and technology use, communication, delivery dates, risk assessment, cost estimates
	■ Publish the results of the analysis

AT&T. *Transition from Development to Production.* Department of Defense (DoD 4245.7-M), September 1985. Techniques for avoiding technical risks in 47 key areas or templates including funding, design, test, production, facilities, logistics, management, and transition plan. Identifies critical engineering processes and controls for the design, test, and production of low risk products.

Bergland, G. D., "A Guided Tour of Program Methodologies," *IEEE Computer,* October 1981. Discusses four software design methods: structured analysis, functional decomposition, data flow design, and data structure design.

Bersoff, Edward H.; Henderson, Vilas D.; and Siegel, Stanley G., *Configuration Management: An Investment in Product Integrity.* Englewood Cliffs, N.J.: Prentice-Hall, 1980. Discusses the principles of configuration control, auditing, and status accounting in software development projects.

Best Practices: How to Avoid Surprises in the World's Most Complicated Technical Process. Department of the Navy (NAVSO P-6071), March 1986. Discusses how to avoid traps and risks by implementing best practices for 47 areas or templates that include topics in design, test, production, facilities, and management. These templates give program managers and contractors an overview of the key issues and best practices to improve the acquisition life cycle.

Boar, Bernard H., *Application Prototyping: A Requirements Definition Strategy for the 80s.* New York: John Wiley & Sons, 1984. Discusses software prototyping as a method for defining the user requirements for a system.

Boehm, Barry W., "A Spiral Model of Software Development and Enhancement," *Computer,* May 1988. Discusses the risk driven spiral model for software development, its focus on early prototyping and simulation, current usage of the model, and advantages of this process model over other software models available.

Boehm, Barry W., "Software Engineering," *IEEE Transactions on Computers, December 1976.* Provides a survey of software engineering technology in several of the life cycle phases and discusses software trends.

Boehm, Barry W., *Software Engineering Economics.* Englewood Cliffs, N.J.: Prentice-Hall, 1981. Gives an in-depth explanation of the COnstructive COst MOdel (COCOMO) for estimating resources, staffing levels and costs of a software development project. Discusses the impact of numerous cost drivers associated with a project such as: product attributes, computer attributes, and personnel attributes.

Boehm, Barry W. and Papaccio, Philip N., "Understanding and Controlling Software Costs," *IEEE Transactions on Software Engineering,* October 1988. Discusses key issues in estimating software development costs.

Booch, Grady. "Object-Oriented Development," *IEEE Transactions on Software Engineering,* February 1988. Discusses key issues in object-oriented development.

Brooks, Frederick P., "No Silver Bullet," *IEEE Computer,* April 1987. Discusses the nature of software development projects and why there are no easy ways (silver bullets) for making large improvements in productivity, reliability and simplicity.

Brooks, Frederick P., *The Mythical Man-Month: Essays on Software Engineering.* Reading, Massachusetts: Addison-Wesley Publishing Company, 1975. Discusses the difficulties of managing large software projects and other software management issues.

Buckley, Fletcher J., *Implementing Software Engineering Practices.* New York: John Wiley & Sons, 1989. Provides guidance and examples for establishing software engineering standards and practices within an organization. Discusses current IEEE and DoD standards.

Cortese, Amy, "Estimating Tools reap 85% Accuracy, Some Say," *Computerworld,* November 14, 1988. Brief discussion of current state of software costs estimation tools.

DeMarco, Tom, *Structured Analysis and System Specification.* New York: Yourdon, 1978. Provides details of structured software design techniques.

DeMillo, Richard A.; McCracken, W. Michael; Martin, R. J.; and Passafiume, John F., *Software Testing and Evaluation.* Menlo Park, California: The Benjamin/Cummings Publishing Company, 1987. Provides a discussion of software testing techniques and an overview of current defense practices gathered from interviews and surveys with military and industrial personnel. Also includes data sheets on numerous testing tools.

Deutsch, Michael S., *Software Verification and Validation: Realistic Project Approaches.* Englewood Cliffs, N.J.: Prentice-Hall, 1982. Discusses software testing methodologies and the use of verification and validation techniques throughout the entire software life cycle.

Fairley, Richard E., *Software Engineering Concepts.* New York: McGraw-Hill Book Company, 1985. Provides a comprehensive description of the entire software development life cycle from concept definition through maintenance. Also includes many good examples of software engineering concepts. Very good section about planning the software development effort.

"IEEE Standard for Software Unit Testing," *ANSI/IEEE Std. 1008-1987,* December 29, 1986. Discusses planning for, implementing and measuring software unit testing.

IEEE Standard for Software Verification and Validation Plans, Std. 1012-1986, November 14, 1986. Provides a complete guide for the development of a standard verification and validation plan throughout the life cycle of a software development project.

Kemerer, Chris F., "An Empirical Validation of Software Cost Estimation Models," *Communications of the ACM,* May 1987. Provides an evaluation of four popular software cost estimation models: SLIM, COCOMO, Function Points and ESTIMACS.

Kirk, Frank G., *Total System Development for Information Systems.* New York: John Wiley & Sons, 1973. Discusses a disciplined method for developing a software system. Includes an activities network which is used as the framework for describing the total system development effort.

"Managing Software Development: Solving the Productivity Puzzle," Course Moderator's Guide, John Wiley & Sons, Inc., 1985. Provides good overview of key points in the software development process with emphasis on the management of a software project.

Meyer, Bertrand, *Object-Oriented Software Construction,* Englewood Cliffs, NJ: Prentice Hall, 1988. Describes how object-oriented software is developed.

Musa, John D.; Iannino, Anthony; and Okumoto, Kazuhira, *Software Reliability: Measurement, Prediction, Application.* New York: McGraw-Hill, 1987. Describes the basics of software reliability measurement, provides the procedures and formulas for applying the measurements, and also details the theoretical background of software reliability.

Myers, Glenford J., *The Art of Software Testing.* New York: John Wiley & Sons, 1979. Describes the various types of software testing and the tools and techniques commonly used. Also provides discussions of test case design, software debugging and program inspections, walkthroughs, and reviews.

Myers, Glenford J., *Software Reliability: Principles & Practices.* New York: John Wiley & Sons, 1976. Describes what needs to be done during each of the various stages of software development to produce reliable software.

Perry, William E., *A Structured Approach to Systems Testing, Second Edition.* Wellesley, Massachusetts: QED Information Sciences, 1988. Comprehensive discussion of software testing throughout the entire life cycle. Includes test, documentation, review and audit checklists and examples.

Peters, Lawrence J., *Software Design: Methods & Technologies*. New York: Yourdon Press, 1981. Discusses the role of and issues of software design. Includes descriptions of tools and techniques used in software design.

Quirk, W. J., ed., *Verification and Validation of Real-time Software*. New York: Springer-Verlag, 1985. Provides an array of articles concerning software development methods and standards.

Smith, David J. and Wood, Kenneth B., *Engineering Quality Software: A Review of Current Practices, Standards and Guidelines including New Methods and Development Tools*. London: Elsevier Applied Science, 1987. Discusses software standards and guidelines in several different countries. Provides checklists for software design, design reviews, programming standards, testing, change control, documentation, and project management.

Vincent, James; Waters, Albert; and Sinclair, John, *Software Quality Assurance: Volume I, Practice and Implementation*. Englewood Cliffs, N.J.: Prentice-Hall, 1988. Discusses the factors of software quality, software metrics and software audits. Provides numerous checklists, forms, and algorithms for assessing and measuring the appropriate factors.

Wallace, Dolores R. and Fujii, Roger U., "Software Verification and Validation: An Overview," *IEEE Software*, May 1989. Discusses methods and standards which support the use of verification and validation throughout the life cycle of a software project.

Wood, Bill; Pethia, Richard; Gold, Lauren Roberts; and Firth, Robert, "A Guide to the Assessment of Software Development Methods," *Technical Report* CMU/SEI-88-TR-8 ESD-TR-88-009, Carnegie-Mellon University Software Engineering Institute, April 1988. Discusses software methods and ways to determine if the methods satisfy a software development organization's needs.

Computer-Assisted Technology

1

Introduction

To the Reader

This part, Computer-Assisted Technology, includes two Transition from Development to Production templates: Computer-Aided Design and Computer-Aided Manufacturing.

The templates, which reflect engineering fundamentals as well as industry and government experience, were first proposed in the early 1980s by a Defense Science Board task force of industry and government leaders, chaired by Willis J. Willoughby, Jr. The task force sought to improve the effectiveness of the transition from development to production. The task force concluded that most program failures were due to a lack of understanding of the engineering and manufacturing disciplines used in the acquisition process. The task force then focused on identifying engineering processes and control methods that minimize technical risks in both government and industry. It defined these critical events in design, test, and production in terms of templates.

The template methodology and documents

A template specifies:

- areas of high technical risk
- fundamental engineering principles and proven procedures to reduce the technical risks

Like a classical mechanical template, these templates identify critical measures and standards. Use of the templates makes it likely that engineering disciplines will be followed.

The task force documented 47 templates and in 1985 the templates were published in the DoD *Transition from Development to Production* (DoD 4245.7-M) manual.[1] The templates primarily cover design, test, production, management, facilities, and logistics. In 1989, the Department of Defense added a 48th template on Total Quality Management (TQM).

In 1986, the Department of the Navy issued the *Best Practices* (NAVSO P-6071) manual,[2] which illuminated DoD practices that compound problems and increase risks. For each template, this manual describes:

- potential traps and practices that increase the technical risks
- consequences of failing to reduce the technical risks
- an overview of best practices to reduce the technical risks

The intent of the *Best Practices* manual is to help practitioners become aware of the traps and pitfalls so they do not repeat them.

The templates are the foundation for current educational efforts

In 1988, the government initiated an educational program, "Templates: Professionalizing the Acquisition Work Force," with courses and books, such as this one, to increase awareness and improve the implementation of the template concept.

The key to improving the DoD's acquisition process is to recognize that it is an industrial process, not an administrative process. This change in perspective implies a change in the skills and technical knowledge of the acquisition work force in government and industry. Many in this work force do not have an engineering background. Those with an engineering background often do not have broad experience in design, test, or production. The work force must understand basic design, test, and production processes and associated technical risks. The basis for this understanding should be the templates since they highlight the critical areas of technical risk.

The template educational program meets these needs. The program consists of a series of courses and technical books. The parts provide background information for the templates. Each part covers one or more closely related templates.

---

[1]Department of Defense. *Transition from Development to Production.* DoD 4245.7-M, September 1985.

[2]Department of the Navy. *Best Practices: How to Avoid Surprises in the World's Most Complicated Technical Process.* NAVSO P-6071, March 1986.

How the parts relate to the templates. The parts describe:

- the templates, within the context of the overall acquisition process
- risks for each included template
- best commercial practices currently used to reduce the risks
- examples of how these best practices are applied

The parts do not discuss government regulations, standards, and specifications, because these topics are well-covered in other documents. Instead, each part stresses the technical disciplines and processes required for success.

Clustering several templates in one book makes sense when their best practices are closely related. For example, the best practices for the templates in this part interrelate and occur iteratively within design and manufacturing. Designers, suppliers, and manufacturers all have important roles. Other templates, such as Design Reviews, relate to many other templates and thus are best dealt within individual parts.

Courses on the templates. The books are designed to be used either in courses or as stand-alone documents. The courses use lectures and other proven instructional techniques such as videotapes, case studies, group exercises, and action plans.

Government and industry program managers should recognize that adherence to engineering discipline is more critical to reducing technical risk than blind obedience to government standards and administrative rules. They should especially recognize when their actions (or inactions) increase technical risks as well as when their actions reduce technical risks.

The templates are a model

The templates defined in DoD 4245.7-M are not the final word on disciplined engineering practices or reducing technical risks. Instead, the templates are a reference and a model that engineers and managers can apply to their industrial processes. Companies should look for high-risk technical areas by examining their past projects, by consulting their experienced engineers, and by considering industry-wide issues. The result of these efforts should be a list of areas of technical risk which becomes the company's own version of the DoD 4245.7-M and NAVSO P-6071 documents. Companies should tailor the best practices and engineering principles described in the books to suit their

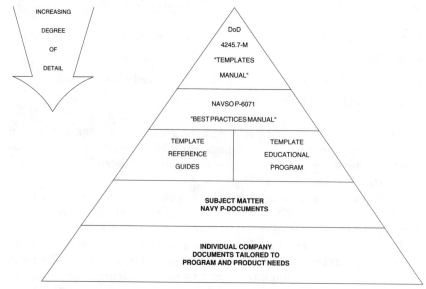

Figure 1 Resources on the acquisition process templates.

particular needs. Several military suppliers have already produced manuals tailored to their processes.

Figure 1 shows where to find more and more details about risks, best practices, and engineering principles. Participants in the acquisition process should have copies of these documents.

This part deals with the application of some of the most powerful tools ever created for the development process. When properly applied, the benefits of computer-assisted technology can greatly increase over-all productivity and quality within the transition from development to production.

Since the investment in computer-assisted technology is large, common-sense planning and foresight are required. Computers are extremely powerful and useful, but they can magnify problems if not properly used.

What Is a Computer-Aided Tool?

A computer-aided tool is a combination of computer hardware and software that assists the user in accomplishing a task. Computer-aided design/computer-aided manufacturing, CAD/CAM, is the name often given to computer-aided tools that assist in the design and manufacturing processes. Computer-aided tools exist for virtually every task in the design and manufacturing process.

History

The modern history of computer-aided design and manufacturing tools began in the 1950s with a series of events at the Massachusetts Institute of Technology (MIT). The first event led to the development of CAD tools. In 1950, a computer-driven display was used to generate simple pictures. This was followed by a system called SKETCHPAD demonstrated in 1963 at the Lincoln Laboratory of MIT. The system, consisting of a cathode-ray tube (CRT) driven by a computer, allowed a user to draw pictures on the CRT screen and manipulate the pictures interactively.[3]

The first CAD systems were basically expensive automated drafting stations. As the cost of computer hardware has decreased and the sophistication of computer software has increased, CAD tools have been used for engineering design and analysis as well as drafting.

The history of CAM began with U. S. Air Force-sponsored research at MIT that led to development of numerical control (NC) machines. A prototype NC machine was demonstrated in 1952. Further research on programming languages for NC machines resulted in the language APT, which stands for Automatically Programmed Tools. APT continues in widespread use today. Although computer-aided manufacturing began as NC, today CAM includes much more, including computer-assisted resource planning and process planning. The most recent developments include the integration of the design and manufacturing functions to form computer-integrated manufacturing (CIM) systems.

Background

In 1980, the use of computer-aided tools was not as widespread or effective as would be expected from a thirty-year-old technology. The cost of computers was high relative to their capability, and this limited the number of workstations available to the engineers in a company. Computers were difficult to use, and this limited their acceptance among engineers. Also, computers did not often communicate among themselves, and this limited their effectiveness. These difficulties often prevented a firm from realizing the potential of computer-aided tools and sometimes caused companies to reject further use of computer-aided tools.

---

[3]Besant, C. B. and Lui, C. W. K. *Computer-Aided Design and Manufacture*. Chichester, England: Ellis Horwood, 1986, pp. 13-15.

Since the early 1980s, the computer has become much more prevalent in industry. Computer-assisted technology has advanced at an incredible pace. Power, performance, and ease of use make it possible for nearly every organization to benefit.

Figure 2[4] reflects how prevalent the computer has become. Projections show that by 1995, the installed base of CAD systems in the United States will be nearly *fifty times* that of 1981.

The computer has become a mandatory tool in today's competitive market. Therefore, a sense of urgency to avoid obsolescence may magnify risks. The templates for Computer-Aided Design and Computer-Aided Manufacturing and this part can help companies realize the benefits of this powerful technology and reduce the technical risks involved.

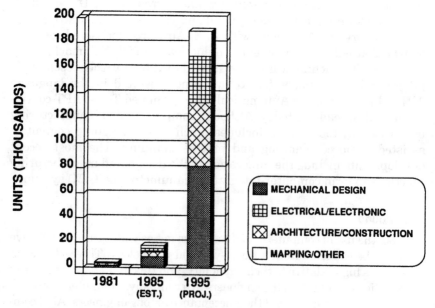

Figure 2 Projected growth of the U.S. CAD base.

[4]U.S. Department of Commerce, International Trade Administration. *A Competitive Assessment of the U.S. Computer-Aided Design and Manufacturing Systems Industry,* February, 1987, p. 21.

What Is Computer-Assisted Technology?

Computer-assisted technology is a way of design and manufacture that integrates the strengths of humans with the strengths of computers to produce a flexible system that responds rapidly to customer requirements.

Typically, the strengths of humans include the ability to:

- analyze new situations
- formulate solutions
- analyze results as reasonable
- interact with other people

Computers are usually better at:

- doing repetitive tasks rapidly
- storing and retrieving information
- displaying results in graphics
- tracking the details of complex processes

If computer-assisted technology is managed and used properly, the benefits to companies and customers can be enormous.

Computer-aided acronyms

Different acronyms refer to various segments of computer-assisted technology. Among the acronyms are:

- CAE: computer-aided engineering
- CAD: computer-aided design
- CAM: computer-aided manufacturing
- CIM: computer-integrated manufacturing
- CASE: computer-aided software engineering

Figure 3 shows how the various segments of computer-assisted technology interact. The figure shows the design and manufacturing process, starting with an idea for a new product and ending with the manufactured product.

The term CAE usually refers to the early part of the design process where features are added to a product to see if the product will meet customer requirements. As the design progresses, CAD tools become more prominent. CAD refers to the physical, i.e., hardware, design of a

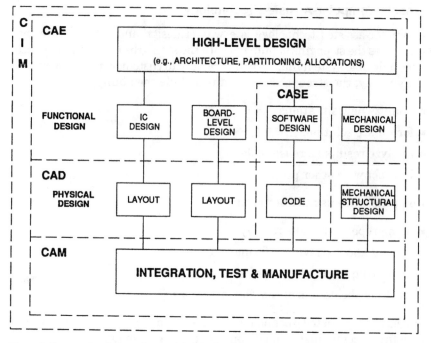

Figure 3 Segments of computer-assisted technology.

product. CAM is concerned with making a product that can be shipped to a customer. CIM refers to the overall integration of the design and manufacturing process. CASE is the equivalent of CIM for software.

CAD/CAE technology can significantly reduce risk in any development effort. CAD systems help design engineers lay out products, select components and materials, and develop a description of the product for use in other phases of the development process. CAE systems generally help analyze the product under specific conditions.

CAM is more than just computer-controlled factory equipment. When integrated with CAD/CAE and other tools, CAM can increase productivity and ease the transition into manufacturing.

CASE requires that specifications for corporate plans, software designs, and software development become fully integrated. CASE replaces paper and pencil with a set of well-integrated computer tools that automate the software development process.

However, computer-assisted technology is more than just automating the design and manufacture of a product. It usually involves reengineering existing processes to take advantage of available technology. Computer-assisted technology also involves continuous evaluation and improvement of the design and manufacturing process, resulting in a continuously improved product. Corporate design policy should require that CAD/CAE be integrated to benefit the entire organization.

Opportunities for Computer-Aided Tools

Historically, hard work and productivity increases were the only things required to keep ahead of the competition. Computer-assisted technology is a means of leveraging capabilities against the competition. It is not the whole answer, but it is a significant part of the solution.

Market pressures

While the complexity of systems has increased at a high rate, useful life has decreased. These trends force companies to use new methods to design and produce systems. The semiconductor industry is an example of an industry under pressure to market products quickly with more functions. Figure 4 shows how semiconductor chip complexity has increased since 1960.[5]

Market pressures and development time interact. If development takes too long, the product will be obsolete when it enters the marketplace. The company's profits will quickly erode.

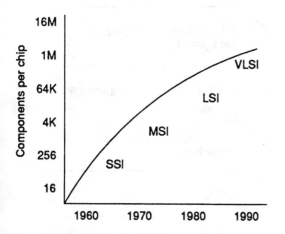

Figure 4 Chip complexity doubles every two years.

---

[5]Ohr, Stephen A. *CAE: A Survey of Standards, Trends, and Tools.* New York: John Wiley and Sons, 1990, p. 5.

Computer-assisted technology offers hope to companies under these market pressures. A company can reconcile demands for more complexity, faster development, and better quality by using computer-assisted technology that is properly integrated with its design and manufacturing processes. Proper integration of tools, including a database with a common description of the product, allows for smoother information flow among design, analysis, and manufacturing and reduced duplication of effort.

In a recent survey, the design and engineering community was polled on attitudes towards design automation. The respondents included companies ranging in size from less than $20 million to larger than $1 billion.

Figure 5 reflects some of the common justifications for implementing computer-assisted technology.[6]

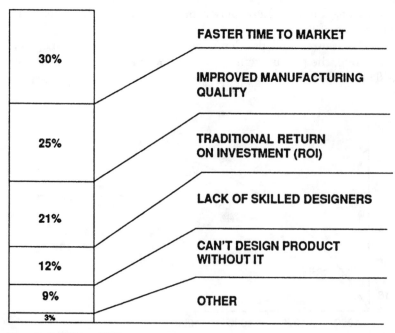

Figure 5 Justifying CAD.

[6]Krause, Irvin and Suchors, Cheryl R. "Design Automation: A Strategic Necessity." *Electronic Business,* vol. 13(8), April 15, 1987, pp. 121-132.

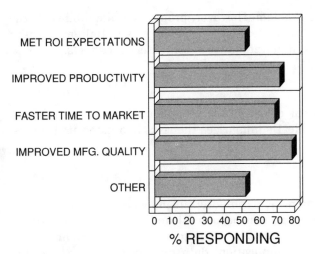

MET ROI EXPECTATIONS

IMPROVED PRODUCTIVITY

FASTER TIME TO MARKET

IMPROVED MFG. QUALITY

OTHER

0 10 20 30 40 50 60 70 80

% RESPONDING

Figure 6 Results from implementing design automation.

Figure 6 reflects feedback from the survey illustrating that companies are meeting their design automation goals.

For example, when Nissan developed the all-new 1990 300ZX sports car, design engineers worked closely with manufacturing engineering, test, and suppliers. This integrated design philosophy was applied in the areas of aerodynamics, structural integrity, suspension, braking, engine design, and transmission design. Computer-aided tools were a critical factor in the successful design of the car. Shuichi Suzuki, Manager of Body Design 300ZX, has said:

> We find computers to be very useful to identify the optimum point of often conflicting requirements, such as rigidity vs. light weight, light weight vs. safety and durability. And because of the accuracy of our computer simulations, we have dramatically reduced the number of cycles in the design process.[7]

Risks and Traps

Quite often, computers are bought without a full understanding of the required needs and areas of application. Many companies want to maintain parity with the competition. Others justify the purchase as a

---

[7]Simanaitis, Dennis. "Interesting Developments." *Road & Track's Guide to the All-New 300ZX.* Special Edition, 1989, p. 45.

cure-all to a chronic problem. However, proper planning and realistic expectations of benefits, costs, and schedule are essential for successful implementation and maintenance of computer-aided tools.

Areas of Risk

There are several areas of risk in the use and implementation of computer-aided tools. Table 1 summarizes the risks and associated consequences.

Improper planning

Upgrading or implementing a computer-aided tool is not simply throwing a switch. Automation can be painful and costly without proper planning. Consideration must be given to such areas as communication, culture, training, application, data management, and maintenance. Also, the system implementation schedule must be well thought-out in advance. Expecting users to welcome a new method of operations with all new tools is unrealistic.

Unrealistic expectations

Many people expect a computer to instantly solve problems. A firm may expect dramatic productivity and quality increases and significant cost

TABLE 1   Risks and Consequences

Risks	Consequences
Use of computer-assisted technology is not properly planned.	■ Use of computer-assisted technology will be ineffective and may reduce productivity and quality.
Firms set unrealistic expectations of benefits.	■ Expectations will not be met, and future improvements may be rejected.
Unnecessary or inappropriate tools are used.	■ Productivity, quality, and worker morale may decrease.
Small parts of the design and manufacturing process are optimized at the expense of optimizing the overall process.	■ Potential benefits of using computer-assisted technology will not be realized.
Users are not trained to use the new technology.	■ Efficiency gained by using computer-aided tools will be offset by inefficient use.
Results of computer analyses are not reviewed carefully.	■ Defects in the product will increase, leading to expensive rework and delayed product introduction.

savings without fully understanding the effort involved. Similarly, if a firm does not establish any historical baseline, quantitative and qualitative evaluations are not possible.

Unnecessary tools

Another critical risk is not understanding how the computer-aided tools fit into the overall design process. An engineer must understand the design process before considering using computer-aided tools. The engineer can then make intelligent decisions about what tools to use and how to use them.

Inappropriate tools

Buying excess capability is another area of risk. Usually many computer-aided tools can do a given function. The engineer may choose the tool that does the most. This may be risky, however, because a more powerful and flexible tool usually requires more user training and experience. The engineer could spend more time learning how to use a tool than the time saved. The engineer should use the tool that best fits the desired function.

Islands of automation

When upgrading or purchasing tools, a firm should consider integration and communication. While each individual tool may be perfectly suited to its respective task, chances for errors are increased every time data has to be reentered for another tool to use. Therefore, avoid tools having incompatible interfaces.

Improperly trained users

An engineer must learn how the computer-aided tool works, or more risk may be introduced into the design process. An engineer is usually trained to avoid risks in the traditional method of solving problems. Use of a computer, however, introduces risks that are not readily apparent.

"Garbage in, garbage out"

An engineer should always confirm the results of the computer simulation and analysis. Developers may believe the results of computer simulations just because they come from a computer. It is essential to understand and document the information that is entered into the model. Similarly, the techniques and expected outputs of the simulation must be understood, or even correct data may produce the wrong solution.

Keys to Success

When computer-assisted technology is properly applied within an organization, all phases of the acquisition process can benefit. However, it is just as easy to incur costly mistakes as it is to reap benefits. The keys to success are few and simple.

Work to meet business goals

The decision to apply computer-aided tools must be driven by the business strategy. A company should first determine the obstacles to meeting the business goals and then decide how computer-assisted technology can help overcome those obstacles. Without a goal driving the use of technology, it will be difficult to get and maintain upper-level management support.

Understand the overall process

Implementers of computer-aided tools should understand the overall design and manufacturing process and the flow of information. Computer-aided tools should be used to optimize the overall process rather than individual tasks. The tools should be integrated so that information flows smoothly to where it is needed.

Enlist management support

Management must support change and encourage new methods and tools. The use of integrated computer-aided tools will affect many people in different organizations. Management support will be critical to getting people with different goals to work together. Management understanding of the changes needed is essential.

Be patient

Benefits are not realized immediately. Expectations must be tempered with the fact that the process of upgrading technology is gradual. Users and organizations must learn to operate in a new environment. The effects of the new technology should be monitored, and problems should be fixed quickly. Procedures for operating in the new environment should be adapted as dictated by experience.

Educate people

Educate users and others affected by the new environment. Let people know what is happening and get them involved early and throughout the implementation. Get their input and discuss their concerns.

Work with vendors

Work closely with vendors as partners. Computer-aided tools are complex systems that require constant attention and maintenance. A vendor should be responsive to problems and help implement new tools and techniques, a continuous process.

2

Procedures

Common hurdles in introducing and implementing computer-assisted technology are corporate culture and perceived problems with job security. Fortunately, these hurdles can be overcome with proper planning. It is not just a matter of choosing the proper hardware and software. Time must be devoted to learning about the technology, seeing how others have implemented the technology, evaluating vendors, developing a strategy for implementation, and training users.

Steps for Success

This process can be divided into seven major steps. These steps are represented graphically in Figure 7.

Step 1 discusses the questions and concerns that a firm should identify prior to determining whether to upgrade or bring in new computer-assisted technology.

Step 2 discusses how to identify needs and develop requirements for computer-aided tools. In this step, it is important to consider both present and future needs.

Step 3 discusses the definitions and applications of the more common tools associated with computer-assisted technology. Those considering which tools best answer users' needs should ask the question: Are there tools that can do what is required?

Step 4 covers how to plan and implement the transition to computer-assisted technology. A smooth transition is desired, and thus it is important to examine how the new technology will fit into everyday operations.

Step 5 explains how to select the tools and vendors to meet the requirements established in Steps 1-4. A trap of technology is the tendency to choose the system with the most "bells and whistles" rather than the system that can best suit present and future needs.

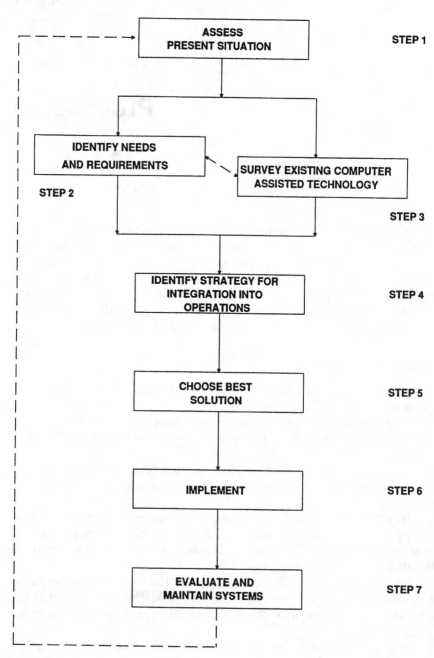

Figure 7 The steps for success.

Step 6 explains how to ensure a smooth implementation by gradually phasing in the new systems. Also critical to implementation is scheduling training so that users can learn the new tools while still being productive with the older methods. Also important is tracking training costs and any changes in productivity. These data will prove useful during later evaluations of the system.

Step 7 emphasizes that throughout implementation and integration it is important to keep accurate evaluations of the system and its users. This helps prevent the system and user skills from becoming obsolete. If trends show that productivity and benefits are slipping, it is possible to begin the process over again.

Step 1: Assess Present Situation

Without proper planning, investing in a technology that is as expensive and complex as computer-aided tools makes no sense. Technology will not make problems disappear and can even serve to magnify existing problems by allowing poor decisions to be made faster. Fortunately, success stories outnumber horror stories. For the organization that wants to update its existing technology base or bring in an entirely new system, the question is: Where do we begin?

The key to starting the process is understanding the opportunities provided by the computer and how to apply the technology for maximum benefit. For example, computer-aided design (CAD) may help an engineer in the process of creating a design, but it does not guarantee the design will be any better. The process of design must be evaluated and improved to take advantage of computer-assisted technology.

Also, computer-assisted technology must be applied wisely to those areas that will provide the most benefits. For example, a British Aerospace study found that the first 5% of design time fixes 85% of the development costs.[8] In addition, drafting, one of the more popular computer-aided tasks, generally represents less than 1% of the total cost of manufacturing.[9] It would not be wise to make a heavy investment in computer-aided tools for manufacturing and drafting while ignoring tools for product design.

It is generally accepted that computer-assisted technology is necessary in today's competitive environment. However, to properly move to-

---

[8]"Concept Development More Critical Than CAD." *Machine Design,* vol. 62(6), March 22, 1990, p. 34.

[9]Krouse, John and others. "How to Successfully Implement Computer-Aided Design and Manufacturing." *CAD/CAM Planning Guide '89,* Supplement to *Machine Design,* vol. 61(15), July 20, 1989.

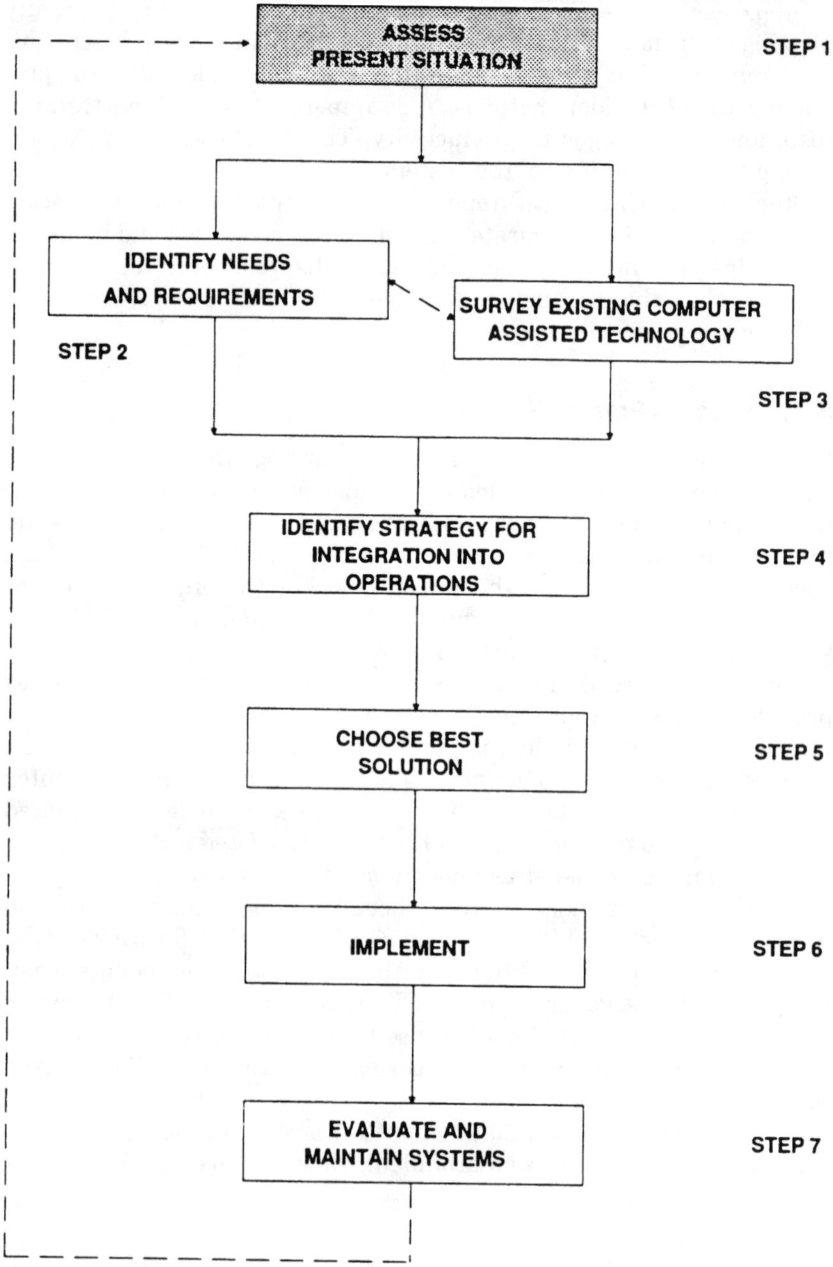

Figure 8 Assess present situation.

wards automation requires asking the right questions. Joel Orr suggests four points to follow when implementing computer technology:[10]

- ask the right questions
- take the right approach and direction
- choose the right people
- change at an appropriate rate

While it sounds simple, this task is very complex. Steps 1-2 describe the process that addresses the first of these points. Success in these activities helps ensure success later. Steps 3-7 continue the process described and address key issues regarding the right approach and direction.

Do we need new computer-aided tools?

The first task is to investigate the current state of affairs. A few key questions can help point out the need for a change in the status of computer-aided tools within an organization:[11]

- Are manufacturing costs constantly decreasing?
- Can the products still be delivered at a lower cost than our competitors' products?
- Can manufacturing goals (e.g., just-in-time) be met without a CAD/CAM system?
- Are we responding quickly enough to customer demands?
- Are design changes, no matter how minor, taking too long to perform?

If the answers to these questions indicate that a change is warranted, it is important to assess the present state of operations. This assessment is important both to understand where opportunities lie and to make informed decisions about potential solutions. This review also helps to identify strengths and weaknesses in existing systems, estimate the effectiveness of existing methods, and identify coordination considerations that must be met by the new system.

---

[10]Orr, Dr. Joel N. "How to Buy the Right CAD Program.", Paper presented at National Design Engineering Show. Chicago: February 26-March 1, 1990.

[11]Vaidyanathan, Pallavoor N. "Considerations in CAD/CAM Implementation." *CAD/CAM Management Strategies.* Eds. Robert M. Dunn and Dr. Bertram Herzog. Pennsauken, NJ: Auerbach Publishers, 1987, p. 2.

What benefits can we expect to see?

Over time, with the proper planning and implementation, some of the expected benefits from computer-aided tools are:

- reduced product cycles
- earlier error detection
- integrated design and analysis
- increased productivity and quality
- coordinated tool and fixture design
- increased material-handling efficiency
- reduced setup and lead times
- reduced manufacturing costs
- competitive products

While the list of benefits is great, it is important to temper expectations with the fact that gains are hardly ever immediate.

Set realistic expectations

It is generally acknowledged that the benefits of computer-aided tools outweigh their costs *when properly implemented*. However, great care must be taken when attempting to justify benefits such as cost reduction and productivity ratios because the initial results can be misleading.[12] Productivity ratio is the ratio of output produced to the input effort. When measuring productivity increases, there is often a tendency to oversimplify the input effort. This is a trap. For example, the efforts associated with purchasing and implementing a system are significant inputs that should be included along with the effort of designing the product.

What are we doing now?

The assessment of the current situation should cover at least the following areas and their relationship to computer-aided tools: the engineering and manufacturing information flow, areas of application, organization, people, work modes, and computers.[13]

A thorough understanding is required of how both engineering and manufacturing use information. It is tempting to try to understand the process through a top-down decomposition that describes what people

---

[12]Stark, John. *Managing CAD/CAM*. New York: McGraw-Hill, 1988, pp. 54-55.

[13]Stark, pp. 84-87.

think the process is. However, more insight can be gained by following the path of the product through the life cycle. This serves to develop an understanding of the information flow and the process as it presently exists.

Activities such as design, process planning, tooling, and machining are areas where computer-assisted technology can be applied. Figure 9 shows how the different activities may interact in the design and manufacturing environment. Each activity should be described in terms of task, information requirements, and present practices. Further, describing the complexity of the work, volume of work, and control procedures can identify additional deficiencies and opportunities.

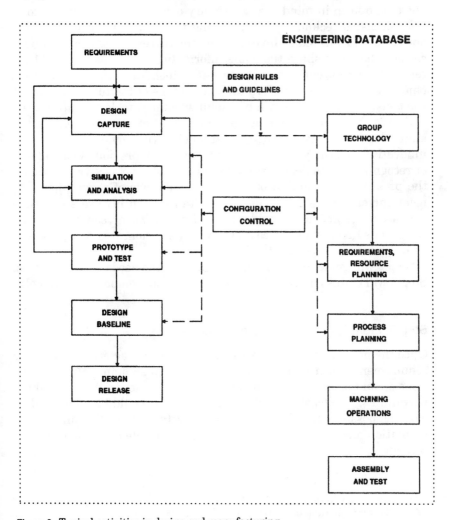

Figure 9 Typical activities in design and manufacturing.

The present organization should be examined from a computer-aided tool perspective. Then the organizations and locations of potential users can be identified. Any factors that may potentially hinder implementation should be noted. All future users should be described in terms of their present skills and their attitude toward new tools and procedures.

What tools and abilities do we have now?

The current use of computers in the organization must be described. In a company that has already introduced computer-aided tools, many first-time stumbling blocks have been removed or identified. It is important to keep in mind that a primary objective of introducing computer-assisted technology is improving the flow of information. This objective is facilitated by investigating the current state of other applications that may share the same information. A company should determine whether current computer-aided tools work to meet business objectives, and whether users take advantage of the full capability of the tools. These answers will be useful when new tools are introduced.

Due to the broad base of opportunities that exist, investigating the state of practice outside the organization is recommended. As a result, innovative ideas may originate in some areas not previously considered or recognized. Outside, neutral experts are able to objectively evaluate the present situation and offer recommendations of requirements to meet current deficiencies. Such consultants often involve a one-time expense that some may consider prohibitive. However, compared to the cost of misdirection or the cost of a wrong solution, this cost proves to be much more reasonable.[14]

Step 2 will guide the reader in determining the needs of the business. Step 3 will discuss how to find information about computer-aided tools.

Step 2: Identify Needs and Requirements

Once the present state of operations has been characterized, the basic requirements for a system can be defined. To accomplish this, questions should be asked to help define or identify needs, wants, and opportunities. The feasibility of implementation should be addressed as early as possible. It is important to understand that after further examination a solution involving new computer-aided tools may not be recommended.

---

[14]Chasen, S. H. and Dow, J. W. *The Guide for the Evaluation of and Implementation of CAD/CAM Systems.* Atlanta: CAD/CAM Decisions, 1983, pp. 24.

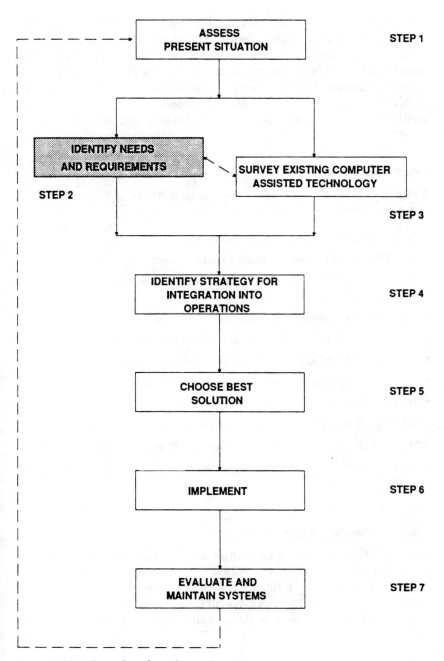

Figure 10 Identify needs and requirements.

The basic objectives here and in Step 1 are to identify opportunities and to develop basic requirements. The requirements should be well understood, documented, and disseminated throughout the organization. Initial efforts serve to gain consensus and to identify conflicts that must be resolved. Managers should explain their objectives, and their involvement and support will promote cooperation later on. More immediate results are some tentative conclusions and fewer alternative solutions.[15][16]

Identify needs

Some questions to ask when investigating needs to be answered by computer-assisted technology are:[17]

- Is the scope of the problem understood?
- What needs to be done to meet business goals?
- Will the applications stand alone or interface with some existing automated tools?
- How will the solution interface with existing automation?
- What are the risks and rewards of the applications?
- What amount of in-house expertise must be developed?
- How well can existing computer systems be adapted?
- Will users be centralized?
- What kind of system can solve the problem?
- How much software exists, and how much must be developed?
- How can present and planned systems be expanded to meet future needs?

What costs can be expected?

Justifying the purchase of a system is hard because the benefits of computer-aided tools are difficult to quantify, and a traditional cost-benefit analysis cannot be applied. It is nearly impossible to accurately determine gains, design improvements, improvements to communication, and all the possible new designs that computer-assisted technol-

---

[15]Stark, pp. 87-89.

[16]Chasen and Dow, p. 48.

[17]Chasen and Dow, pp. 25-26.

ogy can provide. However, the opportunity cost of no implementation will be significant.[18]

Budgeting capital expenses can be difficult. Predictions of such factors as payback, rate of return, and accounting rate of return are usually fuzzy and very dynamic. Some costs are easier to quantify than others. Direct costs are usually straightforward to calculate and include hardware and software purchase, training, and interest payments. However, indirect costs must be considered.[19] [20]

Many costs occur after purchase and have a bearing on how benefits are measured. These costs include maintenance and upgrades, integration, data management, system administration, and productivity losses during implementation. While it is difficult (or even impossible) to quantify these costs, early preparation allows them to be handled effectively when they arise.

One other trap is the rate of change of the technology. System performance per dollar appears to improve daily. It may be tempting to "wait and see" and hope that the answers will become obvious. However, waiting can result in missed-opportunity costs that could far outweigh any savings. Decision making and justification are difficult. Planners who anticipate, evaluate, and solve problems can increase the probability of success.

Develop and document the requirements

When identifying and establishing requirements for tools, get input from the intended user community. Their inputs regarding the system's requirements may be the most valuable. Some users may experience difficulty recognizing how work can be accomplished using the new tools. They should be encouraged to think of new practices that would efficiently use the tools.

The earliest requirements should also focus on the future use of computer-aided tools to facilitate a better understanding of the requirements. Identifying affected work areas and how much automation is necessary will help in identifying applications and phasing in the technology. Expert systems and artificial intelligence should also be considered, but these areas are developing so quickly that it is difficult to evaluate their role.

Ensure that management and engineering requirements match. The final requirements should clearly state why computer-assisted tech-

---

[18]Vaidyanathan, p. 3.

[19]Vaidyanathan, p. 3.

[20]Krouse and others, p. 10.

nology is being upgraded or introduced. They should outline intended uses, projected impact, major assumptions, and potential problems.[21]

The requirements will promote a better understanding of computer-assisted technology if they include:[22]

- a ranked list of functions needed and wanted
- a description of the intended use and flow of information
- a description of interfaces to other systems
- a statement of the work to be done by the system

Possible outcomes

Once the present state has been assessed and needs have been ranked, the search for solutions begins. One of the following is likely:[23] [24]

- A turnkey system will be purchased from one vendor or from a value-added reseller.
- The existing computing facilities will be expanded and augmented.
- A custom system will be developed within the company to meet their exact needs.
- Nothing will be done.

Turnkey system.   A turnkey system is complete; the user turns the key to start using the system. Historically, the trend to turnkey systems has been strong. The traditional turnkey system was provided by one vendor using the vendor's own equipment. More recently, the turnkey system provided by a value-added reseller (VAR) has become popular. The VAR represents several vendors and therefore chooses from a variety of hardware and software. The VAR integrates various computer-aided tools, possibly from several vendors.[25]

VARs can integrate computer-aided tools because of the trend toward standard interfaces. Standards allow tools from different vendors

---

[21]Chasen and Dow, pp. 87-89.

[22]Chasen and Dow, p. 89.

[23]Groover, Mikell P. and Zimmers, Emory W. Jr. *CAD / CAM: Computer-Aided Design and Manufacturing.* Englewood Cliffs, NJ: Prentice-Hall, 1984, p. 462.

[24]Besant and Lui, pp. 382-383.

[25]Krouse and others, p. 45.

to communicate. Companies can choose the best tools regardless of the vendor, as long as the tools use the same standard. Step 3 further discusses standards.

There are many advantages to a turnkey solution, especially considering that most organizations lack the internal expertise to properly integrate and implement on their own. A turnkey system has vendor support for the entire system.

Augmented system. Adding tools to an existing system allows for a great deal of flexibility, but interfacing with the present system may be difficult. In this case, the computing department has a significant role in implementation and decision-making.

Custom system. Developing a custom system can be incredibly complex, time-consuming, and expensive, but this may be the only solution to meet a company's requirements.

The process in Steps 1 and 2 parallel the one outlined in Part 1, Front End Process. With both processes, a need is identified, a description of the system to answer the need is created, feasibility studies and trade-offs are conducted, and clear and understandable requirements are generated.

A firm continues to examine and evaluate solutions until a single alternative is selected and implemented. Steps 3-7 describe this process. The process includes surveying many possible solutions and technologies, identifying implementation issues, preparing requests for vendor proposals, benchmarking candidates, selecting, planning implementation, implementing and integrating, and finally maintaining the system.

Step 3 is often conducted in parallel with Step 2. That is, the firm surveys tools while examining various areas of application. It is often easier to address a tool's suitability when both the task and the tools are examined concurrently.

Step 3: Survey Existing Computer-Assisted Technology

This step helps those responsible for implementing computer-assisted technology survey what is available in computer-assisted technology and match the technology with the applications and requirements developed in Steps 1 and 2.

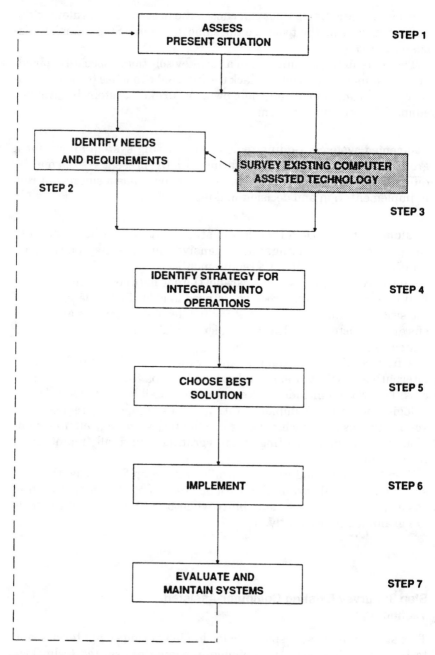

Figure 11 Survey existing computer-assisted technology.

Where to find information

Books, journals, trade shows, other users, and vendors are good sources for information about computer-assisted technology.

Books and journals.   Books offer comprehensive information ranging from introductory to advanced. Journals provide information piecemeal but offer the latest information in the rapidly changing field of computer-assisted technology. Many journals provide a yearly summary of computer-assisted technology in specific fields such as mechanical or electrical design. Advertisements in journals supply vendor names and product information. Journals also provide notices of trade shows and conferences, and some will dedicate most of an issue to discussing the large conferences.

The References chapter in the back of this part lists books on computer-assisted technology and articles from journals that often have information on computer-assisted technology.

Trade shows and conferences.   Trade shows and conferences are a good source for up-to-date information about vendor products and user applications. Often there are three components of a conference: an exhibit by vendors, presentation of papers by users and vendors, and short courses or discussions about specific topics. Vendor exhibits show the latest offerings, and visitors can talk directly with vendors about specific applications. Papers and short courses show what others are doing with computer-assisted technology and can lead to further discussion.

Contacts.   Contacts are another important source of information. Talking with other users can uncover useful points not mentioned in books and journal articles. Contacts may reveal traps encountered in implementing systems and possible escapes. Contacts can be people in your company or in other companies using similar computer-assisted technology.

All possible sources of information should be explored to maximize success and minimize risks of computer-assisted technology.

What information to find

A company interested in upgrading its computer-aided facilities will need a large amount of information before choosing a system and vendor. Some key points to research are:

- Will the tools work together?
- What are the functions and limits of the tools?

Will the tools work together?

Communication among the computer-aided tools may be the biggest hurdle in realizing their full potential. Engineers often use several different tools in the design and manufacturing process. Information produced by one tool is also used by other tools.

Often it is difficult to pass information among the different tools because they use different interfaces. As an analogy, imagine engineers from different countries trying to work together. Assume that none of the engineers speak the same language. Even though each engineer may accomplish a task very well, it will be difficult for the overall project to succeed. To succeed, the engineers must learn a common language, or they must use translators. Such is the problem with computer-aided tools. The computer communication problem requires either a common language or translators.

Standards. Standards make it easier for systems from different vendors to communicate. In principle, two machines using the same standard for communication can exchange information. In reality, this is not always possible because the standards are not sufficiently developed or vendors do not adhere to the standard. Figures 12 and 13 show the result of a transfer between two computer-aided tools using the Initial Graphics Exchange Specification (IGES) standard. Figure 12 shows a component drawing as it left one tool, and Figure 13 shows

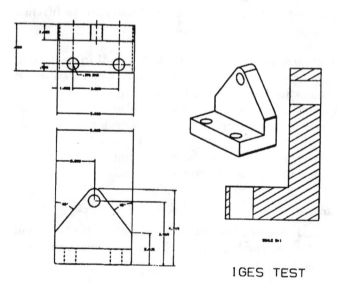

IGES TEST

Figure 12 Original drawing.

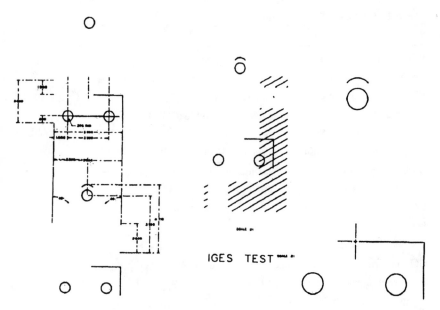

IGES TEST

Figure 13 Drawing after poor IGES transfer.

how it was interpreted by the second tool. Although both tools use IGES, further development would be needed for error-free information exchange.

Many standards exist or are proposed. Table 2 is a list of several standards and formats in various stages of development and application.[26] Two standards that are popular in the United States are IGES and the Electronic Design Interchange Format (EDIF). IGES is an evolving data format for exchanging graphics or geometric information for mechanical design information. EDIF is used for the transfer of electronic integrated circuit information. Another standard under development is the Product Data Exchange Specification (PDES). PDES is more encompassing than IGES. It includes nongeometric data, such as material properties, and administrative information, such as release dates.[27]

---

[26]Adapted from Kuttner, Brian C. and Michael Lachance. "Assessing Standards and Alternative Means of Data Transfer." *Computer-Aided Design, Engineering, and Drafting.* Eds. Robert M. Dunn and Dr. Bertram Herzog. Pennsauken, NJ: Auerbach Publishers, 1986, pp. 1–11.

[27]Stark, p. 128.

TABLE 2   CAD Data Transfer Standards

CAD Data Transfer Standards
Initial Graphics Exchange Specification (IGES)
Product Data Exchange Standard (PDES)
Electronic Design Interchange Format (EDIF)
Government Open Systems Interconnection Profile (GOSIP)
AutoCAD DXF
German Automotive Industry Association VDA
U. S. Air Force Product Definition Data Interface (PDDI)
French Aerospatiale Corporation SET
CAM-I Applications Interface Specification (AIS)
Ford Motor Company Standard Tape
Chrysler Corporation Standard File
GM Corporation Data Exchange Standard (DES)
Intergraph Corporation Standard Interface Format (SIF)
Vought Corporation Standard Data Format (SDF)
Standard Exchange of Product Model Data (STEP)

Engineering database. Having tools that communicate does not ensure that the information will be used effectively to make and maintain a successful product. The probability of success increases if a firm uses tools integrated with an engineering database.

An engineering database is a repository of engineering information needed to design, manufacture, and maintain a product. This information includes design data such as part sizes and shapes, parts lists, materials, and operating conditions. It also includes manufacturing data such as material processing, part inspections, and scheduling. Other information may include field problems and solutions. The purpose of an engineering database is to make accurate and timely information available so that engineers can design and manufacture products.

A database management system manages the engineering database and provides benefits such as elimination of redundant data, access control, and security. Firms use two approaches to make engineering information available from an engineering database. One method uses a single database that stores information for all engineering and production functions. Figure 14 shows this approach.[28]

---

[28]Adapted from Bohse, Michael E. "Integrating CAD and MRP Systems." *Computer-Aided Design, Engineering, and Drafting.* Eds. Robert M. Dunn and Dr. Bertram Herzog. Pennsauken, NJ: Auerbach Publishers, 1986, p. 7.

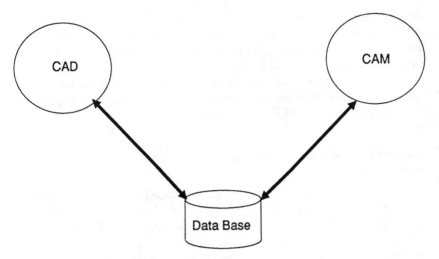

Figure 14 Use of a single database.

The other method uses separate databases for different functions and links the databases together through interfacing systems. Figure 15 shows the multiple-database approach.[29]

A single database simplifies the database management system and ensures that data are not duplicated. In practice, a single database

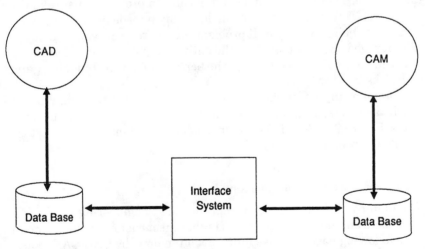

Figure 15 Use of connected, multiple databases.

---

[29]Adapted from Bohse, p. 8.

may become too large and slow to manage and use effectively. Also, it may be difficult to add new systems without redesigning large parts of the database. Multiple databases have the advantage that each database efficiently handles the function for which it was designed. However, interfaces between databases must be designed and implemented, and database management is more difficult. Both approaches have been used successfully in industry.

What are the functions and limits of the tools?

The person who is researching tools should determine the functions of the different tools. Generally, computer-aided tools are designed to do one function well. The tools may also do several related functions. For example, a tool developed to do mechanical design may also do some computer-aided drafting and finite element analysis. Other tools may do computer-aided drafting or finite element analysis better than the tool that does all three functions. However, the tool that does several functions may suffice. Generally it is better to have the fewest tools that meet the needs of the users.

A firm should be aware of problems that may exist in state-of-the-art computer-aided tools. Research on computer-aided techniques is providing a constant flow of new technology, and vendors of computer-aided tools quickly introduce new techniques to gain a competitive advantage. These new products often contain problems that take time to correct. Buying the latest technology may impede productivity while the tool is being debugged. If problems occur frequently, users may get discouraged with the tools, and the entire computer-aided effort may be jeopardized. A firm has to weigh the benefits of new techniques against the risks.

The following pages discuss how computer-aided tools may accomplish various tasks and what to expect when surveying computer-aided tools. Figure 9 in Step 1 shows typical design and manufacturing tasks needed to produce a product.

Computer-aided design/engineering

Computer-aided design and computer-aided engineering (CAD/CAE) tools help engineers design products that meet customer requirements. The following pages discuss functions often done by CAD/CAE tools.

Design capture.   A user enters a description of an idea for a new product into a CAD/CAE system either numerically, graphically, or textually.

Numerical input. For numerical input, the user enters the values of system parameters from a keyboard or a file.

Graphical input. For graphical input, a designer uses symbols to describe the idea. The designer can piece together the symbols to build up the desired system, often by choosing standard symbols. Figure 16, Panels a and b, show typical electrical-component symbols.[30]

Text input. For text input, the engineer describes the behavior and structure of a system using a software programming language. Text input is most often used for electronic design. Figure 16, Panel c, shows a text description of an electrical component. For large electronic systems, text input may consume less time than entry of electronic symbols. Using text, the engineer can more easily simulate the overall circuit before doing detailed design of individual components. With some computer-aided tools, the engineer can direct the computer to convert the text into the equivalent circuit diagram.

After design capture, a model of the physical system is stored in the computer. The engineer manipulates that model to get useful design information. Different engineering disciplines use different models. Electrical designers generally use graphical symbols or programming languages for design. Mechanical designers often use wire-frame, surface, or solids models to represent mechanical components. Only solids models contain both geometrical and material-property information,

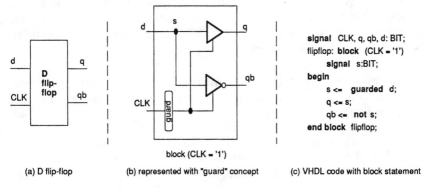

(a) D flip-flop   (b) represented with "guard" concept   (c) VHDL code with block statement

Figure 16 Graphical and textual descriptions.

[30]Barton, David L. "A First Course in VHDL" in *VLSI Systems Design's User's Guide to Design Automation,* Manhasset, NY: CMP Publications, 1988, p. 40.

making them the most intuitive representation of parts. Although the wire-frame model is the most widely used today, the solids model will become more widely used as computers become more powerful.

Simulation and analysis. Computer-aided tools help engineers use *simulation and analysis* to design systems.

Simulation and analysis is the process of applying a stimulus to a model and interpreting the resulting behavior. An engineer modifies the design (that is, the model) until the desired result is achieved. This shortens design time by finding problems before hardware is produced. For example, an engineer may simulate an automobile radiator under various operating temperatures. If the system does not work well, the engineer changes the design so it will be more tolerant to temperature. With this process, most problems should be uncovered before manufacture.

Mechanical engineers use simulation and analysis techniques such as interference and tolerance analysis, variation simulation analysis, stereolithography, kinematic and dynamic analysis, and finite element analysis. Electrical engineers typically simulate digital (logic) and analog circuits with software or with combined software and hardware.

Design rules and guidelines. Computer-aided tools can help avoid problems during manufacture and deployment. The computer can compare a design against rules and guidelines and highlight deviations. Some computer-aided tools automatically position electrical components based on design rules and guidelines. For example, component layout on a circuit board may be optimized by using thermal guidelines that improve reliability. Most tools allow an engineer to override the rules if necessary.

Configuration management. Configuration management is the discipline by which changes are managed. Three main activities are required to successfully manage change:[31][32]

- *Configuration identification*: This defines and identifies every element of the product. As the product matures, this task moves from generation of ideas, through requirements, and finally to modifications.

---

[31]Dhillon, B. S. *Engineering Management.* Lancaster, PA: Technomic Publishing, 1987, pp. 275-277.

[32]Gryna, Frank M. "Product Development" in *Juran's Quality Control Handbook.* 4th ed. Eds. J. M. Juran and Frank M. Gryna. New York: McGraw-Hill, 1988, pp. 13.66-13.67.

- *Configuration control*:   This manages design changes from the initial proposal through implementation.

- *Configuration accounting*:   This documents the status of all changes, both proposed and implemented.

When properly implemented, configuration management facilitates data retrieval, establishes traceability for a design, and reduces redundancies. Computer-aided tools integrated with a configuration management system can track all necessary information. The system can also reduce paper required by storing information on a computer.

As tools become more integrated, shared data become more important. "Engineering data management" or "definition configuration management" can manage data and help maintain data integrity in an integrated environment. These tools store, track, and control both data and process information. Often, they are able to use existing databases, facilitating their implementation and use.

Computer-aided software engineering

To the nonpractitioner, software development can be quite confusing. Part 4, Software Design and Test, discusses in detail the software engineering process. Briefly, the process covers four major phases: requirements, design, implementation, and test. By applying software engineering principles to simpler systems, productivity can be increased as much as 10-to-1. However, as project complexity increases, it becomes harder to realize such benefits. Computer-aided software engineering (CASE) allows for a set of tools to be integrated within the software engineering environment. CASE, like any other computer-aided tool, can increase productivity. However, gains may not appear immediately. Figure 17 illustrates one example of CASE tools' impact on productivity.[33]

CASE dictates that specifications for corporate plans, system designs, and system development become fully integrated. (Note that in software engineering terminology, "system" can either refer to a fully integrated hardware-software system or to software programs alone.) This is accomplished by having the CASE tools and components share the specifications across the corporate planning function, system analysis and design, and system development.[34]

---

[33]Soat, John. "Software Productivity: The Next CASE." *InformationWEEK*, vol. 252, January 8, 1990, p. 24.

[34]Gibson, Michael Lucas. "The CASE Philosophy." *BYTE*, vol. 14(4), April 1989, p. 209.

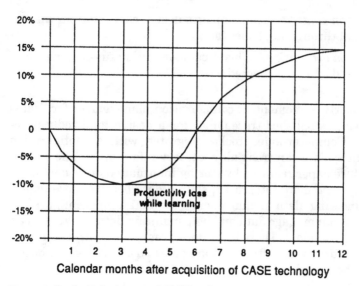

Figure 17   Productivity impact of CASE tools.

CASE has evolved from simple systems analysis and documentation tools to full-function tools that can provide automated support for the entire software life cycle. When choosing a CASE technology, developers should remember that its functionality is the primary criteria. It combines both development methodologies and software tools. The methodologies determine the process and the tools automate it. Ideally, developers can work within a responsive, dedicated environment to develop and maintain software systems.[35]

There are two types of CASE tools. The first type focuses on one task or phase of the software development process. Tools exist for analysis and design, database and file design, programming, maintenance and reengineering, architecture, and project management. The second type of tools offers automated support (to the software development process) and delivers documented, executable software. These tools require careful evaluation when matching platforms, development methodologies, and target systems.[36]

Ideally, a fully integrated CASE environment is similar to the one illustrated in Figure 18.[37]

---

[35]McClure, Carma. "The CASE Experience." *BYTE,* vol. 14(4), April 1989, p. 235.

[36]McClure, pp. 236-237.

[37]Gibson, p. 210.

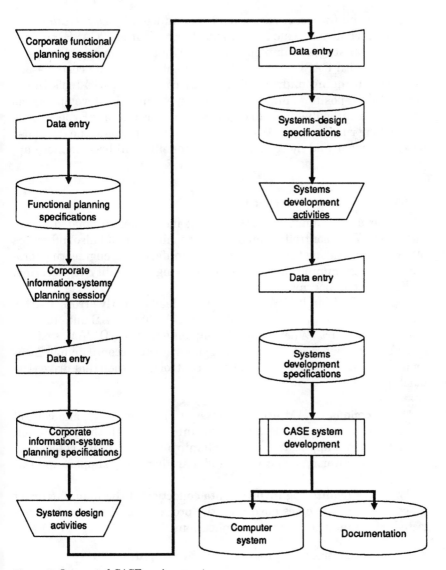

Figure 18 Integrated CASE environment.

In almost any development activity, the "state of the art" and the "state of the practice" can be vastly different. Software engineering is no exception. It can take 15 years for a concept to become accepted practice. CASE can improve the current state of software engineering. However, compared to other computer-aided tools, it remains an immature technology. Most companies that are pioneering CASE expect payoffs in the future. A recent survey conducted by *Datamation* and Price Waterhouse revealed that in most organizations, CASE tools are in the initial stages of implementation. The three main problems experienced with CASE are: industry and sales hype, the costs associated with bringing multiple single-user copies of CASE tools into a single data repository, and frustration experienced by developers who try to tailor inflexible tools to organizational practices.[38] Expectations for the long run are high. As the technology matures and the user base becomes larger and more experienced, benefits will become more apparent.

Computer-aided manufacturing

Computer-aided manufacturing (CAM) is more than automated manufacturing. While controlling production machinery, CAM also influences other manufacturing functions such as manufacturing engineering, production programming, industrial engineering, reliability engineering, and facility engineering.[39]

Applying CAD and CAM separately can lead to costly errors. CAD is used to create a representation of a part. CAM uses CAD data to manufacture the part. When properly implemented, CAD/CAM systems can help engineers design a part, generate the accompanying documentation, and produce the numerical control (NC) code that drives the machine tools.

CAM applications.   CAM uses computers to plan, manage, and control the manufacturing plant through an interface with the production resources. CAM applications often fall into one of two broad categories: computer monitoring and control, and manufacturing support applications.

Computer monitoring involves data collection of the manufacturing process; control involves governing the process based on the data collected. Manufacturing support applications include:[40]

- numerical control part programming by computer

- computer-automated process planning (CAPP)

- computer-generated work standards

---

[38]Statland, Norman. "Payoffs Down The Pike: A CASE Study." *Datamation,* vol. 35(7), April 1, 1989, p. 52.

[39]Datapro Research. *CAD/CAM/CAE Systems.* M07-100. Delran, NJ: McGraw Hill, January, 1989, pp. 401-402.

[40]Groover, and Zimmers, pp. 2-3.

- production scheduling
- material requirements planning (MRP)
- shop-floor control

Traditionally, the data for CAM tasks came from an off-line computer. In a separate task, the information was physically transferred to the tool for processing.

Numerical control (NC). Quite simply, NC operations are processes driven by computerized machine-tool instructions. These processes can be drilling, machining, tool selection, and even assembly. For machining processes, the complexity of the shape that can be produced is determined by the number of axes of the machine tool. Five-axis capability is found on the more sophisticated NC tools.

Just because a part has been designed and analyzed on a CAD/CAE system does not mean it can be machined on an NC tool. Part complexity coupled with high material costs can lead to mistakes that waste both time and money. However, simulation tools can reduce risk by modeling the NC process before any hardware is made. With simulation tools, an object can be machined in the computer. A user programs a simulation tool using part data from the CAD system and data describing the machine tool's characteristics, including the number of axes. The simulation allows early detection of manufacturing problems, design concerns, and verification of tool paths.

MRP and MRP II. Another method of increasing productivity and reducing costs lies in the integration of CAD with material requirements planning (MRP). MRP's major objectives are minimizing inventory and maintaining delivery schedules. MRP focuses on both the current and planned inventories of parts. In tracking inventory, there are four main factors: requirements, available inventory, bill of material, and company plans. Designers provide inputs to MRP by defining parts lists, inventory requirements, and processes. Integration keeps the data flow between functions compatible and relevant.[41]

As the computer became more integrated into the process, an extension of MRP developed. MRP II (or manufacturing resource planning) incorporated more planning functions. The ability to know when to order materials was added. With this new addition, material require-

---

[41]Srch, Richard W. "Developing a Computer-Aided Design and Materials Requirements Planning Interface" in *CAD/CAM Management Strategies*. Eds. Robert M. Dunn and Dr. Bertram Herzog. Pennsauken, NJ: Auerbach Publishers, 1988, pp. 1-4.

ments could be scheduled more efficiently and prioritized if necessary. MRP II is also a more closed-loop function that integrates inventory management, capacity planning, and shop floor control. This closed-loop system can detect problems earlier in the production planning phase.[42] MRP and MRP II are often referred to by the acronym MRP.

Computer-aided process planning.　Process development can also yield productivity and cost benefits. Process planning can reduce production costs by as much as 75%.[43] Computer-aided process planning (CAPP), also known as computer-automated process planning, provides tools that generate and control the data necessary to assemble parts. It links CAD data with assembly data. CAPP studies show:[44]

- a 47% reduction in part throughput on the shop floor
- a 35% improvement in each process planner's efficiency
- a 32% reduction in work-center setup time
- a 7% reduction in total design costs

Group technology: foundation for CAPP.　Key to the successful implementation of CAPP is *group technology* (GT). GT is a systematic approach, more philosophy than technology, that identifies and exploits part similarities. It groups, codes, and analyzes parts and processes by common features. These features can be size, shape, material, tolerance, or processing. The parts are then grouped into families and coded by some scheme, usually numeric. The codes are stored in a database for later retrieval.[45 46]

Figure 19 shows a family of parts sorted by GT methods.[47]

GT lays a foundation for CAPP by promoting standardization in parts and processes. CAPP can then control materials, labor, machinery, and tools. CAPP relies on a CAD model to match the part with the machinery required in production. There are two main types of CAPP

---

[42]Besant and Lui, p. 374.

[43]Granville, Charles. "Computer-Aided Process Planning Systems Come of Age." *Managing Automation,* vol. 37(2), February 1990, p. 60.

[44]Granville, "Computer-Aided Process Planning Systems Come of Age." p. 60.

[45]Sutton, George P. *Computer-Assisted Process Planning.* Business Intelligence Program, Report No. 765, Menlo Park, CA: SRI International, 1988, p. 8.

[46]Granville, Charles. "CAPP Comes to CAD and Manufacturing." *Automation,* vol. 37(3), March 1990, pp. 44-45.

[47]Sutton, p. 9.

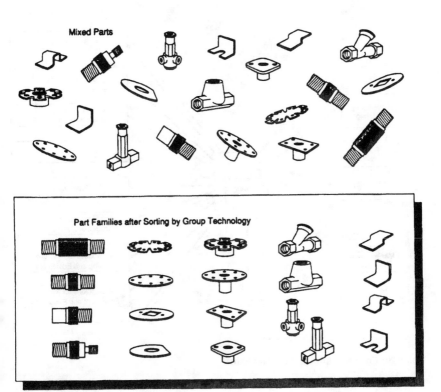

Figure 19 Parts sorted by group technology.

systems: variant process planning and generative process planning.[48][49] Variant process planning, the most common method today, modifies an existing process plan with GT principles of codifying and classifying. Generative process planning is more complex. It uses manufacturing logic, machinery capabilities, standards, specifications, a full part description, and its stored knowledge to find the optimum process plan.[50][51] This is the most advanced CAM application of artificial intelligence.

Figure 20 illustrates process planning's relationship to other manufacturing functions.[52]

[48]Veilleux, Raymond F. and Petro, Louis W. *Tool and Manufacturing Engineers Handbook, Volume 5: Manufacturing Management.* Dearborn, MI: SME, 1988, p. 16-7.

[49]Sutton, p. 8.

[50]Veilleux and Petro, p. 16-7.

[51]Sutton, p. 8.

[52]Sutton, p. 12.

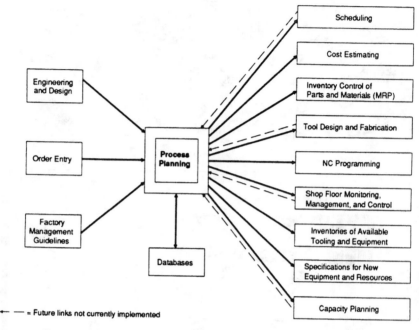

Figure 20 Process planning's relationships.

CAM as it relates to CIM. The notion of "islands of automation" describes a common problem in implementing computer-aided tools. It refers to the problem of introducing computer-aided tools without giving proper thought to integrating the pieces. CAD, CAE, CAM, MRP, GT, NC, and CAPP are all key to a successful implementation of a computer-integrated manufacturing (CIM) strategy.

CIM is an all-encompassing concept that stresses unification of technologies and disciplines. The Computer and Automated Systems Association of the Society of Manufacturing Engineers (CASA/SME) developed the CIM Wheel, a model that represents this ideal of unification. Figure 21 shows the CIM Wheel.

The CIM wheel describes five fundamental topics that have a natural affinity for each other:[53]

■ general business management

■ product and process definition

■ manufacturing planning and control

---

[53]Veilleux and Petro, p. 16-2.

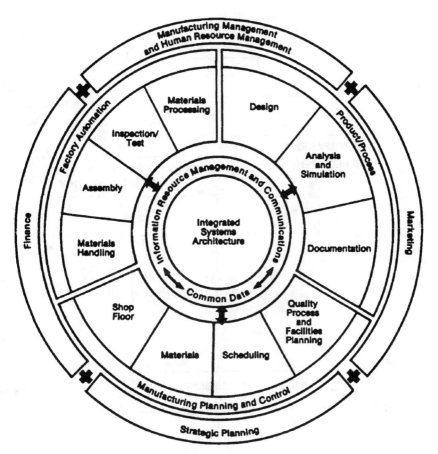

Figure 21 The CIM wheel.

- factory automation
- information resource management

These areas are seen as a family of automated CIM processes, rather than as the islands of automation described earlier.

Step 4 discusses how to integrate CIM into a company's operations.

Step 4: Identify Strategy for Integration into Operations

Need for strategy

Computer-aided tools are improving development costs, schedules, and quality. However, there are potential pitfalls. For example, engineers

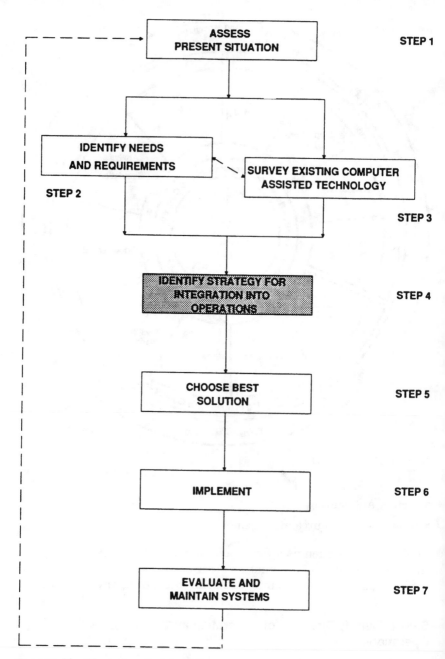

Figure 22 Identify strategy for integration.

and managers often become so enamored with the technology that they forget their objectives. Doing it faster does not always mean doing it better. If wrong things are being done, automation helps do them faster. A strategy for computer-assisted technology helps maintain focus on the objectives.[54] Computer Integrated Manufacturing (CIM) is that strategy.

A view of CIM

An idealized view of CIM was once presented as:

> The basic premise behind CIM is that by totally automating and linking all the functions of the factory and the corporate headquarters, a manufacturer would be able to turn out essentially perfect, one-of-a-kind products: at the lowest possible cost and almost overnight.[55]

While this thought reflects noble aspirations, the fundamental idea of integrated functions and information exchange is sound. The computer allows integration, but integration requires a great deal of foresight and planning. CIM strategy requires a firm to identify what information is needed, where it will come from, where it should go, its format, speed of transfer, method of transfer, and what happens if it does or does not get there.

A firm must reengineer its design and manufacturing processes to take advantage of computer-assisted technology. It is not enough to automate individual manual tasks. The firm and engineers must find the most efficient way of doing things in the new environment. An obstacle to planning for CIM is that not everyone involved with CIM works within the same context. There is often a gap between those who implement automation and those who are directly affected by it.[56] [57] Figure 23 illustrates a typical CIM environment.[58]

---

[54]McGill, Michael E. *American Business and the Quick Fix.* New York: Henry Holt and Company, 1988, pp. 181-200.

[55]Brandt, Richard and Port, Otis. "How Automation Could Save the Day." *Business Week,* number 2935, March 3, 1986, p. 72.

[56]Bozman, Jean S. "How Some Users Were Able to Plant CIM in Their Factories." *InformationWEEK,* issue 065, May 12, 1986, p. 39.

[57]Webb, Michael J. and Felegyhazi, Bill. "CIM in Perspective: The Challenges and Rewards." *Manufacturing Systems,* vol. 8(4), April, 1990, pp. 28-32.

[58]Palecek, Peter, Sutton, George P., and McGinty Weston, Diane. *CIM Market Needs and Opportunities.* Business Intelligence Program, Report No. 744, Menlo Park, CA: SRI International, 1987, p. 3.

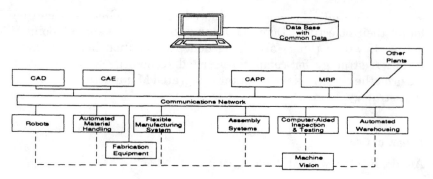

Figure 23 Segments of CIM.

Building a CIM strategy

Companies that have realized significant benefits from CIM have:[59]

- a CIM plan that is driven by the need to integrate all business functions, not just the need to improve efficiency

- a CIM process that simplifies, automates, and integrates business systems and technologies

- a team of CIM people who understand and support the goals and efforts of the organization

Roger Willis of Andersen Consulting identifies five key lessons that support the above:[60]

- Without implementation in support of a business strategy, CIM components will remain islands of confusion.

- Automating a mess does not clean up the mess.

- Competitive advantage is not gained by automation for the sake of technical superiority.

- Integration provides significant results; simplification and automation just supply excitement.

- Implementing CIM requires the right people.

---

[59]Willis, Roger. "The Laws of CIM: Case Studies on Optimizing Manufacturing." *Manufacturing Systems,* vol. 8(2), February 1990, pp. 54-58.

[60]Willis, pp. 54-58.

Justifying CIM

CIM offers many benefits. For example, it helps shorten development time, improve productivity, increase throughput of information, provide more accurate information, and reduce paper.

As with any strategy, there are pitfalls to avoid. Some common pitfalls in CIM justification are:[61]

- relying on traditional cost management and performance measurement techniques
- focusing on incremental investments and short term results
- focusing on hurdles to implementation that are not tied to business strategy
- failing to quantify important benefits and opportunity costs
- not simulating and analyzing alternative scenarios

Key steps to successful CIM

The key steps to CIM are: plan, simplify, automate, and integrate.[62]

Plan.   This step identifies opportunities provided by CIM, a conceptual design of the CIM strategy, a technical architecture for CIM, and an implementation plan. The implementation plan shows how to implement CIM at a reasonable pace, yet still realize benefits quickly.

Simplify.   This step has implications for both engineering and management. This step determines how to: migrate smoothly to CIM, reduce setup, define manufacturing cells, determine material flow, prepare for the appropriate levels of implementation, plan top-down, and implement bottom-up.

Automate.   This step brings in the technologies associated with CIM such as CAD/CAM, automated material handling, MRP, and NC machines. However, as mentioned earlier, it is important to selectively apply these technologies. Group technology (GT) is often an outgrowth of performing the simplify and automate steps in parallel.

---

[61]Wejman, James, Collins, Mike, Darnton, William, and Dalton, G. Reid. "Broadening CIM Horizons." Paper presented at National Design Engineering Show, Chicago, IL, February 26-March 1, 1990.

[62]Wejman, Collins, Darnton, and Dalton.

Integrate.   This step ties new and existing systems into a coordinated, well-managed process that streamlines the flow from product design through delivery. Technologies often implemented here are factory control systems, distributed or direct NC (DNC), local area networks (LANs), and performance measurement systems.

People and CIM

Figure 24 reinforces the idea that people are a key ingredient to a successful CIM strategy.[63]

Three broad categories of people can support the CIM environment:[64] [65]

- Enforcer: This person mandates that tasks and activities associated with CIM are carried out.

- Supporter: This person develops a communication strategy, maintains labor relations, promotes organizational change, and educates personnel.

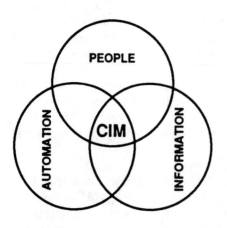

Figure 24 People are a key ingredient in CIM.

---

[63]Wejman, Collins, Darnton, and Dalton.

[64]Wejman, Collins, Darnton, and Dalton.

[65]Gould, Lawrence. "CIM Champions: Grow 'Em or Buy 'Em?" *Managing Automation,* vol. 5(3), March 1990, pp. 20-23.

- Champion: This person is a cheerleader, advocate, sales representative, resource, communicator, manager, and engineer rolled into one. The champion develops enthusiasm and support for change by making everyone feel part of the team.

A firm should avoid the need for an enforcer because of the potential backlash such a person may cause. Supporters and champions require upper-management support and work well together. The champion promotes understanding and change and can provide vision and guidance.

Example of CIM architecture

IBM is working on a CIM architecture that supports business goals and is flexible enough to handle changing manufacturing requirements. It supports existing investments in systems, applications, procedures, and data. It also provides users access to new functions such as integrated data management.[66]

The architecture includes interfaces, functions, and components that will be implemented over time. Five principles guide its development:

- Data Integration: Data are managed so that they can be secured, consistent, and shared wherever needed.

- Open Interfaces: The architecture relies on published interfaces so users have easy access to CIM services. Open interfaces also allow systems to be extended by systems integrators.

- Industry Standards: The architecture supports existing and evolving standards (e.g., OSI, MAP, SQL, PDES, IGES, CALS, and CIM-OSA) so information and integration can be achieved in a multi-vendor environment.

- Platforms: A comprehensive plan for consistency and compatibility between software products and vendors supports application development across major operating environments.

- Protection: Company-wide data sharing is emphasized with a minimal effect on applications. Data from IBM and non-IBM systems can be integrated.

---

[66]Munsinger, John P. "The State of IBM's CIM Architecture." *CIM Review,* vol. 6(3), Spring 1990, pp. 30-33.

IBM has implemented this architecture at many of its own manufacturing sites worldwide.

Parallels between CASE and CIM environments

An objective of CASE is the establishment of the *software factory,* where new, complex, defect-free software can be manufactured. The software factory facilitates software development and production while addressing such topics as organization, unification of tools and methods, planned integration of CASE, and non-technical issues.

Stand-alone CASE tools, much like islands of automation, can be powerful yet ineffective. Few CASE tools effectively talk to one another. In software development, problems include dependence on a programming language, dependence on a specific type of hardware or operating system, portability, and orientation to a single task as opposed to system development. As with CIM, the success of CASE depends on the interconnection of tools and interfaces. Standards support interconnection which is fundamental to the success of the software factory. For CASE tools to be properly integrated, a common database is required, also.[67]

CASE can benefit the software development process in much the same way CIM provides benefits for manufacturing. Many opportunities are the same, but have different applications and disciplines. Among these opportunities are:[68]

- developing a system development model that can be supported across an entire organization
- defining a standard life-cycle framework for development
- connecting tools and methods with the proper support methodologies
- selecting and automating manual methodologies such as design, planning, and project management
- selecting and automating tools that are compatible with the overall framework and support methodologies
- developing a set of integrated computer-aided tools, processes, and methods

---

[67]Weber, Herbert. "From CASE to Software Factories." *Datamation,* 35(7), April 1, 1989, pp. 34-36.

[68]Manley, John H. *Computer Aided Software Engineering (CASE) Foundation for Software Factories.* Report AS85-510-101. Delran, NJ: Datapro Research Corporation, February 1986.

- educating and training everyone to facilitate change within the organization

Step 5: Choose Best Solution

This step discusses how to choose computer-assisted technology that best satisfies the needs of the firm. The process is summarized as follows:

1. Prepare the Request for Proposal (RFP).
2. Choose vendors who will receive the RFP.
3. Evaluate vendor responses and rank vendors.
4. Benchmark computer-assisted technology.
5. Select a suitable vendor and system.

Each step of the process is discussed below.

Prepare RFP

The RFP is a document used to solicit quality proposals from vendors in a format that is easy to evaluate and compare with other proposals. The RFP includes technical requirements for a computer-aided system that meets a firm's needs, but it is not a rigid design document. Typically, the RFP contains information about the firm and its products, the current method of operation, the intended use of computer-assisted technology, and the expected benefits of applying the technology. Also, the RFP includes requirements regarding hardware, software, support, documentation, training, acceptance tests, and maintenance. A table of contents from a sample RFP is found in Chasen and Dow. This suggests what information to include in the RFP.[69]

Choose vendors who receive the RFP

A firm should select about six vendors to whom they send the RFP. The number should be limited so that evaluation of vendor proposals is manageable. Several sources can be used to select vendors or systems: previous experience with vendors, experience of another division or firm, reviews of vendors and systems by outside consultants, and contact with vendors at trade shows.

This list of vendors should be compiled carefully since the success of the computer-aided system hinges on the quality of the vendor. The vendor provides most of the initial training and support to help the

---

[69]Chasen and Dow, p. 124.

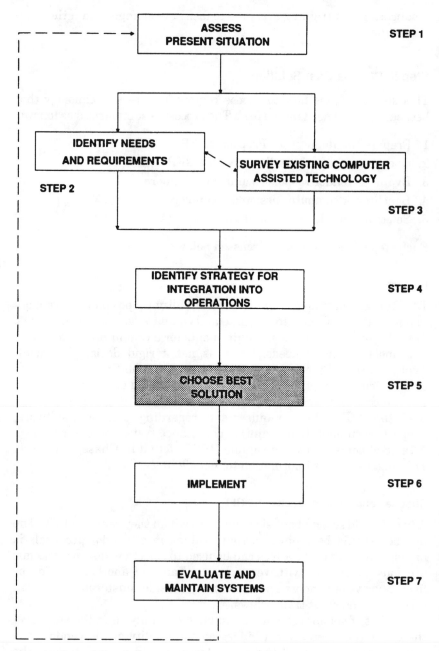

Figure 25 Choose best solution.

firm take advantage of the tools. The vendor works closely with the firm to solve problems with the tools or tailor them. Also, the vendor provides support in the future when problems arise, changes must be made, or new tools are introduced. Partnership with a vendor is important for continued success with computer-assisted technology.

Evaluate vendor responses and rank vendors

A firm should use consistent and objective criteria to evaluate vendor proposals. As proposals come in, reviewers should study them for clarity and completeness and, if necessary, request more information.

The review team should base evaluation on the RFP. A first cut may use criteria that the vendor or system must meet. Table 3[70] shows five vendors, A through E, evaluated against five "must" criteria. Vendor E failed criteria numbers 2 and 5 and was eliminated. The review team continues the evaluation of the other four vendors.

Vendor scoring. Subjective criteria can be made more objective by attaching a numerical score to the evaluation items. One method involves four parts:[71]

- List attributes used for evaluation.

- Assign relative importance to each attribute.

TABLE 3   Initial Screening of Vendor Proposals

	Vendor				
Item description	A	B	C	D	E
1. Facilitates efficient/cost effective production	X	X	X	X	X
2. Facilitates common drawing techniques—overlay capability	X	X	X	X	
3. Possesses progressive display technology—hybrid, storage, or refresh displays	X	X	X	X	X
4. Communicates to remote systems and/or host mainframe	X	X	X	X	X
5. Has computer output microfilm interface	X	X	X	X	
Unsatisfactory rating					X
Satisfactory rating	X	X	X	X	

---

[70]Chasen and Dow, p. 124.

[71]Chasen and Dow, p. 126.

- Evaluate each vendor for each attribute.
- Weight and sum scores.

The Application chapter contains an example of this method.

Benchmark the system

Benchmarking exercises a computer-aided system to evaluate the vendor claims. Two kinds of benchmarks can be used:

- synthetic benchmark
- live benchmark

Synthetic benchmark. *Synthetic benchmarks* are computer programs that have pre-established performance parameters and have been written to operate on many systems. For example, Linpack is a benchmark that checks how fast a computer does calculations by having the computer solve a 100×100 matrix equation. A synthetic benchmark is most useful if it is representative of the applications of the firm. For example, if the firm does matrix math during design, the results of Linpack could serve as one of several benchmarks.

Live benchmark. *Live benchmarks* are samples taken from the users' work that represent all of the work of the firm. This benchmark is preferred for evaluating computer-aided systems. For a live benchmark, the computer system hardware and software should match as closely as possible the firm's configuration. For example, if a firm will have several computer workstations communicating with a remote computer, the benchmark system should be configured similarly. A single (stand-alone) computer operates faster than a communicating (networked) computer and provides an unrealistic benchmark.

Reviewers may benchmark several systems before making a final choice of system. However, benchmarking is costly and time-consuming, so a firm may benchmark only the top-rated system. If the firm is satisfied with the results, the vendor and system are selected. Otherwise the second system is selected for benchmark. This process continues until a choice is made. The evaluation team should choose a system that performs the entire process well, rather than just the individual parts of the process.

Select suitable vendor and system

At this point the evaluation team should summarize the results of the evaluation and benchmark and present the recommendations to management. After management makes a decision to buy, the team can concentrate on implementing the system.

Step 6: Implement

Implementation of computer-aided technology is difficult because those responsible must maintain a continuous, high level of support as they guide users through changes in design and manufacture. A firm can only realize the potential of computer-assisted technology if a firm changes the prevailing culture which often shuns change. Change must become the norm so that new tools and new methods can be integrated into existing operations with minimal disruption.

Implementation process

Implementation can be broken down into several tasks:

- Plan the implementation.
- Prepare for implementation.
- Start using the system.
- Adapt the system to the company's processes.

Most of these tasks are done concurrently and evolve as the company gains experience in the technology.

Plan the implementation.   Those responsible for implementing computer-assisted technology should prepare a high-level plan and a detailed low-level plan for the first year of implementation. This ensures an orderly introduction of the technology. The plans should include the tasks, approximate start and end dates for each task, resources necessary for each task, and persons responsible. It may be difficult to estimate the time to complete a task, but an estimate should be included. Those responsible for implementing tasks should have major roles in the planning process. Top management and middle management should approve the plan. Middle-management review and approval are important since their subordinates will carry out the tasks. Table 4 is a sample of first-year, high-level tasks.[72]

Table 5 is a sample breakdown of one high-level task.[73]

---

[72]Stark, p. 100.

[73]Stark, p. 101.

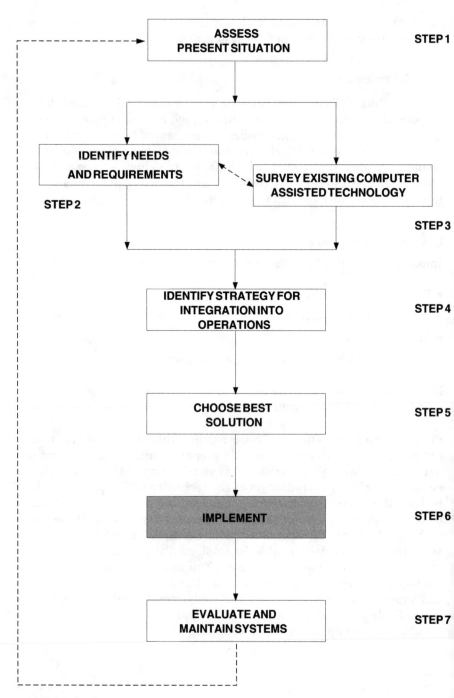

Figure 26 Implement.

TABLE 4   Major High-Level First-Year Tasks

Develop plans for installation-related activities and other activities that will occur during first year's use.

Carry out training of managers.

Prepare purchase orders. Sign contracts. Fix delivery date.

Finalize site location and facility layout.

Prepare for system installation and acceptance.

Develop procedures and standards for system use.

Install.

Adapt system to company—conversions, macros, documentation, etc.

Use system. Communicate experience.

Monitor and review progress. Report progress.

TABLE 5   Detailed Low-Level Tasks Related to Facility Preparation

Develop detailed plans for facility-preparation-related tasks.

Identify potential sites.

Define detailed requirements (e.g., for power, space, environment control).

Select site.

Define detailed site and facility layout.

Build facility.

Order major systems (power, air conditioning, etc.). Order furniture and supplies.

Define facility tests.

Receive and install systems, furniture, and supplies.

Test facility without CAD/CAM system.

Install CAD/CAM system.

Test facility with CAD/CAM system.

Prepare for implementation.   This task puts implementation plans into action. The most critical activity is training those affected by the new environment. Issues related to people may be more difficult than problems related to technology.[74] Training can come from the system vendor, internal sources, independent consultants, colleges and universities, and computer-aided instruction from the system itself. Training is an ongoing task that will help a firm realize the potential of computer-aided tools.

---

[74]Stauffer, Robert N. "Lessons Learned in Implementing New Technology." *Manufacturing Engineering,* 102(6), June, 1989, p. 63.

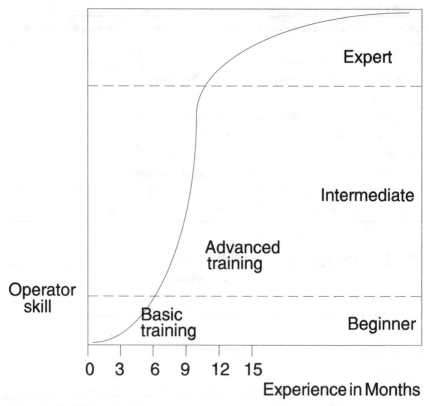

Figure 27 Typical user learning curve.

Figure 27 shows an operator learning curve for computer-aided tools.[75] A user requires different training at each stage in order to reach the expert stage.

Figure 28[76] shows a phased training program. Trainees should get help rapidly to limit lost time and frustration with the new tools. Those indirectly affected by computer-aided tools, such as supervisors, should be trained, also.

One trap is to have the first student train the rest of the users. This is an inefficient way to transfer knowledge and will overload the first student. Another trap is not to treat training seriously, but rather as paid time from work.[77]

---

[75]Besant and Lui, p. 396.

[76]Stark, p. 171.

[77]Chikofsky, Elliot J. "How to Lose Productivity with Productivity Tools." in *Software Development: Computer-Aided Software Engineering (CASE).* Ed. Elliot J. Chikofsky. Washington: IEEE Computer Society Press, 1989, p. 120.

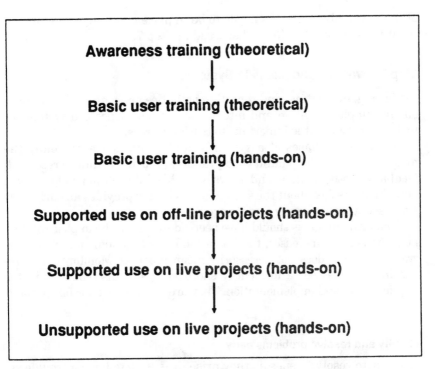

Figure 28 Proper transition of training.

During this task, procedures should be established so that people work efficiently in the new environment.[78] The procedures should be specifically adapted for use with computer-assisted technology and include the design policy discussed in Part 2, Design Policy, Design Process, and Design Analysis. The new procedures should not merely automate the manual tasks since this will not likely lead to the most efficient process.

Start using the system.   The users can start gaining experience and increasing productivity. Initial projects using computer-aided tools should be chosen to ensure some early successes, to maintain user morale, and justify management support. The new tools should not be used on a project with a tight time schedule that will pressure users to skip training. This could lead to inefficient use of the tools and eventual failure.

Adapt the system to the company's processes.   As users gain experience, changes in the processes and tools will be necessary to make better use

---

[78]Stark, pp. 102-103.

of the new tools. A process needs to be in place to highlight and implement these changes. This is discussed in Step 7.

Step 7: Evaluate and Maintain Systems

Reviewing, evaluating, and reporting activities are continuous throughout the implementation and integration process. Once the systems or tools are in place, it is important to track progress.

The earliest phases of a system's implementation are usually the most difficult. To help build good practices, it is important to record all feedback, both positive and negative. This information helps gauge how the users feel about the system and possibly provides an early metric to see whether requirements are being met.

Monitoring progress should have been addressed in the implementation plan. As activity increases, there is a tendency to assume all is well. Increased activity does not necessarily mean progress. Monitoring and reviewing progress serve to identify shortcomings or discrepancies in the requirements and implementation. They are *not* intended for finger-pointing.

Identify and resolve problems early

In order to resolve issues as they arise, it is wise to have a trouble reporting mechanism in place. Typically, a form should be developed that can adequately describe the problem. Users and management should be taught the proper method for reporting trouble and using the form. An archive of troubles and solutions will provide information on the timeliness of resolution. Items on the form should include:[79]

- user name
- application name
- date of the problem
- date solution needed by
- statement of the problem
- justification for resolving the problem
- estimate of resources required
- management approval
- results of solution
- user acceptance

---

[79]Chasen and Dow, p. 185.

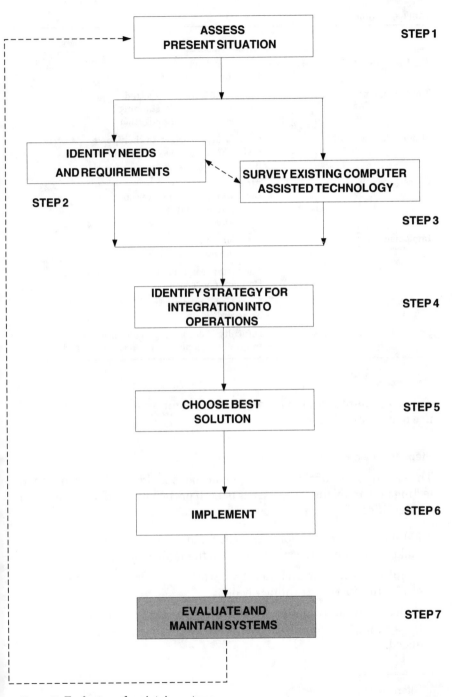

Figure 29 Evaluate and maintain systems.

TABLE 6   Common First-Year Problems

Problem Area	Problem
The CAD/CAM System	The users don't like it It's underutilized
Top Management	Immediate payback expected No foresight or understanding Don't change the organization
Middle Management	CAD/CAM conflicts with their goals CAD/CAM is too risky for real projects Afraid of loss of power
Users	Afraid of loss of job Don't want to be computer operators Once trained, may leave No methodology to use system
Implementation	Start on a high, finish on a low Planners not responsible
Information	The system can't handle all the information Data are not secure
System support	No procedures in place No records kept
System Development	Difficulty of developing in absence of standards Internal developments are not maintained

Table 6 presents sample problems that might arise in the first year of use of a new system or tool.[80]

Steps for review, evaluation, and reporting

The effectiveness of the new system or tool should be measured by collecting and analyzing data. Areas of both proficiency and deficiency can be identified. The major steps in this process include:[81]

- Establish guidelines for a project team to monitor the system and make reports and recommendations to management.
- Monitor the volume and quality of work to either validate user skills or identify further training needs.
- Collect and analyze statistics to evaluate rates of improvement to the system, new applications, and new procedures as they are introduced.

---

[80]Stark, p. 114.

[81]Chasen and Dow, pp. 188-191.

- Keep management involved to facilitate support and corrective actions.

- Develop procedures for backing up data, restoring data, and maintaining security to protect the data and its integrity.

- Respond to changes in corporate needs.

- Review and revise implementation procedures.

- Consider system upgrades for both hardware and software as applications demand.

The more the system is used, the greater the level of user expertise. Keep track of experts in the user community to resolve problems and issues later on. Also, local experts can help train new users.

Software suppliers often provide the source code for in-house maintenance and upgrades. Great care must be taken when making changes to system software. The temptation for improvements to be defined and programmed by the users can lead to several complications. User changes may unknowingly affect other parts of the software. Source code modifications should be handled through the vendor or with a user group and should be conducted with careful configuration control and clear documentation to explain the changes.

In time, the entire system or specific tools may require an upgrade or replacement. To ensure that the objectives and requirements of the users and organization are always met, continuous evaluation is mandated. In Figure 29, the dashed line implies that the process never ends. By repeating Steps 1-7, a firm can continue to profit from the best computer-assisted technology.

Application

This section presents the following case studies and examples to illustrate some of the procedures and principles discussed in the Procedures chapter:

- *Project Nirvana:* Details Sandia National Laboratories' efforts in selecting and evaluating a new CAD/CAE/CAM system.

- *Integrating New and Existing Systems:* Describes Boeing Aerospace's efforts to integrate an automated information system with their existing computer-aided tools.

- *CAD/CAM Justification and Follow-Up:* Details justification and postimplementation results of installing a CAD/CAM system at Simmonds Precision Products' Instrument Division.

- *CAD Tool Comparison Method:* Describes a method used by AT&T for comparing electronic CAD/CAE tools.

- *Generative Process Planning at Lockheed:* Describes the implementation of the GENPLAN system at Lockheed Aeronautical Systems Company.

- *CASE Tool Checklist:* Presents a checklist for examining and evaluating CASE tools.

- *Information Transfer and Database Storage Using Standards:* Presents the architecture and use of standards within Boeing's CAD/CAM Integrated Information Network.

- *Unsuccessful Use of IGES for Communication:* Describes the lessons learned at General Electric during a pilot program to transfer data between different computer systems.

- *Another View of the Process:* Presents a different, yet similar view of the overall process described in this part.

Project Nirvana

Nirvana: A goal hoped for but apparently unattainable

Background

When the people at Sandia National Labs, Albuquerque, NM started to search for a new electronic design and manufacturing automation (EDMA) system, the core team chose a name to reflect both the difficulties and rewards which were anticipated.[82]

Sandia is a part of the Nuclear Weapons Complex (NWC) which is comprised of three laboratories and several manufacturing locations. Most of the data transfer in the NWC occurs between Sandia and its manufacturing affiliate in Kansas City.

Realization of a need.   Recently, the organizations involved in the electronic CAE/CAD/CAM areas at Sandia realized that much of their hardware and software was becoming outdated. In some organizations, it had been a few years since any significant purchases or upgrades had occurred. Also, since the user community was widely distributed, there was no one platform that could be considered standard for their use.

To answer these needs, Sandia decided to put together a "cradle-to-grave" system that could be used by any member of the NWC. One objective was that the system be an overall package that could later be bought on a purchase order (e.g., call the vendor and order a Nirvana system).

Objectives and approach

Figure 30 illustrates the relationships and design transfer links within the electrical CAE/CAD areas in Sandia.[83] To reflect the reality of design transfer, the diagramming symbol for resistors was placed on the links which have been historically difficult.

The goal of the Nirvana core team is to develop an integrated, common electrical CAE/CAD environment (both hardware and software). This environment will support organizational design needs and the bidirectional transfer of design information between design definition, IC design, subsystem design, and manufacturing.

---

[82]From presentations and discussions with the Nirvana core team at Sandia National Labs, Albuquerque, NM.

[83]From presentations and discussions with the Nirvana core team at Sandia National Labs, Albuquerque, NM.

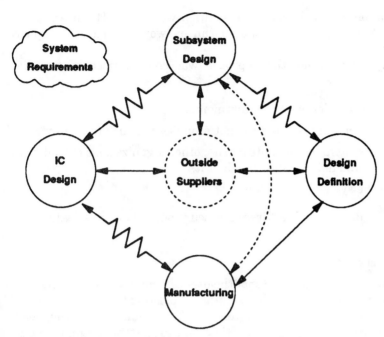

Figure 30 Relationships among CAE/CAD/CAM areas at Sandia.

There is presently an installed user base at Sandia. Several hundred users primarily work on five or six vendor platforms. These users provided suggestions regarding the initial candidate solutions based on their knowledge and experience. A key support factor is that a vice-president at Sandia is a visible champion of the Nirvana project.

One stipulation of the Nirvana EDMA system is that it be sourced from a single vendor. Since it is highly improbable that any system can meet all of their needs, a best-fit will be chosen. Similarly, it has been recognized that it may be necessary to change the present state of operations in order to best fit the new system into Sandia.

Interestingly enough, the core team and its champion simply agreed to the necessity of Nirvana and assumed that the system would show benefits and payback within a reasonable period of time. Nirvana is expected to carry a significant investment for Sandia.

Solution approach

It was assumed that Nirvana would be treated as a hardware independent solution. In this way, Sandia would let the vendors propose the platform on which the software solution would run. However, it was specified that the system should run on workstations and support Unix.[84]

With this in mind, the approach was characterized by four key points:

- Let the users define the requirements.

- Evaluate actual performance, not promises ("fly before we buy").

- Sell the system based on its merits; other organizations should want to use it and not be forced to adopt it.

- Make the system easy to buy.

Figure 31 illustrates the Nirvana procurement plan and schedule.[85]

Establishing the requirements

Since the core team realized that the user community could best identify both their own needs and shortcomings of existing tools, it was decided to let the users define the requirements. The requirements group leaders were from design definition, IC design, subsystem design, manufac-

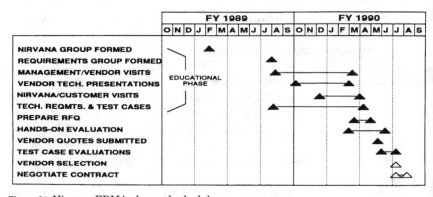

Figure 31 Nirvana EDMA plan and schedule.

---

[84]Unix is a registered trademark of AT&T.

[85]From presentations and discussions with the Nirvana core team at Sandia National Labs, Albuquerque, NM.

turing, and administration. Project subcommittees covered the major subdivisions required within an electrical CAE/CAD environment:

- printed wiring board (PWB) layout
- hybrid micro-circuit (HMC) layout
- IC design
- digital subsystem design
- system capabilities
- mechanical design and analysis
- system integration
- manufacturing
- analog subsystem design
- component database
- product/data configuration management and documentation

Outside help.   Because the subcommittees had difficulty in gaining a consensus, an outside consulting firm was brought in to facilitate the requirements definition process. For two weeks, the representatives of the subcommittees and the consultant met in a hotel near the Sandia facility.

The results of those sessions included:

- developing a common understanding and description of Sandia's design process
- developing a vision of improvements possible with the new system
- defining of over 200 features and support requirements
- developing a scoring strategy that balanced technologies and organizations
- preparing a document which captured all the above as well as:

  project goals and objectives

  problems, issues, and concerns

  policies and procedures required

  additional tasks to be completed

Figure 32 illustrates the Nirvana requirements process.[86]

---

[86]From presentations and discussions with the Nirvana core team at Sandia National Labs, Albuquerque, NM.

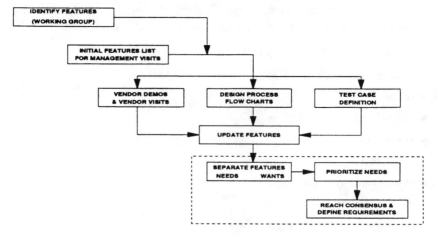

Figure 32 The Nirvana EDMA requirements process.

The system model

Once the Nirvana EDMA system's requirements were established, the core team was able to model the ideal system from process and functional perspectives. Figure 33 illustrates how the Nirvana system's processes fit within an overall electrical CAE/CAD/CAM environment.[87]

Figure 34 shows the functional breakdown within the Nirvana system model.[88]

Important aspects of the RFP

The RFP focused on technical and price proposals. The RFP was a large document consisting of several volumes. The quotation instructions and evaluation criteria descriptions took up six pages. Several key points are summarized below.

General provisions.   Overall instructions included requiring that technical and price proposals be separate documents which did not refer to each other. Also, vendors replying to the RFP had to include a certified balance sheet so that the financial state of the vendor could be evaluated.

---

[87]From presentations and discussions with the Nirvana core team at Sandia National Labs, Albuquerque, NM.

[88]From presentations and discussions with the Nirvana core team at Sandia National Labs, Albuquerque, NM.

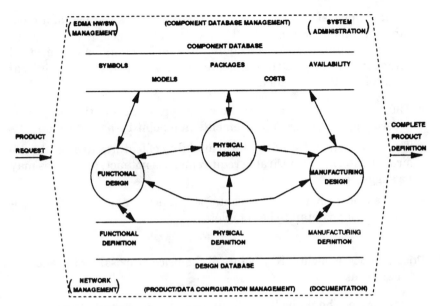

Figure 33 Nirvana EDMA system model.

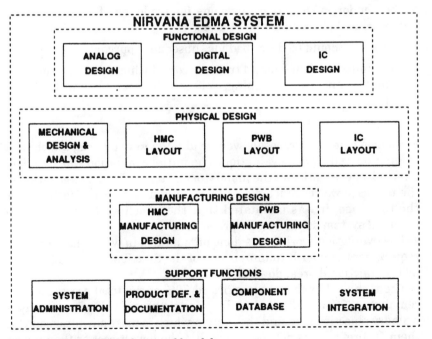

Figure 34 Nirvana EDMA functional breakdown.

Technical proposal content. Technical requirements were summarized in an attachment to the RFP. They were categorized as either mandatory or desirable. To be considered further, vendors had to meet all mandatory criteria. Other key elements regarding the technical content of the RFP included:

- The vendor should indicate how the proposed solution meets or exceeds all mandatory requirements in a point-by-point fashion.

- The vendor should indicate the availability of features to meet the desirable criteria. If a desirable requirement is not met, alternates may be suggested.

- The response should be self-contained, concise, and clear while providing enough detail to be complete.

Price proposal content. Key points of the price proposal had to address such areas as:

- Total prices should be listed for hardware, software, three-year life-cycle hardware maintenance, and three-year life-cycle software maintenance.

- The vendor must make provisions for trade-ins of existing equipment, should Sandia desire to do so.

- Discounts should be described for subsequent purchases.

- Costs must be quoted for all training costs including on-site software training, vendor site training, self-pace instruction, and train the trainer.

Evaluation criteria. Proposals were evaluated in two parts; a technical evaluation and a price evaluation. The initial evaluation was based on a technical evaluation of the mandatory criteria. The demonstration of the test case was conducted on-site at Sandia or a nearby facility. For the test case, it was mandated that the benchmark be run on an identical system as the vendor proposes.

The evaluation scoring was done on a matrix in which each row is a requirement and each column is one of the functional areas identified by the functional breakdown in Figure 34.

The technical evaluation comprised 70% of the total score, and the remaining 30% was devoted to the price proposal evaluation. Sandia's primary objective is to find the system which is most advantageous for them. Therefore it is not necessary that the lowest cost or the highest technical proposal will win the contract.

Issuing the RFP

Sandia wanted to distribute the RFP to as wide an audience as possible. Many vendors were known due to the installed user base already at Sandia. To reach even more vendors, Sandia took out an ad in the *Commerce Business Daily* describing the need for the Nirvana EDMA system.

Benchmarking

In order to properly evaluate the proposed systems, the Nirvana core team decided to make the test evaluations as close to real life as possible. Obvious criteria included measuring the system against the test cases for performance and support functions.

Based on feedback from the user community, a typical Sandia design engineering case was developed. An interesting twist to the test case was making it a process. It was decided that the vendors who would respond to the RFP would have to carry out the exercise on site. Also, having the Sandia team on hand, it allowed for other real-world scenarios to randomly occur. For example, a core team member could suddenly change a requirement or cause a power outage.

These conditions were intended to measure vendor expertise and how capable the system really was. After getting burned before, Sandia was not about to take a vendor's word based on a canned solution to a benchmark test. If the vendor proved capable of successfully completing the demonstration, then the system would score marks for its functional abilities. The vendor would score marks for its ability to provide support in the future.

Comments

As of August 1990, Sandia had narrowed the field down to three vendors and were in the process of evaluating the benchmark scores. An implementation plan had not yet been established. It is intended that a plan be put in place, but the core team feels much of it is conditional on the system chosen. Also, the Nirvana system may result in many changes in practices at Sandia so the implementation plan will be designed to reflect those changes.

Concerning the existing set of designs which are stored in the present set of systems at Sandia, it is assumed that a one-way, one-time translation to the Nirvana system will be all that is required. Depending on the system chosen, this will occur through the use of direct translators or a neutral format such as IGES.

The parallels between the process employed at Sandia and the process described in the Procedures chapter of this part are close.

Integrating New and Existing Systems

Boeing Aerospace of Kent, WA, added more capability to its existing computer-aided tools by working closely with the system vendor to integrate an automated information system with those tools. The goal was to reduce the manufacturing cycle time by improving the flow of information. Welding was the first function converted to the electronic (i.e., paperless) information system. The system is used to create, maintain, transmit, and store documents governing welding.

Problems with paper

The weld design and specifications originate in the Manufacturing R&D organization and are approved by the Quality Assurance Weld Planning Department. Other departments also review the document. The information then passes to the factory floor to the welder doing the work. Several difficulties with paper slowed down the welding process. The paper document had to be routed to various groups in different locations. If the document got lost, it had to be regenerated at a considerable cost of effort and time. Also, on the factory floor, the documents were not always legible, leading to further time losses and quality problems. The electronic flow of information removed many of these difficulties by making the electronic data immediately available from a central database. On the factory floor, the information is up-to-date because it is taken from the central database. Paper copies are legible because they are produced as needed from the computer and not copied from other paper copies.

Reengineering the process

As Boeing implemented the system, they discovered that reengineering could further improve the process. For example, by using graphics on the welding documents, they discovered that many of the welds were similar and some of the information on the document was redundant. The information was changed to eliminate the redundancy.

Continuous improvement

Boeing has plans for further enhancements to the system including expanding the project to other manufacturing processes, adding an expert system to automatically generate welding specifications, and integrating the system with electronic mail systems. Boeing embraces the philosophy that continuous change is necessary to maintain a competitive edge.[89]

---

[89]"Beating the Paper Chase." *Manufacturing Engineering*, vol. 103(6), December 1989, pp. 28-30.

CAD/CAM Justification and Follow-Up

Simmonds Precision Products' Instrument Systems Division is a leading supplier of commercial, military, and aerospace instrumentation and control systems.

This case study will focus on the justification and postimplementation of Simmond's initial CAD/CAM evaluation.[90]

Traditionally, Simmonds based their financial decisions on short-term (e.g., quarterly) earnings results, rather than on long-range strategy. This conservative strategy created a problem when collecting cost data to try to evaluate a CAD/CAM solution based on a longer (two year) plan. For this reason, Simmonds decided to start from scratch to develop their evaluation criteria. Data were collected from user groups and symposia. A list of companies with similar applications and financial constraints was collected. After meeting with representatives from the other organizations, Simmonds realized that trying to quantify the near-term savings opportunities was indeed difficult.

Evaluating potential applications

Simmonds decided to identify applications that were relevant, could perform reliably, and would earn a high rate of return on investment. Four areas of opportunity were identified: printed circuit design, mechanical engineering design, manufacturing, and graphics.

Historical baseline. Printed circuit (PC) design seemed to present the greatest opportunity for high productivity gains. The design group had plenty of well-documented data from manual and automated drafting systems already in place. These data served to provide a good baseline for comparative analyses.

Mechanical engineering design also displayed great potential for productivity. However, the areas of potential application were in areas with little historical detail. With insufficient data to develop a base-lined model, measurable short-term savings were uncertain. This uncertainty was compounded with anticipated training, programming, and database development costs. Simmonds anticipated a large amount of work would be required before savings would begin to show. They decided to be conservative and not create a savings statement.

The areas of manufacturing and graphics were considered to have very limited opportunities. The manufacturing organization chose only to limit their commitment to the area of NC programming. The graph-

---

[90]Van Nostrand, R. C. "CAD/CAM Justification and Follow-Up: Simmonds Precision Products Case Study." *CAD/CAM Management Strategies.* Eds. Robert M. Dunn and Dr. Bertram Herzog. Pennsauken, NJ: Auerbach Publishers, 1988, pp. 1-16.

ics area limited their estimates to only the generation of illustrations for documentation.

Assessing feasibility

The first analyses conducted on CAD/CAM feasibility were based on published reports or vendor data. All the reports enthusiastically endorsed CAD/CAM, citing productivity gains of up to 40-to-1. These data provided the source material for scenarios projecting a payback within two years. Further examination of the reports revealed a lack of consistency and backup data.

Simmonds decided to refer back to its user data and data from five representative companies with mature systems in place. Estimates were prepared for performance requirements of the proposed CAD/CAM system. Table 7 is an example illustrating the proposed labor-hour requirements for PC design.

Preparing and submitting the CAD/CAM justification

Tangible savings had been estimated to be approximately $240,000 per year. Unfortunately, the financial analyses which accompanied the estimate suggested a payback time of 40 months, which greatly exceeded the desired 24-month period. However, the proposal did not address the hidden benefits and the long-range strategic impact of CAD/CAM.

Hidden benefits.   Simmonds was tempted to quantify hidden benefits. A study done at Lockheed suggested that interactive graphics could result in producing more ideas and better designs with 15% fewer design changes. Also, CAD/CAM allows for a high degree of part stand-

---

TABLE 7   Labor Hour Estimates

	Manual (actual)	Computer-Aided Drafting (actual)	CAD/CAM (est.)
Schematics	22	22	16
Feasibility analysis	5	5	1
Design and layout	40	40	20
Fabrication drawing	6	6	2
Assembly drawing	32	6	2
Artwork	40	10	0
Checking	30	30	11
Total	175	119	52

ardization. In one case, CAD/CAM standards resulted in a reduction of 295 part numbers for a single subassembly.

Another hidden benefit is enhanced creativity of the workforce. Frequently, CAD/CAM allows for less-skilled workers to perform more complex tasks. A subsequent benefit of this is that the costs of hiring more experienced staff is reduced.

In the end, Simmonds decided not to attempt to quantify the hidden benefits. The rationale for this decision was simple: if so many hidden benefits could be identified, then an equivalent number of hidden costs could also be present.

Simmonds was frustrated by this offset. Further, it was felt that CAD/CAM was a strategic necessity to insure a competitive place in the market. Simmonds recognized CAD/CAM's abilities to:

- strengthen its position in the marketplace
- produce lower-cost, higher-quality products
- reduce intervals
- generate greater customer confidence
- integrate engineering and manufacturing

Although viewed as a necessity, the apparently poor reconciliation of short-term payback was a cause for concern.

Management's role.   Management intervened and the president of the division encouraged the continuation of the justification effort. A meeting with the chairman of the board of Simmonds was held to determine if CAD/CAM was consistent with corporate strategy.

CAD/CAM consultant.   An expert in CAD/CAM was brought in to audit and evaluate the proposal. The consultant's approach emphasized being conservative and compensating for variables and unknowns. The consultant went on to stress that:

- Savings might not be realized in the first year and a half.
- The majority of savings in the first two years would probably come from areas not yet recognized.
- It was probable that a significant amount of error reduction in both documentation and design would not be measurable.

The consultant helped finalize the proposal and gain approval for the CAD/CAM capital expense.

Implementation factors and tactics

Many of the implementation factors and tactics that affected the actual payback of the CAD/CAM system were:

- balancing training and work assignments such that many applications were begun in parallel
- establishing a full-time, managed, and centralized design group
- maintaining awareness of corporate limitations and expertise
- avoiding (early on) tasks that may have undermined the system's credibility and overall confidence
- implementing incentives to help cover initial overruns during training to ensure support
- dedicating a facility to house the system, thus guaranteeing its dependability

Results

Overall, Simmonds only quantified those savings that could be supported by historical data, rather than simply recording every instance of a savings.

PC design: First year. Objectives for PC design in the near term were saving, on average, 67 labor-hours and $75 in vendor fees per standard design. Simmonds projected six months for training and bringing designers to an intermediate level of expertise. In the first year:

- Training was completed on time.
- Average design time was reduced from 119 to 50 labor-hours.
- Autorouting capabilities exceeded expectations, yielding a 90 to 95% completion rate.
- Four or five feasibility layouts could be performed in the same amount of time previously needed for one.
- It was possible to accurately determine the minimum number of layers required for a design.
- The system allowed for greater compatibility with a larger base of board vendors. This allowed for a cost reduction of $150 per board and turnaround was reduced by three weeks.

PC design: Second year. During the second year, unexpected project loads in the first eight months exceeded the load of the previous eight

months by 220%. Additionally, the new designs required an un-
precedented level of complexity and technical requirements. Overall,
all projects were delivered on time and at or below bid. Specific results
included the following:

- PC board production cycle time was reduced by 100%.
- Production drawings and schematics were produced with greater ef-
  ficiency and fewer errors.
- Productivity gains were calculated at 3.4-to-1.

Mechanical design.   Simmonds initially intended to automate its
existing mechanical design process. Instead, the design process was
restructured, focusing on the design task's building blocks. Several
methods for problem solving were developed which in turn generated
some major productivity improvements:

- Analysis turnaround times were reduced by as much as 20-to-1.
- Chargeable mainframe time was reduced 9-to-1.
- Data conversion errors were virtually eliminated.
- Detailed worksheets were automated.
- The manufacturing fabrication process was improved.

Many of these benefits could not be directly related to the CAD/CAM
system itself. However, Simmonds felt that they were a direct consequence of the implementation process.

Manufacturing.   Savings were projected to be about $18,600 for NC
programming. Due to the increased work load from the PC design area,
however, very little time was spent in this area in the first two years.
Thus, an effective net loss was seen in this area due to training costs.

Graphics.   Initial estimates projected limited savings in the area of
graphics for proposals, publications, and support. However, the benefits were much greater than anticipated and a wide range of applications were developed. Reported results included:

- One project reported savings of approximately $30,000 in costs and
  eight employee-months in direct labor-hours.
- Modifications to printed circuit assembly drawings showed a 6-to-1
  productivity improvement.
- Schematics now took only one iteration from concept to technical
  manual, where previously four were required.

- Software flow diagrams now took 15 minutes with the CAD system. Manual production required between two and six hours.

Error reduction.   Qualitatively, Simmonds found that the CAD/CAM system led to error-free results (given input yields desired output). This fact was even more surprising in light of the increased complexity of designs required in the second year and beyond. Quantitatively, verification errors were reduced by 24%, and the cost of correcting designs was reduced by 32%.

Summary

The increased productivity provided by the CAD/CAM system allowed for the design and drafting staff to remain constant while the work load and the remaining engineering staff increased significantly. An analysis of the savings estimates and the actual results indicates that the CAD/CAM system far exceeded expectations. Table 8 reflects the reconciliation of predicted and actual savings.

Conclusions

On the positive side, Simmonds found that the CAD/CAM system far exceeded expectations. The CAD/CAM group produced consistently high-quality work, was dedicated, and had a lower than average level of problems or absenteeism.

On the negative side, Simmonds discovered that to properly implement CAD/CAM, a full-time applications programmer and instructor is required. In general, such experts are rare and expensive. Hardware restrictions limited the expansion of the system user base.

Overall, Simmonds concluded that CAD/CAM's strategic role was soon overshadowed by its use and value as a fundamental tool.

TABLE 8   Predicted vs. Actual Savings

	Predicted	Actual
Gross Savings	$240,000	$546,000
Service Contract		($ 39,000)
Training		($ 41,000)
Library and Procedure Development		($ 12,000)
Net Savings		$454,000
After-Tax Savings		$227,000
Basic System Amortization	50%	70%

CAD Tool Comparison Method

Motivation

Recent trends in the area of electronic design automation, both in vendor improvements and vendor growth, led AT&T to begin to examine some of the tools used for printed circuit board (PCB) design. Additional pressures from increased emphasis on productivity and time to market added to these pressures.

A committee was formed to create a strategy and plan for a cost-effective CAD environment which would support AT&T's product realization process. This activity was divided into three phases: compare tools for technical advantages, analyze the advantages of the different tools, and pick the best CAD/CAE tools for the users. This case study will discuss the first phase of the process.

Due to the proprietary nature of most of the background material, AT&T and vendor-specific information are not included.

Designs

Five designs, representing a typical sampling of the work required, were proposed to check the abilities of the systems. The board designs ranged as follows:

- logic: analog, digital, analog and digital
- PCB: two through ten layers, varying dimensions
- components: through-hole, surface-mount, mixed
- complexity: total number of component and connector pins

Evaluation criteria

A primary goal in the comparison was to determine how effectively each system supports the PCB design process. Based on feedback from designers, assessment was to be based on a task-oriented as well as a function-oriented comparison. The following are the criteria used to rate and compare the respective systems for the design set described above:

- time

  total elapsed time

  system execution and wait time
  user think time

- ease of use
  amount and ease of user input
  system messages (length and completeness)
  graphics
  user control and flexibility
  number of errors and re-tries (did the system do what was expected?)
- learning effort
  on-line help
  user documentation
  relationships to ease of use
- quality and design for manufacturability (DFM)
  adherence to electrical design rules and constraints
  adherence to DFM/design for test (DFT) design rules and constraints

Comparison activities

Table 9 illustrates how the results of the comparisons are tabulated. For each task, the tools are rated in the four areas described above. The ratings are based on a scale of 0 to 10 (0 = worst, 10 = best).

Activities which are not covered in this phase of the task included archiving, backplane specific tasks, component libraries, documentation, hybrid circuits, integration of CAD and CAE, and future plans. These items are planned for the second phase of the process.

Conclusions from comparisons

It is quite possible that for a given task, one tool may be clearly superior to another. Other times, their scores may be tied. Determining a clear-cut winner is not easy. Careful trade-offs must be conducted, with the user community's inputs as the primary decision criteria.

Ambiguous results are not uncommon and serve only to reinforce the fact that a single-vendor turnkey solution may not be the best choice. An integrated system comprised of different tools may best meet the needs of the users and the company.

Generative Process Planning at Lockheed

As discussed in Step 3 of the Procedures chapter, group technology (GT) and computer-aided process planning (CAPP) are two major implementation areas for CAM. At Lockheed Aeronautical Systems Company's Marietta, Georgia (LASC-G) facility, generative process planning and GT are widely used. LASC-G designs and produces

TABLE 9   CAD Tool Comparison Score Sheet

Design Task	Time		Ease of Use		Learning Effort		DFM/Quality	
	Tool A	Tool B	Tool A	Tool B	Tool A	Tool B	Tool A	Tool B
Interactive Placement								
Automatic Placement								
Partitioning, Swapping								
Placement Checking								
Interactive Interconnect Editing								
Automatic Routing								
Interconnect Checking								
Test Feature Creation & Checking								
Metal Optimization								
Soldermask Creation & Checking								
Nomenclature Creation & Checking								
Solder Paste Creation & Checking								
Debugging Plots								
Manufacturing Outputs								

cargo aircraft such as the C-130 Hercules, the C-141 StarLifter, and the C-5 Galaxy.[91]

LASC-G has implemented GENPLAN, a proprietary software package that allows for rapid generation of work instructions from engineering drawings. GENPLAN generates operation sheets for such items as: machined parts, wiring, plastics, composites, sheet metal, tubing, extrusions, fiberglass, subassemblies, decals, blankets, nameplates, and laminated shims. The primary input to the GENPLAN system is a proprietary GT code describing the part.

The codes for a new part are entered into the system that automatically generates a skeletal plan associated with the process logic in the GT code. Planners can then edit or add data to the plan. The resulting operation sheet is reviewed by quality assurance and sent to time standards engineering.

---

[91]*Report of Survey Conducted at Lockheed Aeronautical Systems Company, Marietta Georgia.* Best Manufacturing Practices Program. US Navy RM&QA(OASN), August 1989.

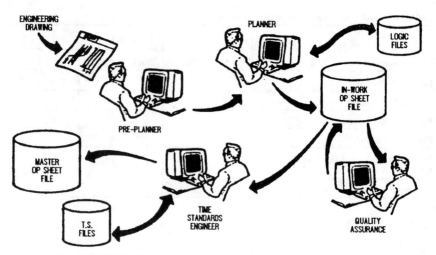

Figure 35   The GENPLAN flow.

Figure 35 illustrates the GENPLAN system flow.

The information in a plan typically consists of headers, material requirements, and operation instructions. Variations of the plans are created for fabrication, assembly, and tool orders.

GENPLAN also has many sophisticated file management features. Included among them are the following:

- plan databases set up to allow for global updates

- planner and review identification automatically handled by the system

- tools in place to ensure that planners complete work in a timely fashion

- tool orders produced for tool design, manufacturing aids, fabrication and assembly tooling, and NC tape

GENPLAN provides varying levels of automation across many manufacturing areas. Some areas are automated such that planners are only required to review and sign off on a plan. The system's reliability is high, experiencing only short periods of downtime during its implementation. Completed plans are also entered into an operation sheet database to ensure reliable access to planning data.

CASE Tool Evaluation Checklist

Table 10 shows a fairly comprehensive checklist from *BYTE* designed to offer some structure when evaluating CASE tools.[92]

---

[92]Adapted from McClure, Carma. "The CASE Experience." *BYTE*, vol. 14(4), April 1989, p. 240.

TABLE 10   CASE Tool Evaluation Checklist

General Information	Diagramming Support	Methodology Support	Error Checking
▫Vendor name	▫Data flow	▫Yourdon	▫Syntax
▫Product name	▫Control flow	▫DeMarco	▫Consistency
▫Date introduced	▫Decision table/matrix	▫Gane/Sarson	▫Completeness
▫Number of copies	▫Hierarchical tree	▫Bachman	▫Requirements traceability
sold	structure	▫Chen	▫Quality assurance
▫Price	▫Structure chart	▫Merise	**Type of Toolkit**
Hardware Platform	▫Warnier/Orr	▫Orr	▫Planning
▫Personal computer	▫State transition	▫Jackson	▫Analysis
▫Workstation	▫Pseudocode	▫Ward/Mellor	▫Design
▫Mainframe	▫Screen layout	▫Hatley	▫Database design
Graphics	▫Dialogue flow	▫Object-oriented	▫Real-time design
▫Color	▫Report layout	▫SADT	▫Code generator
▫Mouse	▫Data structure	▫Stradis	▫Programming
▫Windows	▫Entity relationship	▫Method-1	▫Framework
Life-cycle Support	▫Logical records	▫LSDM	**Prototyping Support**
▫Planning	▫Booch	▫Other	▫Screen painter
▫Design & analysis	▫Petri nets	**CASE Repository**	▫Report painter
▫Implementation	▫Other	▫Host-based	▫Functional model
▫Maintenance	**Code Generation**	▫PC-based	▫Simulation
▫Project manage-	▫Skeleton program	▫DBMS architecture	**Target System**
ment	▫Complete program	▫Reports	▫On-line
Reengineering	▫Language	▫Change/version	▫Batch
Support	▫On-line program	control	▫Transaction processing
▫Static analyzer	▫Batch program	▫Audit trail	▫Real-time
▫Redocumentation	**Other Requirements**	▫Download	▫Embedded
▫Restructure	▫ _____	▫Logical partitioning	
▫Reverse engineering	▫ _____	▫Consolidation	
▫Dynamic analyzer			
▫Converter			

Information Transfer and Database Storage Using Standards

The Boeing Commercial Airplane Company used standards to facilitate transfer of data among various CAD/CAM systems and an engineering database. Figure 36 shows the architecture of the Boeing system. Boeing calls the information system the CAD/CAM Integrated Information Network (CIIN). The outer circles are the various computer-aided systems used for design and manufacturing. The inner circle is the engineering database with data stored in a neutral format. The wedges are the translators that allow communication between the computer-aided systems and the database. The engineering database can communicate with other databases through a geometric database management system (GDBMS). This project was

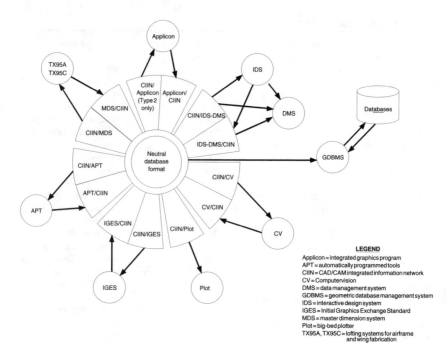

Figure 36 Boeing's CAD/CAM integrated information network.

successful partly because adequate resources were allocated for the life of the project.[93]

Unsuccessful Use of IGES for Communication

General Electric Company undertook a pilot program to use the IGES standard for transferring data between computer-aided systems from two different manufacturers. The data to be transferred included annotated designs and wire-frame models. One computer-aided system was scheduled to have an IGES processor available in time for the project, although some custom changes would be needed. The other computer-aided system would not have an IGES processor in time for the project. General Electric decided to implement its own IGES processor for this system.

[93]Beeby, William D. "The Heart of Integration: A Sound Data Base." *IEEE Spectrum,* vol. 20(5), May 1983, pp. 44-48.

Problems with translation

General Electric had problems with data translation between the two computer-aided systems. Most of the problems were due to incompatibilities between the two systems. For example, some of the annotations on the designs were not recognized by the system receiving the data. Many of these problems could be solved by changing the translator, but some problems had to be resolved manually. Another serious problem was that too much time was required to do the data translation and transfer. Both of these problems were enough to offset any efficiency gained in connecting the systems, so the IGES translator was postponed. Instead GE went with direct translators to transfer data between the two systems. This was successful partly because of the experience gained from the IGES project.

Lessons learned

General Electric learned several lessons from the IGES translation project. The resources required to do such a job in-house are large. Some information required by the company to do such a job involves proprietary vendor information, and may not be readily available. Vendors are better suited to do the work, and an adequate translator can result if a company works closely with the vendor.[94]

Another View of the Process

It is interesting to compare how others view the process of implementing computer-assisted technology. Several sources try to simplify the process of acquiring a CAD system by laying out the basic steps to structure the process (e.g., Stark, p. 83). Dr. Joel N. Orr, a CAD/CAM consultant, offers what he refers to as the "Steps of Creeping Commitment."[95] Dr. Orr explains that we are constrained by our paradigms when it comes to computers. We get caught in the trap of doing a task via the same old process with brand new tools. He sees the analogy of "beating the engine with a buggy whip to get a car to go faster."

Dr. Orr goes on to explain that there really is no coherent standard to mechanical design engineering. As a result, many poor decisions are often made when implementing CAD. The steps that are presented at-

---

[94]Giguere, Marshall E. and Kennicott, Philip R. "Transferring Annotated Design Data with IGES: General Electric Case Study." *Computer-Aided Design, Engineering, and Drafting.* Eds. Robert M. Dunn and Dr. Bertram Herzog. Pennsauken, NJ: Auerbach Publishers, 1984, pp. 1-20.

[95]Orr, Dr. Joel N. "How to Buy the Right CAD Program", Paper presented at National Design Engineering Show, Chicago, IL, February 26-March 1, 1990.

tempt to bring some order to a confusing and complicated process. Creeping commitment refers to the idea that a gradual process that includes the proper planning that can greatly increase the odds of success.

The Steps of Creeping Commitment

1. Find a champion.
2. Describe how you currently do things.
   - Identify the what, who, how long, etc.

3. Postulate how you will do these things with the new technology.

   - This is a big step, so determine how much change you can stand.

4. Perform a reality check on Steps 2 and 3.

   - Get feedback and advice of many.

5. Create a mapping of now to then.

   - This step is crucial. Try to determine the number of steps that can minimize the "bloodshed" at each step.

6. Assign values to improvement.
7. Describe the required system functionally.

   - Describe the system in terms of statements like: "I need a system that does this...."
   - Do not concern yourself with performance specifications of the equipment.
   - Consider full automation.

8. Set up a pilot with measurable goals.
9. Compile a list of vendors who *seem* to meet your needs.
10. Check out the vendors.

    - This involves checking their reputation, history, and financial strength.

11. Conduct interviews with users of current systems.

    - Try to find users doing what you want to do.

12. Conduct demos of the capabilities of "surviving" vendors.

    - Ask the vendor "Can you do what I want?"

13. Gather proposals and price quotes.
14. Evaluate and select system based on a successful pilot.

Chapter

4

Summary

All companies involved in the design and manufacture of products use computers to some extent. Many of these companies, however, are not getting the full benefit of these tools because their computers are not properly integrated into the design and manufacturing process. Many factors can hinder a company in its quest for better quality products at a lower cost in a shorter time. In some cases, the previous manual processes are not adapted for use with the new technology. In other cases, the latest computer is bought to help speed up a process that is not really understood. Sometimes the company will give up continued use of automated tools because the purchased technology does not live up to their unrealistic expectations. Sometimes, not enough resources are committed to implementation beyond the capital expenditures. These are just some of the many risks that have been mentioned throughout this part.

To realize the full benefit of computer-assisted technology, a company must carefully follow a process that will minimize the risk of failure. This process begins with a commitment to allocate sufficient resources to properly integrate computer-assisted technology into the everyday functions of the company. The process does not end when power is turned on to the new tools. Rather, the process of evaluating and integrating new tools continues as long as the company exists. Companies that make computer-assisted technology an integral part of operations often become leaders in their field, reaping financial benefits.

In an article discussing modern tire technology, the point was made that even for such mundane items as tires, it is possible to design, analyze, and manufacture with the very latest technology has to offer. For as in any mature industry, competitive advantage requires investment in both time and money. The article concluded by describing the efforts of the tire builder:[96]

---

[96]Simanaitis, Dennis. "Technology Update: Tires." *Road & Track,* 41(11), July 1990, p. 104.

This artful portion of the process is performed by that most elaborate of computer-assisted devices—a real person. As computer-cozy as technology has become, there's still room—and need—for craftsmen.

A comforting thought, this.

This part has laid out a process for selecting and integrating computer-assisted technology into the design and manufacturing process. It can be adapted to the needs of individual companies. Figure 7 shows the basic process. This process, however, must be under the umbrella of corporate commitment for success to be possible.

The Application chapter presented case studies illustrating several parts of the Procedures chapter and their results. There are many successes in the use of computer-assisted technology despite all the possible risks. The successes can serve as examples and encouragement for companies embarking on the process. The failures can also serve to highlight the traps inherent in such a complex process.

Summary of the Procedures

Table 11 summarizes the key points of the Procedures chapter.

TABLE 11 Summary of the Procedures

Step 1: Assess Present Situation	
Procedures	Supporting Activities
Assess need for computer-aided tools.	■ Review current state of affairs to determine if change is needed. ■ Identify strengths and weaknesses in existing systems and methods. ■ Identify coordination considerations for new system.
Set realistic expectations.	■ Determine how benefits will be measured. ■ Remember that gains are not immediate.
Assess current situation.	■ Examine information flow and areas of application. ■ Identify potential users.
Describe current tools and abilities.	■ Detail current use of computers. ■ Investigate state of practice outside organization. ■ Seek expertise from outside, neutral expert.

Step 2: Identify Needs and Requirements	
Procedures	Supporting Activities
Identify needs.	■ Understand scope of problem. ■ Ensure computer-aided tools are consistent with business goals. ■ Plan for future needs.
Determine costs.	■ Quantify as many direct costs as possible. ■ Prepare and allow for hidden costs. ■ Get the right people involved in the problem-solving process.
Develop and document the requirements.	■ Understand, document, and disseminate requirements ■ Ensure management is involved and supportive. ■ Consider requirements for future applications. ■ Consider integration requirements.
Be aware of possible outcomes.	■ Remember that solutions can take several forms: turnkey system (single vendor or VAR) augmented current facilities, custom system, or no system.

Step 3: Survey Existing Computer-Assisted Technology	
Procedures	Supporting Activities
Look for information.	■ Read books and journals. ■ Attend trade shows and conferences. ■ Use peer contacts.
Identify necessary tool information.	■ Identify ease of interfacing individual tools. ■ Consider differences in data-transfer standards. ■ Ensure proper use and management of engineering databases. ■ Identify tool functions and limits.
Survey computer-aided tools	■ Survey areas of application (e.g., mechanical, electrical, software, and manufacturing). ■ Survey tool classifications (e.g., CAD, CAE, CASE, CAM). ■ Consider configuration management scheme at all times. ■ Consider integration at all times.

TABLE 11   Summary of the Procedures (Continued)

Step 4: Identify Strategy for Integration Into Operations	
Procedures	Supporting Activities
Have a realistic view of CIM.	■ Identify what information is needed, where it will come from, where it should go, its format, method of transfer, and what happens if it does or does not get there.
Prepare for CIM strategy.	■ Recall lessons learned and pitfalls. ■ View CIM within the context of a strategy for the entire business.
Follow key steps for successful CIM.	■ Ensure understanding, foresight, and planning.
Plan.	■ Identify a conceptual design for CIM, a technical architecture, and an implementation plan.
Simplify.	■ Provide for a smooth migration to CIM.
Automate.	■ Bring in and selectively apply technologies associated with CIM.
Integrate.	■ Tie together new and existing systems into a coordinated, well-managed process. ■ Get the right people involved early.

Step 5: Choose Best Solution	
Procedures	Supporting Activities
Prepare RFP.	■ Use format that makes it easy to evaluate and compare vendors. ■ Include detailed technical requirements.
Choose vendors that will receive RFP.	■ Limit number to keep proposal evaluation manageable.
Evaluate vendor responses and rank vendors.	■ Base criteria on requirements in RFP. ■ List attributes used for evaluation. ■ Assign relative importance to each attribute. ■ Evaluate vendors on each attribute. ■ Weight and sum scores.
Benchmark systems.	■ Evaluate vendor claims for each system. ■ Be aware that benchmarking can be costly and time consuming.
Select suitable vendor and system.	■ Summarize results. ■ Present recommendations to management.

Step 6: Implement	
Procedures	Supporting Activities
Establish the implementation process.	■ Plan the implementation. ■ Prepare for implementation. ■ Start using system. ■ Adapt system to company's processes.

Step 7: Evaluate and Maintain Systems	
Procedures	Supporting Activities
Identify and resolve problems early.	■ Put a trouble-reporting mechanism in place. ■ Teach proper methods of trouble reporting to users and management. ■ Archive trouble reports and associated solutions.
Evaluate, review, and report.	■ Identify areas of proficiency and deficiency. ■ Continuously evaluate state of affairs.

Barton, David L. "A First Course in VHDL." *VLSI Systems Design's User's Guide to Design Automation,* Manhasset, NY: CMP Publications, 1988, pp. 40-47. Describes several features and uses of the VHSIC Hardware Description Language (VHDL). Discusses how the language can be used by designers in high-level and low-level design of circuits.

Bathe, Klaus-Jurgen. *Finite Element Procedures in Engineering Analysis.* Englewood Cliffs, NJ: Prentice-Hall, 1982. An advanced text on derivation and use of finite-element analysis.

"Beating the Paper Chase." *Manufacturing Engineering,* vol. 103(6), December 1989, pp. 28-30. Records Boeing's success in replacing paperwork with electronic information transfer in its Kent, WA manufacturing facility. This is a good example of how a company worked with a vendor to continuously upgrade its integrated computer-aided system.

Beeby, William D. "The Heart of Integration: A Sound Data Base." *IEEE Spectrum,* vol. 20(5), May 1983, pp. 44-48. Discusses several issues concerning the integration of a database with the design and manufacturing operations. Issues include reengineering of processes for efficient automatic data transfer, single vs. multiple databases, and difficulties encountered in setting up a database.

Beeby, William and Collier, Phyllis. *New Directions Through CAD/CAM.* Dearborn, MI: SME, 1986. Explains the techniques, advantages, and processes of CAD/CAM. Provides some historical detail as well as some case material to describe CAD/CAM's impact on industry.

Bennett, Robert E. and others. *Cost Accounting for Factory Automation.* Montvale, NJ: National Association of Accountants, 1987. Details the results of a study on the cost accounting problems associated with automating manufacturing processes.

Besant, C. B. and Lui, C. W. K. *Computer-Aided Design and Manufacture.* New York: John Wiley & Sons, 1986. Contains a discussion on the history of computer-aided design and manufacturing.

Bohse, Michael E. "Integrating CAD and MRP Systems." in *Computer-Aided Design, Engineering, and Drafting.* Eds. Robert M. Dunn and Dr. Bertram Herzog. Pennsauken, NJ: Auerbach Publishers, 1986, pp. 1-12. Discusses the use of design and manufacturing databases. Compares the advantages and disadvantages of single and multiple databases.

Bozman, Jean S. "How Some Users Were Able to Plant CIM in Their Factories." *InformationWEEK,* May 12, 1986, pp. 39-56. Presents examples of successful CIM and how barriers were overcome during its implementation.

Brandt, Richard and Port, Otis. "How Automation Could Save the Day." *Business Week.* March 3, 1986, p. 72-74. Describes the ideals and problems facing the "factory of the future" in the United States.

Burnett, David S. *Finite Element Analysis.* Reading, MA: Addison-Wesley, 1987. Introductory text on derivation and use of finite-element analysis. Includes an introduction of the basic concepts as well as detailed explanations of the analysis method.

Chasen, S. H. and Dow, J. W. *The Guide for the Evaluation and Implementation of CAD/CAM Systems.* Atlanta:*CAD/CAM* Decisions, 1983. Provides a detailed and comprehensive description of the process of evaluating and implementing computer-aided systems.

Chikofsky, Elliot J. "How to Lose Productivity with Productivity Tools" in *Software Development: Computer-Aided Software Engineering (CASE).* Ed. Elliot J. Chikofsky. Washington: IEEE Computer Society Press, 1989. Discusses several areas of risk in using computer-aided tools.

"Concept Development More Critical Than CAD." *Machine Design,* vol. 62(6), March 22, 1990, p. 34. Discusses the differences between concept-development packages and traditional CAD.

Crandall, Stephen H. *Engineering Analysis.* New York: McGraw-Hill, 1956. Surveys numerical procedures for solving mathematical equations describing engineering systems. Derives system equations for many different kinds of engineering problems.

Datapro Research. *CAD/CAM/CAE Systems.* M07-100. Delran, NJ: McGraw Hill, January, 1989. Provides an overview and definitions regarding the state-of-the-art in CAD/CAM/CAE.

Date, C. J. *An Introduction to Database Systems.* Reading, MA: Addison-Wesley, 1983. Discusses several aspects of database systems including distributed databases, integrity, concurrency, and security.

Department of Defense. *Transition from Development to Production.* DoD 4245.7-M, September 1985. Describes techniques for avoiding technical risks in 48 key areas or templates in funding, design, test, production, facilities, logistics, management, and transition plan.

Department of the Navy. *Best Practices: How to Avoid Surprises in the World's Most Complicated Technical Process.* NAVSO P-6071, March 1986. Discusses how to avoid traps and risks by implementing best practices for 48 areas or templates, including topics in funding, design, test, production, facilities, and management.

Dhillon, B. S. *Engineering Management: Concepts, Procedures, and Models.* Lancaster, PA: Technomic, 1987. Discusses the importance and use of value engineering and configuration management.

Dietz, Daniel. "Stereolithography Automates Prototyping." *Mechanical Engineering,* vol. 112(2), February 1990, pp. 34-39. Defines and describes the concept of stereolithography. Also discusses state-of-the-art capabilities and future applications.

Dow, James W. "Preparing Technical Specifications for CAD/CAM Systems." in *CAD/CAM Management Strategies.* Eds. Robert M. Dunn and Dr. Bertram Herzog. Pennsauken, NJ: Auerbach Publishers, 1984, pp. 1-15. Describes in detail those items that should be included in a Request for Proposal (RFP). Dow is one of the coauthors of *The Guide for the Evaluation and Implementation of CAD/CAM Systems.*

Elliot, Stephen W. and Hydak, S. J. "Meet the Smart Factory of the '90s." *Manufacturing Engineering,* vol. 103(4), October 1989, pp. 60-63. Discusses how a variety of computer systems were integrated at a GM assembly factory. The resulting system was designed to be flexible so that new equipment could be added as technology or requirements change.

Faina, David G. "Reducing the Odds." *Manufacturing Engineering,* vol. 103(6), December 1989, pp. 23-24. Describes variation simulation analysis and how Westinghouse used this method to improve product design.

Gibson, Michael Lucas. "The CASE Philosophy." *BYTE,* vol. 14(4), April 1989, p. 209-218. Explains CASE philosophy in terms of corporate plans, systems design, and systems development. Also details many benefits of CASE.

Giguere, Marshall E. and Kennicott, Philip R. "Transferring Annotated Design Data with IGES: General Electric Case Study" in *Computer-Aided Design, Engineering, and Drafting.* Eds. Robert M. Dunn and Dr. Bertram Herzog. Pennsauken, NJ: Auerbach Publishers, 1984, pp. 1-20. Discusses the difficulties that GE had in using IGES to transfer design information and the lessons they learned from the trial program.

Gould, Lawrence. "CIM Champions: Grow 'Em or Buy 'Em?" *Managing Automation,* vol. 5(3), March 1990, pp. 20-23. Provides support to the notion that a champion can greatly increase the odds of success. Also describes desired traits of a champion.

Granville, Charles. "CAPP Comes to CAD and Manufacturing." *Automation,* vol. 37(3), March 1990, pp. 44-45. Discusses the benefits of integrating CAD and manufacturing through group technology (GT) and computer-aided process planning (CAPP).

Granville, Charles. "Computer-Aided Process Planning Systems Come of Age." *Managing Automation,* vol. 37(2), February 1990, p. 60. Provides a brief description of CAPP and its associated benefits.

Groover, Mikell, P. and Zimmers, Emory W., Jr. *CAD/CAM Computer-Aided Design and Manufacturing.* Englewood Cliffs, NJ: Prentice-Hall, 1984. Discusses design and manufacturing processes and how the computer can be used in those processes.

Gryna, Frank M. "Product Development." in *Juran's Quality Control Handbook.* 4th ed. Eds. J. M. Juran and Frank M. Gryna. New York: McGraw-Hill, 1988. Discusses the process of translating user needs into a set of product design requirements for manufacturing. Includes detailed discussions on basic concepts, planning issues, configuration management, and improving effectiveness.

Johnson, Pete. "Software vs. Hardware Models for System Simulation." *VLSI Systems Design, Design Automation Guide,* 1988, pp. 50-56. Discusses trade-offs between software and hardware simulations.

Juran, J. M. and Gryna, Frank M. Eds. *Juran's Quality Control Handbook.* 4th ed. New York: McGraw-Hill, 1988. Describes the process of new product development, including procedures for defining design requirements, setting up configuration management systems, and designing experiments for trade studies.

Krause, Irvin and Suchors, Cheryl R. "Design Automation: A Strategic Necessity." *Electronic Business,* vol. 13(8), April 15, 1987, pp. 121-132. Presents the results of a survey conducted by Coopers & Lybrand and *Electronic Business* regarding design automation. Analysis of the survey's results indicated that while it is possible for a company to realize its strategic objectives, there is still room for improvement. Management involvement and support is also stressed as a key to success.

Krouse, John, and others. "How to Successfully Implement Computer-Aided Design and Manufacturing." *CAD/CAM Planning Guide '89* Supplement to *Machine Design,* vol. 61(15), July 20, 1989. Describes issues associated with successfully implementing CAD/CAM. Topics include planning, benefits and costs, CAD/CAM basics, technology, purchasing decisions, and applications.

Kuttner, Brian C. and Lachance, Michael. "Assessing Standards and Alternative Means of Data Transfer." in *Computer-Aided Design, Engineering, and Drafting.* Eds. Robert M. Dunn and Dr. Bertram Herzog. Pennsauken, NJ: Auerbach Publishers, 1986, pp. 1-11. Discusses why there are many different standards for data transfer. Discusses an alternative to data transfer. Includes a case study of how one company transferred data using translators rather than standards.

Machlis, Sharon. "How to Bring CAD to Life." *Design News,* vol. 46(6), March 26, 1990, pp. 54-60. Presents a view of stereolithography as desktop manufacturing. Discusses the pros, cons, and limits of the technology.

Manley, John H. *Computer Aided Software Engineering (CASE) Foundation for Software Factories.* Report AS85-510. Delran, NJ: Datapro Research Corporation, February 1986. Describes CASE as an integral part of realizing the software factory. Provides a discussion on software engineering principles, benefits of CASE, and obstacles to implementation.

McClure, Carma. "The CASE Experience." *BYTE,* vol. 14(4), April 1989, pp. 235-246. Explains the basic differences in the types of CASE tools. Provides some ideas on where to begin implementation and supporting case material.

McGill, Michael E. *American Business and the Quick Fix.* New York: Henry Holt and Company, 1988. Explains many of the myths and problems facing modern American business and its managers.

Messenheimer, Susan and Weiszmann, Carol. "Quality Software Quest." *Software Magazine,* vol. 8(2), February 1988. pp. 29-36. Describes several current software engineering methodologies and recent approaches to the software development task.

Munsinger, John P. "The State of IBM's CIM Architecture." *CIM Review,* vol. 6(3), Spring 1990, pp. 30-33. Details the principles, concepts, and components of IBM's CIM architecture.

Nolen, James. *Computer-Automated Process Planning for World-Class Manufacturing.* New York: Marcel Dekker. 1989. Provides a comprehensive survey of the state-of-the-art in process planning. Explains opportunities for CAPP and promotes its use. Also comes with two demonstration programs of CAPP software for personal computers.

Ohr, Stephen A. *CAE: A Survey of Standards, Trends, and Tools.* New York: John Wiley and Sons, 1990. Provides a good, current review of computer-aided tool use and application. Mainly slanted toward electrical (logic) design.

Orr, Dr. Joel N. "How to Buy the Right CAD Program," Paper presented at National Design Engineering Show, Chicago, IL, February 26-March 1, 1990. Presents the "Steps of Creeping Commitment," a framework to help ensure the right approach will be followed in implementing CAD.

Palecek, Peter, Sutton, George P., and McGinty Weston, Diane. *CIM Market Needs and Opportunities.* Business Intelligence Program, Report No. 744. Menlo Park, CA: SRI International, 1987. Discusses the drive to integrate CAD, CAE, CAPP, and CAM by way of CIM. Also presents an analysis of the CIM market and some of its associated opportunities.

Petre, Peter. "How GE Bobbled the Factory of the Future." *Fortune,* vol. 112(11), November 11, 1985, pp. 52-63. Discusses GE's failed attempt to become a one-stop supermarket of automation products. Points out that companies must thoroughly evaluate vendors prior to selecting a system from them.

Pfeil, W. and others. "The Application of LQR Synthesis Techniques to the Turboshaft Engine Control Program.", Paper presented at *AIAA/SAE/ASME 20th Joint Propulsion Conference,* Cincinnati, OH, June 11-13, 1984. Discusses the use of modeling and simulation in the design of an advanced engine control system for a modern helicopter.

Priest, J. W. *Engineering Design for Producibility and Reliability.* New York: Marcel Dekker, 1988. Presents an overview of engineering concepts, engineering methodologies, and design practices that have been proven in industry to improve producibility, reliability, and quality.

Prun, Jon. "Maneuvering in the NC Minefields." *Machine Design,* vol. 62(2), January 11, 1990, pp. 107-112. Presents several key questions to ask when evaluating NC software.

Report of Survey Conducted at Lockheed Aeronautical Systems Company, Marietta Georgia. Best Manufacturing Practices Program. US Navy RM&QA(OASN), August 1989. Presents the results of a survey by the Best Manufacturing Practices program that reviews and documents the best practices and industry-wide problems at Lockheed Aeronautical Systems Company, Marietta Georgia.

Shearer, Lowen J. and others. *Introduction to System Dynamics.* Reading, MA: Addison-Wesley, 1971. A good introductory text on system modeling and simulation.

Simanaitis, Dennis. "Interesting Developments." *Road & Track's Guide to the All-New 300ZX.* Special Edition, 1989, p. 45. Presents a detailed discussion regarding the development effort on the 1990 Nissan 300ZX sports car.

Simanaitis, Dennis. "Technology Update: Tires." *Road & Track,* vol. 41(11), July 1990, pp. 99-104. Describes the use of CAD and CAE in modern automobile tire development.

Soat, John. "Software Productivity: The Next CASE." *InformationWEEK,* vol. 252, January 8, 1990, pp. 22-24. Compares earlier implementations of CASE with anticipated applications in terms of measured and expected benefits.

Srch, Richard W. "Developing a Computer-Aided Design and Materials Requirements Planning Interface" in *CAD/CAM Management Strategies.* Eds. Robert M. Dunn and Dr. Bertram Herzog. Pennsauken, NJ: Auerbach Publishers, 1988. Discusses the requirements for interfacing CAD and MRP systems. Identifies and explains CAD and MRP systems, their relationship, and integration considerations.

Stark, John. *Managing CAD/CAM* New York: McGraw-Hill, 1988. Discusses management aspects of evaluating and implementing computer-aided technology. Does not get bogged down in technical aspects.

Statland, Norman. "Payoffs Down the Pike: A CASE Study." *Datamation,* vol. 35(7), April 1, 1989, pp. 32-33, 52. Documents the results of a survey conducted on CASE tools. Explains that productivity payoffs and expectations are not met immediately.

Stauffer, Robert N. "Lessons Learned in Implementing New Technology." *Manufacturing Engineering,* vol. 102(6), June 1989, p. 63. Interviews several companies that have successfully implemented computer-aided technology. Highlights the factors that contributed to the success.

Sutton, George P. *Computer-Assisted Process Planning.* Business Intelligence Program, Report No. 765. Menlo Park, CA: SRI International, 1988. Describes the importance of CAPP to the development process. Discusses current practices, process planning technology, the market for CAPP, and implementation considerations.

Sutton, George P. *CIM Options and Implementation.* Business Intelligence Program, Report No. 735. Menlo Park, CA: SRI International, 1986, p. 12-13. Reports on the challenge of CIM to the corporation, provides a framework for integrated manufacturing, and discusses issues in implementation.

U.S. Department of Commerce, International Trade Administration. *A Competitive Assessment of the U. S. Computer-Aided Design and Manufacturing Systems Industry.* February, 1987. Summarizes the history and forecasted outlook for the U.S.'s CAD/CAM market. Analyses of trends in growth, revenues and applications are also presented.

Vaidyanathan, Pallavoor N. "Considerations in CAD/CAM Implementation" in *CAD/CAM Management Strategies.* Eds. Robert M. Dunn and Dr. Bertram Herzog. Pennsauken, NJ: Auerbach Publishers, 1987. Identifies the need for CAD/CAM technology, system selection considerations, potential problems, and benefits.

Van Nostrand, R. C. "CAD/CAM Justification and Follow-Up: Simmonds Precision Products Case Study" in *CAD/CAM Management Strategies.* Eds. Robert M. Dunn and Dr. Bertram Herzog. Pennsauken, NJ: Auerbach Publishers, 1988, pp. 1-16. Presents a complete case study of a corporation's investigation, acquisition, and implementation of a CAD/CAM system.

Veilleux, Raymond F. and Petro, Louis W. *Tool and Manufacturing Engineers Handbook, Volume 5: Manufacturing Management.* Dearborn, MI: SME, 1988. Provides a comprehensive discussion on manufacturing management as it relates to such topics as strategic planning, cost estimating, management philosophy, CIM, and quality.

Webb, Michael J. and Felegyhazi, Bill. "CIM in Perspective: The Challenges and Rewards." *Manufacturing Systems,* vol. 8(4), April, 1990, pp. 28-32. Discusses the opportunities, obstacles, implementation considerations, and benefits provided by CIM.

Weber, Herbert. "From CASE to Software Factories." *Datamation,* vol. 35(7), April 1, 1989, pp. 34-36, 52. Explains the need for software factories, and how CASE is required in order for them to be successful.

Wejman, James, Collins, Mike, Darnton, William, and Dalton, G. Reid. "Broadening CIM Horizons." Paper presented at National Design Engineering Show, Chicago, IL, February 26-March 1, 1990. Presents a discussion on several aspects of successful CIM: the technical architecture, gaining the competitive edge, and cost justification.

Willis, Roger. "The Laws of CIM: Case Studies on Optimizing Manufacturing." *Manufacturing Systems,* vol. 8(2), February 1990, pp. 54-58. Explains five common "laws" associated with CIM. Includes case studies.

Design for Testing

1

Introduction

To the Reader

This part includes two templates: Design for Testing (DFT) and Built-In Test (BIT).

The templates, which reflect engineering fundamentals as well as industry and government experience, were first proposed in the early 1980s by a Defense Science Board task force of industry and government leaders, chaired by Willis J. Willoughby, Jr. That task force sought to improve the effectiveness of the transition from development to production. It concluded that lack of understanding of the engineering and manufacturing disciplines used in the acquisition process causes most program failures. It then focused on identifying engineering processes and control methods that minimize technical risks in both government and industry. It defined these critical events in design, test, and production in terms of templates.

The template methodology and documents

A template specifies:

- areas of high technical risk

- fundamental engineering principles and proven procedures to reduce the technical risks

Like a classical mechanical template, these templates identify critical measures and standards and can impose engineering discipline on suppliers to the DoD.

The task force documented 47 templates and in 1985 the templates were published in the DoD *Transition from Development to Production*

(DoD 4245.7-M) manual.[1] The templates primarily cover design, test, production, management, facilities, and logistics. In 1989, the Depart-ment of Defense added a 48th template on Total Quality Management (TQM).

In 1986, the Department of the Navy issued the *Best Practices* (NAVSO P-6071) manual, which illuminated DoD practices that compound problems and increase risks.[2] For each template, this manual describes:

- potential traps and practices that increase the technical risks
- consequences of failing to reduce the technical risks
- an overview of best practices to reduce the technical risks

The intent of the *Best Practices* manual is to help practitioners become aware of the traps and pitfalls so they do not repeat them.

The templates are the foundation for current educational efforts

In 1988, the government initiated an educational program, "Templates: Professionalizing the Acquisition Work Force," with courses and books, such as this one, to increase awareness and improve the use of good engineering practices.

The key to improving the DoD's acquisition process is to recognize that it is an industrial process, not an administrative process. This implies a need to change the skills and technical knowledge of the acquisition work force in government and industry. Many in this work force do not have an engineering background. Those with an engineering background often do not have broad experience in design, test, or production. The work force must understand basic design, test, and production processes and associated technical risks. The basis for this understanding should be the templates since they highlight the critical areas of technical risk.

The template educational program meets these needs. The program consists of a series of courses and technical books. The books provide background information for the templates. Each book covers one or more closely related templates.

---

[1]Department of Defense. *Transition from Development to Production.* DoD 4245.7-M, September 1985.

[2]Department of the Navy. *Best Practices: How to Avoid Surprises in the World's Most Complicated Technical Process.* NAVSO P-6071, March 1986.

How the parts relate to the templates.   The parts describe:

- the templates, within the context of the overall acquisition process
- risks associated with each included template
- best commercial practices currently used to reduce the risks
- examples of how these best practices are applied

The books are based on best commercial practices and use examples from both government and commercial sectors. They do not discuss government regulations, standards, and specifications, because these topics are well-covered in other documents and courses. Instead, the books stress the technical disciplines and processes required for success.

Clustering several templates in one book makes sense when their best practices are closely related. For example, the best practices for the templates in this part interrelate and occur iteratively within design and manufacturing. Designers, suppliers, and manufacturers all have important roles. Other templates, such as Design Reviews, relate to many other templates and thus are best dealt within individual books.

Courses on the templates.   The books are designed to be used either in courses or as stand-alone documents. The courses use lectures and other proven instructional techniques such as videotapes, case studies, group exercises, and action plans.

The template educational program will help government and industry program managers understand the templates and their underlying engineering disciplines. They should recognize that adherence to engineering discipline is more critical to reducing technical risk than strict adherence to government military standards. They should especially recognize when their actions (or inactions) increase technical risks as well as when their actions reduce technical risks.

The templates are models

The templates defined in DoD 4245.7-M are not the final word on disciplined engineering practices or reducing technical risks. Instead, the templates are references and models that engineers and managers can apply to their industrial processes. Companies should look for high-risk technical areas by examining their past projects, by consulting their experienced engineers, and by considering industry-wide issues. The result of these efforts should be a list of areas of technical risk and best practices that becomes the company's own version of the DoD 4245.7-M and NAVSO P-6071 documents. Companies should tailor the best practices and engineering principles described in the books to suit their particular needs. Several military suppliers have already produced manuals tailored to their processes.

Figure 1 shows where to find more details about risks, best practices, and engineering principles. Participants in the acquisition process should have copies of these documents.

Testability is "a design characteristic [that] allows the status (operable, inoperable, or degraded) of an item to be confidently determined and the isolation of faults within the item to be performed in a timely manner."[3] Without testability, system performance suffers and maintenance cost grows. Testability succeeds best when system planners treat it as a key design objective from project conception through system acquisition.

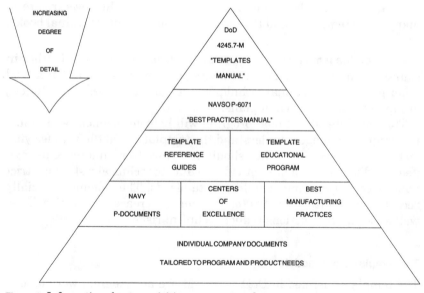

Figure 1 Information about acquisition-process templates.

---

[3]Rome Air Development Center. *A Guide for Contractor Program Managers: How You Can Produce and Deploy an Improved Weapon System Diagnostic Capability—A Systems Engineering Process.* Rome, NY: RADC, 1990, p. xi.

Many manufacturers build testability into their products. Makers of cellular telephones, personal computers, and other new commercial products are taking the lead. In the 1990s, testing will be part of front-end design and manufacturing, and "design for testability" will be the creed of electronic engineers.[4]

As the military services recognize the need to improve system testing, they are demanding, as part of acquisition, designed-in testability.

Testing is done throughout the life of a system, starting with its design. Subsystems, prototypes, production-line subassemblies, and the complete system are tested to determine status and to locate defects. During operation, status testing influences mission success, and the quality of diagnostic testing affects the duration of repairs and thus system availability.

Testability can control system cost for the manufacturer. "One company [that] went to testability," says George Heilmeier, senior vice president and chief technical officer of Texas Instruments, "slashed testing costs by a factor of five while boosting market performance 10 times."[5]

Testability can control system cost for the user because an easily fixed system minimizes repair costs and maximizes system availability. That is one reason MIL-STD 2165 (*Testability*) requires suppliers to pay equal attention to testability and mission objectives.

Poor testability can be expensive. For example, at an F-16 tactical fighter wing, maintenance people spent over 13,600 hours removing nondefective parts.[6] This is equivalent to twenty people needlessly replacing items for six months. The cause was inadequate diagnostic capability, in this case *false alarms,* and lack of skill.

Many people contribute to system testability—requirement writers, test engineers, and manufacturers—but designers are key. Designers must weigh pressures to achieve performance at lowest cost against testability requirements that add weight, bulk, and cost without directly improving performance.

During the 1990s, system developers in government and industry will:

---

[4]Special Report on Testing. "Credo for a New Decade: Design for Testability." *Electronic Engineering Times,* August 7, 1990, p. 23.

[5]Special Report on Testing. "Credo for a New Decade: Design for Testability." *Electronic Engineering Times,* August 7, 1990, p. 26.

[6]Rome Air Development Center. *A Guide for Contractor Program Managers: How You Can Produce and Deploy an Improved Weapon System Diagnostic Capability—A Systems Engineering Process.* Rome, NY: RADC, 1990, p. xi.

- design more capability into fewer new systems
- extend the useful life and capabilities of existing systems
- use fewer technicians to maintain existing and new systems
- integrate testing elements internal and external to the system
- emphasize deployment of systems instead of employment of personnel as training becomes more difficult and costly

This part presents best practices for testability and a systems-engineering approach to designed-in testability.

Integrated Diagnostics

Currently, the leading concept in testing military systems is *integrated diagnostics.*
    Integrated diagnostics is a

> structured design and management process to achieve the maximum effectiveness of a...system's diagnostic capability by considering and integrating all related pertinent diagnostic elements.[7]

The integrated-diagnostics process includes design, engineering, testability, reliability, maintainability, human factors, and ongoing system support. Its goal is a cost-effective way to detect and find all faults in systems and equipment.
This part covers two key elements of integrated diagnostics addressed in DoD 4245.7-M: Design for Testing (DFT) and Built-In Test (BIT). DFT is a method for anticipating system testing needs from the earliest stages of development. BIT, which is a DFT tool, means putting the hardware and/or software needed to test a system inside the system.

Testing Caveats

Customers want performance, quality, and reliability at the lowest cost. To this end, engineers must do trade-off studies to decide what combination of performance, cost, producibility, quality, and reliability will make the most successful system. Although simple in concept, cost analysis demands both knowledge and forecasts of a wide spectrum of disciplines. Cost analysis calls on the engineer to envision future changes and innovations that may affect the system's cost. Costs, in

---

[7]Rome Air Development Center. *A Guide for Contractor Program Managers: How You Can Produce and Deploy an Improved Weapon System Diagnostic Capability—A Systems Engineering Process.* Rome, NY: RADC, 1990, p. xi.

this context, include designing, testing, purchasing, operating, and maintaining a system.

Designers should also consider to what degree testability adds to complexity, which increases the risk of a premature failure. That is part of the trade-off studies described below.

Design for Testing

Testing and inspection are essential to the production and operational environments. System design must allow for access for inspection and various types of automatic test equipment (ATE) during production and operation.

In the past, testability was not part of a system's earliest design goals, and the resulting poor testability hurt system performance. System testability can be improved by treating it as a design process that starts in the conceptual phase and continues through the acquisition process.

DFT ensures that test capabilities are put in the equipment to support testing at all levels (e.g., component, board, subsystem, and system). DFT goals include economical manufacture and ease of maintenance.

Table 1 summarizes risks and consequences that can be avoided by Design for Testing.

TABLE 1   Risks and Consequences Addressed by Design for Testing

Risks	Consequences
Testability requirements for production are defined after design release	Redesign needed Test throughput cannot support an efficient production rate Production test equipment is delayed Production starts late
Testability requirements are unclear	Poor testability increases the risk during production that the unit will fail at the next assembly or system level Poor testability decreases system availability in the field and increases the need for spares
Testing yields false alarms	Good systems are unavailable for missions Good systems are unnecessarily "repaired" during production and operation

TABLE 1   Risks and Consequences Addressed by Design for Testing (Continued)

Risks	Consequences
Testing yields false test-OKs	Defective systems pass production-line screens, causing costly rework Defective systems compromise missions
Product is designed only for performance testing	Cost and time increase for product-acceptance testing and fault isolation
Government-furnished equipment lacks testability	Diagnostic capability of the system suffers
Design is mission-focused, and automatic test equipment (ATE) is ignored	ATE is not compatible with prime system or not available Mean time-to-repair (MTTR) increases
Development-team members do not interact	Designers focus on subsystems and modules, neglecting the system Designers omit needed test capabilities
System manager lacks experience with earlier systems	Testing designs do not capitalize on prior experience New system gets new ATE Maintenance centers must support old and new ATE

Risk: Testability requirements for production are defined after design release

Testability added after the fact to a completed design can cause these undesirable consequences:

- increased development cost because of additional design efforts
- incompatibility between the modified system and pre-existing test equipment
- delayed availability of production test equipment
- decrease in production rate because testing takes too long
- difficulty relating production-system performance with development-system performance because tests differ
- inability to access test points on densely populated printed-circuit boards
- poorer system performance, especially in aircraft, because of the size and weight of components added to support tests
- failure to meet testability goals because of weight or space limitations of aircraft, submarines, and tanks
- A well-planned development that uses DFT avoids these risks by establishing a corporate design policy on testability and having pro-

duction-test people work with designers during trade-off studies and design reviews.

Risk: Testability requirements are unclear

Without clear test specifications, designers may overlook key aspects of system testability, such as production and maintenance tests. Other needs, such as operational availability and mean-time-to-repair, may not be translated into testability requirements.

If a subsystem test used during production fails to show latent defects, those defects will appear at later assembly stages or in the field. Repair and replacement are then harder, may cost more, and may jeopardize the mission.

DFT considers testability, maintainability, and supportability requirements for a given system. It includes trade-off studies for cost-effective use of BIT, ATE, and manual testing. A DFT plan can reduce the need for spares by increasing the ease and accuracy of diagnostics and repair.

Risk: Testing yields false alarms

Production lines for complex systems rely on subsystem tests to reveal faults so workers can repair them before assembly in larger systems. If those tests falsely report failures, the production line loses time and resources. The suspect subsystem leaves the assembly line for repairs and more tests or may directly delay system production while workers prove that the alarm was false.

In the field, false alarms may delay the start of missions, abort missions in progress, and tie up storage facilities with to-be-repaired systems.

Risk: Testing yields false test-OKs

When production-line testing fails to show defective subsystems, those defects surface later. If found prior to system use, repairs may require disassembly. If defects first appear during use, lives could be lost.

False test-OKs are particularly dangerous for system status reports.

Risk: Product is designed only for performance testing

Performance testability will increase production costs because extra hardware is usually required to do performance testing. It will not guarantee maintainability of the product under field conditions because performance testing by itself does not identify faulty replaceable subsystems. DFT builds production testing, acceptance testing, and diagnostic (fault isolation) testing into the system.

Risk: Government-furnished equipment lacks testability

If government-furnished or off-the-shelf equipment is untestable or difficult to test, any system that depends on it is compromised. Just as a chain's strength is limited by its weakest link, the testability of a system and its maintainability are no better than its least testable and maintainable subsystem.

Even if government-furnished or off-the-shelf equipment is testable, uncertainty about its maintainability may cause contractors to overdesign to compensate for suspected weakness.

Designs that use government-furnished or off-the-shelf equipment may not allow needed access or control for testing if contractors lack details about those needs.

Risk: Design is mission-focused, automatic test equipment (ATE) is ignored

If designers focus on the system and its mission, testing needs, particularly the need for ATE, may not be met. Designers may forget to create the system's interfaces to existing ATE, or development of new, needed ATE may start too late.

DFT ensures that designers plan for ATE early to achieve compatibility between the system and ATE.

Risk: Development-team members do not interact

Diagnostics engineers do not always interact effectively with other designers and engineers. Currently, for example, about 40% of the designers say that diagnostic requirements are verbally communicated to them.[8] Human-factors experts are not consulted.

This risk can result from a tendency of designers to stress subsystem design, losing sight of the overall system and its mission.

Some contractors use a system design group of experienced engineers who ensure system testability by giving system test requirements to all design team members. For example, the Electronics Division of General Dynamics, which builds testing equipment for mil-

---

[8]Aeronautical Systems Division of Wright-Patterson AFB. *Generic Integrated Maintenance Diagnostics Task 12.* Draft of February 19, 1990, p. D-22.

itary systems, publishes a "Design Requirements Bulletin" that forces designer to put on paper everything that will be needed.

A systematic approach to test design should include open and effective communications.

Risk: System manager lacks experience with earlier systems

Sometimes the manager of development of a new system has little prior experience with similar, existing systems. That manager may be unaware of testability problems that plagued earlier, similar systems. He or she may not know that effective testability techniques and procedures have evolved.

A manager without preconceptions may create innovative testability schemes, but, for complex systems, the cumulative wisdom of earlier developments is a safe starting point and standard of excellence.

A manager new to a system may want to develop new, system-specific ATE, but reuse of existing ATE has advantages. Reuse avoids development of the new ATE, and depots that maintain the new system avoid the logistics and expense of procuring and maintaining two types of ATE.

Built-In Test

Built-in test (BIT) uses internal system hardware, software, or both to test the system or subsystems. A common example of BIT is the startup routine found in today's computers. It often uses internal microprocessors and self-test software to isolate failures. This part describes BIT for both electronic and nonelectronic systems.

(Some experts reserve "BIT" to mean *manually initiated* tests that use the system's internal hardware and software. A test that the system does *automatically,* either scheduled or in response to some internal state, is called a built-in self-test [BIST]. On the other hand, some advocate using "BIST" for electronic systems and "BIT" for mechanical systems. Experts in the field have not reached consensus.[9] This part uses the phrase "built-in test" [BIT] for any test capability that relies on internal hardware or software.)

Benefits of BIT include:

- reduced requirements for external test equipment
- fewer interfaces between the system and the external world

---

[9]Konemann, Bernd and Wagner, Kenneth. "Practical BIST: Old and New Tools in Concert." *IEEE Design & Test of Computers,* August 1989, p. 16.

- less damage from improper invasive inputs
- more accurate testing because BIT is non-invasive and does not add noise to the measurement
- reduced skill level of maintenance personnel
- rapid troubleshooting and reduced downtime
- verification that the equipment works (improved status-monitoring and readiness)
- reduced life-cycle cost

Many designs lack enough BIT to isolate failures to a field-replaceable assembly. One common result is trial-and-error replacement of good assemblies with bad ones, increasing downtime and costs of replacement assemblies.

Table 2 summarizes risks and consequences that can be avoided by BIT.

Risk: BIT is designed independently of system design

If BIT designers and system designers are independent and do not communicate, their designs may be incompatible and require redesign. They may fail to share space, processing power, memory, and power sources, making the system costlier and less reliable.

System designers need to tell BIT designers the system's reliability objectives so they can design BITs that are more reliable than the system they test. That reliability objective is achievable because generally BIT components experience less stress than the system. If BITs are less reliable than the system, users who experience false alarms and false test-OKs will not trust either indication, rendering the test useless.

TABLE 2   Risks and Consequences Addressed by Built-In Test

Risks	Consequences
BIT is designed independently of system design	BIT fails to support operational and maintenance needs
	BIT's mean time between failure exceeds system's
BIT is designed after the fact	BIT design is costly and less effective
Production people are not consulted on BIT	BIT is not effective in the factory

Trade-off studies are not complete if system designers and BIT designers do not discuss conflicting goals and find solutions.

When system design is independent of BIT design:

- BIT often does not adequately support operational and maintenance needs
- additional costs and extra time may be needed to retrofit BIT

Risk: BIT is designed after the fact

When BIT design is not incorporated into the initial system design, BIT is less effective and more costly.

Best practice is to include BIT in system design from the start of design. Designers need to plan for field, production, and maintenance testing. Early BIT design avoids delays that could result from unavailable parts or incompatible system design.

Risk: Production people are not consulted on BIT

BIT should be designed for production, factory, operational, and depot testing. Targeting BIT only for operations may not benefit other phases, especially prototype, production, and maintenance testing. Representatives of production organizations should therefore help plan BIT.

Keys to Success

By applying DFT and BIT, contractors and the customer can avoid risks and improve the acquisition process.

Keys to success include:

- ensuring the adequacy, completeness, and consistency of test requirements
- understanding test objectives and adapting acquisition strategies accordingly
- integrating life-cycle, environmental, and mission-profile considerations into the testing philosophy
- ensuring that maintainability (e.g., easy fault detection and isolation) is designed into the system
- conducting trade-off studies to evaluate test alternatives

- using lessons learned to improve testability by improving the design process
- developing testability rules and guidelines to ensure that all designs include access for testing
- employing testability analysis tools
- involving manufacturing and maintenance groups during the initial design

2

Procedures

Recognizing the importance of testability in complex systems, this part outlines up-to-date design, manufacturing, and testing procedures. Manufacturers should consider adopting a corporate policy that reflects these practices.

For example, Unisys Corporation Computer Systems Division managers are committed to testability of all equipment. Company policy requires that factory test procedures be established before full-scale development. Programmer maintenance stations for automatic test equipment are designed concurrently with prime (system) equipment. Testability is also a part of design reviews.[10]

A common fallacy is that ingenuity in the design of manufacturing tests and inspection equipment can compensate for deficiencies in the testability of the hardware. If testability is missing in the original design, it is costly and difficult to do later in the design process.

This chapter presents procedures for planning, designing-in, and building-in testability. It does this in three steps as Figure 2 shows:

Step 1 describes how to plan for testability

Step 2 describes how to design for testability and is based on the DFT template.

Step 3 describes how to put BIT concepts into the system by following the BIT template.

---

[10]Best Manufacturing Practices (BMP) Review Team. *Best Manufacturing Practices, Report of Survey Conducted at Unisys Corporation Computer Systems Division, St. Paul, MN.* Washington, DC: Office of the Assistant Secretary of the Navy (Shipbuilding and Logistics), November 1987, p. 5.

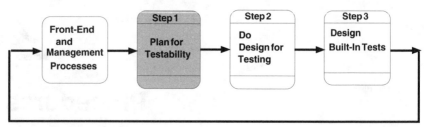

Figure 2 Steps for achieving testability.

Case Study

A Personal Computer

Throughout this part a typical personal computer (PC) serves as a simple, familiar example of points made in the text.

This PC consists of three major elements: system unit, display, and keyboard. The system unit houses the system board, floppy drives, hard drives, and power supply. The display is a television-like screen that shows the data entered from the keyboard and the system unit's interaction with the internal and external interfaces, such as the floppy drive. The keyboard is attached to the system unit by cable.

The system board (often called the motherboard) has five functional areas: microprocessor with supporting elements, read-only memory, random-access memory, input-and-output channel, and sockets for plugging in extra boards.

The read-only memory holds programs or data that the computer needs when the power is turned on. The information in the read-only memory cannot be changed by users and does not disappear when the PC is turned off. The data in random-access memory can be changed by users. Those data vanish when the PC loses power.

An input-and-output channel is a path for communication between the microprocessor and external devices. The sockets accept extra circuit boards that allow the microprocessor to control disk drives, displays, printers, etc.

Step 1: Plan for Testability

Planning for testability puts test capability into the design *early* to improve prototype, production, acceptance, operational, and maintenance testing.

Prototype testing

The new system and its subsystems experience the most intensive testing during prototyping. The prototype's subsystems may be new, and designers may have assembled them in a novel way. Prototype testing validates the system design, confirms system performance, and shows where improvements are needed.

The high cost of post-prototype design changes motivates contractors to test prototypes of the production process and the operational system thoroughly. The same costs should lead contractors to plan enough prototype testability and testing during an early concept stage.

Because prototype testing can avoid later high costs, prototype testers should concentrate on test thoroughness and not worry about test cost and redundant testing.

Production testing

Efficiency is key to production testing. The contractor is motivated to use the least expensive means to do the fewest, shortest tests of isolated and assembled subsystems because those production tests slow the production rate and increase the per-system production costs. The motivation to limit production testing is balanced by the need to find subsystem flaws prior to assembly and ensure that the system passes acceptance tests and satisfies its eventual users.

These conflicting goals challenge system designers to plan for production testability. Strategies for production-line testing can vary. For example, a complex subsystem, such as an aircraft engine, should pass exacting tests before it joins other subsystems, especially if the assembled subsystem is harder to repair. Simple subsystems, such as disc-brake calipers, may be easy to test when the whole system is assembled and fail too seldom to warrant individual, preassembly testing.

Testability designers must choose a strategy for each needed production test. They must try to predict the fewest, quickest, lowest-cost tests whose results ensure a satisfactory product. They must make subsystems accessible, controllable, and observable by user-friendly low-cost external test equipment and built-in tests that reliably provide those results.

Acceptance testing

Acceptance tests confirm basic system performance. They verify that assembled subsystems meet system requirements, not whether each subsystem works. They should be fast, few in number, low cost, and easily performed by the customer.

Because acceptance tests duplicate some of the production tests, designers may do little testability planning that is specific to acceptance testing. (Acceptance testing may include actual system operation, which is not usually part of production testing. No special testability planning is needed to operate the system.)

Operational and maintenance testing

Post-acceptance testing includes performance and operational testing in the field and (typically) three levels of maintenance testing. The three maintenance levels are *organizational testing,* which is done in the field by the operating unit, *intermediate testing,* which is done in regional shop facilities that support several operating units, and *depot testing,* which is done by the manufacturer or in industrial-type facilities. Table 3 shows how testing capabilities grow as maintenance moves from an operations site to a depot.

Maintenance locations tend to be fixed for Air Force planes, which return to base for repairs. The Navy puts maintenance centers on ships to support shipboard systems.

Without advanced planning, test capabilities are added in response to observed needs. For example, after the system fails repeatedly or catastrophically, designers, in desperation, add tests.

Test capabilities added in response to observed failures tend to be expensive because designers may need to change many elements of the original design. Further, test-capability needs may surface late in the project if proof of need relies on system failure.

The best practice is to foresee the need for test capabilities and to design them in the system from the start. MIL-STD-2165 (*Testability*) can help contractors design testability into products.

The government can help by emphasizing key aspects of the design project, including diagnostics, to the contractors. A government repre-

TABLE 3   Maintenance Levels

Maintenance Level	Performing Organization	Capabilities
Organizational	Operating unit	■ Inspection ■ Service ■ Handling ■ Modification
Intermediate	Centrally located facilities	■ Support of operating units ■ Repair ■ Modification ■ Test of components that require shop facilities
Depot	Industrial-type facilities, including contractor plants	■ Overhaul ■ Repair ■ Test capability beyond intermediate and organizational levels

sentative should also coordinate with the program's designated systems engineer.

Objectives of the testability plan

The testability plan:

- translates performance requirements into testability requirements for each level of test (prototype, production, acceptance, and post-acceptance)

- ensures that testability requirements are met for each level of maintenance

- involves test experts with the system design team to ensure an economical, testable design

- improves testability throughout the system's life cycle

- identifies operational constraints that affect testability

- identifies technical advances that may reduce (the need for) special test equipment, technical manuals, staffing, skill needs, as well as diagnostic costs, spares, and system downtime

- identifies existing and planned diagnostic resources outside the system and their limitations

- identifies testability problems observed in similar systems and avoids those problems

- agrees on *fault coverage,* which is the percentage of expected failures that available tests can detect and is an accepted measure of testability[11]

- agrees on the fault-isolation level, which is the smallest subsystem to which faults are traced

- agrees on the false-alarm rate, which is the number of incorrect failure indications per unit of operation time

- requires the testing of a prototype system

The system requirements that support testability should not constrain contractors. They should allow the contractor to choose any alternatives that meet the testing needs, preferably using minimum additional hardware to maximize reliability.

---

[11]Cheng, K. T. and Agrawal, V. D. *Unified Methods for VLSI Simulation and Test Generation.* Boston: Kluwer Academic Publishers, 1989, p. 15.

Objectives of the Testability Plan

For the PC:

- User friendliness is crucial to encourage users to use the built-in test capability. For example, the error messages should use simple English and avoid technical jargon.

- Users need simple displays that show the results of built-in tests. Dashboard displays for automobiles do this job well.

- New subsystems require extensive prototype testing.

- Production testability means that failure-prone subsystems that are difficult to repair after assembly, such as a hard-disk drive, should be testable just prior to assembly.

- Testability allows the user to assess the general status of major subsystems.

- Testability allows the technician easy access to circuit boards and other subsystems.

- Documents describing problem-correction procedures must be simple and concise.

Getting started

To start planning for testability, the prime contractor should:

- identify the organization responsible for implementing the plan for testability

- establish contacts and data interfaces among organizations responsible for elements of the testability plan

- develop a process that integrates testability with other requirements

- ensure that each subcontractor's test practices are consistent with overall system or equipment requirements

- plan to review, verify, and use test data

MIL-STD-2165 (*Testability*) requires contractors to set up an organization that is responsible for testability. That organization submits a formal plan showing how it will do the seven tasks spelled out in the standard. The tasks, which range from program planning to detailed design and analysis, include testability demonstrations. Task 201—testability requirements—calls for state-of-the-art built-in and external testing.

PC Case Study

Getting Started

During PC design, representatives of production, sales, customer service, and repair should discuss testability requirements with designers. Because PCs are used by unsophisticated users, PC testability should address the needs of two different technical levels: novice user and trained technician.

The owner of a nonworking PC needs enough testability to decide whether to:

- attempt do-it-yourself repair
- call for telephone support by a technician
- call for on-site service by a technician
- drop off or ship the PC for service at a service location

The technician called by a user needs to decide, based on user-described symptoms and the results of user-initiated tests, whether to ship user-replaceable parts, dispatch a technician, or ask the user bring the PC to a repair center. The technician who works directly on the PC needs enough testability to decide whether to fix the PC locally or ship to the manufacturer.

Step 2: Do Design-for-Testing for Each Subsystem

As systems become more complex, testing, i.e., monitoring system status and isolating failed subsystems, becomes more important, complex, and difficult. For example, early Motorola microprocessors lacked built-in tests (BITs), but as microprocessors became more complex, testing the many new functions required more engineers and more time. Motorola solved its testing problem by using a formal design strategy, like DFT, that puts test capabilities in the earliest phases of new designs.[12]

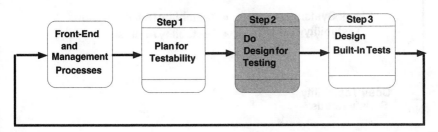

Figure 3 Steps for achieving testability.

---

[12]Daniels, R. G., and Bruce, W. C. "Built-In Self-Test Trends in Motorola Microprocessors." *IEEE Design and Test,* April 1985, p. 64.

System tests can use external test equipment or tests can be *built-in.*
Design for Testing can improve both kinds of tests. (Step 3 focuses on
Built-In Test.)

Figure 4 shows how to do DFT as a series of activities. The following
subsections explain these activities in more detail.

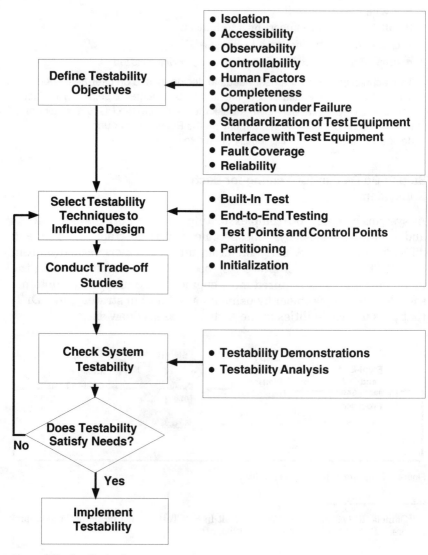

Figure 4 Design for testing.

Define testability objectives

Testability objectives and constraints should be defined early to ensure testability for qualification, acceptance, production, operation, and maintenance testing. The following are candidate objectives and constraints for testability.

Isolation. In testing, the term *isolate* has two meanings, both important:

- identify the faulty subsystem, as in "Testers *isolated* the data-link failure to the fiber-optic splice." With this meaning, *isolation* is the process of finding the failed subsystem. Equipping a system with line-replaceable units only makes sense if diagnostic tests can *isolate* failures to a specific unit.

- before testing a subsystem, quarantine it by cutting or otherwise neutralizing its external interfaces to avoid interactions with the rest of the system. For example, lamps in a series Christmas-tree string can be *isolated* (unplugged) one by one and tested.

Subsystem isolation (quarantine) is a technique that sometimes helps isolate (find) troubles. Isolation before test is especially important when subsystems interact in complex ways, because otherwise interactions between the subsystem under test and other subsystems can cause misleading results.

Designers of complex systems can help maintenance crews do their job by simplifying the isolation (quarantine) of subsystems for test and replacement. Isolation may be mechanical, e.g., unplugging a printed-circuit board, or nonmechanical, e.g., built-in firmware that tests a microprocessor may order the microprocessor to ignore all input, thus achieving virtual isolation.

PC Case Study

Isolation

The user can do some trouble isolation if trouble appears. Defective PC subsystems can be found by built-in self-test during power-up, user diagnostics, subsystem (such as keyboard) testing, and subsystem swap. For example, swap the nonworking unit's boards with new boards.

PC functions are modularized to ease isolation of troubles. That is, the system, disk drive(s), display, keyboard, mouse, and data interfaces are physically self-contained and connect with the rest of the PC through simple interfaces that disconnect for testing and replacing individual modules.

During power-up, the screen shows whether modules have passed or failed built-in testing. If no display is connected or the display is defective, the PC beeps. The number of beeps shows the problem, e.g., memory or display. (This method is flawed because users may forget the relation between beeps and subsystems. A display, independent of the PC display, showing the failed subsystem is better.)

Using diagnostics software, the user can select and test field-replaceable modules. The test-one-module option shows which modules are eligible for automatic testing. During testing, the screen displays test progress and information on user options. The test-status messages are: testing, passed, failed, aborted, and not done. The screen may show other data specific to the module under test. All data-destroying tests display a message requesting confirmation of the desire to test. If errors occur, display messages describe the failure and suggest which part to replace. One flaw with this method is that to do customer diagnostics, the floppy drive must be operational.

Using the customer-diagnostics results as the starting point, the technician can isolate further. Logic diagrams simplify the circuitry for easy troubleshooting.

Accessibility. *Accessibility* is the physical ability to reach a system's parts or subsystems. Accessibility makes it possible to test and replace defective parts. Test access may require special pins or fittings, but it is generally better to use existing interfaces to minimize subsystem connectors.

Memory devices may require direct access if their tests use long input-and-output (I/O) sequences that intervening devices would slow.

Limited access makes faults difficult to find and can force maintenance people to replace major subsystems. For example, if a B-1B subsystem loses electric power, the fault could lie in the power assembly that supplies power, the subsystem, or the wiring linking them. Because the B-1B offers few points to monitor power, repair crews usually replace the whole power assembly rather than its failed subassemblies.[13]

Sometimes designers provide a means to monitor that is physically inconvenient, such as an oil dipstick that is hard to reach or read. For example, B-1B designers provide test access through connectors, but those connectors are hard to reach in some areas of the B-1B bomber. Because repair personnel cannot easily see, test, or handle those inaccessible connectors, B-1B performance suffers.[14]

---

[13]Aeronautic Division of Wright-Patterson AFB. *Generic Integrated Maintenance Diagnostics (GIMADS) Task 12,* Draft of February 19, 1990, p. D-224.

[14]Aeronautic Division of Wright-Patterson AFB. *Generic Integrated Maintenance Diagnostics (GIMADS) Task 12,* Draft of February 19, 1990, p. D-211.

Equipment changes must not interfere with accessibility, e.g., by hiding test-access points or their identity.

PC Case Study
Accessibility

The accessibility of the circuit-board modules in the system unit is poor. Those modules have labels for identification, but getting to them often requires disconnecting and moving the keyboard and display. Then the repairperson finds and removes screws to release the system-unit cover. Accessing the hard-disk and floppy-disk drive(s), which are housed in the system unit, is also difficult.

The display, keyboard, and mouse are external to the system unit and easily accessed.

Observability. *Observability* is the ability to monitor a subsystem through its outputs. It includes perceptions by human senses (e.g., sight) and instrument sensing of electrical, thermal, mechanical, or chemical conditions.

For an electronic circuit, observability is "the ease of determining at the primary outputs of the circuit what happened at the outputs of a component."[15]

Often the output connections of an integrated circuit mounted in the middle of a circuit board are inaccessible or too small to monitor directly and safely. Those outputs may have no direct exterior connection and connect only to other interior components. A key measure of observability is how easy it is to indirectly tell the state of those outputs, e.g., by deduction from the states of the board's external interfaces.

PC Case Study
Observability

The microprocessor, random-access memory, read-only memory, and interface sockets for other boards communicate through a shared data channel or data bus.

The data channel provides observability, but because it is housed in the system unit and difficult to access, observability suffers.

The front panel provides some observability:

---

[15]Stephenson, J. H., and Grason, J. "A Testability Measure for Register Transfer Level Digital Circuits." *Proceedings of the Sixth International Symposium on Fault-Tolerant Computing* (FTCS-6), Pittsburgh, PA, June 21-23, 1976, p. 102.

- An access indicator for each floppy-disk drive that lights when the computer reads from or writes to the floppy disk.
- An access indicator for each hard-disk drive that lights when the computer reads from or writes to the hard disk.
- A display-on indicator that lights when the video display is on.
- A power-on indicator that lights when the power is on.

Controllability. *Controllability* is the ability to affect the operation of a system or subsystem by using its inputs. Control is often required to start system functions during tests. System responses are the test results.

For an electronic circuit, controllability is the ease of putting the circuit in desired states by adjusting its controls and changing its input voltages and currents.

PC Case Study

Controllability

The microprocessor, random-access memory, read-only memory, and interface sockets for other boards also share a *control channel or control bus.*

The control channel provides control, but it is housed in the system unit, difficult to access, and inconvenient.

The front-panel reset switch provides some controllability. For example, it causes system reset (restart) without turning off system power.

Human Factors. Usability is the *human factors* of testability. Tests that technicians cannot do have little value, yet test designers sometimes overlook technician needs.

For example, most technical manuals for use inside aircraft are too bulky. Some have 72-inch foldouts. Because technicians cannot conveniently consult the manuals, they make mistakes, and procedures take too long.[16]

Manufacturing, operations, and maintenance personnel with skills appropriate to their roles should find testing procedures easy, interesting, and effective. They must know what, when, where, and how to test. They need to know what extra documents and/or equipment they need and where to find them.

If procedures are not self-guiding, users should have step-by-step guides at hand. If required, formal user training and test procedure reviews must be planned and offered to technicians.

---

[16]Aeronautic Division of Wright-Patterson AFB. *Generic Integrated Maintenance Diagnostics (GIMADS) Task 12,* Draft of February 19, 1990, p. D-242.

When new diagnostics replace existing ones, users familiar with earlier procedures may need special training to help them unlearn past methods.

Completeness.   Testing should be *complete* enough to ensure mission success. A test is sometimes so essential to mission success that it is more an operational feature than test. For example, because chargeable batteries can be damaged by overcharge, real-time built-in status tests that monitor battery voltage, current, and temperature are essential to the charging feature.[17]

Tests of one subsystem can often protect other subsystems. For example, partial failure of a three-phase power supply can damage electronic equipment powered by that source. Thus a real-time status test that shuts down the three-phase power supply when one phase fails can protect dependent electronics.[18]

PC Case Study
Human Factors and Completeness

PC manuals guide users in step-by-step procedures. Diagnostics software offers menus that help users find failures.

The PC's testability is complete enough to help a user choose among repair options and help a trained technician decide whether to fix it on site or ship to the factory.

Operation under failure.   If the system fails, system tests should suggest alternative modes of operation and the expected degree of mission success for each alternative.

Standardization of test equipment.   As test equipment grows in importance and cost, its reuse becomes important. The following standards help promote test-equipment reuse:

- standard test-equipment commands for test functions
- standard test languages (e.g., ADA, ATLAS)
- standard interfaces between test equipment and computers

---

[17]Aeronautic Division of Wright-Patterson AFB. *Generic Integrated Maintenance Diagnostics (GIMADS) Task 12.* Draft of February 19, 1990, p. D-231.

[18]Aeronautic Division of Wright-Patterson AFB. *Generic Integrated Maintenance Diagnostics (GIMADS) Task 12.* Draft of February 19, 1990, p. D-5.

- standard environments and tools for system development
- test-equipment software that runs on standard computers
- standard architectures that cover a spectrum of price and performance

Experience from other standardization efforts shows that when the number of competing standard architectures is low, the support for them is high, and most resources are spent on developing hardware and software for those architectures.

Interfaces with testing equipment. Designers must provide compatible interfaces for external test equipment that supports the system. These interfaces should be accessible, standard, and small in number.

For example, any subsystem that depends on automatic test equipment (ATE) for testing should be compatible with existing or planned ATE. When ATE and BIT are designed separately, ATE is sometimes incompatible with BIT, making it difficult to do production and field tests.

Any subsystem that will be screened for failure under environmental stress must be compatible with environmental stress screening (ESS) equipment, such as temperature-cycling devices. ESS is discussed in Part 3, Parts Selection and Defect Control.

PC Case Study

Operation under Failure

The PC's modular design allows its use even when subsystems fail. For example, if the floppy-disk drive fails, the user can do any activity that does not involve that drive. Failure of the data ports does not interfere with local, intra-PC features.

The PC has standard connectors that can link it with standard test equipment.

Fault coverage. *Fault coverage,* a widely accepted measure of system testability, is the number of failures that available tests can detect, expressed as a percentage of possible failures. The higher the fault coverage, the better the system's testability. As fault coverage increases, however, the expense and duration of testing also may increase, so designers must do trade-off studies. (Refer to Part 1, Front End Process, for more information on trade-off studies.)

Designers should use failure models and the chosen fault-coverage objective to help them choose system test needs. The fault-coverage target may evolve during system design.

Reliability.   Testing is reliable if it accurately shows system status without seriously degrading system performance. Testability, whose aim is to improve system performance, is seldom seen as a potential threat, yet there are examples of a test capability jeopardizing the system.

While some error in test results is tolerable, testing inaccuracy harms system performance at two extremes:

- False alarms
- False test-OK

A *false alarm* is an apparent indication of failure, when, in fact, no failure exists. A *false test-OK* occurs when testing intended to reveal a failure implies the system is fine, but the failure has occurred. In reliable systems, false test-OKs occur less frequently than false alarms because they require the system and the failure-detection test to fail.

Testing inaccuracy at either extreme hurts customer satisfaction. In the case of military systems, false alarms and false test-OKs can compromise missions. For example, a false alarm can cancel a mission that the system could actually perform. False test-OKs encourage missions that probably cannot succeed. For that reason, false test-OKs may be more life-threatening and costly than the more-frequently-occurring false alarms.

Because unreliable testing hurts system performance, testing reliability should exceed system reliability. In other words, testing, which is designed to enhance system performance, should not *limit* system performance. For example, if a test capability having mean time between failures (MTBF) of 1000 hours is put in a system having a 2000-hour MTBF, the new system's reliability is close to 1000 MTBF because false alarms and false test-OKs hurt performance as surely as real failures of the system.

Because false alarms may result from intermittent but real problems, they need careful study to ensure they do not indicate life-threatening or system-threatening conditions.

Table 4 lists possible causes of false alarms and false test-OKs.

In a contractor survey, many false alarms resulted from electrical overstress caused by transient voltages from the power supply that reach sensitive circuits.[19]

The M-1 tank offers an example of false test-OK. During operation of the M-1 tank, the main nuclear, biological, and chemical system mal-

---

[19]Aeronautic Division of Wright-Patterson AFB. *Generic Integrated Maintenance Diagnostics (GIMADS) Task 12*, Draft of February 19, 1990, p. D-159.

TABLE 4   Some Causes of False Alarms and False Test-OKs

Error	Possible Causes
False Alarm	■ Stresses on subsystems that do the testing, such as heat, humidity, vibration, and power-supply transients ■ Poor test-system design, such as monitoring vibration instead of temperature ■ Incorrect alarm thresholds that cause tests to "fail" subsystems that meet specifications
False Test-OK	■ Test-system failure ■ Poor test-system design ■ Incorrect alarm thresholds that cause tests to "pass" subsystems that fail to meet specifications ■ Unexpected failure modes of the system

functioned and there was no indication of failure.[20] The malfunction-indication lights on the control panel did not light. A faulty resistor in the control panel caused this problem. Studies showed that the resistor fails at low temperature.

Select testability techniques

The prime contractor should consider using the following design techniques to improve system testability and reduce risk.

Built-in test.   Device-level and system-level BIT are discussed in Step 3.

End-to-end testing.   *End-to-end* testing is performed by generating known inputs and comparing the system's outputs to simulated results or previous test results of the system. For example, a system that checks landing-gear position could be fed inputs that mimic landing-gear malfunction to see whether the subsystem's output is an appropriate warning. This tests one aspect of the end-to-end performance of this landing-gear checker.

The end-to-end test is normally considered the final test of a system. To be effective, it should be run in real time in the operational environment.

End-to-end tests can verify performance of systems and subsystems that have distinct inputs and distinct outputs. For example, end-to-end tests may be more appropriate for a warning system than for a tank.

---

[20]Aeronautic Division of Wright-Patterson AFB. *Generic Integrated Maintenance Diagnostics (GIMADS) Task 12,* Draft of February 19, 1990, p. D-503.

End-to-end testing is intended to reduce the number of tests needed to verify that a system is operational. If an end-to-end test fails, then more-detailed tests, such as measuring an amplifier's gain, would be done.

Test points and control points. A common method of designing for testability uses test points and control points. *Test points,* which generally provide external measuring devices with non-intrusive access to system elements, show the status of those elements. For example, a car's oil dipstick is a test point for the lubrication system. *Control points* allow external control of elements in the system. For example, a car's idle-adjustment screw is a control point that mechanics use to set engine speed during tune-up.

In electrical subsystems, a test point is often circuitry added to a node whose status is otherwise difficult to measure. Test-point nodes generally provide key system status information.

A control point is often circuitry that makes it easy to place a node in a state that is otherwise difficult to achieve. The node is chosen so that controlling its state causes observable changes that characterize system performance or status. In some cases, changing the state of the node enables system adjustments. In other cases, controlling the node can fix the system or reset it to a desired state.

Careful selection and use of test points and control points ensure that they will not adversely affect the system. Sometimes many test points and control points can be combined in one interface, thus limiting the number of access avenues that could hurt system performance.

Test- and control-point selection is important for *redundant systems,* which automatically replace a failed subsystem with a backup. Overall tests of a redundant system, such as end-to-end tests, may show normal operation even though a subsystem has failed. Because it is difficult to isolate problems in redundant systems, they should have enough test points and control points to separate them into individual, nonredundant subsystems so those nonredundant subsystems can be tested.

Partitioning. *Partitioning* divides the system, e.g., by function, printed-circuit board, or mechanical subsystem. Initially, partitioning may group related functions. By dividing and conquering, partition of a complex system into (physically) separable modules or subsystems can simplify repair, providing that test points and control points permit tests that isolate failures to modules. Partitioning allows easier production, test, maintenance, and repair because only the failed module needs to be removed, replaced, and repaired. Testing modules independently reduces assembly and maintenance costs.

The costs of partitioning include:

- operational delays caused by module-to-module interfaces
- coordination of module design groups to achieve module integration
- cost and potential unreliability of the module interconnection system

PC Case Study
Partitioning

PCs are partitioned into the following modules:

- power supply—provides power for the system board, keyboard, disk drives, expansion boards, and other system components
- peripheral device bays—provide two locations for the following peripherals: 3.5-inch floppy-disk drive, 5.25-inch floppy-disk drive, and cartridge-tape unit
- system board—provides plug-in locations for expansion for many system components, including numeric co-processor and memory modules
- expansion slots—provide plug-in locations for expansion boards

Initialization. *Initialization,* i.e., setting the starting values of controllable parameters, usually precedes testing because test results may depend on the state of the system at the start of the test. Therefore, testability usually includes the ability to initialize the system.[21]

In electronic circuits, initialization, which often includes reconfiguration of subsystems, starts with a reset signal or short reset sequence.

Initialization provides these advantages:

- reduces test time and test errors by eliminating manual steps otherwise needed to ready the system for testing
- changes the existing, possibly unknown state to a known state

PC Case Study
Initialization

A PC can be initialized by three methods:

- power on—runs full repertoire of built-in hardware and software tests
- front-panel reset—checks and initializes memory
- keyboard reset—stops the application program without checking and initializing the memory

---

[21]Priest, J. W. *Engineering Design For Producibility and Reliability.* New York: Marcel Dekker, 1988, pp. 248-249.

All three methods stop the application program without saving the data in random-access memory. Memory initialization sets the contents of each memory location to zero, which can be important in some applications.

Conduct trade-off studies

Trade-off studies for testability weigh different ways of testing the system and other methods that improve system availability. For an inexpensive and easily replaced subsystem, like an engine's distributor cap, replacement may prove less costly than exhaustive testing. Designers can specify subsystem replacement per calendar interval, hours of use, or symptoms, if that strategy is more effective than testing. The goal is to find the best way to test during manufacturing, acceptance, operation, and repair—or otherwise increase system availability.

Using prototypes, computer models, or other means, designers should compare the above test techniques, singly and in combination, to pick the most effective set of tests at reasonable cost.

Designers also must decide between BIT, ATE, and manual testing early in the design. This is the time to consider supplementing tests with other availability-enhancing features, such as redundant subsystems. Designers should ensure that the tests they choose will satisfy system needs during full-scale development and production.

Contractors can sometimes modify and make testable government-furnished or off-the-shelf equipment that lacks testability. For example, a contractor may wire internal nodes to an external connector to provide access, observability, and control. The contractor must weigh the savings in first costs against the cost and risk of adding testability to inherently untestable equipment.

Figure 5 shows a way to conduct trade-off studies.

Check system testability

Gaps in system testability may surface during production, operations, or maintenance.

To assess system testability before trying the tests in the field, customers or contractors can do *testability demonstrations* or *testability analyses*.

For example, the Hughes Missile Systems Group (MSG) requires testability assessment. In addition, the Hughes MSG's design-for-testability philosophy recommends incorporating testability criteria before designs are frozen.[22]

---

[22]Best Manufacturing Practices (BMP) Review Team. *Best Manufacturing Practices, Report of Survey Conducted at Hughes Aircraft Company, Missile Systems Group, Tucson, AZ*. Washington, DC: Office of the Assistant Secretary of the Navy (Shipbuilding and Logistics), August, 1988, p. 6.

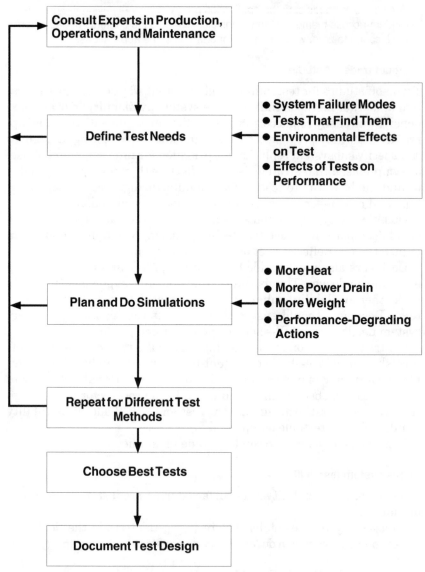

Figure 5 How to do trade-off studies.

Testability demonstrations. Diagnostic demonstrations show the progress and status of testability. To do a comprehensive and complete diagnostics demonstration, all operational diagnostics tools must be available, including support equipment, training, technical manuals, and other applicable diagnostic tools.

Scheduling and obtaining resources for the demonstration are important in any project. Government program managers should be aware of demonstration plans and ensure that such plans are included in the earliest top-level planning documents.

Testability analysis. *Testability analysis* is a theoretical assessment of system testability, i.e., it does not exercise the system or its test tools. Generally, the system architecture and subsystem properties provide the data for the testability analysis.

Various independent auditors can do testability analyses. For example, the RAC testability analysis shows how well the system design meets the requirements of controllability, observability, and accessibility.[23] It is an accepted commercial and government method for measuring the testability of a system design. The result may suggest changes and serves as a baseline for gauging improvements.

Alternatively, testability-analysis tools are available.[24] Some compute a figure of merit for testability.[25]

Testability-analysis tools typically use logic modeling for electronic, hydraulic, electromechanical, and mechanical systems.

Testability-analysis tools can be automated or manual. In general, automated tools yield quantitative testability analyses while manual tools yield qualitative testability analyses.

Step 3: Design Built-In Tests

Built-in test (BIT) helps monitor the operation of the system, generate stimuli, detect faults, and isolate failed subsystems without the need for external equipment. A BIT can be as simple as a pressure gauge that is part of a tire or as complex as a microprocessor that checks the subsystems of a main-frame computer.

The power of a BIT increases if it can be controlled externally, e.g., by automatic test equipment (ATE). Thus designers of BITs should include interfaces to ATE.

Whether to test a subsystem using BIT depends on:

- the importance of the test to system and mission success
- the likelihood that BIT will find the failure

---

[23]The Reliability Analysis Center of the DoD Information Analysis Center, Rome, NY does testability analysis.

[24]DETEX Systems Incorporated of Orange, CA offers Systems Testability Analysis Tool (STAT), Weapon Systems Testability Analyzer (WSTA), and Computer-Aided Fault Isolation Tool (CAFIT).

[25]Kovijanic, P. G. "Single Testability Figure of Merit." *Proceedings of the 1981 IEEE Test Conference,* October 19-21, 1981, pp. 521-529.

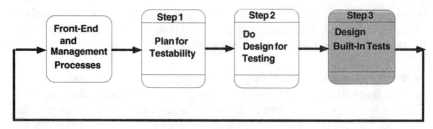

Figure 6 Steps for achieving testability.

- the availability of external test equipment (in the factory and field)
- the cost of self-test

For example, if an autopilot test is critical to mission success and the choice is between a bulky external tester that will not fit in the cockpit or a few extra built-in integrated circuits, designers should choose BIT. If autopilots rarely fail or if missions usually succeed even if the autopilot fails, the benefits of BIT may not pay the cost of BIT.

Figure 7 shows how to add BIT to a system.

Propose built-in tests

Built-in tests generally begin as proposed solutions to testability needs. A comprehensive system analysis, such as DFT, shows which tests to build in, and the process of detailed BIT design and development begins.

The trade-off study outlined in Step 2, Design for Testing, may yield a complete test strategy that includes a list of needed BITs. The simulation studies done during the trade-off study may suggest how to design and develop those BITs.

Plan BIT. The prime contractor, in cooperation with the subcontractors and the government, should schedule milestones in the development of the BITs. Those milestones are generally the activities outlined in Figure 7.

The schedule should allow for time delays. For example, a subsystem may be redesigned and the new design may need new BITs.

Write BIT requirements. The prime contractor should consult subcontractors to obtain lists of BITs recommended for their respective subsystems.

The prime contractor, drawing on data from the DFT trade-off studies, subcontractor suggestions, and customer ideas, should specify

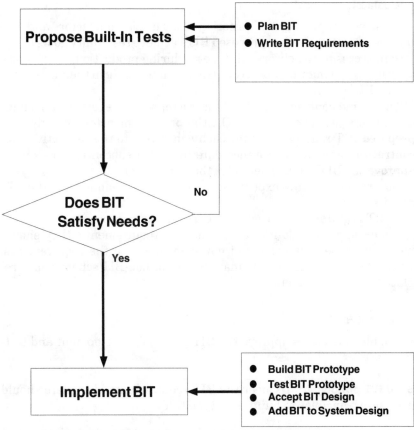

Figure 7 Design built-in tests.

needed BITs for each subsystem and for the complete system. Requirements will include test points, control points, partitioning, ATE interfaces, and BIT efficiency.

BIT efficiency includes the probability that the BIT will detect failure, the probability of isolating the defective subsystem, and the average time needed to complete the test. For example, requirements for BIT efficiency may be:

- 95% detection of all failures
- 80% isolation to replaceable subsystem
- 10 minutes to complete the BIT
- less than 0.1% false alarms

Evaluate BIT proposal

To ensure that proposed BITs satisfy their needs, contractors and customers estimate whether proposed BITs achieve testability goals of isolation, accessibility, observability, etc., during production, acceptance, operation, and maintenance. (Contractors later confirm their estimates with a BIT prototype.)

The prime contractor and subcontractors may develop the BIT requirements jointly as a team. Or, the prime contractor may give the proposed BIT requirements for each subsystem to the respective subcontractors and the acceptance date by which subcontractors should approve the BIT requirements for their subsystem.

Government testing experts also may wish to comment on the BIT requirements.

If BIT requirements satisfy key parties, design proceeds.

If subcontractors do not accept the BIT requirements, they should help the prime contractor and government revise the requirements. Unsatisfactory requirements may change, or non-BIT schemes may replace those requirements.

Implement BIT

When all contractors approve the BIT proposal, development and testing starts.

Build BIT prototype. To meet the BIT requirements, designers should use these principles to build prototype BITs:

- Use current BIT methods, such as *boundary scan,* which gives access to an electrical circuit's test points through a built-in, dedicated *bus* or *data highway.*[26][27] The Air Force requires boundary scan for its advanced tactical fighter and the Army calls for it in the LH helicopter.

- Use standard designs if possible.

- Avoid hurting system performance.

- Have inexperienced users try the prototype, aided by user-training and test guides, to confirm human-factors aspects of the proposed design.

- Test BIT prototype.

---

[26]Maunder, C. M. and Tulloss, R. E. *The Test Access Port and Boundary-Scan Architecture.* Los Alamitos, CA: IEEE Computer Society Press, 1990, p. 12.

[27]Kinnucan, P. "Boundary Scan Makes an Impact on Military ATE." *Military & Aerospace Electronics,* September 1990, p. 37.

Contractors should measure the effectiveness of their BIT designs, e.g., by simulating failures and noting whether the BIT prototypes achieve efficiency targets.

The prime contractor should coordinate BIT-prototype testing for tests that cross subsystem boundaries and involve several subcontractors.

If a BIT prototype fails its efficiency objective, the prototype may be at fault, the BIT may need redesign, or the customer and contractors may agree to change the efficiency goal.

Accept BIT design. After BIT prototypes prove that the BIT designs achieve the needed testability, those designs are documented and receive final approval.

Add BIT to system design. Approved BIT designs become part of the system when contractors add them to their respective subsystem designs.

As with any design, production realities and field experience may require changes that improve the BITs.

Overall system testability is achieved when BIT and all other DFT techniques, such as environmental-stress screening, automatic test equipment, and user-support materials, have been studied, tried, and made part of the system.

PC Case Study

Built-In Test

The PC has built-in tests that check modules whenever the PC is turned on. During start-up, the PC performs a power-on self-test to check the condition of the system before the PC makes itself available for use.

The following shows a typical start-up display of the PC.

Resident Diagnostics Version x.x

CPU (i286)	Pass
ROM Module	Pass
DMA Timer	Pass
DMA Control	Pass
Interrupts	Pass
640-Kb RAM	Pass
RT Clock	Pass
Fixed Disk	1 Present
Floppy Disks	2 Present

If the display is properly installed, the power-on self-test checks the PC, and the PC screen shows the results. If the PC detects no errors, its speaker beeps once. If the PC detects an error and if the display is missing or defective, the speaker gives a long beep followed by one or more short beeps. The number of short beeps identifies the problem to a trained technician.

The PC offers the following menu of user-selectable diagnostic tests.

Diagnostic	Description
Configuration Check	Lists the system configuration and allows change for testing.
Setup Utility	Sets time and date, identifies configuration errors, and provides hardware configuration options.
Test All Modules	Automatically sequences the tests to test all modules in the configuration.
Test One Module	Tests a specific hardware module.
Set Test Options	Defines the test process.
Error-Log Utilities	Defines the format of test results and where they should be recorded.

Chapter

3

Application

This section illustrates the methods and techniques of this reference guide by discussing an M-1 Abrams tank case study and typical lessons learned from other military projects.

M-1 Abrams Tank

Introduction

Applied to diagnostics, "Do it right the first time" means designing the system's diagnostic capability as system design progresses. Unfortunately, designing the diagnostic capability and following through to deployment are becoming more challenging. Improved diagnostics and system redesign will continue to be the solutions to diagnostic problems. The following example of the M-1 Abrams tank illustrates some of these points.[28]

Deficiencies in the diagnostic capability of the M-1 tank led to an integrated-diagnostics improvement program—a positive example of improving the diagnostic capability after-the-fact.

Dissatisfaction with the diagnostic capability of the fielded M-1 was not limited to maintenance personnel—technology and cost issues also concerned Army management.

To develop and make the necessary improvements, a Joint Working Group (JWG) for Integrated Diagnostic Improvement was formed with a charter to develop a systems-engineering approach to improve and integrate the M-1 tank's diagnostic capability. The JWG first established integrated-diagnostic improvement objectives and then developed a road map to achieve the objectives. The objectives were:

---

[28]Rome Air Development Center. *A Guide for Contractor Program Managers: How You Can Produce and Deploy an Improved Weapon System Diagnostic Capability—A Systems Engineering Process.* Rome, NY: RADC, 1990, pp. A-14 to A-22.

- resolve the M-1 tank diagnostic problems
- identify cost-effective solutions that complement short-term fixes
- ensure that the diagnostic concept supports future enhancements
- communicate lessons learned to contractors and combat personnel

M-1 diagnostic problems

The JWG defined the scope of their effort and also identified the following four areas of M-1 diagnostic problems.

Vehicle diagnostic problems.   The vehicle lacked an integrated diagnostic system:

- limited BIT capability (See Table 5)
  absence of diagnostic capability in the mobility system
  absence of BIT for sensors

TABLE 5   M-1 Tank Built-In Tests

M-1 Tank Subsystem	Automatic Fault Detection	Operator-Initiated Fault Detection
Fire Control and Gun Stabilization	- Display in gunner's sight shows failed subsystem - Panel indicates defective LRU in thermal-imaging system - Computer run finds faulty LRU in wind sensor - Remaining faults isolated to four or five LRUs	- Special test equipment isolates faults in remaining subsystems to single LRU
Armament	- Manual-loading continuity tester test condition of firing circuit - Faults detected by firing lock-out	- Special test equipment isolates faults in remaining subsystems to single LRU
Hull Electronics	- No built-in tests	- Special test equipment isolates faults to single LRU
Chassis Power and Actuation	- Engine and transmission fault indications - No built-in tests for track and suspension	- Special test equipment isolates faults in remaining subsystems to single LRU
Fire Suppression	- No built-in tests	- Special test equipment verifies status periodically

- lack of standard diagnostic connector assemblies
- inability to isolate a fault to a single line-replaceable unit (LRU) or cable
- lack of data recording capability
  no record of intermittent faults

  no failure history
- poor LRU, cable, and connector assembly
- limited sensors for vehicle transmission
- limited space for new hardware

Special-test-equipment problems.   The following problems afflicted special test equipment (STE):

- STE failed environmental requirements
- STE display was hard to read in direct sunlight
- STE self-test took too long

Test-program-set problems.   The test program sets:

- required cumbersome cable connections
- lacked user documentation
- examined failure at component level rather than system level
- stopped failure isolation after first fault was identified
- lacked an efficient method of recovering from operator error
- prohibited mechanic from bypassing unnecessary test steps
- did not use known fault data (e.g., from BIT)
- failed to meet required test-duration limits
- misinterpreted test messages
- did not provide test measurements to mechanics

Problems with technical manuals.   The following problems were identified with technical manuals:

- manuals were not user-friendly because of quantity and excessive cross-referencing
- procedures were cumbersome and time-consuming

- manuals were not suited for field use
- manuals were written only to lowest skill level
- procedures for changing the manual were time-consuming and difficult

Personnel-training problems. Despite training:

- mechanics lacked theoretical knowledge
- mechanics lacked advanced training and review

Integrated-diagnostics improvement

After developing a list of M-1 diagnostic problems, the JWG identified 24 diagnostic improvement programs. To provide a comprehensive and cost-effective plan, the JWG suggested developing diagnostics requirements based on operational requirements, criticality of component repair, and components repair echelon. In addition, the group developed the integrated-diagnostics method shown in Table 6.

To fit the M-1 production schedule, the JWG segregated diagnostic improvement into short-term fixes and long-term solutions. The JWG limited short-term fixes to already-funded diagnostic programs. The JWG addressed problems such as standard test connector design with the long-term solutions.

TABLE 6   Integrated-Diagnostics Method

Maintenance Level	Procedure
Unit Level	<ul><li>Develop diagnostic alternatives that meet requirements</li><li>Examine whether current diagnostic approach can be improved</li><li>Determine acquisition, logistic, and support cost of each alternative</li><li>Select optimum diagnostic approach</li></ul>
Intermediate Level	<ul><li>Determine direct-support electrical-systems test-set manpower and workload</li><li>Determine requirements of embedded diagnostics and tester-independent support</li><li>Determine causes of high no-evidence-of-failure rates</li><li>Recommend solutions</li></ul>

M-1 diagnostics system

The JWG ranked the battle criticality of tank subsystems as follows:

- mobility
- firepower
- surveillance
- survivability
- communications

They concentrated their diagnostic improvement efforts on the top three subsystems.

To diagnose problems, the JWG developed a system that:

- obtains results of built-in diagnostics (e.g., trouble lights, status meters) to guide later stages of problem analysis
- locates catastrophic failures by sight, touch, and smell
- uses a break-out box and multimeter to measure signals and isolates faults by comparing measurements with normal values
- uses STE to find defective subsystems if all other methods fail

The JWG made specific and detailed recommendations for carrying out the above steps.

Summary

The systems-engineering method applied to the M-1 Abrams diagnostics improvement program resulted in solutions that affected all diagnostic elements. The solutions focused on the integration of the diagnostic elements to provide a comprehensive and cohesive diagnostic capability within existing constraints.

Typical Lessons Learned

The following four lessons in the area of diagnostics were learned during the development of the B-1 bomber and the F-111 aircraft.[29] The lessons reinforce design-for-testing topics covered in this reference guide.

---

[29]Rome Air Development Center, *A Guide for Contractor Program Managers: How You Can Produce and Deploy an Improved Weapon System Diagnostic Capability—A Systems Engineering Process.* Rome, NY: RADC, 1990, pp. A-1 to A-11.

The purpose of "lessons learned" is to help managers of future projects avoid past mistakes.

Haste hurts testability

Development of the B-1A bomber began in 1971 and continued until 1977. In 1977 the B-1A project was canceled. Renamed the B-1B project, it resumed in March 1982. To make up for the lost time between 1977 and 1982, it was decided to do the full-scale development and production programs concurrently. This approach led to production problems and poor diagnostics at the deployment phase.

Emphasis determines success

The history of the F-111D fighter aircraft shows that whether a test design is successful can depend on how much the customers or contractors emphasize the test.

For example, BIT usually serves two functions: one is to advise the operator about the status of a system and the other is to serve as a diagnostics aid to maintenance personnel. The F-111D status-reporting BIT performed well, but the diagnostic protion of BIT did not work properly. The reason is that, during F-111D development, designers emphasized operations testing rather than diagnostic testing. Simulation of F-111D operation in the laboratory was fruitful and usually resulted in the elimination of failures, but BIT diagnostic problems remained. This was not because refining diagnostics in a laboratory is any harder than simulating mission conditions. The emphasis on diagnostics was not there.

Maturation needs

One definitive B-1B lesson is the need for a period of system operation to mature operational and amintenance tests. During maturation, emphasis is placed on testability analysis, reliability, and maintainability.

A critical step of maturation is system testing under field conditions to refine the test method and fault limits.

It is sometimes hard to predict how much maturation is needed. For example, it was expected that 70 flights during full-scale development would suffice to mature the B-1B diagnostics. During full-scale development, it was learned that 70 flights were not enough. Then a plan was developed to use 468 Strategic Air Command sorties in 1985 and 1986. The system under development did not fly 468 sorties over that time period and the program was extended through November 1987. Additional aircraft deliveries and an increase in sortie rate produced a total of 1,060 sorties during that period. By then enough data had been gathered to indicate a more-than-acceptable level of performance. It is estimated that at least 400 to 500 sorties are required to mature an on-board test system.

Once bitten...

A partially working diagnostic system can cause maintenance technicians to lose confidence in that diagnostic and stop using it. Later, it is hard to convince them that the diagnostic has been improved.

For example, B-1B maintenance technicians who were exposed to inaccurate diagnostic methods never fully accepted improved versions. They had used early, imperfect diagnostics that called for replacing parts even though the system appeared and was normal. Because they lacked confidence in the diagnostics, they changed no parts, even when later, more-accurate diagnostics demanded part replacement. Change only occurred when the system failed, and progress toward avoiding system failure was slow.

According to field data, bases that used the poorest version of the diagnostics continued to have the highest system failure rate. People on those bases probably had low confidence in the diagnostics and used them infrequently.

4

Summary

This reference guide covers two templates: Design for Testing (DFT) and Built-In Test (BIT).

DFT is a development strategy that ensures adequate, low-cost testing throughout the life of a system by anticipating diagnostic and status-reporting needs at project conception.

BIT is a particular test method that ensures testability by using a systems's internal hardware and software.

DFT and BIT should be used for complex systems because they lower system costs and increase system availability.

Table 7 outlines procedures and supporting activities for DFT and BIT.

TABLE 7 Summary of Design-for-Testing Procedures

Step 1: Plan for Testability	
Procedures	Supporting Activities
Understanding testing needs during system life	■ Plan prototype testing that is extremely thorough ■ Plan production-line testing that is fast, costs little, finds flaws, and avoids false alarms ■ Plan acceptance testing that confirms basic system performance ■ Plan operational and maintenance testing that supports the diagnostic needs of organizational, intermediate, and depot levels of maintenance.
Define objectives of the testability plan	■ Translate performance requirements into testability requirements for each level of maintenance ■ Meet those requirements ■ Involve test experts with the system design team to ensure a low-cost and effective test capability ■ Improve testability throughout the system's life cycle

TABLE 7   Summary of Design-for-Testing Procedures (Continued)

Step 1: Plan for Testability (*Continued*)	
Procedures	Supporting Activities
Define objectives of the test ability plan	■ Identify operational constraints that affect test-ability ■ Identify technical advances that may decrease need for test equipment, technical manuals, staffing, skills, diagnostics, spares, and system downtime ■ Identify existing and planned diagnostic resources outside the system and their limitations ■ Identify testability problems observed with similar systems and avoid them ■ Choose fault coverage, the percentage of expected failures that available tests can detect ■ Agree on the false-alarm rate, the number of incorrect failure indications per unit of operation time ■ Plan to test a prototype system
Getting started	■ Identify the organization responsible for implementing the plan for testability ■ Establish contacts and data interfaces among organizations responsible for elements of the testability plan ■ Develop a process that integrates testability with other requirements ■ Ensure that each subcontractor's test practices are consistent with overall system or equipment requirements ■ Plan to review, verify and use test data

Step 2: Do Design-for-Testing for Each Subsystem	
Procedures	Supporting Activities
Define testability objectives	■ Isolate faults to defective subsystems ■ Isolate subsystems before testing ■ Consider accessibility ■ Consider observability ■ Consider controllability ■ Consider human factors ■ Check for completeness ■ Consider operation under failure ■ Standardize test procedures, test interfaces, and external test equipment ■ Consider interfaces with test equipment, such as ATE and ESS ■ Establish fault-coverage guidelines ■ Consider reliability of testing and minimize false alarms and false test-OKs

TABLE 7   Summary of Design-for-Testing Procedures (Continued)

Step 2: Do Design-for-Testing for Each Subsystem (*Continued*)	
Procedures	Supporting Activities
Select techniques for testability	■ Consider built-in test ■ Consider end-to-end testing ■ Consider the use of test points and control points in system design ■ Consider partitioning ■ Consider initialization
Conduct trade-off studies	■ Define the system's test needs throughout the system's life ■ Consult experts in production, operations, and maintenance to satisfy their inspectability and test needs ■ List system failure modes and tests that detect them ■ List effects of the environment on the proposed tests ■ List effects of the proposed tests on system perfor mance ■ Evaluate cost and benefits of BIT, ATE, and ■ manual testing ■ Plan and run needed simulations ■ Repeat for different combinations of test techniques ■ Pick the best tests ■ Document the test design
Check system testability	■ Do testability demonstrations ■ Do testability analyses

Step 3: Design-In Built-In Tests	
Procedures	Supporting Activities
Propose built-in tests	■ Consider built-in tests from DFT trade-off studies ■ Plan BIT design ■ Write BIT requirements
Evaluate BIT proposal	■ Give BIT requirements to subcontractors ■ Revise BIT requirements, if needed
Implement BIT	■ Build BIT prototype ■ Test BIT prototype ■ Accept BIT design ■ Add BIT to system design

Abramovici, M., Breuer, M. A., and Friedman, A. D. *Digital Systems Testing and Testable Design*. New York: Computer Science Press, 1990. Provides a comprehensive and detailed treatment of digital-systems testing and testable design. Is an excellent reference and source of information for engineers interested in test technology: integrated-circuit designers, system designers, test engineers, and developers of computer-aided design (CAD).

Agrawal, V. D. and Seth, S. C. *Test Generation for VLSI Chips*. Washington, DC: IEEE Computer Society Press, 1988. Describes various topics in electronic testing such as VLSI testing, types of testing, methods of testing, test development process, fault modeling, test evaluation, testability analysis, design for testability, and automatic test equipment.

Aeronautic Division of Wright-Patterson AFB. *Generic Integrated Maintenance Diagnostics (GIMADS) Task 12*. Draft of February 19, 1990. Narrates lessons learned from some diagnostics areas known to place heavy maintenance burden on present systems. Covers: fault indications that cannot be duplicated, faults that are difficult or costly to isolate to a replaceable item, effect of power-supply quality of electronic failures, and mechanical-system diagnostics.

Best Manufacturing Practices (BMP) Review Team. *Best Manufacturing Practices, Report of Survey Conducted at Hughes Aircraft Company, Missile Systems Group, Tucson, AZ*. Washington, DC: Office of the Assistant Secretary of the Navy (Shipbuilding and Logistics), August 1988. Describes Hughes Missile Systems Group policy, practices, and strategy in the functional areas of design, test, production, facilities, logistics, management, and transition planning. Categorizes individual practices as they relate to critical-path templates of the DoD 4245.7, *Transition from Development to Production*.

Best Manufacturing Practices (BMP) Review Team. *Best Manufacturing Practices, Report of Survey Conducted at Unisys Corporation Computer Systems Division, St. Paul, MN*. Washington, DC: Office of the Assistant Secretary of the Navy (Shipbuilding and Logistics), November 1987. Describes Unisys Corporation Computer Systems Division policy, practices, and strategy in the functional areas of design, test, production, facilities, logistics, management, and transition planning. Catalogs practices and relates them to critical-path templates of DoD 4245.7, *Transition from Development to Production*.

Cheng, K. -T. and Agrawal, V. D. *Unified Methods for VLSI Simulation and Test Generation*. Boston: Kluwer Academic Publishers, 1989. Discusses various topics in electronic testing. Includes: the testing problem, testing techniques, logic simulation, other simulation methods, fault analysis, test-generation approaches for combinatorial and sequential circuits, and several methods of test generation.

Daniels, R. G., and Bruce, W. C. "Built-In Self-Test Trends in Motorola Microprocessors." *IEEE Design and Test*, April 1985, pp. 64-71. Provides an insider's account of a decade of built-in self-test development at Motorola.

Department of Defense. *Acquisition of Major Defense System.* Directive 5000.1, Washington, DC, July 1971. Establishes policies, practices, and procedures consistent with concepts and provisions of defense acquisition-program procedures. Covers management of major and non-major defense acquisition programs.

Department of Defense. *Transition from Development to Production.* DoD 4245.7-M, September 1985. Describes techniques for avoiding technical risks in 48 key areas or templates that include: design, test, production, facilities, logistics, and management. Identifies critical engineering processes and controls for design, test, and production of low-risk products.

Department of Navy. *Best Practices: How to Avoid Surprises in the World's Most Complicated Technical Process.* NAVSO P-6071, March 1986. Describes how to avoid traps and risks by implementing best practices for 48 areas of risk (templates) that include: design, test, production, facilities, logistics, and management. Provides program managers and contractors with an overview of the key issues and best practices to improve systems and their acquisition.

Joint DARCOM/NMC/AFLC/AFSC/Commanders. *Built-In Test Design Guide.* Washington, DC, 1981. Provides and overview of the different approaches and requirements available to the designer and the acquisition manager. Discusses standard methods for evaluation of these approaches. Is an excellent reference for personnel responsible for design for test, maintainability, reliability, and logistics support.

Kinnucan, P. "Boundary Scan Makes An Impact on Military ATE." *Military & Aerospace Electronics,* September 1990, pp. 37-40. Describes the importance of boundary scan in military applications. Explains the Joint Test Action Group's boundary-scan scheme (IEEE 1149.1 standard). Provides examples of weapon systems currently using boundary scan.

Konemann, Bernd and Wagner, Kenneth. "Practical BIST: Old and New Tools in Concert." *IEEE Design & Test of Computers,* August 1989, pp. 16-17. Introduces the special issue of *IEEE Design & Test of Computers* dedicated to built-in test from a practitioner's view.

Kovijanic, P. G. "Single Testability Figure of Merit." *Proceedings of the 1981 IEEE International Test Conference,* Philadelphia, PA, October 27-29, 1981, pp. 521-529. Derives a single figure of merit for testability that is based on TESTSCREEN, a design- automation tool that helps logic designers study the testability of comples electronic circuits. Shows agreement between this fast testability estimator and testability obtained from an established system for test generation and fault simulation.

Maunder, C. M. and R. E. Tulloss. *The Test Access Port and Boundary-Scan Architecture.* Los Alamitos, CA: IEEE Computer Society Press, 1990. Discusses the trends in product and test technology that motivated the development of IEEE Std 1149.1. Introduces the standardization of test-support features for integrated circuits and loaded printed-wiring boards. Presents an abbreviated, tutorial verison of IEEE Std 1149.1 circuitry. Offers examples of IEEE Std 1149.1 applications.

McCluskey, E. J. "Built-In Self-Test Techniques." *IEEE Design and Test,* April 1985, pp. 21-28. Surveys structures used to do the built-in self-test (BIST) function. Describes various techniques used to convert system bistables into test-scan paths. Describes the addition of bistables associated with the I/O bonding pads so that the pads can be accessed via a scan path.

McCluskey, E. J. "Built-In Self-Test Structures." *IEEE Design and Test,* April 1985, pp. 29-36. Describes structures that support self-test features such as generating test patterns and analyzing circuit responses. Offers various techniques that convert the system bistables into test-scan paths. Presents various linear-feedback shift-register designs for pseudorandom or pseudoexhaustive input-test-pattern generation and for output signature analysis.

Military Standard 2165: *Testability.* Discusses selective implementation of system-level diagnostic strategies, partitioning to enhance fault isolation, initialization of circuitry under test control, controllability, observability, parts selection, test-point placement, and built-in test.

Parker, Kenneth P. "The Impact of Boundary Scan on Board Test." *IEEE Design & Test of Computers,* August 1989, pp. 18-30. Describes boundary scan for production board testing and its advantages. Explains how boundary scan, even thoug it is not a replacement for automatic test equipment, can increase accuracy and decrease test complexity and test cost.

Priest, J. W. *Engineering Design For Producibility and Reliability.* New York: Marcel Dekker, 1988. Describes various topics in the area of testing. Includes: developmental testing and reliability growth, TAAF, integrated test-and-evaluation strategy, environmental and design-limit testing, life testing, accelerated life testing, operational field testing, economics of testing, testability design requirements, levels of testing testability approaches for electronic systems, and design techniques for testability.

Raytheon. *Transition from Development to Production.* Managment Manual 956010, February 1985. Discusses the 48 templates. Includes: technical risks, money phasing, design, testing, production, transition plan, facilities, and logistics. Combines elements from both *Best Practices* (NAVSO P-6071) and *Transition from Development to Production* (DoD 4245.7-M).

Rome Air Development Center. *A Guide for Contractor Program Managers: How You Can Produce and Deploy an Improved Weapon System Diagnostic Capability—A Systems Engineering Process.* Rome, NY: RADC, 1990. Provides guidance for industry personnel in circumventing known problem areas of diagnostics. Problem areas include: engineering, field, human factors, machine type, weapon systems, and support systems.

Rome Air Development Center. *A Guide for Government Program Managers: How You Can Produce and Deploy an Improved Weapon System Diagnostic Capability—A Systems Engineering Process.* Rome, NY: RADC, 1990. Provides guidance for government personnel in circumventing known problem areas of diagnostics. Problem areas include: engineering, field, human factors, machine type, weapon systems, and support systems.

Runyon, S. "Special Report on Testing—Credo for a New Decade: Design for Testability." *Electronic Engineering Times,* August 7, 1989, pp. 23-25. Describes the testing challenges of the 1990s and explores the realities of doing design for testability.

Runyon, S. "Special Report on Testing—Designers Wait for Tools to Forge Design-to-Test Links." *Electronic Engineering Times,* August 14, 1989, pp. 46-50. Discusses efforts to link the design and test processes.

Runyon, S. "Special Report on Testing—Standards, Software Lend Design Status to Test." *Electronic Engineering Times,* August 21, 1989, pp. 80-86. Describes the contributions of software and standards to system testability.

Stephenson, J. E. and Grason, J. "A Testability Measure for Register Transfer Level Digital Circuits." *Proceedings of the Sixth International Symposium on Fault-Tolerant Computing* (FTCS-6). Pittsburgh, PA, June 21-23, 1976, pp. 101-107. Describes a testability measure for digital circuits that predicts ease of generating tests for such circuits. Shows the measure is easy to calculate and helps design for testability.

Williams, Thomas W. and Parker, Kenneth P. "Design for Testability—A Survey." *Proceedings of the IEEE,* vol. C-31(1), January 1982, pp. 98-112. Describes the basics of design for testability. Reviews testing and reasons for testing. Discusses in detail different techniques of design for testability. Includes techniques that apply to recent technologies and techniques used in industry.

Zorian, Y. and Jarwala, N. "A Structured Approach to Complex VLSI Testing Using BIST and Boundary Scan." *ATE Instrumentation Conference West,* 1990, pp. 69-72. Proposes built-in self-test and boundary-scan strategies to solve complex VLSI system problems. Describes how to partition a VLSI system into different structural modules, various regular structures, and random-logic blocks. Describes specific BIST algorithms that achieve high fault coverage in each of these modules. Explains the use of boundary scan to control and access all on-chip test resources.

Configuration Control

Chapter

1

Introduction

To the Reader

This part, Configuration Control part, is intended for use by program managers and design, manufacturing, and project management professionals involved in governmental and commercial product development efforts to develop overall Configuration Management (CM) policies, procedures, and metrics for measuring CM program effectiveness. This part is not narrowly focused on any one CM function. For more information and guidance on specific CM topics, consult the references listed in Chapter 5.

Although the title of this template is Configuration Control to be consistent with the existing template terminology, the overall CM discipline is covered herein. Configuration identification, control, status accounting, audits, and subcontractor CM are all essential in effective CM program implementation. These elements combine to facilitate product development and the transition from development to production through a comprehensive CM program.

Configuration Management (CM) is a discipline that organizes and implements, in a systematic fashion, the process of documenting and controlling system configuration.[1] The CM discipline is essential in reducing technical risk and assuring design integrity during the transition of any product design program from development to production. CM is essential for overall system/product quality and supportability.

---

[1] *Best Practices: How to Avoid Surprises in the World's Most Complicated Technical Process,* Department of the Navy, (NAVSO P-6071), March 1986.

CM Strategy

CM should be part of an overall strategy that incorporates the *Transition from Development to Production* templates[2] that are the foundation for this series of books. Together with the *Best Practices,*[3] the templates provide the engineering discipline and road maps needed to achieve a successful, low-risk transition of products from development to production.

Ultimately, implementation of CM will reduce costs associated with redesign, factory rework, and support and will contribute to the overall success of the procurement effort.

Rationale

Implementing CM properly and early in the development process provides design traceability. By freezing and documenting the design at various points, baseline configurations can be established and controlled to prevent unauthorized changes. This ongoing configuration management provides adequate data for effective program management and permits final planning for production, installation, maintenance, and support. Good configuration management can be helpful in reducing the number of different configurations in operational use, by preventing unnecessary changes and by retrofitting operational systems as necessary. System designers must understand their responsibilities in supporting CM; training courses are essential in this regard.

During the transition from development to production, CM ensures that configuration management procedures are an integral part of the production process. Thus, the as-built hardware or software Configuration Item (CI), typically an end-use functional item, will exhibit all the functional and physical characteristics of the approved configuration identification (the total documentation package for the CI). CM must also be applied throughout the life cycle of the equipment to avoid degraded operational availability and higher support costs during system deployment.

CM objectives

The overall objectives of CM are to:

- Ensure that the delivered system meets all the functional and physical requirements of the approved configuration identification

---

[2]*Transition from Development to Production,* Department of Defense (DOD 4245.7-M), September 1985.

[3]*Best Practices,* Department of the Navy, March 1986.

- Ensure that the documentation is accurate to ensure repeatable performance, quality, and reliability for future procurements of the same system.

The overall objectives for hardware and/or software CIs (HWCIs/CSCIs) can be further refined to:

- Identify all documentation required for product design, development, fabrication, software coding, and test
- Maintain accurate and complete descriptions of the approved product configuration, including descriptions of each material, part, subassembly, and assembly of the product (including drawings, parts lists, specifications, test procedures, and operating manuals)
- Maintain traceability of the as-built system and its components to these descriptions
- Maintain accurate, complete control and accounting of all changes to CI descriptions and to the actual as-built CIs.

The cost benefits gained from an effective CM Program far outweigh any additional efforts. The resulting design is more complete and cohesive, lowering costs associated with redesign and rework. Products perform according to specification requirements, with lower customer maintenance costs. *Remember: These objectives will only be met when experienced people run an effective CM Program. A poorly run CM Program only hinders progress.*

CM Defined

CM applies technical and administrative direction and surveillance to:

- Identify and document the functional and physical characteristics of a CI
- Control changes to those characteristics and to the associated documentation
- Record and report change processing and implementation status
- Plan and conduct configuration audits to ensure that the functional and physical characteristics of the as-built CI match those required by the approved configuration documentation.

An effective CM Program must continue through all five acquisition phases of a product's life cycle: the concept exploration phase, the demonstration/validation phase, the full scale development phase, the production phase, and the operation and support phase. The timing

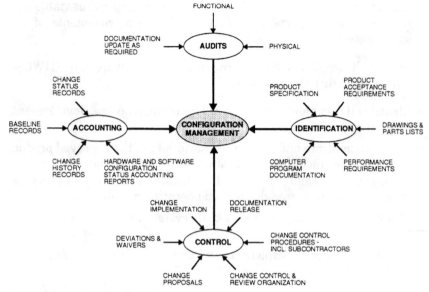

Figure 1 Overview of configuration management.

and sequence of various CM activities and events are detailed in Chapter 3, Application.

All configuration identification, including specifications, drawings, schematics, diagrams, associated parts and data lists, software files, and software design documents, will be under configuration control. This control will be *internal* until baselines are established and approved by the customer; then, all documentation associated with a baseline will be under *formal* configuration control, requiring customer approval or concurrence in classification, for each proposed change.

Internal CM

Internal CM must begin early to optimize selection of CIs and to organize and monitor the development process. Internal CM is similar to formal CM, except that formal customer approvals are not required. Change order packages are prepared and processed internally, rather than formal Engineering Change Proposal (ECP) packages which require customer approval. Internal CM is described in more detail in Chapter 3, Application.

Formal CM and baselines

Formal CM is mandated by the customer after customer-approved baselines are established. For example, formal CM begins at contract award for any existing system specifications called out in the contract. However, the system specification may not always be provided with the contract; occasionally the contractor is required to develop or update this specification. These specifications and the system-segment specifications, if any, constitute the *functional baseline*. During the Design Review process, *allocated baselines* for hardware and software are established. This baseline usually consists of hardware development and software requirement specifications. A *product baseline* is established when the functional and physical configuration audits are successfully completed. After each baseline is established, the associated documents are brought under formal configuration control. Formal CM is designed to meet the customer's specific requirements as mandated in the contract. An experienced CM organization, an adequate CM Plan, and subcontractor control are critical to formal CM.

Organization

An experienced Configuration Manager administers the CM Program. Proper staffing by qualified personnel is essential to ensure effective interface with design and manufacturing.

The CM function is most effective when there is strong management support and CM is delegated to a separate CM organization with sufficient authority and personnel to enforce its policies. This minimizes possible conflicts of interest. For example, in organizations where CM is subordinate to either engineering or manufacturing, decisions affecting CM are often based on what is needed to satisfy the short-term cost or schedule requirements rather than overall CM Program goals. Simply put, in this arrangement, CM policies are frequently compromised for short-term program goals.

Even though, ideally, CM is part of a separate organization, the Configuration Manager should report functionally to the Program Manager. Although CM is separated from engineering, designers should be trained in CM policies and procedures. They should work with the Configuration Manager to ensure that CM policy is followed.

CM plan

A comprehensive CM Plan is a key element for an effective CM Program. The CM Plan must be concise, yet thorough in addressing CM policies and procedures for the entire product life cycle. The CM Plan must also define subcontractor, contractor, and customer interfaces. It must be coordinated with other organizations, and these organizations must agree to abide by it. Lastly, the customer usually approves the CM Plan *before* it is official.

Subcontractor control

The contractor is responsible for applying CM policies and procedures to subcontractors, as applicable. All applicable CM requirements should be documented in subcontractor Statements of Work (SOWs) and Purchase Orders (POs). The contractor may require subcontractors to prepare and submit a separate CM Plan. This CM Plan may be integrated into the contractor's CM Plan.

Major factors in subcontractor selection should be a compliant CM Program and the ability to support the contractor's CM requirements. Pre-contract surveys and fact-findings conducted during the proposal phase should verify capability. Then, periodic audits should be conducted after award to verify compliance.

Keys to Success

Remember: A successful CM Program satisfies all its objectives at minimum cost to both the contractor and the customer.

KEYS TO SUCCESSFUL CONFIGURATION MANAGEMENT

Define CM goals, scope, and procedures in the CM Plan *early and completely.*

Integrate CM into the System Engineering Process.

Select CIs carefully for optimum CM, support, and program management.

Implement an *orderly and thorough baselining* process.

Evaluate and process changes *promptly.*

Describe changes *completely.*

Coordinate closely with key elements of the project team, including design, logistics, and manufacturing.

Empower a *strong Configuration Control Board* (CCB) and chairperson.

Account for configuration *status accurately, thoroughly,* and *in a timely manner.*

Initiate a *comprehensive configuration audit* program.

Cultivate a *cooperative* and *responsive buyer* by striving for customer satisfaction.

Design for *minimum labor* requirements.

Encourage timely feedback from the field to keep better track of deployed system configurations.

Initiate and implement a comprehensive CM training program for applicable personnel.

A list of major DOD (Department of Defense) and other Government standards frequently followed for CM is shown in Chapter 5, References. These standards should not be applied directly before the need for tailoring is determined. Suggestions for tailoring of these DOD standards and specifications are included in Chapter 3, Application. Industry standards or company procedures may be used for commercial application of CM.

Procedures

Seven major steps are required to implement an effective Configuration Management program, as shown below. These steps are described in detail in this chapter.

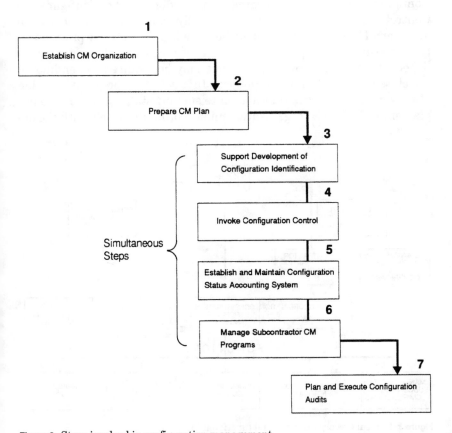

Figure 2 Steps involved in configuration management.

Step 1: Establish CM Organization

The CM organization structure for a design and development program is usually established during the proposal phase. A Configuration Manager is selected to direct the program and to serve as the central contact point for all CM matters. The CM organization should become active during the proposal phase, with the Configuration Manager participating in decisions and planning which affect CM.

Optimally, the Configuration Manager is administratively part of a central CM organization, although he/she reports functionally to the Program Manager. This separate, overall CM organization minimizes possible conflicts of interest. In order for this organization to be effective, company procedures which define CM responsibilities relative to projects should be developed and enforced. This will provide the leverage needed by a Configuration Manager to prevent project managers from ignoring CM protocol.

A typical CM organization structure is shown in Figure 3. The Configuration Manager coordinates the effort of the Configuration Control Board (CCB) which is typically chaired by the Program Manager. The CCB is the final authority within a company program on the disposition of proposed changes. After formal baselines are established, customer approval of a change may be required, depending on its classification and total impact. In addition, the Configuration Manager should ensure that adequate support staff is available to implement CM. The CM organization must coordinate closely with the de-

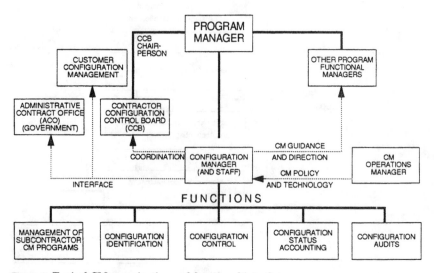

Figure 3 Typical CM organization and functional interfaces.

sign and production organizations for an effective CM program. Chapter 3, Application, discusses how to establish a CM organization in more detail.

Step 2: Prepare CM Plan

The Configuration Manager should prepare the CM Plan during the proposal phase or immediately upon contract award. A comprehensive CM Plan is a key element of an effective CM Program.

The customer usually prescribes the format and content of the CM Plan. The CM Plan must be concise, yet thorough. It must address all functional interfaces with other disciplines, CM objectives and organization, and policies and procedures for the major CM activities, including managing subcontractor CM Programs, as shown in Figure 4. Copies of any forms used for CM must be included. The CM Plan must be coordinated with other organizations, especially program management, and these organizations must agree to abide by it.

The CM Plan should address any tailoring done to customer required standards and specifications. Also, with the onset of design and manu-

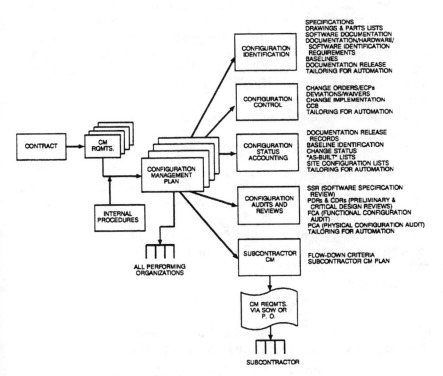

Figure 4 Overview of typical CM plan.

facturing automation, the Plan should describe how CM requirements will be met in an automated environment. In addition, the Plan should define:

- CIs specific to the program
- Documents that will constitute the various baselines
- CM Milestone Schedule, including dated baselines
- Procedures for controlling changes to baselines
- Procedures and tools for the configuration identification, control, status accounting, and audit functions

Step 3: Support Development of Configuration Identification

What is configuration identification?

Configuration identification is the process of documenting product performance, design, qualification, fabrication, and acceptance requirements. Configuration identification is a key element of CM, essential for controlled development, production, and maintenance. It may also refer to the actual approved documentation, as well as documentation numbering and product marking. Change identification is also an element of configuration identification.

The product configuration identification includes:

- Any technical documentation provided by the customer with the contract
- The additional technical documentation prepared/revised by the contractor and subcontractors and approved by the customer

Usually, configuration identification is continuously modified throughout the product life cycle as changes are approved for incorporation. Product documentation becomes more detailed as design and testing progress.

CM support for configuration identification ends only when:

- The last change is documented in the approved technical documentation package and incorporated into the product and
- The last item is built and properly identified.

Purpose

Configuration identification defines the functional (performance) and physical (design) characteristics of the hardware and/or software in sufficient detail to select, develop, test, evaluate, produce, accept, operate, and maintain it.

When properly applied, configuration identification:

- Provides the basis for effective configuration control, status accounting, and technical reviews and audits
- Protects design integrity during implementation of required changes or during maintenance in the operational and logistic support phase

Selecting configuration items

Selecting hardware and/or software CIs is the first step in the configuration identification process. CIs may vary widely in complexity, size, and type, for example, from an entire ship system down to a circuit card assembly. The customer may define top-level CIs in the contract. Additional CIs are selected through the contractor's system engineering process; normally, all such CIs must be customer approved.

A CI is an aggregation of hardware or software, or any of its discrete parts, which satisfies an end-use function and is designated by the customer for application of CM.

CIs provide the basis for:

- Requirements documentation
- Status accounting
- Reviews and audits
- Testing and reporting
- Baselining of hardware and software
- Change control.

Efficient CM depends on wise selection of Hardware CIs (HWCIs) and Computer Software CIs (CSCIs). See Figure 5 for a typical breakdown of hardware and software CIs. Properly selected CIs should:

- Be capable of independent development
- Allow incremental development and testing
- Support Work Breakdown Structure (WBS) and subcontractor efforts

Also, any aggregation of hardware or software which controls a ma-

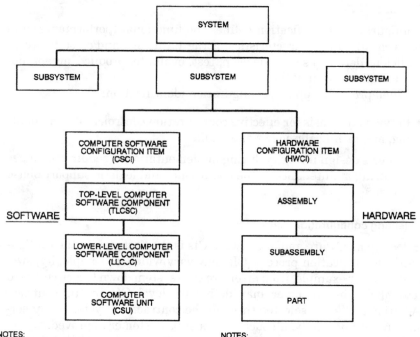

Figure 5 Typical breakdown of hardware/software CIs.

jor interface with another system(s) should also be selected as a separate CI.

Selecting too many CIs causes an unnecessarily increased management burden through too many specifications. The specifications tend to become redundant in statement of requirements, therefore causing an excessive number of engineering changes, difficulty in keeping specifications current and accurate and increasing development time and cost. On the other hand, selecting too few CIs results in an inadequate specification set for good management control, difficulty in determining a maintenance approach, and therefore, poor supportability.

CM and logistics personnel should participate in CI selection to ensure optimal CM and logistic support.

Numbering and marking of HWCIs and CSCIs should be according to contract requirements and the approved contractor CM Plan.

TABLE 1   CI Selection Guidelines

Is the item a critical high risk and/or a safety item?

Is it readily identifiable with respect to size, shape, and weight (for hardware)?

Is it newly developed?

Does it incorporate new technologies?

Does it have an interface with hardware or software developed under another contract?

With respect to form, fit, or function of the item, does it interface with other CIs whose configuration is controlled by other entities?

Is there a requirement to know the exact configuration and status of changes to it during its life cycle?

A useful CI selection checklist is given in Table 1.[4]

Role of CM in configuration identification

The responsibilities of the Configuration Manager with regard to configuration identification include those listed below.

Participating in the system engineering process to ensure that configuration items are selected and developed for optimum CM and support.

Identifying, in conjunction with design/development engineering, the documentation that will constitute the functional, allocated, and product baselines.

Ensuring that drawings and specifications are prepared according to contract requirements.

Requesting customer assigned drawing numbers and computer software identification numbers in accordance with contract requirements, and assigning these numbers or company numbers, as appropriate.

For nomenclature requiring customer approval, requesting the nomenclature in accordance with contract requirements and ensuring proper use on documentation and the product.

For customer provided serial numbers, requesting the serial numbers in accordance with contract requirements and controlling their use.

---

[4]*Configuration Management Practices, Appendix XVII,* Department of Defense (MIL-STD-483A), June 1985.

Assigning drawing titles according to contract requirements and using these titles properly.

Ensuring that drawings and specifications provide proper instructions for marking and labeling the product (hardware/software) and for resolving any problems before baselining.

Coordinating the baselining process, providing advice, participating in design reviews, and documenting the baselines and release records accurately in the Configuration Status Accounting System.

Monitor any subcontractor development of configuration identification.

Phases of configuration identification

Normally, there are three formal phases of configuration identification:

- Functional Configuration Identification (FCI)
- Allocated Configuration Identification (ACI)
- Product Configuration Identification (PCI)

Functional configuration identification.   The FCI is the *approved* system specification and/or system segment specification(s).

The FCI establishes functional, performance, design, and test criteria for the design, development, and testing of a complete product or system.

Normally, the customer supplies the system/segment specification with the contract. Sometimes a preliminary system specification is provided with the contract to be finalized by the contractor before the final specification is approved. If the contract is for a single HWCI or CSCI, note that the FCI can be a development specification.

The initially approved FCI constitutes the Functional Baseline (FBL); the FBL plus any approved changes constitute the *approved* FCI.

Allocated configuration identification.   The ACI begins as part of the system engineering process. System requirements are allocated to HWCIs and/or CSCIs in newly created lower-level specifications. This should be documented in a requirements traceability matrix showing where in the lower-level specifications each higher-level requirement is allocated.

System requirements should be traceable both downward and upward between the upper-level and lower-level specifications.

Next, development specifications are prepared for the HWCIs and CSCIs. (These are sometimes called "design to" specifications.) Interface Control Documents (ICDs) are prepared, as required.

The initially approved ACI constitutes the Allocated Baseline (ABL); the ABL plus any approved changes constitute the *approved* ACI.

Product configuration identification. The PCI is the third phase of configuration identification. During this phase product documentation is developed, including:

- Product specifications, software source listings, and design documents
- Any process specifications
- Any material specifications
- Detailed drawings and parts lists
- Test and acceptance specifications

Process and material specifications are prepared *if* material and/or processes are program-unique or state-of-the-art.

Functional Configuration Audits (FCAs) and Physical Configuration Audits (PCAs) are conducted on the first article. If successful, the customer approves the documentation as the initial PCI.

The initially approved PCI constitutes the product baseline (PBL); the PBL plus any approved changes constitute the *approved* PCI.

Baselines

A baseline is a set of formally designated configuration identification approved or accepted at a specific time in the developmental process.

It is the departure point from which all design and performance changes will be controlled.

Typically, there are three formal baselines, as mentioned earlier:

- FBL
- ABL
- PBL

The ABL can be optional, depending on customer requirements and whether there are lower-level hardware and/or software CIs.

Sometimes informal or internal baselines are established to provide additional management visibility and traceability; for example, a test baseline or a design baseline may be established.

The approved baselines, plus any approved changes, constitute the current product/system configuration identification. See Figure 6,[5] below, for an overview of a typical design review and baselining process for a Government contract. A similar process may be followed for commercial application, paralleling any proposed or required customer design review activity.

Specifications and engineering drawings

This template does not contain guidelines for preparing various types of engineering drawings and specifications. That information is provided in the standards and specifications listed in Chapter 5, under References for Government contracts.

Commercial drawings, would use either industry standard or company standard format(s), as required by the customer.

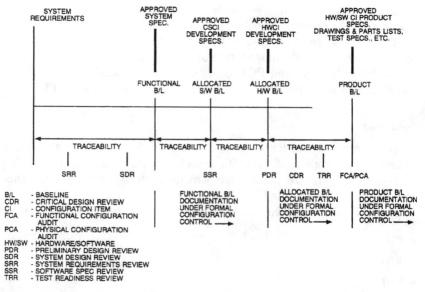

Figure 6 Overview of typical design review and baselining process.

---

[5]Dean, J. W., *Advanced Configuration Management* (*Seminar Guide*), Spring 1989.

Drawing types.   Typically, there are three basic types of drawings developed on Government contracts, which are similar in detail and purpose to commercial drawings:

- *Conceptual design drawings* contain engineering design information sufficient to evaluate an engineering concept. They may contain enough information to fabricate developmental hardware.

- *Developmental design drawings* provide sufficient data for the analysis of a certain design approach and for the fabrication of prototype hardware for test.

- *Product drawings* provide the necessary design, engineering, manufacturing, and quality requirements for procurement and fabrication of production items. These drawings should enable a qualified manufacturer to produce many items identical functionally and physically to the original design, without additional design data or aid from the original designer.

These drawing types provide a natural progression from design inception to production. Government standards for additional types of drawings are listed in the References Chapter.

Commercial drawing types vary according to industry and company standards. *However, during product development, evolving commercial specifications and drawings should follow a logical sequence for traceability.*

Role of CM in software configuration identification

Normally, the responsibilities of CM as regards software configuration identification include:

- Identifying the documentation that establishes the functional, allocated, and product baselines and the developmental configuration, defined below

- Identifying the documentation and computer software media containing code, documentation, or both to be placed under configuration control

- Identifying each CSCI and its corresponding Computer Software Components (CSCs) and Computer Software Units (CSUs)

- Identifying the version, release, change status, and other identification details of each deliverable item of software

- Identifying the version of each CSCI, CSC, and CSU to which the corresponding software documentation applies

- Identifying the specific version of software contained on each deliverable medium, including all changes incorporated since the previous release.[6]

Developmental configuration for CSCIs

For internal CM during CSCI development, a developmental configuration is established, consisting of the software and associated technical documentation that define the evolving CSCI configuration. Normally, software development for Government contracts proceeds through the following steps:

1. The Software Development Plan and System/Segment Design Document are prepared. Once approved, these documents become part of the FBL for the overall system.

2. The Software Requirements Specification (SRS) and Interface Requirements Specification (IRS) are prepared for each CSCI. Once approved, these documents become the ABL.

3. A Software Design Document (SDD) is then written for each CSCI to further define the design. The SDD for each CSCI should be incorporated into the developmental configuration for that CI prior to delivery to the customer. Sometimes the SDDs are initially presented to the customer at a Preliminary Design Review (PDR). *If they are received successfully at PDR, the SDDs are the beginning of the developmental configuration.* During development, any updates to the SDDs should be incorporated into the developmental configuration for the affected CSCI.

4. The Software Test Plan (STP) and preliminary Interface Design Document (IDD) are then prepared and placed under configuration control prior to delivery to the customer. Any updates to these documents should also be placed under configuration control. If the IDD and any updates to the SDD are received successfully at Critical Design Review (CDR), these documents are incorporated into the developmental configuration.

5. The Software Test Descriptions (STDs) for each CSCI are prepared and placed under configuration control prior to initial delivery to the customer.

6. Source code and associated listings for each CSU should be incorporated into the appropriate developmental configuration and should come under internal configuration control upon completion of successful test and evaluation, prior to software integration.

---

[6]*Defense System Software Development,* Department of Defense (DoD-STD-2167A), February 1988.

7. During CSC integration and test, any updates to the SDDs and source code listings for each successfully tested and evaluated CSC should be incorporated into the appropriate developmental configuration.

8. CM should identify the exact version of each CSCI to be delivered upon successful completion of CSCI testing. This data must be documented in a Version Description Document (VDD) for each CSCI.

9. After successful completion of CSCI testing, the updated design documents, plus source code listings, are combined to form the software product specification ("C5-spec"), also known as the as-built specification.

10. Once the FCA and PCA are successfully completed, the Software Product Specification (SPS) for each CSCI is incorporated into the PBL after it is authenticated by the customer. *At this point, the developmental configuration for each CSCI is no longer needed and ceases to exist.*

Commercial software development effort should also follow a similar, logical sequence for traceability.

Product marking

As part of the configuration identification task, CM procedures prescribe the marking of as-built hardware/software CIs. The Configuration Manager must ensure that the configuration end items are marked according to the contract requirements as tailored by the approved CM Plan.

Usually, a product marking is an identification plate or software media label including:

- The item's approved nomenclature
- The item's part number
- The item's serial number
- Applicable manufacturer information and used-on information
- CSCI/CSC number, if any.

On some lower-level items, however, there is only enough room to stamp or etch the part number and serial number directly on the item. An example is a circuit card assembly. Another method used on such items is to tag the item and bag it.

The customer or contractor serial number should always be applied on a product and its repairable subassemblies.

Specific guidelines for assigning part numbers and marking and labeling product should be in accordance with the required standards for Government applications.

Software marking and labeling must meet customer requirements and facilitate storage and retrieval of specific versions/items of software.

For Nongovernment applications, hardware/software can be marked according to commercial conventions and any customer requirements.

For numbering documentation, the Configuration Manager determines whether the numbering scheme should use significant or nonsignificant numbers.

- Significant numbers attempt to tie related items together by applying specific meaning to numbers or groups of numbers.

- Nonsignificant numbers are assigned in sequence without trying to number related items similarly. For example, the numbers 697310 and 697311 may be two totally unrelated items. This is the most flexible numbering scheme.

Whichever is used, the numbering scheme should be clear and meet the customer's needs and contract requirements.

Documentation for a DOD program is usually numbered according to certain standards for specifications and drawings. However, for some Government programs, the customer provides drawing and specification numbers. Documentation numbering for a commercial contract is done in accordance with company/industry standards.

Nomenclature and drawing/specification titles are normally assigned according to Government requirements or commercial conventions. Government programs typically require customer approved nomenclature for identification plate drawings and the identification plates themselves.

Re-identification of parts.   Sometimes changes require re-identification of parts for safety or interchangeability reasons. The logic flow diagram in Figure 7, below, shows how to determine whether re-identification is necessary.

Step 4: Invoke Configuration Control

Configuration control is the systematic preparation, evaluation, coordination, approval, or disapproval of proposed changes and the implementation of all approved changes to the configuration of a CI after establishment of its baselines.

Change should not be discouraged, but be considered normal in the design/development process. Through the baselining process described earlier as part of Step 2, departure points are established from which future changes may be made. When properly managed through effective configuration control, change can be a positive force by optimizing

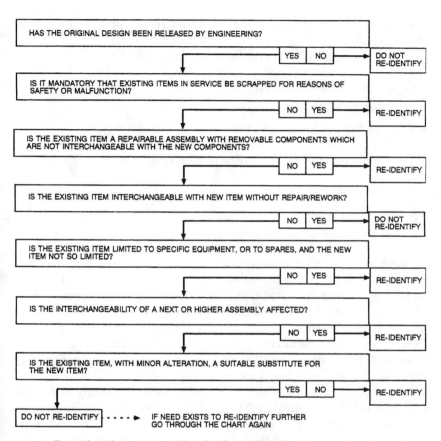

Figure 7 To re-identify or not to re-identify after making changes.

management, technical, and cost considerations for a program. Effective configuration control also ensures system supportability for both Government and commercial contracts.

A product becomes progressively more defined as it proceeds through the various development and production stages, becoming more expensive to change as it matures due to broader impact. The degree of configuration control increases as each baseline is established.

Configuration control requirements specified in a contract can be tailored to better suit CM program needs using the tailoring guidelines in the Applications Section.

Baseline management

Configuration control is applied to both internal and customer approved baselines as they are established during the development process. Achieving the proper timing of this control is crucial to the success of a program.

Premature application of configuration control:

- Causes an excessive number of changes
- Increases the cost of development
- Reduces the productivity of the development effort.

Late application of configuration control results in:

- Poor management control
- Schedule slippages
- Costly production and field changes.

Customer approved baselines are changed only via customer approved ECPs. ECPs are prepared according to contract requirements as tailored by the approved CM Plan. Therefore, during development and prior to PBL establishment, ECPs are prepared for customer approval only for those changes which impact the FBL or ABL. Otherwise, development changes which impact *internal program baselines only* do not require ECPs. Such changes can be documented and processed using a company's internal procedures.

The evolving internal baseline for software is commonly called the developmental configuration. Though there is no like term for evolving hardware baselines, any internally released hardware documentation constitutes an internal baseline. The evolving baselines are established incrementally, as drawings and specifications are internally released. For this type of hardware and software documentation and media, internal change approvals are required to determine program cost and schedule impacts.

Configuration control for evolving baselines is typically the contractor's responsibility: the contractor usually defines the internal configuration control procedures. However, the customer sometimes requires the contractor to maintain such traceability. For example, for some Government contracts, certain software design documents must be placed under *internal* configuration control prior to customer approval.

When the hardware and/or software development effort is completed, a first article is fabricated for performance of the FCAs and PCAs for each CI. Upon successful completion of these audits, the PBL is established. Internal baselines then cease to exist; all future changes must be approved by the customer prior to implementation.

Internal change analysis is basically required to:

1. Identify total change impact

2. Identify costs and schedule for redesign effort.

Manufacturing impact must be considered in the analysis.

Remember: Good internal configuration control during the development phase is the foundation for effective CM after PBL establishment.

Role of the Configuration Control Board

The program CCB is responsible for evaluating and approving or disapproving proposed changes to product configuration and ensuring proper implementation of the approved changes. The board includes members from all appropriate functional program areas.

CCB membership and responsibilities.   The exact membership of the CCB depends on product complexity and type and the product life cycle as dictated by the contract. The membership of the CCB can change over time to keep pace with changing program status and requirements. A typical CCB usually consists of members from the following major functional areas:

- Program Management
- Hardware/Software Development Engineering
- System Engineering
- Interface Control (if appropriate)
- Manufacturing Engineering
- Test and Evaluation
- Contracting
- Purchasing
- Integrated Logistics Support
- Reliability and Maintainability
- Configuration Management
- Operational Engineering
- Human Engineering and Safety
- Quality Assurance
- Subcontractor Representation, as required.

A typical program CCB and board/member responsibilities are shown in Figure 8. However, every CCB will not have members from these functional areas. Some CCBs will have members from other functional areas. For example, risk management may be important enough for a particular program that a separate CCB member from risk management is needed. Also, integrated logistics support frequently includes training, product usage documentation (technical manuals), spare repair or replacement parts ordering and storage (provisioning),

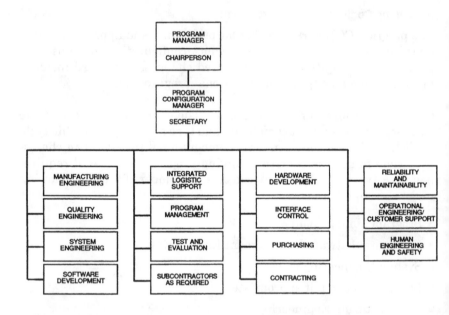

Responsibilities of CCB:
- Ensures total program impact has been identified for a proposed change
- Evaluates technical merit of a proposed change and verifies need for change
- Pays particular attention to cost and schedule impact
- Classifies or verifies classification of a change (Class I or Class II)
- Evaluates:
  — Internal change requests (and approves/disapproves them)
  — Engineering Change Proposals (ECPs)
  — Requests for Deviations/Waivers in product
- Final decision for submission of proposed change or Request for Deviation/Waiver to the customer rests with the CCB except customer requested change proposals
- Monitors change implementation

Responsibilities of CCB Members:
- CCB Chairperson:
  — Approves/disapproves proposed change based upon recommendation of each CCB member

- Configuration Manager (Secretary):
  — Serves as Secretary of CCB
  — Receives and records proposed change package and tracks change status
  — Ensures that proposed change is in good order and distributes to CCB members
  — Schedules CCB meetings and distributes agendas
  — Prepares CCB Directives
  — Prepares and distributes minutes of CCB meetings (minutes require chairperson approval)
  — Provides advice and coordinates change proposal preparation (or prepares proposal)
  — Monitors release of updated documentation and change implementation
- All other CCB members:
  — Coordinate proposed change with their particular functional area and determine impact
  — Recommend approval or disapproval of change proposal to Chairperson and provide rationale for the recommendation
  — Commit their respective organization to a course of action related to a change
  — Review documentation being released to ensure adequacy for their area

Figure 8 Typical configuration control board.

repair service activity, logistics support analysis, etc. One or more of these functions may be important enough for a particular program that a separate CCB member is appointed from that area. Security representatives or system users may also be important representatives on a CCB. The Configuration Manager should tailor the CCB for the program.

The CCB chairperson (usually the contractor Program Manager) obtains the views and recommendations of board members, as well as specific details in any impacted functional areas. He/she then approves or disapproves the proposed change or defers the decision. The CCB secretary (usually the program Configuration Manager) coordinates review of the change package, schedules CCB meetings and prepares agendas, and records the views and supporting data for each board member (issued as meeting minutes), as well as the chairperson's decision. Such decisions are issued as CCB directives. For changes affecting internal baselines only, the chairperson's decision is final; however, for changes affecting formal customer approved baselines, the change must be approved by the customer CCB before implementation is authorized.

Determining total change impact.   The CCB as a whole is responsible for convening, as required, to determine the combined impact of changes and to approve or disapprove proposed changes. If the volume of changes is consistent, regularly scheduled CCB meetings should be held to process changes efficiently. Special CCB meetings may be called to review urgent changes. CCB members should give a high priority to CCB tasks and meeting attendance. In determining whether total change impact has been identified, the CCB checklist[7] in Table 2 may be used.

Software configuration control boards.   Sometimes, separate Software Configuration Control Boards (SCCBs) are established, if required by the contract or deemed necessary to handle a high percentage of software related changes. The functions of the SCCB are similar to the CCB, except that:

- Changes usually result from software problem reports.

- The SCCB usually assigns or approves priority levels to software problems for resolution.

---

[7]Adapted from Samaras, T. *Configuration Management Deskbook,* Advanced Applications Consultants, Inc., 1988.

TABLE 2   Change Impact Analysis Checklist

1. How does this change affect total CI performance?

2. Are other projects or CIs in the system affected by the change?

3. How does the change affect the reliability and maintainability of the CI?

4. Can the change be implemented by manufacturing?

5. Is special tooling required to implement the change? If so, how much will it cost and how long will it take to make this tooling?

6. Will previously delivered CIs have to be returned to the factory for rework, or will they be retrofitted in the field?

7. How will the CI change affect the delivery schedule?

8. How will the change affect system life-cycle cost?

9. What existing documentation (drawings, specifications, manuals, and procedures) requires change?

10. How are the CI weight, size, balance, stability, and power consumption affected by the change?

11. Is the safety of operations and/or maintenance personnel affected?

12. Does the change affect spare parts and assemblies?

13. Is the service life of the CI affected?

14. Are repair and maintenance made more difficult by the change?

15. Will the magnetic and radio interference characteristics of the CI be changed?

16. Will the mechanical and electrical installation of the CI be affected?

17. Will the first article or qualification CI have to be modified?

18. Will parts procurement be a problem?

19. Will completed assemblies/subassemblies without the change have to be scrapped?

20. What is the effectivity of change incorporation (What serial numbers of the CI will incorporate the change during production)? Which items must have the changes made to them?

21. Will additional tests be required? What are their schedule and cost impact?

22. Will item interchangeability, substitutability, or replaceability be affected?

23. Will suppliers of CIs be changed?

24. Will changes be required to skills, manning, training, biomedical factors, or human engineering design?

25. Will compatibility with other equipment, trainers, or training devices be affected?

26. How will the change impact the environment?

- Any potential impact on hardware or the total system must be identified and coordinated with the program CCB.

- Software changes impacting customer approved baselines must be classified as Class I or Class II and referred to the CCB for review.

- The SCCB should ensure that an appropriate corrective action process is followed for resolving software problem reports.

The change control process

Every CM Program should have in place a consistent and proven change control process to integrate customer requirements and internal company procedures.

An effective change control, or change management, process tracks a product's evolving design during development and ensures supportability and traceability to system requirements.

Steps involved with a typical change control process can be summarized as follows:

1. Requesting the change and verifying the need for it.
2. Developing, testing, and documenting the change solution.
3. Evaluating the proposed change for total program impact and classifying the change.
4. Approving the change internally (through CCB).
5. Obtaining customer concurrence in classification for Class II changes (if required).
6. Obtaining customer approval for Class I changes (if required).
7. Distributing authorized changes for implementation.
8. Verifying and recording change implementation.

An example of a typical change control process is shown in Figure 9. This flow chart shows who is responsible for each step of the process. The individual steps of this process are explained in more detail in the following paragraphs.

Requesting a change. A change must be officially requested to initiate the change control process. Anyone in the program organization may request a change by:

- Completing a Change Request Form
- Obtaining approval from immediate supervision
- Submitting the request to the CM organization.

Any required supporting documentation should be attached to the

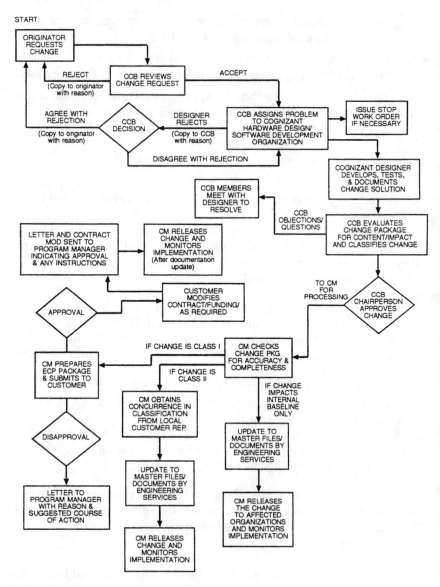

Figure 9 The change control process.

form. A sample Change Request Form is included in the Appendix. Change Request sources include:

- People of all levels, from an electronics assembly tester to a program functional area manager

- Personnel from manufacturing, software development, quality assurance, testing, reliability engineering, and other areas
- Subcontractors or vendors.

The customer may also request a change via letter or contract modification, directing the contractor to further investigate the change and/or prepare a change proposal.

Once received, CM personnel review the request; make the initial classification; assign a change request number; and record its number, title, and date of receipt. CM then submits the request to the CCB for evaluation.

Evaluating a change request. The CCB proceeds through the following steps in evaluating a Change Request, as shown in Figure 9.

1. Determine if the change is necessary.
2. Make a preliminary assessment of total change impact in all functional areas.
3. Approve/reject the Change Request.
4. Assign the request to the cognizant design organization for development of a solution and method of implementation for the change (if approved), *or*
5. Return the request to the originator with reason for rejection.

For complex changes, the CCB may appoint an engineering committee to evaluate the change request first, to ensure that no technical impacts have been overlooked. The committee then makes its recommendation to the CCB.

Change requests should only be submitted for changes which impact internal or formal baselines. If the affected design has not yet been released internally *or* formally approved by the customer, the CCB does not get involved with change analysis or incorporation. In such cases, internal documentation may be generated within the engineering organization to document the decision and background information.

Developing a solution/method of implementation for a change. The cognizant hardware designer or software developer is tasked with developing a solution and recommending a method of implementation for the change. If, upon further analysis, the designer determines that the change is not feasible or unnecessary, he/she then sends it back to the CCB with an explanation. The CCB may agree with or override the decision of the designer.

Otherwise, the designer proceeds with the task at hand. He/she is responsible for the following steps:

1. Developing a solution
2. Proving that the solution will work through testing, as required
3. Recommending a method of implementation based on his/her technical expertise and coordination with manufacturing and support personnel
4. Documenting all findings and recommendations
5. Submitting this package to the CCB secretary.

Evaluating a change package.   The CCB secretary (normally the program Configuration Manager) distributes copies of the change package to each CCB member for advance review. All members should come to the meeting familiar with the change package and its impact on their respective functional areas.

When the total program impact has been determined, the CCB makes the final change classification, and the CCB chairperson must approve or disapprove the change.

Handling of expedited changes is discussed in the Application chapter under Metric 4: Improving Change Process Cycle Time.

Change classification.   Changes must be classified properly for effective change management. There are basically three different types of changes which may be processed:

- Internal baseline changes
- Class I ECPs
- Class II ECPs.

Descriptions of these change types and differences in their evaluation/approval processes follow. In addition, deviations and waivers are described, both of which permit temporary departures from contractual requirements during production.

Internal baseline changes

Internal baseline changes are changes during product development which impact any internally established baseline design or test requirements. No formal customer approved baselines are impacted.

Since the functional and allocated baselines are not impacted by internal baseline changes, customer approval of these changes is not required. However, since internally established baselines are impacted, the CCB must evaluate these changes and approve or disapprove them for implementation. The main reason for CCB involvement is that the requirements affected by these changes are the basis for related development; any changes to the foundation requirements could significantly impact the schedule and cost of the total development effort. The only exceptions to CCB approval prior to change implementation are for expedited changes, which get post-release CCB review.

Once the PBL is established, the previous internal baselines are no longer needed and cease to exist. Changes after this point are classified as Class I or Class II ECPs.

Class I ECPs

Class I ECPs are any changes which: 1) impact the FBL, ABL, or PBL in any way or 2) impact contractual provisions such as cost, schedule, or warranties.

For Government contracts, the following factors are usually considered to determine if a change should be Class I.

- The FCI or ACI is affected to the extent that one or more of the following requirements would be outside specification limits or tolerances:

  Performance

  Reliability, maintainability, and survivability

  Weight, balance, and moment of inertia

  Interface characteristics

  Electromagnetic characteristics

  Other technical requirements of the FBL or ABL.

  NOTE: Minor clarifications and corrections to the FBL and ABL should be made only as an incidental part of the next Class I ECP. A separate Class I ECP would not be cost-effective for these minor changes.

- A change to the PCI, once established, will affect the FCI or ACI, as described above, or impacts one or more of the following items:

  Government Furnished Equipment (GFE)

  Safety

  Deliverable operational, test, or maintenance computer software associated with the HWCI or CSCI being changed

  Compatibility or interoperability with interfacing hardware or software CIs, support equipment/software, spares, trainers, or training devices/software

  Configuration of the hardware/software being changed such that retrofit action is required

Delivered operation and maintenance manuals for which adequate change/revision funding is not provided in existing contracts

Preset adjustments or schedules affecting operating limits or performance to such extent as to require assignment of a new identification number

Interchangeability, substitutability, or replaceability of CIs and of all subassemblies and parts except those of nonrepairable subassemblies

Approved sources of CIs or reparable items at any level defined by source control drawings

Skills, manning, training, biomedical factors, or human engineering design.

- One or more of the following contractual factors are affected:

Total cost to the Government including incentives and fees

Contract guarantees or warranties

Contractual deliveries

Scheduled contract milestones.[8]

Types of changes: Commercial application of CM

In commercial applications, various types of changes include the following changes:

- Changes which correct inoperative conditions, extremely unsatisfactory operating conditions, safety hazards, or a nonconformance to the product technical specification. Change extent and method, location, and schedule for change implementation must be defined for these changes. Both the technical and business aspects of the change must be considered. Documentation associated with the change must also be addressed. Such changes are usually covered at full cost by the contractor.

- Changes which introduce new features or enhanced service capabilities to customers. Such changes are usually billable to customers. These change packages should describe the recommended customer application.

---

[8]Adapted from *Configuration Control,* G-33 Data and Configuration Management Committee Education and Training Task Group (EIA Configuration Management Bulletin No. 6-4).

- Changes which propose implementation of minor product improvements, such as component changes; facilitate manufacturing; or reduce the product cost to the contractor. These are sometimes call "manufacturing convenience changes." These change packages should be based on overall cost-effectiveness, considering cost-benefit trade-offs versus any rework/scrap expense incurred. These change packages should also describe the urgency and method of implementation in the manufacturing facility.

Commercial equivalents of Class I ECPs may be prepared in letter form by industry for their customers. On Government contracts, if design or performance specifications are impacted, submittal of Specification Change Notices (SCNs) with the Class I ECP is usually required. If subcontractors, associate contractors, or the customer have drawings or specifications affected by a contractor's change on a Government contract, a Notice of Revision (NOR) must be included in the change package.

Non-developmental item changes

For Non-Developmental Item (NDI) acquisitions, configuration control of form, fit, and function data is typically preferred to control of full design disclosure engineering data. Control of the latter is more expensive, often without any added benefits. Long-term product supportability is typically the customer's main concern when determining the level of configuration control for NDIs.

Class I ECP justification categories.   Companies are required by Government customers to assign a change justification category to each Class I change. Necessary or beneficial engineering changes are defined by one of the categories listed in Table 3.[9] (For changes falling into more than one of these categories, the most descriptive one should be used.) For changes submitted to the Government, a predefined code is used to indicate which category is applicable. Nongovernment contracts may also benefit from the use of categories/codes.

Class I ECP types.   There are two basic types of Class I ECPs: preliminary and formal. The type used depends on the circumstances relating to schedule, cost, required research and development, and other factors.

A preliminary ECP provides a technical summary and justification of the change and discusses major impacts on other functional areas.

---

[9]EIA CM Bulletin CMB6-4, *Configuration Control,* pp. 41-42.

TABLE 3   Class I ECP Change Justification Categories

1. *Record only changes* are Class I changes because they affect form, fit, function, or logistics; however, they are within the scope of the contract. Record only ECPs normally do not require customer approval. Example: Update of spares provisioning documentation resulting from a previously approved Class II ECP which changed part numbers.

2. *Interface changes* are Class I changes because they are required to correct deficiencies in interface design by eliminating interference or incompatibility at an interface between CIs or their components.

3. *Compatibility changes* are Class I changes because they correct design deficiencies discovered during functional checks or installation and checkout, which *must be corrected for the system to work.* These changes are considered to be within the scope of the existing contract.

4. *Deficiency changes* are Class I changes because they are generated to correct deficiencies discovered after extensive testing has been successfully completed. The safety, interface, and compatibility categories exemplify various types of specific deficiencies. A change to correct a deficiency should be classified in one of these more descriptive categories, if applicable. However, if not, the deficiency change category should be chosen.

5. *Operational or logistics support changes* are Class I changes due to the impact on logistics support. Such an ECP proposes design changes which would improve the operational capabilities or logistics supportability of the item being changed. The benefits of such an ECP should outweigh the costs.

6. *Production stoppage changes* are required to prevent a significant slippage in an approved delivery schedule. For example, if a specific part is no longer commercially available, either redesign effort is called for to find a substitutable part which is available, or manufacturing must set up to produce the part in-house. These changes are Class I changes because they have significant schedule impact.

7. *Cost Reduction changes* propose to make changes to the design or to manufacturing processes which would result in net total cost savings to the customer. To arrive at the net cost, the company must determine cost savings for units ordered but not yet built, less the costs of updating previously delivered units and for changes affecting the logistics support of the units.

8. *Safety related changes* correct design deficiencies which cause hazards damaging the item itself and/or injuring users or maintenance personnel. Such a change would eliminate the hazard(s) through a complete safety analysis. The analysis would identify the type of hazard and predicted effects on the items/personnel and would determine the probability of occurrence of the hazard.

9. *Value engineering changes,* as opposed to cost reduction changes, result in net life cycle cost saving to ordered units and future procurements of the same unit. A value engineering change is more than a short-term cost reduction for units already on order. The change package shows the total cost impact of the change, including long-term maintenance costs. Both customer and contractor share in the cost savings.

A preliminary ECP may be submitted for one of the following reasons:

- To allow the customer to authorize further research and development expense for a specific change or to evaluate alternative change proposals
- To briefly outline an emergency or urgent change by telephone or telex when time does not permit preparation of a formal ECP
- To obtain customer approval for a software change prior to detailed code development required for the change. (In this way, the company can preclude wasted time and expenditures by ensuring customer approval beforehand.)[10]

A formal ECP provides complete, finalized engineering data and a total impact analysis (including price and schedule) to support change approval and funding by the customer.

Class I ECP priorities.   Class I ECP priorities are assigned to inform the customer how quickly an ECP should be processed, approved, and implemented. If several Class I changes have been submitted for approval, the priority classification informs the customer that certain ECPs require more expeditious evaluation than others. The three priority classifications in Table 4[11] are generally used.

The assigned priorities must translate into recommended internal preparation and processing times. Each change must be evaluated individually to determine the appropriate processing time to meet program requirements. In a letter accompanying each urgent or emergency Class I change submittal to the customer, a target processing time (or response time) should be requested of the customer. The letter should explain that expedited processing is necessary for program requirements to be met. It is also essential that company project managers determine their own internal target processing times for Class I changes, and assure adherence to them once they are established.

Class II ECPs

Class II ECPs are any changes during the production phase which do not meet Class I change criteria.

---

[10]Ibid., p. 43.

[11]EIA CM Bulletin CMB6-4, *Configuration Control,* p. 44.

TABLE 4   Class I ECP Priorities

Emergency priorities are assigned when a delay in implementation of a change could:
- Threaten national security
- Threaten the user company's existence or cause significant financial damage to the customer (commercial application)
- Pose a serious hazard to personnel or equipment. (In such cases, the equipment is usually removed from service temporarily, due to the seriousness of the hazard.)

Urgent priorities are assigned when a delay in change implementation could:
- Seriously compromise the operational effectiveness of the equipment/software or of the user company
- Result in a potentially hazardous condition which could cause equipment damage or injury to personnel. (A potentially hazardous condition compromises operator safety, but continued use of the equipment is usually permitted if the operator knows there is a hazard and takes appropriate precautions.)
- Adversely impact schedule and/or cost due to an interface problem.
- Reduce potential savings which could be realized through expedient implementation of a Value Engineering or cost reduction change.

Routine priorities are assigned when emergency and urgent priorities are not applicable to a change.

During production, many changes impact the baselined drawings and specifications (the PCI) without falling within the Class I change criteria. These changes are called Class II changes. Class II changes are frequently used to effect minor producibility changes and drawing clarifications. A Class II change is within the scope of the existing contract. It results in updating the approved product configuration identification and incorporating the change.

Prior to establishing the PBL, there are no Class II changes—only changes processed internally or Class I changes to baselined specifications. When items have been delivered, any changes which require retrofit of existing items would be Class I changes.

The format for documentation of a Class II change may vary from company to company, depending on customer requirements. Usually, the Government will approve the use of the contractor's internal change request form for Class II ECPs. Whatever form is used, it should include the information listed below as a minimum:

- Identification of the originating activity

- Name and part number of the affected item

- Name and part number of next higher assembly

- Serial number of the item affected, if any

- Description of the engineering change

- Reason for making the engineering change
- Identification of contract(s) affected by the change
- Appropriate signatures for approval and concurrence in classification.[12]

Most projects have a large number of Class II manufacturing convenience ECPs during the production phase. Class II ECPs are reviewed by CCB members who provide their recommendations at regularly scheduled CCB meetings. Then the CCB Chairperson, or his designee, signs his approval on the change form and CM obtains concurrence in classification from the local customer representative, if required by the contract. The change is then released for incorporation into the documentation and distributed to affected organizations for implementation.

Commercial equivalents of Class II ECPs may be prepared by industry for their customers.

Obtaining change approval

Once a change has been evaluated, impacted, classified, and approved by the CCB, the change package is forwarded to the Configuration Manager for processing. The Configuration Manager then checks the package for accuracy and completeness. Since the CCB has determined the impact of the change on each functional area, the Configuration Manager has all the information he/she needs to prepare a formal Class I ECP (if the change has been classified as a Class I change by the CCB). Once prepared, the Class I ECP is submitted to the customer for review and approval/disapproval.

No Class I change can be implemented without prior customer approval.

For Class I changes, the customer notifies the contractor of the approval via a contract modification. The contractor is notified of disapproval via a letter from the customer.

For Class II changes, the Configuration Manager first checks the package for accuracy and completeness. If the change is complete and the authorized forms for Class II ECPs have been used, the Configuration Manager then obtains concurrence in classification from the authorized local Government representative. No approval other than internal CCB approval is normally required for Class II changes. Copies of Class II changes may be provided to the customer for record purposes.

---

[12]EIA CM Bulletin CMB6-4, *Configuration Control*, p. 45.

Implementing approved changes

When a change is approved, CM personnel ensure that the documentation is properly updated to reflect the approved change. CM then releases the change to affected organizations for incorporation into the equipment and/or software files. The Configuration Manager ensures that the Quality organization receives the final, approved change package and documentation, so that their inspections will verify incorporation of the change.

Complete change status and supporting data, from change initiation to implementation, are entered into the Configuration Status Accounting System. This system is normally a database tailored to meet project requirements for tracking status of changes and other CM related data. (See Step 5, Establish and Maintain Configuration Status Accounting System, for more information.)

Software configuration control process

The software configuration control process follows the same basic principles as the hardware configuration control process. However, for software-intensive programs, the methods used may differ slightly. A typical software configuration control process is shown in Figure 10.[13] A single form may be used as both the Software Problem Report and the Software Change Request for improved process efficiency.

Strong interfacing between the hardware CCB and software CCB is essential for many programs today due to the extensive use of processor control, embedded computer systems, distributed processing, and hybrid characteristics of firmware. The CCB interface may be accomplished by having a software CCB member on the hardware CCB, a hardware CCB member on the software CCB, joint CCB approval, or joint implementation directives. There is currently a trend toward a single CCB.[14]

Requests for deviation

Prior to manufacture, purchase, or test of an item, if a contractor considers it necessary to depart temporarily from the mandatory requirements of the specifications or drawings, a Request for Deviation is originated.

A deviation is processed instead of an ECP when the approved, documented design is considered superior to the proposed deviation. If the

---

[13]EIA Bulletin No. 6-4.

[14]From J. W. Dean's seminars, Spring 1989.

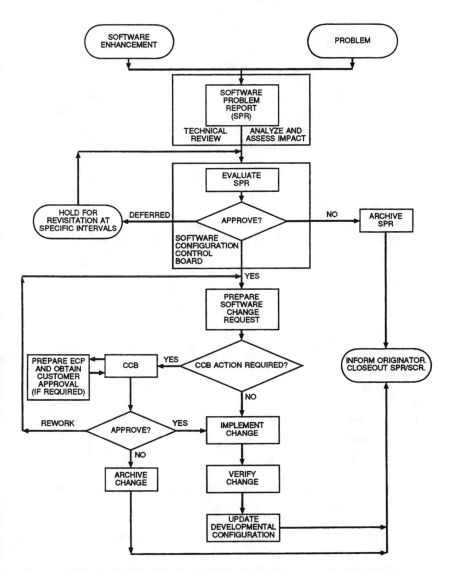

Figure 10 Typical software configuration control process.

deviation is considered superior to the approved baseline requirements, the deviation should be converted to an ECP. No item should be delivered which contains a known departure from a baseline unless a request for deviation or waiver has been approved beforehand. However, use of Preferred Parts Lists (PPLs) and Qualified Products Lists (QPLs) may allow approved part substitutions which do not require a request for deviation.

Examples of Requests for Deviations:[15]

Hardware—A deviation might be requested to allow use of an equivalent material to be used for electrical insulation because the specified material is not available.

Software—A deviation might be requested to allow postponing the availability of a specified software function to a later version of a CSCI by deviation from the software requirements specification. This will allow earlier qualification testing of critical parts of the software.

Categories of deviations. Deviations, like Class I ECPs, are assigned to a category as an aid in reflecting the impact of the temporary change and to guide the evaluation process.

The three categories of deviations are minor, major, and critical.

A *major deviation* is a departure from requirements which impacts:

- Health
- Performance
- Interchangeability
- Reliability
- Survivability
- Maintainability
- Durability
- Effective use or operation
- Weight and size
- Appearance (when a factor)

A *critical deviation* is a departure from requirements which impacts safety. Critical deviations should not be considered or approved unless unusual circumstances exist.

A *minor deviation* is a departure from requirements which is not classified as major or critical.

For some Government contracts, product defects are classified in accordance with a Classification of Defects (CD) listing, which classifies certain defects in the product as minor, major, or critical. When a CD exists, the company should follow its guidelines for categorizing deviations, in addition to those listed above.

Content of deviations. Each request for deviation should contain the following information:

---

[15]Examples from EIA Configuration Management Bulletin No. 6-4.

- Name and address of originating activity
- Contract number
- Name and part number of affected item
- Number of specification or drawing for which the nonconformance is requested
- Category of the deviation
- Description of the deviation
- Quantity of items involved
- Schedule impact
- Previous deviations, if recurring
- Logistic support impact
- Cost impact[16]

Requests for waiver

A Request for Waiver is submitted if a unit or software file is found to depart from baselined requirements during manufacture, test, or inspection but is considered suitable for use as is or after repair by an approved method. The contractor may obtain approval to deliver such a unit by processing a Request for Waiver.

Categories of waivers.   The three categories of waivers are minor, major, and critical, just as for deviations.

Major waivers request acceptance of an item(s) which does not conform to requirements involving:

- Health
- Performance
- Interchangeability
- Reliability
- Survivability
- Maintainability
- Effective use or operation
- Weight
- Appearance (when a factor)

---

[16]EIA CM Bulletin CMB6-4, *Configuration Control,* p. 48.

Critical waivers request acceptance of an item(s) which does not conform to safety requirements.

Minor waivers are waivers which are not categorized as major or critical.

If a CD exists for the project which describes certain defects as minor, major, or critical, its guidelines should be followed, in addition to those above for categorizing waivers. If an Acceptable Quality Level (AQL) is contained in an approved specification or other document, its guidelines on acceptable quantities of minor or major defects should be used in categorizing waivers. For example, if a sample lot of items being waivered contains a number of minor defects which equal or exceed the number requiring rejection of the lot, the waiver should be classified as minor.

Content of waivers.   Each request for waiver should contain the following information:

- Name and address of the originating activity
- Contract number
- Name and identifying number of item
- Name and identifying number of affected parts
- Number of the specification, drawing, or document to which the waiver is requested
- Description of the nonconformance
- Category of the nonconformance
- Quantity of items involved
- Schedule impact
- Identifying number of inspection or test plan
- Effect on logistic support materials
- Corrective action required
- Estimated cost reduction by using item as opposed to obtaining a new item.[17]

Deviation and waiver approval process.   Requests for major or critical waivers or deviations are normally submitted for approval through the same channels as Class I ECPs. Requests for minor waivers or devi-

---

[17]EIA CM Bulletin CMB6-4, *Configuration Control,* p. 50.

ations are normally submitted for approval similarly to Class II ECPs, except for projects which have a properly constituted Material Review Board. In this case, Material Review Board procedures are used *in lieu* of minor waivers.

All Requests for Deviation or Waiver should be expedited through the evaluation process and approved or disapproved as quickly as possible. Some companies request no more than 30 calendar days for review of major or critical deviations and waivers after receipt by the customer. Minor deviations and waivers typically should be approved or disapproved within 10 work days after receipt by the reviewers.

Recurring deviations and waivers. Sometimes deviations and waivers are submitted more than once to resolve the same problem on additional units. In such a case, probably a change should be processed. Eliminating further recurrences of the same deviations or waivers should take priority over resubmitting those same deviations or waivers.

Step 5: Establish and Maintain Configuration Status Accounting System

Configuration status accounting is the recording and reporting of information needed to manage configuration control effectively, including a listing of approved configuration identification, the status of proposed changes to configuration, and the implementation status of approved changes.

Automated configuration status accounting systems should be tailored for specific projects. "Write" access to the database should be limited to certain authorized personnel to prevent unauthorized entries. "Read" access may be made available to those who have a need for the information but are not authorized to make changes to the database, including the customer.

Factors to be considered in the design of an automated configuration status accounting system include:

- System complexity
- Applicable baselines
- Number of HWCIs
- Number of CSCIs
- User requirements
- Production quantity

- Logistics support requirements
- Type and frequency of reports for the customer
- Type and frequency of internal reports
- Interfaces and compatibility with other databases
- Level of detail required
- Data elements required
- Projected cost
- Program schedule/duration

Configuration status accounting records and reports have many different titles; however, they will all fall into one of the generic record/report types listed below:

- Current configuration item lists (hardware/software)
- Latest configuration identification listings
- Approved baseline documentation
- History of changes
- Status of internal change requests
- Effectivity of changes
- All documentation affected by a change
- Status of ECPs
- Status of deviations and waivers
- All items affected by a deviation/waiver
- Change implementation status
- Configuration verification lists
- As-built lists by part number and serial number
- Site configuration data, including retrofit information
- History of delivered item configurations
- Parts usage lists
- Generation breakdowns
- Engineering release records

Examples of some configuration status accounting reports are shown in Figure 11.

The Configuration Manager should ensure that reliable, accurate information sources are established to provide the as-built and change implementation data.

APPLICATION LIST REPORT

APPLICATION LIST REPORT NO 40839-E1	ITEM NAME DATE	MICROCIRCUIT 12/09/89	AL SM-C-803441-1 PAGE 1	REV

* · · · · · ASSEMBLY INFORMATION · · · · · · * NEXT ASSEMBLY	PROJECT NAME	DIV	QTY	* · C/I INFORMATION · * CONTROL ITEM	DIV	* · · · · END ITEM INFORMATION · · · · * END ITEM	QTY	PROJECT NAME	STATUS
C5000511	TPQ-36 PROD	C	3.0	C5000511	C	SM-D-801724	3.0	TPQ-36 PROD	
C5000511	TPQ-36 PROD	C	3.0	C5000511	C	SM-D-803100	3.0	TPQ-36 PROD	
C5000511	TPQ-36 PROD	C	3.0	C5000511	C	SM-D-803107	3.0	TPQ-36 PROD	
C5000511	TPQ-36 PROD	C	3.0	C5000511	C	SM-D-803297	3.0	TPQ-36 PROD	
C5000511	TPQ-36 PROD	C	3.0	C5000511	C	SM-D-803964	3.0	TPQ-36 PROD	

CURRENT CONFIGURATION LIST

INDENT LEVEL	PART NUMBER	AUTH	CHG CL	CHANGE NUMBER	RELEASE DATE	DESCRIPTION PFX CHANGE ID NO	DC	NO. SHT	H/R IND	LOC SEQ DDT	NEXT ASSEMBLY
4	1612120-100	PC		B00	05/21/79	SELF TEST, RF	02	0001	Y	D	1651332-100
4	1612120-100	RC		B00	05/21/79	SELF TEST, RF	02	0001	Y	D	1651339-100
5	1612120-200	RC		D00	05/21/79	SELF TEST, RF	02	0005	R	F1J2	1612120-100

CHANGE PROCESSING STATUS

ECP NO.	SPEC/DWG NO. AFFECTED	TITLE OF CHANGE	CCB ACTION	ACTIONEE	ACTION REQUIRED	DUE DATE
98583	1651430-300	UPDATE W/L	APPD	TAYLOR	PREPARE ED/DCN DD1 00144	0628
98585	1651431-200, 300	UPDATE W/L	APPD	TAYLOR	RELEASED DDT 00264	0628
98625	161252, 3	RELOCATE HEAT EXCHANGER	APPD	WENTWRTH	ED/DCN RELEASED DDT 98625	0629
98626	1651315-100	CORR. NOTE 10	REJ	SISEMORE		0601
98627	1651314-100	CORR. NOTE 10	REJ	SISEMORE		0701

Figure 11   Configuration status accounting reports.

The configuration status accounting reports are only as reliable as the data they contain.

Options should be considered to maximize the accuracy and efficiency of the status accounting task. These options include use of a single entry database and standardized report formats. Report format, content, and frequency should be discussed in the program CM Plan and agreed to by the customer.

Software configuration status accounting

Software configuration items rely on accurate, effective integrated configuration status accounting to record changes that are otherwise invisible. Typically a software file control system used by software engineers during development is integrated with the configuration status accounting system. Also, the software library and status accounting databases are often integrated.

A software configuration status accounting system should be designed to:

- Control changes to source code and documentation
- Record the individual making the change

- Record the date and reason for making the change
- Provide version description to the unit level
- Automatically assign new version numbers upon successful compilation of files
- Aid in origination, tracking, and resolution of software problem reports and change requests during software integration and test.

Existing configuration status accounting software

There are currently several commercially available automated configuration status accounting packages which will meet most status accounting needs. Individuals interested in such software should contact one of the numerous CM consultants for this information.

Whether a project chooses to develop its own automated configuration status accounting system or buy an existing package, care should be taken to ensure that all status accounting needs are met. Designing an internal system for limited use can be costly: it is recommended that commercially available software be used for limited application.

Step 6: Manage Subcontractor CM Programs

As described earlier under Step 2, CM Plan Preparation, the contractor must define and impose those CM requirements and procedures which are applicable to subcontractors in the subcontractor SOWs and POs. This is a joint effort between project management, engineering, and logistics.

Applicable CM requirements and procedures must be flowed down to subcontractors via SOWs and POs and monitored to the extent necessary for the contractor to fulfill contractual obligations.

The subcontractor SOW defines the level of control by the contractor, depending on the criticality of the supplier effort and complexity of the item being procured. The prime contractor should screen customer-imposed CM requirements for a project to avoid flowing down unnecessary, expensive CM requirements to the subcontractor.

The contractor manages subcontractor CM Programs through the following process.

1. Select suppliers based partly on their proven, compliant CM system.
2. Impose specific design and manufacturing requirements via the PO and accompanying SOW.

3. Establish a CM point of contact at the subcontractor location.
4. Verify that any required design documentation or deliverable prepared by the subcontractor meets requirements.
5. Review and process all subcontractor change requests, deviations, and waivers in an expeditious manner.
6. Obtain from the point of contact any data required for configuration status accounting.
7. Predefine and impose any required subcontractor participation in configuration audits.
8. Monitor subcontractor compliance to CM requirements in conjunction with the Quality organization through periodic site inspections and audits.

The following information is relative to management of subcontractor CM Programs.

- Major subcontractors are normally required to submit CM Plans to the contractor.

- Any requirements placed on a subcontractor may be changed only through an official PO modification.

- All correspondence with subcontractors must be channeled through the purchasing organization, including change implementation authority.

- Audits of equipment/software/documentation produced by the subcontractor may be conducted at the subcontractor's facility (source inspection) or the contractor's facility.

- Subcontractors should be represented on the prime contractor CCB, as required.

- The program Configuration Manager may be a member of the subcontractor CCB to provide linkage to the program CCB.

For subcontractors doing development work, the subcontractor's internal CM organization is responsible for performing all required CM functions for their product, including (as applicable):

- *Configuration identification,* including reviewing subcontractor documentation for compliance, ensuring proper product identification and marking, and baseline maintenance.

- *Configuration control,* including preparation and processing of deviations, waivers, and ECPs and coordination of subcontractor CCB activities. Interface with the contractor for change approval is essential.

- *Configuration status accounting,* including provision of records on engineering release and baselines, incorporation of approved changes, verification of change incorporation by item part number/serial number, items affected by deviations and waivers, and as-built lists.

- *Coordination of/participation in audits* as required.

The subcontractor CM organization must retain close interface with the contractor CM organization to ensure that any CM issues requiring contractor approval are resolved as quickly as possible. A subcontractor CM point of contact is recommended.

Proposed Class I and Class II ECPs, deviations, and major or critical waivers are forwarded by the subcontractor CM office to the contractor for disposition. If a subcontractor Material Review Board is authorized by the contractor, minor waivers may be approved internally by the board.

For NDIs which are vendor proprietary, CM requirements should include[18]:

- CM requirements limitation to item fit, form, and function only

- Existence of internal item/document identification and document issue system

- Existence of internal configuration control system

- Provisions for proposing changes to contractor procurement drawing/specification

- Contractor approval of changes to maintenance-significant items

- Requirement for certificates of conformance

For subcontractors performing manufacturing effort, CM requirements should include:

- A requirement for conformance to the contractor furnished baseline list and documentation

- Existence of a system to issue and control contractor furnished documentation

- Method of transmitting subcontractor change proposals, waivers, or deviations to the contractor

- Method for contractor issuance of changes to the subcontractor

---

[18]J. W. Dean, *Basic Configuration Management,* Seminar, 1988.

For products which require adherence only to industry or military standards, rather than a complete documentation package, no contractor CM requirements should be imposed. However, quality inspections of such items, whether at the source (subcontractor) or at the contractor's receiving area, should verify that the item meets the requirements of the specified standard.

Step 7: Plan and Execute Configuration Audits

A Configuration Audit is the verification of a CI's conformance to functional (performance) and physical (design) requirements as contained in customer approved specifications and drawings and other contract requirements.

Configuration Audits are formal meetings between the customer and contractor held at the contractor's facility. (Sometimes audits are held at the subcontractor facility, if the CI being audited was built there.) Their purpose is to audit a first article CI to determine if it meets functional and physical requirements, prior to authorizing production of a quantity of the item.

There are basically two types of audits:

- *A Functional Configuration Audit (FCA)* is conducted to verify that the functional requirements of the specifications are accurately reflected in the performance of the configuration item. Test and analysis data should verify achievement of performance required by the contract.

- *A Physical Configuration Audit (PCA)* is conducted to verify that the as-built CI matches the design requirements of the conditionally approved engineering drawings, lists, software design documents, and product specifications. The PCA also verifies that conditionally approved test procedures are adequate for acceptance of production units of the CI by quality personnel.

The FCA includes:

- A review of test plans and procedures
- A review of test results for compliance
- A listing of any required tests not performed
- A listing of any deviations and waivers for the item being audited.

The PCA includes:

- A review of HWCI product specifications
- A review of drawings against hardware
- A review of drawings for completeness
- A review of manufacturing instructions
- A listing of deviations and waivers
- A listing of outstanding changes
- A review of FCA minutes
- A review of CSCI product specifications
- Verification that CSCI conforms to the software design documentation[19]

Upon successful completion of the FCA and resolution of any action items, the PCA is scheduled. Upon successful completion of the PCA and resolution of any discrepancies or action items, the conditionally approved technical documentation, including HWCI/CSCI product specifications, is formally approved by the customer as the PBL. Any future changes proposed to this baseline must be reviewed and dispositioned by the customer prior to implementation.

Software physical configuration audits

For CSCIs, the PCA includes an audit of the computer program specifications, flow charts, listings, manuals/handbooks, and other documents. The software will be audited to verify that its inputs, outputs, processing interfaces, and physical construction are accurately and adequately described in its technical documentation. Verification of the software in this manner ensures that it is supportable and maintainable in operational use by programmers and technicians other than those individuals who actually developed the programs.

CM role in configuration audits

The Configuration Manager provides a support role in the conduct of design reviews. However, he/she must assume the *lead role* in the planning and conduct of configuration audits. Table 5 summarizes the Configuration Manager's role in configuration audits. The Configuration Manager should conduct a dry run for each audit in which the technical documentation to be used for the audit is compared to the hardware/software, and all errors are corrected prior to the formal audit.

---

[19]Based on J. W. Dean's seminars, Spring 1989.

TABLE 5   The Configuration Manager's Role in Configuration Audits

Preparation of FCA/PCA plans to be approved by customer

Preparation and validation of data package

Establishment of contractor technical support team

Handling of audit logistics, such as meeting space, scheduling equipment availability, security clearances, and other requirements

Scheduling and coordination of dry run of audit

Preparation of list of incorporated and outstanding changes

Invoking a moratorium on changes during the audits

Acting as recorder, or possibly chairperson, of the audit

Obtaining agreement with the customer on action items

Preparation and distribution of audit minutes

Follow-up on audit action items and ensure mutual resolution before closing out

Configuration audit plans

Whether or not audit plans are required by the customer and contract, a plan should be prepared for the conduct of each audit (functional and physical). These plans should be agreed upon between the contractor and the customer.

An audit plan should include:

- Date, time, and location of audit
- Any security clearance requirements
- List of technical support team members
- List of CIs to be audited
- Level to which the audit will be conducted
- Specific documentation to be made available and the depth of that documentation
- Documentation media (paper, microfilm)
- Audit agenda and procedures
- Requirements for audit minutes
- Requirements for resolving action items (How, when?)
- Criteria for audit approval
- Any information requested from the customer prior to audit
- A description of working space and administrative assistance to be provided for the customer

Application

This chapter describes how to apply the previously outlined procedures for CM to specific product procurements. It provides a CM timeline showing various milestones and CM tasks for a typical project. Also, this chapter includes:

- Metrics for measuring CM performance
- Traps, case studies, and lessons learned on actual CM programs
- Suggestions on bidding CM and staffing a CM organization
- Tips for tailoring contract requirements to support varying programs.

CM can be a useful project management tool, if properly implemented. This chapter points out some traps to watch for, as well as how to avoid these pitfalls. CM should be conducted as described in the Procedures Chapter; however, CM contract requirements should be tailored to fit the specific application. Key factors to consider include product complexity, the amount of hardware or software, the type of contract, and other variables.

Once a CM program is in place, CM performance should be continually analyzed using Process Quality Management Improvement. Internally developed metrics, applicable to a specific product acquisition, should serve as a roadmap to measure the performance and effectiveness of the CM function.

Time Phasing of Configuration Management

Figure 12 presents a timeline detailing CM milestones and tasks during a typical product acquisition life cycle. Time is represented from left to right in terms of various CM-related reviews and audits, which dictate when formal baselining activity will occur. The timeline events are further discussed in the following paragraphs.

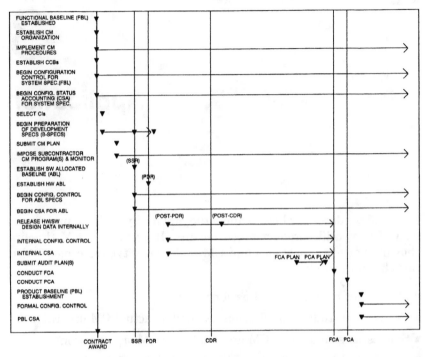

Figure 12 Timeline for implementation of configuration management.

Although it is not shown on the timeline, CM should get involved during the proposal phase, when much preliminary design work is done. If the Configuration Manager does not have input into engineering decisions, such as selection of configuration items, the CM program has already been put at a disadvantage, and program costs may well have been escalated.

Initial CM activities

Typically, the FBL (system specification) is established upon contract award and consists of existing specification(s) imposed by the customer as part of the contract technical requirements. However, occasionally only a preliminary system specification exists, and this specification must be modified/completed by the contractor. In such cases, the FBL is not established until System Design Review (SDR).

Also, at contract award:

- The CM organization is established, and CM procedures implemented.

- Members are appointed to the CCB, and the FBL is placed under formal CM, including configuration control and status accounting.

- The Configuration Manager participates in the selection of CIs prior to preparing the development specifications.

- A CM Plan is normally prepared within 1 month of contract award, and it is subject to approval by the customer on most Government contracts.

Included in the CM Plan are applicable CM requirements and procedures to be used by subcontractors in performance of the CM function, as well as how the prime contractor plans to oversee and interface with the subcontractors regarding CM. Once the CM Plan is in place, the subcontractors are required to implement CM according to that Plan and to continue the CM function throughout the life of the contract.

CM during development

During preparation of the development specifications, the Configuration Manager ensures that the specifications are being prepared in accordance with contract requirements. Prior to PDR, the Configuration Manager's involvement increases significantly in preparation for the review. CM supports development of the data package for PDR by ensuring that it is complete and logically organized and that the specifications meet contract requirements for format and content.

Successful completion of the PDR, including customer approval of the development specifications, establishes the hardware ABL. The Software Specification Review (SSR) establishes the ABL for software. The Configuration Manager now places these baselined specifications under formal CM, including maintaining configuration status and control throughout the remainder of the product life cycle.

Between establishment of the ABL and successful completion of the configuration audits (establishment of PBL), design documentation is released internally through CM and placed under internal CM.

In a group design effort where many design engineers are working together and each has his/her own individual design responsibilities, the work is interdependent. Therefore, the critical path for completing the development effort must dictate the design release schedule, which is developed by Engineering.

The Configuration Manager should monitor the predetermined internal release schedule as closely as possible, notifying the Project Manager if a schedule slippage occurs. The design is normally released incrementally according to the development schedule. Once incremental internal design releases begin, the released documentation must be placed under internal CM, including configuration control and status accounting, until it is formally baselined.

Planning and execution of configuration audits

Upon successful completion of CDR, the product design documentation is conditionally approved to allow fabrication of the first article CIs to be audited. The Configuration Manager prepares and submits the functional and physical configuration audit plans, respectively. This should be at least 3 to 4 months prior to the actual audits, to allow time for customer approval and acceptance of the plans. The Configuration Manager should make sure that these plans are tailored to the product, yet meet all contract requirements.

The FCA is conducted first, with CM and Quality having lead roles in its execution. Engineering, test, and data management have supporting roles in FCA execution. Once the audit is successfully completed and the Configuration Manager, in conjunction with the Quality organization, has ensured the resolution of any resulting action items, the date for the PCA is confirmed. CM also has the lead role in the conduct of the PCA.

Upon acceptance of the configuration item and its associated product documentation, the PBL is established, and all configuration changes are thereafter either Class I or Class II changes. Formal configuration control and status accounting is implemented for all customer approved documentation beginning at this point and continuing throughout the duration of the contract.

Suggestions for Tailoring Contract CM Requirements

In order to meet contract requirements efficiently and effectively, CM requirements should be tailored for specific projects. Tailoring of contract CM requirements prevents wasted time and effort on unproductive activities which provide little, if any, benefit to a program. Tailoring also ensures that project CM procedures are in place and will be effective in meeting requirements.

Pitfall: Do not oversimplify CM requirements in the tailoring process. Certain tasks may be underbid, and cost overruns will come out of profits.

The following suggestions apply to tailoring CM programs.

- Carefully review *all* parts of the Request for Proposal (RFP) for CM requirements. They typically appear in several segments: CM section of the SOW, CM attachment, general provisions, special provisions, quality assurance part of the SOW, program management section of the SOW. Determine if requirements stated in the various parts of the RFP are in agreement or in conflict. Resolve any conflicts. This is the basic data for tailoring recommendations.

- Carefully review CM requirements determined in Step 1 against the type of contract that will ensue: Demonstration and Validation, Full-Scale Development, Production, Operational Support. Recommend appropriate tailoring where the CM requirements are excessive for the type of contract.

- Review any contract specification requirements to determine if less restrictive types of specifications could be recommended as suitable for the program at a reduced cost.

- Review any drawing requirements stated in the contract to determine if they are appropriate for the type of contract and acquisition phase. If not, recommend tailoring to proper drawing type. If integral part lists are required by the RFP and the contractor has computer-generated parts, attempt to tailor to computer-generated lists. *This can be a large cost factor if not resolved.*

- Establish a flexible document release system (specifications and drawings) that can accommodate both contractor and customer document numbers and release data. The system must also be capable of releasing both traditional paper and automated design system files. The contractor internal CM operating plan defines the tailoring of the system for specific contracts.

- Establish a change processing system that has media and procedures appropriate to the project phase under control: on-the-spot manufacturing floor documentation and approval for "make work, make fit" changes on the first build of an item; expedited and routine processing; temporary change procedures for initial integration and test; application to automated design/manufacturing systems; etc. Similar procedures for software should be established.

- Tailor CCB membership to the requirements of the product and type of contract. Membership does not have to be the same for every program. A determination of what is needed for complete change impact analysis should be the guide for CCB establishment.

- Choose or design a status accounting system with a report generator feature. This will enable the contractor to easily tailor status accounting report format and content to program requirements.

- Document all tailored standards, tools, and procedures in the project CM Plan. Once approved, the CM Plan protects the contractor during configuration audits by having a contractual document which states approved departures, modifications, or deviations from customer specifications and standards.

- Use the contract CM Plan to show tailoring for application of customer CM requirements and/or contractor CM systems to automated design and manufacturing systems. It is important to address automation in the CM Plan, since existing coverage for this type of media in specifications and standards is minimal.

- Review SOW requirements for applicable baselines against the type of contract. The SOW may result in the establishment of a premature baseline.

- Review any specifications and standards in the RFP for a tailoring appendix. Many of these documents have them and will provide ideas for tailoring.

- Software CM programs are usually *upscoped* internally from contract and specification requirements, because the requirements contained in the current military software development specifications broadly address software CM. In upscoping program procedures, the contractor should ensure that the CM Plan and Software Development Plan include enough detail to put the procedures in place, without self-imposing too many restrictions.

- Obtain a draft copy of the RFP for review, if possible. Make CM tailoring recommendations to the customer at that time.

Bidding and Staffing a Configuration Management Program

To bid the CM effort effectively, CM requirements for a project must first be determined. Then, the following steps must be taken.

1. Create a detailed listing of all CM tasks to be performed as outlined in the Procedures Chapter.
2. Determine when these tasks must be performed.
3. Determine the quantity and level of personnel required to perform each task.
4. Determine the amount of time needed to perform each task.
5. Determine any required resources to perform tasks.

Parameters for bidding CM

To arrive at these estimates, consider the following parameters:

- Estimated number of documents/files to be controlled
- Expected changes to process
- Drawing and specification requirements

- Meetings to be attended
- Type and number of reviews to support
- Type and number of audits to plan and conduct
- Travel and living costs
- Configuration status accounting *system* requirements
- CM status accounting tools to be modified or purchased
- Configuration status accounting *report* requirements
- Product identification and marking requirements
- Special configuration control requirements
- Training costs for new CM personnel
- CM training costs for project personnel
- Subcontractor CM program coordination
- Estimate of quantity of drawings and lines of code
- Engineering release procedures
- Number and type of functional interfaces for CM
- Unique or special buyer or project requirements
- Project security requirements
- Number of hardware and software CIs
- Program duration and schedule
- Level of CCB involvement and SCCB support by CM
- Use of combined or separate software libraries
- Level of automation required by customer/planned by project
- Anticipated level of detail for CM Plan
- Complexity and type of product (also extent of commercial off-the-shelf hardware/software used)
- CALS (Computer-aided Acquisition and Logistic Support) requirements, for Government contracts

If historical CM costs for similar programs are available, they should be used to develop bids for future CM programs. Otherwise, existing rule of thumb guidelines may be used in estimating CM costs[20] as illustrated in Figure 13.

---

[20]From J. W. Dean's seminars, Spring 1989.

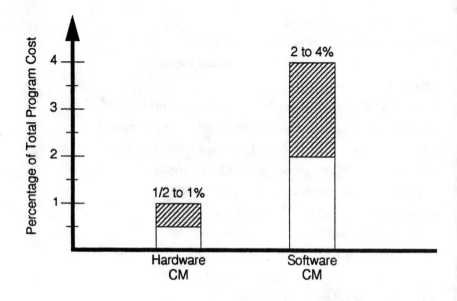

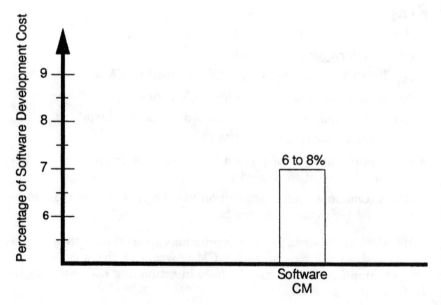

Figure 13 CM effort as a percentage of total program cost/software development cost.

Suggested qualifications for CM personnel

In recent years, extensive work has been done to define more clearly the background and qualifications for CM personnel. However, in the final analysis, the complexity of a product and contract requirements dictate CM function staffing.

Sample job descriptions developed in an EIA CM workshop,[21]are shown in Figures 14, 15, and 16.These job descriptions are for a CM assistant (clerical effort), a Configuration Manager, and the company CM Operations Manager. A company should tailor these qualifications to fit its particular needs.

Job Code

Page 1 of 1

TITLE: Configuration Management   ___ EXEMPT   ___ SUPERVISORY

Assistant (Entry Level) X NON-EXEMPT   X NON-SUPERVISORY

SCOPE

Assist CM and engineering personnel in the implementation and coordination of CM unique procedures and disciplines of Configuration Identification, Change Control, Status Accounting and Audits

DUTIES

1. Operate CRT to input and change data files for status accounting.
2. Log, track and report status of proposed design changes.
3. Record the receipt of engineering change requests and coordinate with originator to obtain additional information as required.
4. Review proposed design changes and determine contract affected and equipment levels affected information.
5. Assist CM personnel in schedule of Configuration Change Control Board meetings and follow-up actions.
6. Assist CM personnel in planning, organizing and conducting informal and formal configuration audits.
7. Assist CM personnel in compiling data for design reviews, special reports, etc.

REQUIREMENTS

1. High school graduate or equivalent with work experience in manufacturing or engineering environment desired.
2. Data entry skills and blueprint reading abilities desired.
3. Perform a variety of routine and repetitive tasks with frequent interruptions in an accurate and timely manner.

Figure 14 Job description for CM assistant.

---

[21]*Data and Configuration Management Workshop,* Sponsored by the G-33 and G-34 Committees, Electronic Industries Association, September 1984.

```
                                                                   Job Code
                                                                Page 1 of 1
      TITLE: Configuration Manager    X  EXEMPT          ___ SUPERVISORY
                                     ___ NON-EXEMPT    X  NON-SUPERVISORY
```

SCOPE

Manages the interpretation and coordination of contractually specified Configuration Management requirements and cost effective implementation of required Configura-tion Management disciplines of Configuration Identification, Change Control, Status Accounting and Audits to assure delivery of quality, supportable products within the Company and Government specifications, budgets and schedules.

DUTIES—(These duties to be performed in accordance with
contract and company requirements

1. Review RFP's to identify CM and contract data requirements, prepare cost estimates and implementation plans.
2. Provide guidance in selection and identification of hardware and software configuration end items.
3. Plan, establish and implement company and contractual baseline and schedules.
4. Provide guidance in preparation of contract specifications.
5. Coordinate CM requirements with interfacing organizations and customer representatives.
6. Support informal and formal design reviews.
7. Prepare subcontractor CM requirements.
8. Establish controls for release and control of engineering drawings and associated documents.
9. Organize and conduct the Configuration Control Board for review of proposed hardware and software changes.
10. Process proposed hardware and software Engineering Changes Proposals (ECPs) and Requests for Deviation or Waiver.
11. Establish and control internal and contractual configuration status accounting system.
12. Plan, organize and conduct informal and formal configuration audits.
13. Provide training and development of support CM personnel.

REQUIREMENTS

1. Bachelor's degree in management or technical field with four to six years related experience or equivalent within an engineering environment.
2. Minimum five years CM experience.
3. Comprehensive knowledge and skills to interpret technical documentation (engineering drawings, contract documents, specifications, etc.)
4. Thorough knowledge of Government/Industry CM requirements and policies.
5. Thorough knowledge of functional interrelationships among participating elements (i.e., Contracts, Engineering, Manufacturing Operations, Quality, Logistics, etc.)
6. Effective written and verbal communications skills are essentials.
7. Working knowledge of automated data systems.

Figure 15 Job description for configuration manager.

<table>
<tr><td></td><td></td><td style="text-align:right">Job Code</td></tr>
</table>

		Job Code
		Page 1 of 1

TITLE: Configuration Management   X EXEMPT   X SUPERVISORY

        Operations Manager   ___ NON-EXEMPT   ___ NON-SUPERVISORY

SCOPE

Establish and implement company CM policies and procedures. Provide technical and administrative direction and surveillance to identification of company products, control of design changes, change status accounting and configuration audits. Budget and allocate CM resources in support of contract and company requirements.

DUTIES

1. Establish and implement company CM policy procedures.
2. Allocate CM resources including facilities, systems, equipment and personnel.
3. Responsible for the training, evaluation and career development of CM personnel.
4. Coordinate and interface with other functional managers.
5. Evaluate and assure the quality of CM documentation and the effectiveness of the CM program.
6. Manage and direct assigned personnel.
7. Perform customer interface/liaison in CM matters.
8. Support informal and formal reviews.
9. Establish guidelines for informal and formal configuration audits.
10. Establish and control change status accounting system.
11. Establish control and release policies and procedures for design documents and contract data.
12. Establish policies and procedures for Configuration Control Board activities.
13. Establish policies and procedures for planning, establishing and implementing company and contract baselines.
14. Establish guidelines for processing proposed hardware and software contract modifications (e.g., Deviations, Waivers, STR's, etc).
15. Develop computer support systems to meet company CM requirements.
16. Plan for future requirements to deliver CM data in an automated format.

REQUIREMENTS

1. Bachelor degree in management, a technical field, or equivalent experience.
2. Eight to ten years experience in CM in both Government (DoD) and commerical environments.
3. Knowledge of automated data systems and interfaces.

Figure 16 Job description for CM operations manager.

These descriptions represent desired qualifications; however, availability of experienced CM personnel is limited. For this reason, an effective CM training program is usually needed.

Certain general qualities should be characteristic of any Configuration Manager. Some of these qualities are:

- Comprehensive knowledge of the CM discipline
- Familiarity with customer/project CM requirements
- Ability to organize effectively
- Ability to manage and direct the efforts of CM support staff
- Technical background
- Aggressiveness with diplomacy and tact
- Knowledge of common documentation requirements
- Good written and verbal communication skills
- Attention to detail
- Knowledge of the product
- Familiarity with company procedures and key people
- Understanding of project functional interrelationships
- Ability to foresee CM related problems and to work to alleviate them.

Using Lessons Learned for Configuration Management

The following paragraphs describe lessons learned and tips for implementing each of the four major functional areas of CM. These suggestions are summarized here, because mistakes in these areas often adversely impact a CM program. Many were also discussed in the Procedures Chapter.

Configuration identification

- *Integrate CM into the system engineering process.* CM should be included in all engineering decisions affecting the program from day one.
- *Manage the design release process.* This includes control of automated Computer Aided Design/Computer Aided Manufacturing (CAD/CAM) database drawing/specification files. Ensure that all associated with development understand CM procedures for design release.

■ *CM should operate all software libraries.* Software developers should not be allowed to operate the libraries. This ensures that file versions are logged, labeled, stored, and impounded for history, traceability, and protection from unauthorized changes.

■ *Strong CM interfaces with all program functions should be established early* to ensure that all cost, schedule, and technical data with a potential impact on CM is relayed to CM in a timely and accurate manner. This includes as-built data from manufacturing.

■ *Do not lose baseline configuration with automated systems.* Not all of these systems have archiving features.

■ *Provide for configuration variations without new drawings,* if possible. Add dash numbers to the part number, e.g. "-5", for varying item functions and configurations.

■ *CM should have sole authority to assign* part numbers, nomenclature, and serial numbers, drawing titles, and drawing/specification numbers.

■ *CM should drive the baselining process.*

Configuration control

■ *Tailor* configuration control to the program, and ensure that the entire change process is defined and documented.

■ *Automate change processing* as much as possible for consistency and efficiency.

■ *Distribute* Change Requests and agenda far enough *in advance* of CCB meetings to allow members adequate time to review.

■ *Each CCB member should analyze change requests / packages* for impact prior to the CCB meeting.

■ *Each member should attend CCB meetings* or send an alternate who has the authority to make decisions or commitments for him. If a member from an area deems beforehand that the change(s) in question do not impact his/her area, he/she need not attend. However, the member should still contact CM to let the CCB know *before* the meeting.

■ Ensure that *all necessary technical documentation is available* at each CCB meeting.

■ *Prepare a complete CCB directive* to authorize or disapprove changes.

■ *Prepare and distribute CCB meeting minutes* in a timely manner.

- *A tape recorder* is suggested for maximum CCB efficiency.

- *Establish company standard criteria* for part number changes early.

- *Ensure total program impact is identified* in change proposals.

- *CM should assign ECP classification* recommended by the CCB.

- *Ensure that the CCB establishes change classification* once a change has been analyzed for impact.

- *Establish a separate Software CCB* for programs with extensive hardware and software development effort.

- *Ensure strong hardware / software CCB interface* by utilizing joint CCB approvals and implementation directives.

- *Include a software CCB member on the hardware CCB, and a hardware CCB member on the software CCB.*

- Ensure effective interface with the Interface Control Working Group (ICWG). *Have a member of the ICWG sit on the program CCB.*

- *Avoid premature baselines,* which result in excessive ECPs.

- *Avoid late baselines,* which result in high retrofit costs.

- *Avoid changes with a high degree of risk* to a program.

- *Coordinate approval of Class I ECPs, major waivers and deviations* with customer to expedite approval

- *Establish procedures for handling emergency change requests.*

- *Use standard codes* to identify change disposition, including implementation status of approved changes.

Configuration status accounting

- *Design the status accounting system around all data elements* required by the contract and tailored requirements.

- *Set up computer interfaces* with design, manufacturing and project databases to ensure communication of all required data to CM.

- *Tailor existing status accounting systems* as necessary for each specific program.

- *Consider user needs and constraints* in designing a configuration status accounting system, and plan system enhancements based on user feedback.

- *Ensure timely and accurate entry of all CM data* into the database, and identify functions responsible for input. (For example, manufacturing personnel may be responsible for entering as-built data if they can access the CM database directly.)

- *Ensure timely and complete submittal of all required configuration status accounting reports,* both external (customer) and internal (other organizations).

- *Standardize report formats* to the maximum extent possible

- *Design or choose a status accounting system* which allows "write" access for authorized data entry and CM personnel but only "read" access to others.

- *As-built lists should be included* in the database.

- *Integrate software libraries and status accounting systems* when possible.

- *System flexibility and vendor support* are vital if a commercial status accounting product is used.

Configuration audits

- *CM* must take *lead role* in audits.

- *Prepare thorough audit plans* and coordinate customer approvals.

- *CM should handle all administrative arrangements,* such as space requirements, and availability of first production article of equipment.

- *CM should coordinate* formation of a technical support team.

- *CM should organize and conduct a dry run* of the audit.

- *Ensure that all technical documentation is available,* including all incorporated and outstanding changes.

- *Obtain agreement with customer on action items* and follow up to ensure timely action item close-out.

- *Prepare audit minutes* or report.

- *Try to restrict hard-copy drawings to Lowest Replaceable Unit (LRU) level,* with rest of drawings accessible in microfilm.

- *Consider union requirements* in moving equipment being audited, as well as removing any subassemblies.

- *Define* with customer the *audit approval criteria.*

- *Brief the customer audit team* on drawing and numbering system.

- *Keep audit from being a design review.*

- *Avoid disassembling equipment* if possible. Disassembly sometimes means retesting. (Use dental mirrors to inspect small components in inaccessible locations.)

- *Provide generation breakdown* of system/product.

- *Develop an attendance list* for customer and contractor.
- *Resolve all audit action items* to mutual satisfaction.

Case Studies: A Look at CM Implementation

The following case studies of actual CM programs have been compiled to illustrate the importance of CM tasks and the results if appropriate CM guidelines are not followed.

Case Study 1
Importance of Configuration Identification

The crew of a space shuttle had difficulties in retrieving a satellite because of interference between the grapple and an antenna. Much effort was required by the astronaut to bring the satellite into the shuttle cargo bay for repair. Retrieval of the satellite was very questionable for a time, and maximum space walk time was almost exceeded.

Problem:	Technical documentation for the satellite did not show the final location for the antenna which was determined during final test. The design of the grapple used for retrieval of the satellite was based on the original antenna location rather than the actual final location.
Lesson Learned:	Actual configuration identification of delivered systems should be documented even though, as in the case of satellites, repair and/or support after launch is not considered in the realm of possibility. This should be accomplished by a function similar to a PCA, and affected specifications and drawings should be updated as required. Basically, this is a function of *configuration identification*.

Case Study 2
Configuration Control Example

A hardware change was processed through the CCB with an estimated cost of $2000. The CCB approved the change without adequate consideration of the grossly underestimated cost. Manufacturing did a re-estimate of the implementation costs and found that the actual costs were $200,000, two orders of magnitude greater than the original estimate. The change package was returned to the CCB for reconsideration and was rejected. The ultimate disposition of the change might cause one to ask, "Was the change really needed in the first place?"

Problem:	Lack of thorough change analysis by the CCB with respect to the cost of a change.

Lesson Learned: Not doing a thorough change impact analysis because of the time and cost involved is false economy, as shown in this example. Each member should analyze the change for total impact in their area of responsibility, particularly in the area of cost. The Configuration Manager should coordinate this activity, backed by a strong CCB Chairperson. This effort is part of the *configuration control* task.

Case Study 3
Practical Configuration Status Accounting

A certain program is 18 months into the development contract, design reviews have been conducted, audits are approaching, and the Configuration Status Accounting System to be used is still being developed internally at a cost far exceeding the original estimate *and* allocated funding. Testing now reveals that there are significant problems with the design, and the schedule is tight.

Problem: No comprehensive database exists which accurately documents formal baseline specifications or internally baselined design data. Therefore, no verifiable data package exists for design reviews and configuration audits. The status of certain changes is unknown because the large number of changes cannot be adequately tracked manually. Clear traceability to system requirements is nonexistent. In short, the program manager does not know what he has.

Lessons Learned: Establish a comprehensive configuration status accounting system *early* for traceability and easy access to accurate configuration data. This greatly facilitates preparation for reviews and audits and even routine design activity. Also, companies should seriously consider using existing commercially available configuration status accounting software packages. Several such products are available which run on platforms and operating systems ranging from personal computer to mainframe applications.

Case Study 4
Planning for Configuration Audits

The customer for a certain program subcontracted their audit effort to a local subcontractor not familiar with the profuct being audited. This company sent 16 people to the contractor's facility with calipers and micrometers. The audit was not organized effectively, and since there were too many people involved, it lasted for two weeks. This resulted in a much higher cost than originally estimated.

Problem: Lack of thorough planning for configuration audits.

Lesson Learned: Regarding configuration audits, CM should always prepare a comprehensive configuration audit plan which:

1. Requires the customer to submit a list of attendees in advance and recommends a certain number of Government participants
2. Clearly defines the audit steps, as well as handling of post-audit corrective actions.

Also, CM should always accomplish a dry run review of items to be audited prior to the audit.

Case Study 5
Part Number Re-identification

Changes were made to electronic assemblies which made them noninterchangeable (either functionally or logistically) for all configurations of a system. Unfortunately, the part number was not changed, and therefore, the noninterchangeable assemblies were stored in the same parts bin since storage is by part number. Therefore, a large number of systems built did not pass test, delivery schedules were missed, and chaos ensued. The problem was traced to the part. *The company finally voluntarily stopped production and delivery to a number of customers for more than 6 months in order to correct the configuration identification (specifications and drawings). The loss to the company was in the millions, and the damage to the company reputation was inestimable.*

Problem: Failure to change the part number of several electronic assemblies when interchangeability (function and/or logistics) was affected. The assemblies were used in a number of systems, each under a different program manager, and each program was making this type of change without adequate change coordination.

The reason given by program personnel for not changing part numbers was the *cost* of preparing documentation for the new part number; however, they failed to recognize that their action or lack thereof would result in noninterchangeable assemblies bearing:

1. A common part number
2. A common bin in manufacturing stores
3. A common National Stock Number.

They assumed that the revision letter change could segregate the noninterchangeable items, which was grossly incorrect. As production proceeded, systems received assemblies from stores which were incorrect for their various applications, even though the assembly part number was as correct according to the parts list.

Lessons Learned: There can be no greater false economy than not changing part numbers when required. CM should train per-

sonnel so that the need for changing, and the consequences of not changing, part numbers is understood. Part
number change criteria should be a major item in the
change impact analysis done by CCB members, and part
number change should be directed by the CCB as a part of
change implementation.

Metrics: Measuring CM Effectiveness

The following examples of CM metrics may be used by the CM practitioner as a guide for:

- Assessing effectiveness of a CM program
- Developing internal metrics applicable to a particular product procurement

Metrics 1 and 2: Measuring effectiveness of internal CM during development

Metrics 1 and 2 compare two similar projects, each worth approximately $100 million. The first graph compares the number of changes
per month for each project during various program stages. Project X,
which did not practice internal change control, is designated by the
solid line. Project Y, which implemented a comprehensive internal
change control procedure early, is designated by the dashed line.

As can be seen in the first graph, Project X processed relatively few
changes during development, but saw a significant increase in changes
beginning after the FCA and PCA and into the production phase.
Changes for Project Y were processed mostly during development, with
the level of changes declining with the start of production.

The second graph illustrates the cumulative cost of changes (in millions of dollars) during various contract stages. Obviously, Project X,
without an internal change control procedure during development, experienced much higher costs, beginning during first article testing and
into production. The costs for Project Y, on the other hand, remained
lower, peaking just prior to the audits and decreasing steadily thereafter into production.

These metrics illustrate that unless designers are trained in internal
CM procedures and CM is present during development:

- Changes are haphazardly made without being documented
- Full impact of changes are not realized
- Traceability is lost

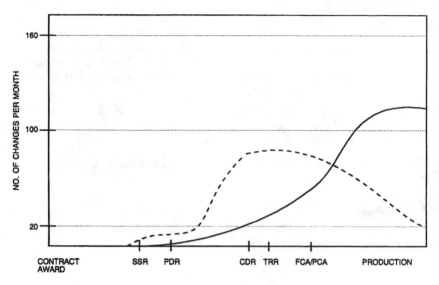

Metric 1 Measuring effective internal CM prior to product baseline establishment.

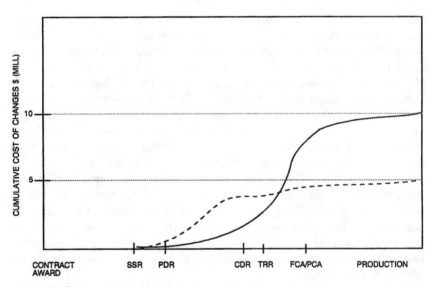

Metric 2 Dollars spent during various stages of development.

As seen by the second graph, this lack of early internal configuration control can increase project costs significantly. *The earlier problems are corrected through controlled changes, the less expensive changes will be.*

Metric 3: Efficiency of software development with applied CM

Metric 3 illustrates various steps in a typical software development process with the person-days of effort required to accomplish each task. The project represented is for the development of 20,000 lines of code. According to equations in *Software Engineering Economics,* by Barry W. Boehm,[22] this project would occupy about 5 people for 12 months. The task would break down as follows:

Task	Without CM Tool (Person-days)	With CM Tool (Person-days)
Coding	500	500
Maintaining Source Documentation	35	1
Module Building and Testing	170	120
Dependency/Wrong Module Factors	30	5
Integration Debugging	200	170
Change Control	65	5
Total	1000	801

These numbers are given both for development without an effective automated software CM tool, and development with such a CM tool, *combined with comprehensive CM procedures.* The following assumptions apply to the development of this breakdown.

- The distribution of time spent for development is based loosely on percentages and case studies from *Software Engineering Economics.*

- Nearly all of the source code documentation can be automated with an effective computer based CM tool.

- Automatic module building and elimination of compiler switch errors, module dependency errors, and unnecessary compilations will save approximately 30 percent of building and testing time.

- Most system dependency and wrong-module errors can be eliminated

---

[22]B. W. Boehm. *Software Engineering Economics,* Prentice-Hall, Inc., 1981.

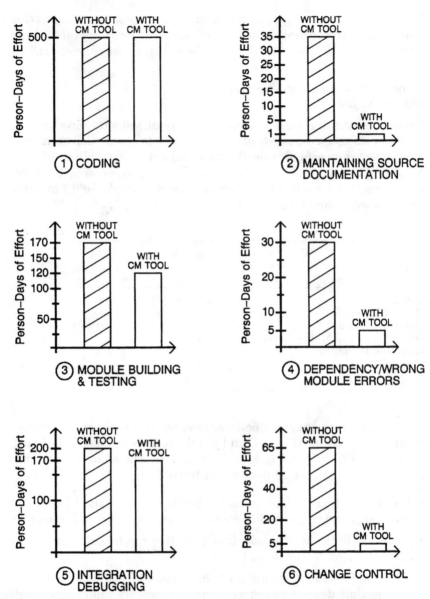

Metric 3 Software development with and without CM tool for a certain project.

by using an automated, intelligent "Make" system, which integrates software while retaining version control and using an advanced module librarian function.

- Integration and debugging time can be reduced by at least 15 percent when automated, comprehensive source code version control is used.

- An effective CM tool will automate most change control requirements.

With these assumptions, on a medium sized project, about 200 person-days will be saved during development.[23] *This equates to about $60,000 considering programmer/administrator pay and overhead.* In effect, about 15 percent of development time will be eliminated. Automated CM also saves time during the maintenance and upgrade phases, because knowledge about system design, organization, and construction is retained indefinitely.

Companies may use this estimate to develop similar metrics for other software development projects to determine how automated software CM tools reduce their development costs.

Metric 4: Improving change process cycle time

Metric 4 illustrates how the *change processing time decreases as the cumulative number of changes processed increases* for a typical project. These numbers will vary between projects, depending on internal CM procedures, product and change complexity, schedule, and customer requirements. Configuration Managers should develop internal metrics for specific projects.

For urgent minor waivers, deviations, and Class II changes which may stop production if not approved rapidly, procedures for on-the-spot change origination and approval may be used, thus allowing an optimal 1-day turnaround on urgent change requests/proposals. Drawing amendments, generated by the manufacturing engineer on the plant floor, have the following format:

Was	Is
1. Description of current configuration	Description of proposed change
2. Any illustrative sketches (hand-drawn)	Sketch (hand-drawn)

---

[23]B. W. Boehm. *Software Engineering Economics,* 1981.

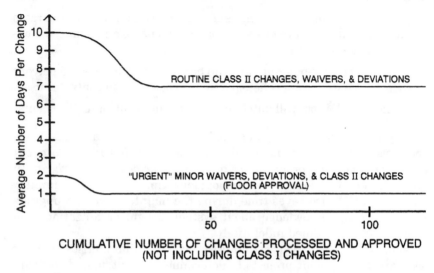

Metric 4 Change processing time vs. cumulative changes processed for a typical project.

Normally, these amendments or "floor change orders" are sketched neatly on an 8-inch by 10-inch sheet of paper and then submitted to the appointed local CCB member for approval. The Quality organization will inspect product per these amendments. Before authorizing use of floor change orders, specific guidelines should be established to govern their use. For example, some companies authorize use of this procedure for "make work, make fit" changes on first-build items only. Also, the number of authorized amendments for any given drawing should be limited. Some companies allow five amendments before the drawing must be officially updated by drafting/engineering services. The floor changes normally get a post-release CCB review as soon as possible after initial floor approval.

As can be seen from the graph, routine Class II changes, waivers, and deviations also show a decrease in processing time, as the cumulative number of changes increases. The limiting factor is the frequency of scheduled CCB meetings, which are often held biweekly or weekly. (For the sample project, meetings were held biweekly.) If routine changes need to be processed more rapidly, the Configuration Manager should schedule more frequent CCB meetings.

Metric 5: Measuring volumes of changes at various program stages/milestones

Metric 5 illustrates sample volumes of changes for a project having approximately 3000 drawings to control. Internal change control is prac-

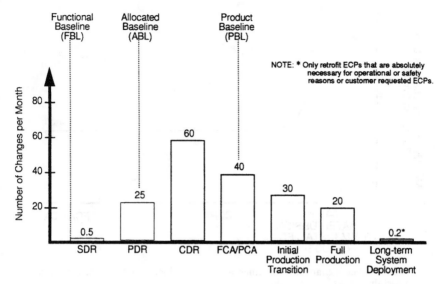

Metric 5 Volume of changes at various program stages/milestones for a certain project.

ticed for the evolving design documentation during development. The product consists of complex digital signal processing equipment unique for its military application.

As the graph shows, the number of changes per month peaks at CDR, tapering slightly at the audits and initial production and reaching lower levels during production and deployment. This distribution of changes corresponds roughly to the data contained in Metric 1 for a project practicing effective internal change control.

Configuration Managers may use these figures as rough guidelines for establishing internal project-specific metrics.

Metric 6: Effect of design automation on error level

This metric shows the number of design changes caused by minor design errors during development, both *before* and *after* installation of an automated design system for this project. As can be seen from the graph, the monthly level of these changes dropped from about 25 to approximately 8 changes per month in a relatively short period of time. This particular project experienced a reduction in "minor error" changes of 68 percent through the use of an automated design tool.

Research has shown that design automation reduces the number of drawing error changes. Built-in checks and logic flows may alleviate

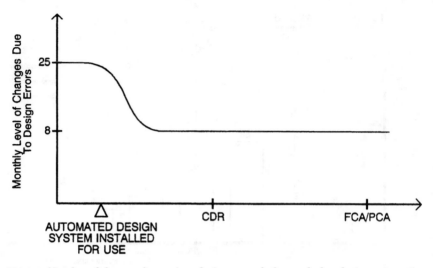

Metric 6 Number of changes from minor design errors before and after design automation for a certain project.

any tolerance and dimensioning problems, as well as problems in electrical design. Specific tools are available for different applications, whether mechanical, electrical, or software. The Configuration Manager should be involved in the selection of automated design packages.

Pitfall: *Careful planning beforehand* facilitates transition to the use of computer-aided design and engineering tools. An effective tool, if poorly implemented, will hinder progress and may increase development costs.

CM, in conjunction with engineering, may establish similar internal project-specific metrics to measure the effectiveness of automated design tools.

Summary

Effective implementation of the CM discipline is essential, both during development and in the transition from development to production. Properly operated, a CM program for any product acquisition ensures that:

- Products meet requirements
- Costs for redesign and factory rework are minimized
- Traceability to system requirements is retained

In short, repeatable performance and design requirements in present and future procurements of the product will be guaranteed at lower costs through effective implementation of the four elements of CM: identification, control, accounting, and audits (see Figure 17[24]).

Follow best practices

The potential benefits of applied CM only result if best practices are followed. Examples of these best practices are:

1. Obtain total management commitment and support for the CM program.
2. Select qualified, dedicated personnel to staff the CM office and ensure that they are trained in CM.
3. Prepare a comprehensive CM Plan outlining policies and procedures to be used, and obtain customer approval.
4. Tailor contract requirements to fit CM program needs.
5. Establish a schedule upon contract award for significant CM milestones and activities, and follow it.

---

[24]From J. W. Dean's Seminars, 1988.

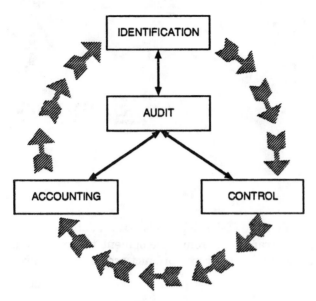

Figure 17 CM element relationships.

6. Ensure that all functional areas interfacing with CM are trained in CM policies and procedures.
7. Ensure that any program automated design, development, or manufacturing tools are integrated effectively into the CM process.
8. Monitor and support the development of configuration identification to ensure that content, format, and marking/labeling instructions correspond to program CM requirements.
9. Establish project-specific configuration control policies and procedures, and monitor their compliance.
10. Establish clear, concise CCB goals and responsibilities.
11. Ensure that each functional area is appropriately represented on the CCB to determine the total program impact for proposed changes.
12. Design or choose a configuration status accounting system which will accurately and efficiently meet project needs.
13. Plan in advance with the customer for FCA and PCA. CM should assume the lead role.
14. Develop internal project-specific metrics to monitor the effectiveness of the CM program, and strive for continuous process quality improvement.
15. Integrate CM into the systems engineering process from day one.
16. Provide CM training to other program interfaces.

Summary of Procedures

Effective CM programs do not happen by accident. They are the result of careful planning, following the step-by-step procedures contained in this part (tailored for specific applications) and adhering to the principles in Best Practices and the Transition from Development to Production templates. The steps presented in the Procedures Chapter are summarized below.

What	Who	Key Points
Step 1: Oganization		
Select Configuration Manager.	Project Manager	Select qualified CM Program Manager.
Select CM support staff.	Project and CM	Ensure adequate level and quantity of staff.
Define functional relationships.	Project and CM	Establish these interfaces according to project needs.
Step 2: Prepare CM Plan		
Define contract CM policies and procedures.	Configuration Manager	Review customer RFP.
Perform necessary tailoring.	Configuration Manager	Define project and internal needs. Obtain customer approval.
Prepare CM Plan.	Configuration Manager	Include functional interfaces, policies, and procedures.
Obtain project and customer approval.	Configuration Manager	Interface with customer and incorporate customer comments, as necessary.
Step 3: Support Development of Configuration Identification		
Select Configuration Items.	Engineering, CM, Logistics, and Customer	Use CI selection guidelines (see p. 561).
Review all documentation for compliance as it is developed.	CM, Engineering, and Quality	Ensure contract content and format requirements are met. Provide guidance to engineering, as necessary.
Assign drawing titles and numbers.	CM	Follow contract requirements and approved CM Plan.
Establish and maintain baselines.	CM, Customer, Engineering, and CCB	CM coordinates baselining process with engineering/CCB; customer approval at design reviews establishes baselines. Internal baselines may also be established. CM controls changes to all baselines.

What	Who	Key Points
Step 3: Support Development of Configuration Identification (*Continued*)		
Release documentation as it is baselined.	CM	Release both internal and external baselined documentation. Distribute to subcontractors, project, engineering, and manufacturing, as necessary.
Assign serial numbers and nomenclature.	CM	Follow contract requirements and approved CM Plan.
Ensure proper product marking.	CM and Quality	Follow approved CM Plan.
Step 4: Invoke Configuration Control		
Establish CCB.	CM, Project Manager	Select members from each functional area. Each member should have the authority to commit his organization to the decision/course of action.
Define change control process and implement it.	CM, Project, and CCB approval	Tailor process to project.
Control changes to FBL and ABL.	CM, Customer, and CCB approval	Customer must approve these Class I ECPs.
Control internal baselines.	CM	Control changes to internally released design data.
Control changes to the PBL.	CM, CCB, and Customer approval/ concurrence	Only Class I changes require customer approval; Class II changes require CCB approval *and customer concurrence in classification.*
Control deviations and waivers.	CM and CCB	Minor waivers and deviations require CCB approval; major and critical waivers and deviations require customer approval. Minor waivers may be dispositioned by the Material Review Board, if one exists, rather than by the CCB.
Step 5: Establish and Maintain Configuration Status Accounting Systems		
Determine status accounting requirements.	CM, Project input	Tailor to RFP requirements and consider project needs.
Design/choose a status accounting system.	CM, Project input	Choose a system that meets requirements, without excess. A computer programmer may be needed for system design.

What	Who	Key Points
Step 5: Establish and Maintain Configuration Status Accounting System (*Continued*)		
Install, implement, and maintain the system.	CM	Services from the company's Management Information System group may be required for installation and interfaces. Either the Configuration Manager or a separate system administrator working under the Configuration Manager should maintain the system.
Set up interfaces with other project data bases when cost-effective and beneficial.	CM, Project input	Design for single entry, if possible; avoid redundant data entry. Possible interfaces include depot, material requirements planning, data management, logistics, project, engineering, manufacturing, etc.
Ensure timely and accurate data entry.	CM	Train users and data entry clerks and monitor data input, as necessary.
Submit required reports on time.	CM	Follow contract requirements for quantities and content, as tailored by the approved CM Plan.
Step 6: Manage Subcontractor CM Programs		
Define CM requirements applicable to subcontractors.	CM, Project input	Tailor contract requirements, as necessary.
Flow down applicable CM requirements through POs and/or SOWs.	CM, Purchasing, Project input	Review by the Legal Department is advised.
Define/approve subcontractor CM procedures.	CM, Project input	Review internal subcontractor procedures for consistency. Subcontractor CM procedures should be reflected in the Subcontractor CM Plan.
Monitor subcontractors for compliance with CM requirements.	CM, Quality	Conduct periodic audits at subcontractor's facilities.
Step 7 : Plan and Execute Configuration Audits		
Determine audit requirements.	CM, Customer	Meet contract requirements, as tailored by the approved CM Plan.
Schedule each audit tentatively.	CM, Project	Work with project on scheduling audits.

What	Who	Key Points
Step 7: Plan and Execute Configuration Audits (*Continued*)		
Prepare and submit audit plans.	CM, Project Management approval	Submit audit plans in advance, even if not required by the contract, to obtain customer agreement on audit procedures/logistics.
Obtain customer approval of audit plans.	CM	Interface with the customer, and incorporate comments, as required.
Ensure that all logistical details for the audit are covered.	CM	Plan for space, equipment availability, availability of technical support personnel, etc.
Plan and execute a dry run for each audit to be conducted.	CM, with Project support	An audit dry run will help ensure that the actual audit runs smoothly.
Conduct audits.	CM and Project, Engineering, and Quality support	Conduct the FCA first. Upon successful completion of the FCA, conduct the PCA.
Prepare and submit audit reports.	CM	Meet contract requirements, as tailored by the approved CM Plan.
Resolve audit action items on time and thoroughly.	CM, Project, Quality	CM should work with quality to ensure timely resolution of action items by appropriate personnel.

Summary of Application

The Application Chapter of this part discusses how to apply the procedures for CM to specific product procurements. Factors to be considered for successful CM implementation include:

- Scheduling major CM activities effectively
- Bidding and staffing the CM function accurately and training new CM personnel
- Tailoring contract requirements for a project, addressing the level of automation
- Using lessons learned and classic case studies from other programs
- Developing internal metrics for tracking CM effectiveness

The Application Chapter contains a timeline for CM implementation, guidelines for bidding and staffing CM, suggested qualifications of a configuration manager, suggestions for tailoring contract requirements to specific project needs, case studies on actual CM programs, and examples of metrics which can be developed to measure CM effectiveness.

CM can be a useful project management tool, if properly planned and implemented. CM should be conducted as described in the Procedures Chapter; however, CM contract requirements should be tailored to fit the specific application. Key factors to consider include product complexity, the amount of hardware or software, the type of contract, and other variables.

Chapter

5

References

Babich, W. A. *Software Configuration Management: Coordination for Team Productivity.* Reading, Massachusetts: Addison-Wesley, 1986. Discusses modern-day problems encountered during software development and how proper application of software CM can alleviate these problems.

Bersoff, E. H., Henderson, V. D., & Siegel, S. G. *Software Configuration Management: An Investment in Product Integrity.* Englewood Cliffs, NJ: Prentice-Hall, Inc., 1980. Discusses the four elements of software CM and how they can ensure product integrity.

Boehm, B. *Software Engineering Economics.* Englewood Cliffs, NJ: Prentice-Hall, Inc., 1981. Discusses the economic justification for having applied CM, as well as design reviews and peer inspections, early in the developmental phase in order to detect, resolve, and fix errors while the cost is lower.

Buckle, J. K. *Software Configuration Management.* London: Macmillan Press, 1982. Describes policies and procedures for the four major functions of configuration management as applied to software. Also describes peer reviews and inspections, software testing and validation, and software CM tools.

Department of Defense. *Drawing Practices.* (DoD-STD-100C). Describes the requirements for preparation and revision of engineering drawings and associated lists prepared for the government. This is the companion document to MIL-T-31000. Information about types of drawings is given in Chapter 200.

Department of Defense. *Engineering Drawings and Lists.* (DoD-D-1000). Prescribes the requirements for engineering drawings and associated lists to be prepared by contractors. *This document has been superceded by MIL-T-31000, Technical Data package.*

Department of Defense. *Transition from Development to Production.* (DoD 4245.7-M), September 1985. Describes strategies for minimizing technical risks in 47 template areas including management, design, test, production, logistics, and transition planning.

Department of the Navy. *Best Practices: How to Avoid Surprises in the World's Most Complicated Technical Process.* (NAVSO P-6071), March 1986. Discusses how to avoid traps and risks during the transition from development to production by implementing best practices for 47 templates that include topics in design, test, production, logistics, facilities, and management. These templates give project managers and contractors an overview of the key issues and best practices to use throughout the product acquisition life cycle.

Eggerman, W. V. *Configuration Management Handbook.* USA: TAB BOOKS, 1990. Describes policy and procedures for the four major areas of CM in a modern development and production environment.

Electronic Industries Association. *Configuration Control (Preliminary).* (Bulletin No. 6-4), Washington, DC: Electronic Industries Association. Recommends policy and procedures for application of configuration control on both government and commercial projects.

635

Electronic Industries Association. *Guideline for Transitioning Configuration Management to an Automated Environment.* (Bulletin No. CMB7-2), Washington, DC: EIA, March 1991. Focuses on automation of the four major CM functions: identification, control, status accounting, and audit in the modern development and production environments.

Electronic Industries Association. *Computer Resources, Data, and Configuration Management Workshop.* (Workshop Report), Washington, DC.: EIA G-33 Committee, September 1984. Summarized the results of the eighteenth annual Configuration and Data Management workshop in the areas of acquisition streamlining impact, career recognition in CM, software engineering and management, and software quality assurance.

Military Specification 83490: *Specification Types and Forms.* Describes the types and forms of specifications which can be prepared by contractors.

Military Standard 973: *Configuration Management (Preliminary).* Describes CM policy and procedures as related to the four major functional areas of CM: identification, control, status accounting, and audits. This document, when finalized, will replace MIL-STD-480, MIL-STD-481, MIL-STD-482, MIL-STD-483, and the audit portion of MIL-STD-1521. The design review portion of MIL-STD-1521 will be moved to MIL-STD-499. MIL-STD-973 will also replace the CM portion of DoD-STD-2167A.

Military Standard 480: *Configuration Control: Engineering Changes, Deviations, and Waivers.* Describes CM requirements for configuration control procedures, ECPs, deviations, and waivers which are normally mandated on government contracts.

Military Standard 482: *Configuration Status Accounting Data Elements.* Contains approved data elements for configuration status accounting which are sometimes required on government contracts. Even when not required, these data elements can provide useful guidelines.

Military Standard 483: *Configuration Management Practices (USAF).* Describes program configuration management practices for U.S. Air Force contracts. Provides useful guidelines for the selection of configuration items.

Military Standard 490: *Specification Practices.* Describes requirements often mandated on government contracts for preparing, controlling, and changing contract specifications. Types of specifications are defined in Section 3.1.3.

Military Standard 1456: *Contractor Configuration Management Plans.* Describes an approach for the organization, content, and preparation of CM plans. This standard is often invoked by the government in the RFP in order to have an approved plan in place at the beginning of a contract.

Military Standard 1521: *Technical Reviews and Audits for Systems, Equipments, and Computer Software.* Details on reviews and audits typically required on government contracts, e.g., Critical Design Review, Functional Configuration Audit, Physical Configuration Audit, etc.

Military Standard 2167: *Defense System Software Development.* Describes CM requirements for software applications which are often mandated on government contracts. Also discusses many other aspects of software development.

MIL-T-31000 (Military): *Preparation of Technical Data Packages.* Describes the requirements for engineering drawings and associated lists, and other elements of technical data packages. It establishes four categories of drawings: conceptual, developmental, production, and commercial.

National Aeronautics and Space Administration. *Apollo Configuration Management Manual.* Washington, DC: January 1970. Details configuration management policy and procedures to be used on all Apollo programs as mandated by NASA.

Samaras, T. *Configuration Management Deskbook.* North Babylon, NY: Advanced Applications Consultants, Inc., 1988. Describes modern methods for application of the CM discipline for ensuring that products meet customer requirements. Covers all aspects of configuration identification, configuration control, configuration status accounting, and configuration audits. Also discusses software CM peculiarities and methods, management of subcontractor CM programs, bidding configuration management effort, and preparing comprehensive CM plans. This book is highly recommended by CDM Consultants.

6

Glossary

ABL   Allocated Baseline

ACI   Allocated Configuration Identification

ACO   Administrative Contract Office

AQL   Acceptable Quality Level

CAD/CAM   Computer Aided Design/Computer Aided Manufacturing

CALS   Computer-aided Acquisition and Logistic Support

CCB   Configuration Control Board

CD   Classification of Defects

CDR   Critical Design Review

CI   Configuration Item

CM   Configuration Management

CSA   Configuration Status Accounting

CSCI   Computer Software Configuration Item

CSC   Computer Software Component

CSU   Computer Software Unit

DoD   Department of Defense

ECP   Engineering Change Proposal

EIA   Electronics Industries Association

FBL   Functional Baseline

FCA   Functional Configuration Audit

FCI   Functional Configuration Identification

GFE   Government Furnished Equipment

HWCI   Hardware Configuration Item

ICD Interface Control Document

ICWG Interface Control Working Group

IDD Interface Design Document

IRS Interface Requirements Specification

LLCSC Lower-Level Computer Software Component

LRU Lowest Replaceable Unit

MIL-STD Military Standard

NDI Nondevelopmental Item

NOR Notice of Revision

PBL Product Baseline

PCA Physical Configuration Audit

PCI Product Configuration Identification

PDR Preliminary Design Review

PO Purchase Order

PPL Preferred Parts List

QPL Qualified Products List

RFP Request for Proposal

SCCB Software Configuration Control Board

SCN Specification Change Notice

SDD Software Design Document

SDR System Design Review

SOW Statement of Work

SPR Software Problem Report

SPS Software Product Specification

SRS Software Requirements Specification

SRR System Requirements Review

SSR Software Specification Review

STD Software Test Description

STP Software Test Plan

TDP Technical Data Package

TLCSC Top-Level Computer Software Component

TRR Test Readiness Review

VDD Version Description Document

WBS Work Breakdown Structure

Design Reviews

Design Reviews

Design reviews are extremely important because, in limiting technical risk and assessing design maturity, they are of proven benefit to development projects and DoD procurement. They are often misunderstood, misused, and underutilized in the design process. Other design review problems:

- Viewed as mandated "dog and pony shows" for government clients or senior management

- Thought useful only if they comply with military standards

- Omitted due to lack of time and budget (low-level technical reviews)

- Expected to present only "good news;" "bad news" is postponed due to anticipated, but often unrealized, solutions

- Focused on success or adhering to a schedule (rather than finding unmet requirements)

- Expected to cover too much material at too fast a pace with too little preparation

Design Review Strategy

Design reviews should be part of an overall strategy that incorporates the *Transition from Development to Production* templates[1] that are the foundation for this series of books. Together with the *Best Practices,*[2] the templates provide the engineering discipline and road maps needed

---

[1]*Transition from Development to Production,* Department of Defense (DoD 4245.7-M), September 1985.

[2]*Best Practices: How to Avoid Surprises in the World's Most Complicated Technical Process,* Department of the Navy, (NAVSO P-6071), March 1986.

to conduct a successful, low risk transition of products from development to production. (See the partial template diagram below on this page.)

Design reviews are an integral part of a successful transition because they make it likely that design alternatives, problems, and errors will surface early, before they become expensive to fix. This will reduce costs and contribute to the overall success of the DoD procurement effort.

Rationale

The scheduling of good design reviews early in the design process allows you to assess design maturity and make sure the design is on track. If corrective action is necessary, it can be applied when it will be most beneficial: before design change is expensive to implement.

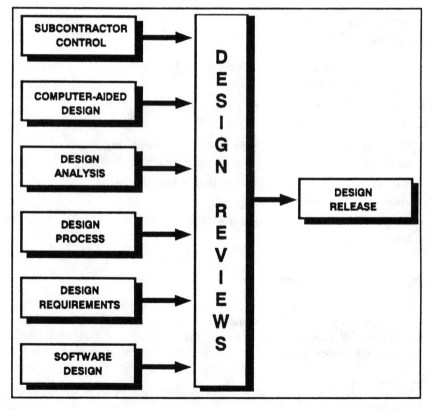

The Design Review template is related to six others: Subcontractor Control, CAD, Design Analysis, Design Process, Design Requirements, and Software Design all feed into Design Reviews; the Design Release is the outcome.

Design reviews can be a simple, effective technique for improving quality, containing cost, and staying on schedule. In fact, design reviews should occur throughout the product life cycle to eliminate errors and technical risks in hardware, software, and firmware. Planning should include both government contract-mandated and internal design reviews.

Design review objectives

The *overall objective* of a design review is to ensure that the product will fulfill its requirements. Many design reviews fail to meet this objective because they lack specific plans, discipline, and resources to perform a truly objective evaluation of the new product and processes.

With the overall objective in mind, the main purposes of the design review are to:

- Assess design maturity.
- Clarify the design requirements and evaluate the design for compliance.
- Challenge the design and related processes.
- Bring additional knowledge and experience to bear on the design and processes.
- Shorten development time, reduce changes, and prevent mistakes and omissions.
- Determine what issues remain to be solved and what effort will be required.
- Ensure the product is testable, manufacturable, and usable.
- Ensure the product is safe and reliable.
- Communicate requirements, design concepts and descriptions to other departments.

The benefits gained during the product's life cycle outweigh the additional costs of conducting design reviews. Fewer errors reach the next stage of development, and thus lower costs and less time are needed to correct them.[3]

Properly conducted design reviews pay for themselves. Products are ready for use sooner, with lower maintenance costs to the manufac-

---

[3]Boehm, B. "Software Engineering," *IEEE Transactions on Computers,* Vol. C-25(12), December 1976. This study shows that costs typically increase as faults are detected later in the process. What would cost one dollar to fix at requirements would cost five dollars to fix during detailed design, ten dollars to fix at design implementation, fifty dollars to fix during system test, and one hundred dollars after delivery.

turer and the customer. Products perform accurately and according to user specifications. With work done correctly, designers can spend time on new work, rather than on reworking old designs.

Good design reviews should be conducted for all internal contractor reviews and contract-mandated reviews. It is the responsibility of the contractor to verify that subcontractors of major components also follow these same practices.

Remember: These objectives will only be met when experienced people conduct timely and thorough reviews to ensure that existing problems are brought to the surface. Poorly conducted reviews seldom benefit anyone and tend to postpone the inevitable realization of project flaws.

Design Reviews Defined

Design reviews are formal evaluations to determine design maturity and to ensure the design is technically adequate and will meet requirements for performance, quality, cost, and availability. During reviews, a technically competent team searches for errors and design problems and may propose design alternatives.

All documents, material, and analyses regarding the design are to be reviewed. These documents can include requirements, results from tradeoff studies and design analyses, wiring board schematics and layout, interface specifications, testing plans, code, and functional specifications, to name a few. The review process actually includes a series of meetings and activities between meetings that should be planned in advance. The nature of the review process depends upon the type of the products being reviewed: hardware, firmware, or software.

Rather than occurring on pre-established dates, design reviews should occur at the end of each step of the design process before a commitment is made to proceed to the next design stage. The timing and sequence of design reviews are covered in chapter 3.

Internal reviews

Internal design reviews should be held at the prime and subcon tractors' sites. These internal reviews should anticipate the man dated reviews; they should strive to eliminate some of the problems that crop up in the government contract-mandated reviews. There are more details about the internal reviews in chapter 3, Application.

Contract-mandated reviews

In addition, certain design reviews are mandated by government contracts to assess technical progress (see Table 1 on page 645). The pri-

TABLE 1   Major Contract-Mandated Reviews

Review	Description
System Requirement Review (SRR)	Review system requirements and specifications with the customer to ensure they are clear, concise, accurate, and verifiable and to obtain agreement.
System Design Review (SDR)	Review the high-level system design and architecture with the customer; review system partitioning and requirements allocations and obtain customer agreement.
Software Specifications Review (SSR)	Review the software requirements for clarity, consistency, and completeness to obtain customer agreement.
Preliminary Design Review (PDR)	Review hardware and software prototypes for functional flow, requirements allocations, thermal analysis, power, packaging, reliability, maintainability, and manufacturability.
Critical Design Review (CDR)	Review the working models of hardware and software to ensure functionality, interfaces, and supportability to assess readiness for release to manufacturing.
Preproduction Reliability Design Review (PRDR)	Review failure reports, failure analysis reports, corrective actions, and retests to assess whether the system's design is mature enough for transition to production.

mary objective of contract-mandated design reviews also varies. Preliminary Design Reviews (PDR), for example, concentrate on whether the design will meet the stated requirements, while Critical Design Reviews (CDR), which occur later, focus on whether the design can be released to manufacturing. These reviews are described briefly below and more fully in MIL-STD- 1521 (*Technical Reviews and Audits*) and MIL-STD-2167 (*Defense System Software Development*).

Types of reviews

Formal *design reviews* are scheduled when the design is at the appropriate level of maturity. Design reviews are used for architecture, high-level plans, requirements documents, specifications, overview functional documents, analysis results and test results. The review team typically reviews the information (documents, drawings, test data, research notes, analysis results and the like) in more general terms, without examining every written word, drawn line, etc.

A *design inspection* is a type of review that encourages team members to examine every detail in the design information being discussed. The purpose is to find errors, not to suggest unrequired improvements. The more complex the design, the more likely an inspection is needed. Information that is subjected to inspection: interface specifications, functional design specifications, circuit board schematics and layout, custom devices (such as very large-scale integrated circuits—VLSI), backplanes, equipment, source code, firmware code, unit test plans, and release documents.

Here are some typical activities that are included in the design review process:

- *Planning activities* such as selecting participants, scheduling future activities, and setting criteria.

- The *design overview review* is a meeting that gives project members an opportunity to review the new design and offer suggestions for improvement. It also introduces the new product and design approach to the reviewers. Topics can include: functional operation, system configuration, wattage values for thermal and power analyses, stocklists, and failure rates.

- A *walk-through* meeting helps the reviewers more quickly understand the complex operations. This allows reviewers to do a more thorough and effective search for errors. It should immediately follow the design overview review.

- *Electrical inspection* determines whether the system functions properly, performs logic operations correctly, and meets requirements. This inspection includes individual activities leading up to, and including, the inspection meeting. The inspectors report on errors found and checklist item completion.

- *Physical inspection* (held after the electrical inspection) verifies that the design meets standards developed in the physical design checklists. Some possible topics: discussion of prototype data, preliminary component placement, parts list, thermal analyses, inputs and outputs, and other special requirements and documentation.

- *Manufacturing inspection* is a final manufacturability check before the design is released to the factory. Some topics: final drawings, debugging, user documentation, and testing guidelines.

2

Procedures

Design reviews assess design progress, maturity, and rigor. To be effective, design reviews need high-level management support and a well-defined, disciplined structure. The support involves allocating time and money to conduct reviews. The discipline comes from following the processes described in *Best Practices* and *Transition from Development to Production*.

The diagram on the next page gives an overview of seven design review steps explained in detail here. The seven steps are:

- Before a review:

  Step 1: Plan

  Step 2: Familiarize
- **During** the review process:

  Step 3: Individual Review

  Step 4: Team Review
- After the review:

  Step 5: Resolve

  Step 6: Follow-up

  Step 7: Improve

At the bottom of the substep box is the person responsible.

Step 1: Plan

Choose team members

First, project management selects the review leader who then helps select the other team members. Each member should be at approximately the same technical and management level to avoid any feelings of dominance for any particular point of view during review meetings.

Steps for Effective Design Reviews and Inspections

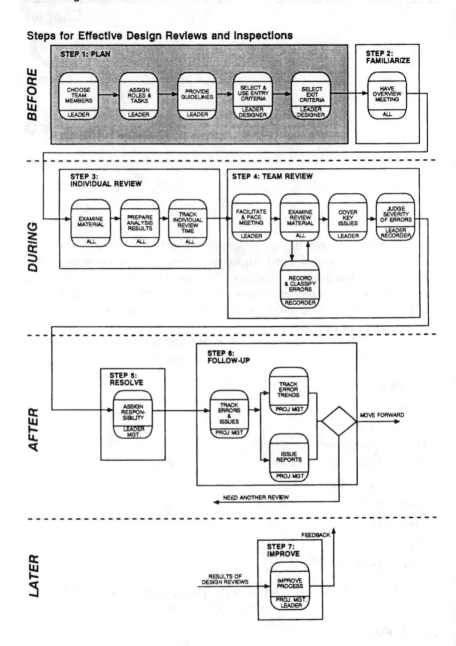

They should also be at approximately the same level as the person whose material they are reviewing. It is important that management support the review process and decisions reached by the review team, but keep a "hands-off" approach. For most projects, three to seven participants can find errors impartially and efficiently.

Guidelines for choosing essential reviewers vary from project to project. Here are some rules of thumb:

- Choose peers of the designer (individuals with a similar or greater level of technical expertise as the designer).

- Choose experienced designers who are working on the same project or have worked on a similar or related project.

- Choose people who worked in earlier phases of the project (a previous version, for example) or people who will work on later phases of the project (test, manufacture, maintenance, deployment, for example).

Assign roles and tasks

The review leader assigns roles to the review team members. The following are typical roles:

Review leader	Takes the leadership role, guides meetings, helps to control the review process, and makes sure all key issues are covered. During meetings, the leader assists in judging the severity of errors.
Designer	Develops the material to be reviewed. The designer should not also be assigned the role of leader or reviewer.
Reader	Paraphrases the document during review meetings; during inspections, the reader reads the code or document line-by-line. A reader is not assigned for all reviews.
Recorder	Writes down comments and corrections during a review meeting. Documents action items.
Reviewers	Perform the actual review and present their findings during the review meeting.

On small teams, the same person may act as review leader and recorder. When the material is familiar and not overly complex, a small team with well-defined roles and tasks can be effective. All participants except the designer act as reviewers in addition to other roles.

Since the review leader plays a major role in ensuring a successful, impartial, and accurate outcome, many projects in industry use dedicated, full-time review leaders to help ensure consistency and continuity. Other projects rotate the review leader duties, with more experienced people taking the duties initially.

In addition to assigning roles, the review leader determines a number of review tasks to be completed and assigns these tasks to various participants. For example, the leader may determine that the review material must be studied for the following: proper interfaces, adequacy of documentation, and adherence to standards. The reviewers must give special attention to these details when conducting individual reviews and report their findings during the review meetings.

The appropriateness of particular assigned tasks depends on the material to be reviewed and the level of design maturity.

Provide guidelines

The review leader provides guidelines on the appropriate amount of individual review time, types of analyses to be performed, meeting times, and meeting pace. These guidelines vary with the size, complexity, and type of material (e.g., requirements, circuit schematics, test plans).

NOTE: Individual projects usually have overall project guidelines. These will include design review guidelines.

Guidelines for analyses to be performed. Types of analyses to be performed during individual review will vary with project and design maturity. See Chapter 3, Application.

Guidelines for individual review time. For some material, such as software code, the appropriate individual review time is fairly easy to quantify because there is an accepted unit of measure and historical data relating individual review time to the number of errors detected. For example, for code inspections, a useful metric is the number of hours of individual review per lines of code.[4]

For material not easily quantified, such as complex circuit boards, the review leader or other expert may estimate the appropriate individual review time. The estimate should include enough time to use analytical tools such as those available with ComputerAided Design (CAD).

Use historical data, if available, to improve individual review time estimates. It is not easy to quantify estimates; the quality of the review and analysis is more important, at any rate.

For example, to review a complex six-layer circuit board, some project teams recommend forty to eighty hours individual review time to inspect the hardware functionality and initialization routines, and ten to forty hours to inspect the electrical timing, interfaces, connectors, etc. Appropriate times for other types of inspection (e.g., test, interfaces, or certification) range from ten to eighty hours.

---

[4]Studies using this metric at AT&T conclude that one hour review for each fifty lines of code is effective, one hour per one hundred lines is marginal, and one hour for more than one hundred lines is questionable.

Meeting time and pace. Plan meetings that best fit the needs of the material to be reviewed. Schedule meetings and distribute material in advance to allow individual review. Two-hour sessions are best; longer sessions risk loss of concentration. Schedule enough two-hour meetings to avoid having to cover too much material at too fast a pace. The more complex the material, the more likely it is that multiple two-hour meetings will have to be scheduled.

Select and use entry criteria

The review leader and the designer select entry criteria to specify when the material is ready for a review. The review leader and the designer check to make sure the entry criteria are satisfied; only after the criteria are satisfied is a review meeting scheduled. Entry criteria vary with the nature of the project and design maturity. A review should be held when it is felt the material is ready for review. They are usually based on design rules that the project team has established. The design rules are a concise set of important criteria for good design practice. Areas of consideration include: logic, interfaces, device and physical design, analog, test, manufacture, and reliability.

Examples of entry criteria for hardware:

Review or Inspection Type	Entry Criteria
Electrical design of circuit card or module	Worst-case timing and voltage parameters set Maximum junction temperatures set at 110°C
Physical design of circuit card or module	Prototypes' performance tested Process technology assessed Device availability listed
Structural design of enclosures for military electronics	Minimum allowable resonant frequency determined Trade studies done on overall weight and shock safety
Design of cylindrical pressure vessels for underwater systems	Failure criteria for O-rings set Producibility analyzed

Select exit criteria

The project team selects exit criteria according to the review type and project standards. Exit criteria help the review team determine in Step 6 whether the material being reviewed is acceptable.

Examples of exit criteria for hardware:

Review or Inspection Type	Exit Criteria
Electrical design of circuit card or module	Timing and voltage parameters satisfy worst-case requirements Junction temperatures are under 110°C
Physical design of circuit card or module	Performance meets requirements Consistent with manufacturing process (producible) Device availability fits schedule
Structural design of enclosures for military electronics	Minimum resonant frequencies are within allowable limits Trade study results presented
Design of cylindrical pressure vessels for underwater systems	Failure criteria meets requirements Pressure vessels are producible within cost

In the individual reviews (Step 3), reviewers check for adherence to exit criteria. During the review meeting, the review team makes sure the outcome agrees with the exit criteria.

Step 2: Familiarize

Have an overview meeting

The review leader holds an overview meeting. In this early meeting, the designer presents complex or unfamiliar material to the review team members. Overview meetings may have a wider audience than reviews since their purpose is to educate or bring people up to date, not to find errors or technical flaws. Managers may be included in the overview meeting to ensure their support of personnel assigned to the review team. The amount of time and effort required of reviewers must be made clear to both management and team members.

With especially complex and detailed material (e.g., schematics for a new VLSI device), consider a second overview meeting in which the material is presented in detail to make sure the team understands the intended functions. Then the team can agree on how much individual review time will be needed, how much material should be evaluated in each session to ensure the appropriate pace, and how this additional time will fit into their existing workloads. Communicate these agreements to all review team members and management.

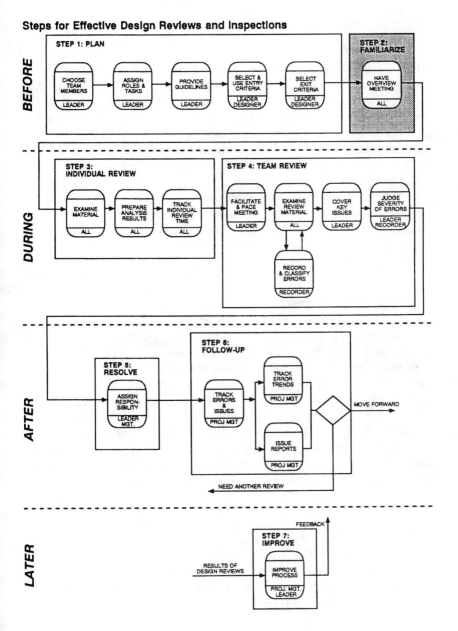

Steps for Effective Design Reviews and Inspections

Step 3: Individual Review

Examine material

All members of the review team use the guidelines (provided in Step 1) on appropriate individual review times, effort, and analyses while studying the material to find as many errors, areas of risk, and design flaws as possible. The review team also assesses the level of design maturity. The most effective reviews occur when all team members have spent adequate time in individual review. Better individual review also leads to an increased number of errors detected and may also lead to better design alternatives.

If the schedule prevents rigorous examination of all the material, examine the problem areas thoroughly, rather than all the material superficially. Assign each person a specific part of the material to examine thoroughly, concentrating on the tasks and types of errors assigned. Collectively the team will find most of the errors.

Prepare analysis results

Analysis tools such as Computer-Aided Design (CAD) tools should be used wherever possible in the individual reviews. Reviewers are likely to do this analysis individually in their own work space. Example of such analyses are: tracing circuit layout, checking routes and paths, simulating various functions, performing thermal analysis, performing mechanical analysis, checking physical design, and testing events. Results from CAD tools can be applied to various review tasks to allow a great deal of flexibility in using "what-if" and "sanity check" analyses of related design issues. Individual review time may not necessarily be reduced. The benefit is the ability to perform more complete analyses. Interpretation, understanding, and reporting of results are often more complete.

Track individual review time

Each team member should track the time they spend on various review tasks in individual review. The review leader keeps track of total time, the size of the project, and the complexity of the review material. This information will be used later in a process measures (Step 7: Improve).

Step 4: Team Review

Facilitate discussion and pace meeting

Team review occurs in the review meeting. The review leader brings meeting objectives, the agenda, guidelines, analysis results, and material to be

Steps for Effective Design Reviews and Inspections

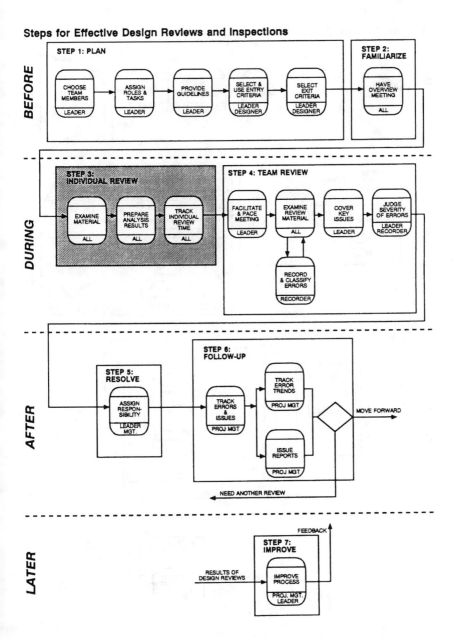

reviewed to the meeting. The leader uses these materials to help keep the meeting on track, address all issues, and set the appropriate pace.

The review leader should encourage all participants to focus on issues to resolve, rather than on justifying decisions, arguing about particular approaches, or criticizing individuals. The review leader sets an

Steps for Effective Design Reviews and Inspections

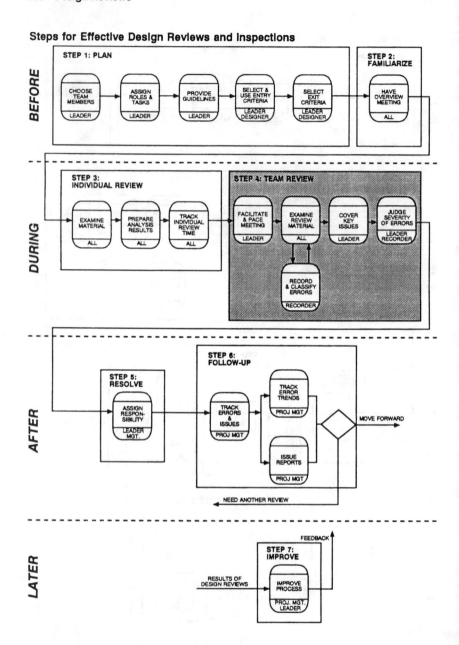

example for the other participants by not arguing or criticizing, and by arbitrating disputes, keeping the discussion on track, and deferring the hunt for solutions until after the meeting.

A successful review meeting produces a clear set of detected errors and issues to be addressed.

Examine material

Examine the material and/or analytical results at the pace set in the guidelines. The appropriate pace varies with the complexity, familiarity of the material, and the project's error history—in much the same way as a driver's speed varies with familiarity with the route and weather.

Cover key issues

The review leader also makes sure all key issues are covered. An objective approach to finding and listing errors will pay off much more than attempting to hide them for the sake of keeping to the schedule and avoiding blame for mistakes.[5]

Record and classify errors and issues

During the review meeting, the recorder writes down issues and classifies errors by type and class using the format prescribed by the guidelines. Type refers to the general category: e.g., interface, documentation, layout, testability. Class refers to whether the error indicates something is missing, wrong, or extra. Some errors are more difficult to categorize. For example, a logic problem on a device may be due to a failure to specify requirements properly.

The recorder creates the list of issues, action items, and errors. The review leader uses the list of issues as a starting place for assigning responsibility (Step 5: Resolve). The review leader will also check the classification after the meeting to make sure it is consistent with the project's format and standards.

Judge the severity of errors

The recorder and review leader are responsible for using a preestablished scheme to classify the severity of errors and open issues. Different organizations use different severity assignment schemes. Below is a typical set of definitions errors and open issues.[6]

Critical defect: A critical defect is likely to result in hazardous conditions for individuals using, maintaining, or depending upon the product. It is also likely to prevent performance of the tactical function of the major item of which it is a part: ship, aircraft, tank, missile, or space vehicle.

---

[5]Freedman, D. P. and Weinberg, G. M. *Handbook of Walkthroughs, Inspections, and Technical Reviews.* Boston: Little, Brown, 1982. Contains details on effective meetings.

[6]Paraphrased from MIL-STD-105, *Sampling Procedures and Tables for Inspection by Attributes.*

Major defect: A major defect is likely to result in failure or a material reduction of the usability of the product (as opposed to preventing performance of the major item, as above).

Minor defect: A minor defect is not likely to reduce the usability of the unit. It may also be a departure from established standards having little bearing on the use of the unit.

Open issue: May be used temporarily if the severity cannot be assigned without additional investigation.

Another classification system uses *Criticality 1* to refer to defects that will cause loss of life or mission failure and *Criticality 2* for defects that involve only a portion of the mission.[7]

Examples of error type and class:

Review Type	Description of Error	Type	Class	Severity	Reason
Hardware	Qualify EX1F with read.	Logic	Wrong	Major	Input gate missing input and circuit would not work as required.
	IC96 timing is wrong.	Logic	Wrong	Major	Incorrect timing will cause malfunction.
Firmware	Provide integration ratios for the alarm.	Requirements	Missing	Major	Requirement not met.
	Say where tolerances can be found.	Document	Missing	Minor	Missing information was not part of requirements.
	What should data link be initialized to?	Requirements	Missing	Open Issue	More information needed.
Software	Change "3 sec" to "cannot exceed 2.5 sec."	Document	Wrong	Minor	Incorrect information; does not affect requirements.
	Add "digroup is added & not restored."	Functionality	Missing	Major	Missing function causes ambiguity in low-level design and a possible coding error; would cause a change request.
	Specify "odd" range.	Document	Missing	Minor	Does not affect requirements.
	Add "translation to a new channel number."	Functionality	Missing	Major	Will generate a change request; does not meet requirements.

---

[7]MIL-STD-785, *Reliability Program for Systems and Equipment Development and Production.*

Step 5: Resolve

Assign responsibility

Review leader uses the list of classified errors and issues (from Step 4) as a starting point in this step. Then, with input from management and other team members, the review leader assigns responsibility for re-

Steps for Effective Design Reviews and Inspections

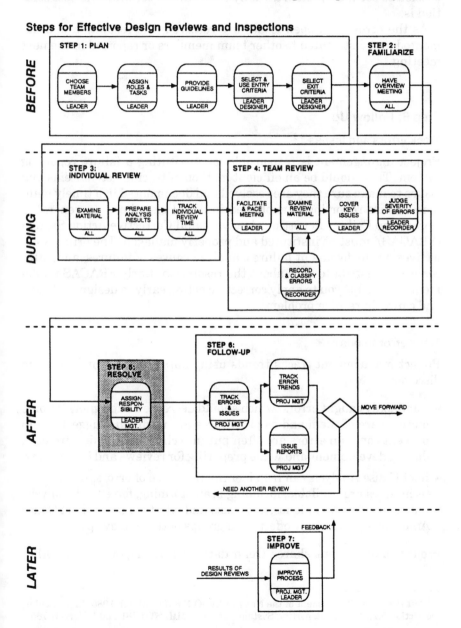

solving errors and open issues. The responsibility may be assigned to the designer, other team members, or an expert outside of the review. These assignments are formally documented and tracked. The review leader and the person assigned agree on an expected resolution date. Resolving open issues usually requires more information or discussion to see if a problem actually exists and if so, what the best solution is.

As the errors and open issues are resolved, the resolution is documented and distributed to other team members or reported in the next meeting.

Step 6: Follow Up

Track errors and issues

Project management is responsible for instituting a failure reporting system. This should be an efficient method of tracing action items and issues to make sure they are resolved and documented in a timely manner. Failure reporting is explained in detail elsewhere. In general, a closed-loop Failure Reporting, Analysis, and Corrective Action System (FRACAS)[8] must be instituted and properly managed. The primary objectives are to document failures, analyze causes of failures, and, after corrected, distribute data about the resolution. Early FRACAS implementation helps you to apply corrective action early in design evolution and can help resolve problems.

Track error trends

Project management tracks trends using data from recent reviews to discover:

- Commonly caught errors, which should decrease as designers become more aware of them and as procedures are modified. Suggestions to prevent common errors are often put into checklists, which are used during development and while preparing for reviews and inspections.

- Root Cause Analysis can help identify the cause of errors. A variety of techniques are available, including brainstorming, force field analysis, critical incident techniques, and cause-and-effect diagrams. Root Cause Analysis is particularly effective in groups of three to five people.

Project management shares trend data with management and other

---

[8]*Best Practices,* Department of the Navy (NAVSO P-6071), March 1986. In particular, see section 5.2, "Failure Reporting System." See also MIL-STD-781 and MIL-STD-785.

Steps for Effective Design Reviews and Inspections

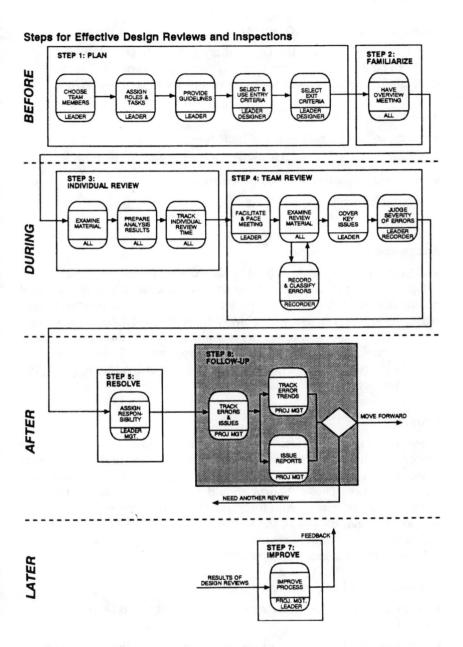

project members to explore possible causes, solutions, and preventive measures. Sharing trend information also improves communication among project members and other groups.

Error traceability should be emphasized throughout design and development to prevent design ambiguity and prevent common errors.

Examples of Errors Often Overlooked

Classify errors to show the types likely to be detected early and those likely to be detected in later phases. Vigorous reviews and inspections throughout the design stages help prevent these common errors:

- neglecting extreme cases
- failing to consider all paths of logic flow
- failing to check if plans for error recovery meet all the requirements
- using variables or data inconsistently
- making incorrect assumptions

An example of error trends

The graph below[9] shows error trends from a typical large system development project. Notice that the first release had fewer faults found during the initial phases of the project and more in later phases. In contrast, the second release, much larger than the first, found more errors early in the project and fewer in later phases. These data indicate that if more emphasis is placed on finding errors early, fewer will become evident in later phases. This information provided valuable input into future project planning.

Issue reports

Prepare summary reports for management and others. The review leader usually writes reports on particular review meetings; project management usually writes the summary reports on the results of a series of reviews. These reports should include error summaries and trends. Reports can be counterproductive if they focus on individual mistakes. When complete, the summary reports are circulated to project members and management.

Include in the summary reports:

- Assessment of design maturity.
- Summaries of errors detected by type, severity, and by subset of features.
- Outcome of reviews: percent accepted, accepted with minor revisions, not accepted and why.
- Analysis of the trends of error types and severity.

---

[9]American Telephone and Telegraph, New York, NY. 1989.

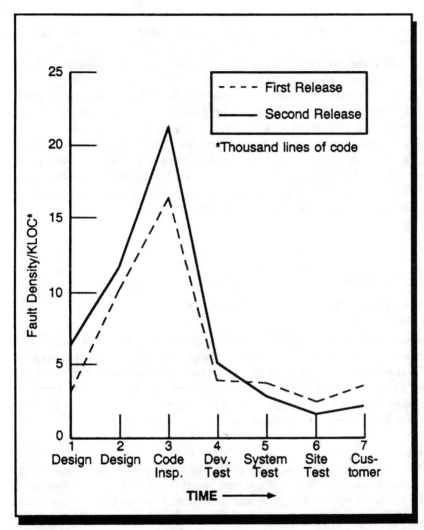

Fewer faults were found during the initial phases of the first release; more in later phases. The second release, where more errors were found earlier, had fewer faults in later phases.

- Suggestions for additional training, clarification of design review guidelines, and adjustment of schedules to allow additional individual review and meeting time.

Is another review needed?

The review team compares exit criteria and meeting objectives (from Step 1) with the meeting results to decide whether to accept the material as is with minor revisions, accept pending major revisions, or reject the material. (If rejected, the material is corrected and reviewed again.) The decision is based on the level of design maturity, the number of errors and their severity, the number of open issues, and whether all exit criteria and objectives have been met.

Some other reasons to hold another review: If a review finds fewer faults than expected for the size and complexity of the material and if individual review effort was too slow or the meeting pace was too fast, consider scheduling another meeting. Also consider meeting again if the number of errors found was higher than expected for material of that size and complexity. Errors are a good predictor of more errors.

If another meeting is to be held, the review leader issues appropriate notices to the review team and management.

Step 7: Improve

Use outcome and process measures for fine tuning

To assess the value of reviews and inspections, the review leader and project management use outcome measures as well as process measures. Use these measures to evaluate what parts of the process are working and what parts need to be improved. Outcome measures are the criteria against which process measures are evaluated.

The definitions and examples below help explain:

Outcome measures.   These are direct measurements of a product's effectiveness and quality. As such, they have a validity that is independent of other forms of measurement (such as process measures). The problem with using outcome measures alone is that they are not available until the review process is over. By then it is too late to make a significant change.

Examples of outcome measures.   Useful outcome measures include:

- cumulative faults found during system test, field tests, and after release

- number of changes or modification requests after the design is released for manufacturing

Process measures.   Process measures, on the other hand, give immediate feedback on how successful the process has been. They pre-

Steps for Effective Design Reviews and Inspections

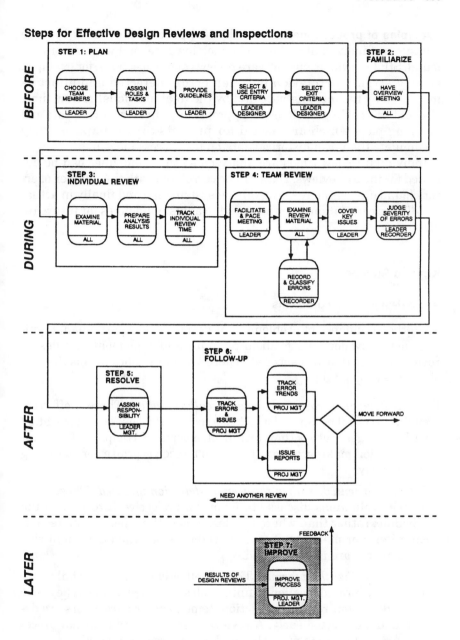

dict outcome measurements so that adjustments to the process can be made. Process measures must be used in conjunction with outcome measures and they must be carefully chosen so that data will correspond with the data of an outcome measurement you are interested in.

Examples of process measures. Useful process measures include measures of individual review time, meeting pace, and when errors are found. Studies evaluating software reviews and inspections, for example, show a strong correlation between errors detected and effort as measured by the meeting pace and by the total individual review time. These studies offer guidelines for appropriate individual review rate and meeting pace which are useful for projects to customize as they accumulate their own historical data on the process measures.

Individual review times are analyzed so that guidelines can be adjusted for future reviews. Data can be correlated with the number of errors found during the reviews, during system test, and finally in actual use. These correlations are then used to adjust guidelines for future reviews.

Keys to Success

Avoid design review problems

With the abundance of problems that design review participants typically note, how does one go about planning and implementing a design review process that will succeed? There are four major reasons that design reviews can fail:

- *Success-oriented, rather than technically-oriented.* These ineffective evaluations fail because of a lack of a technical focus. They are staffed with managers rather than technical experts. Participants fear punishment for making errors; they don't receive praise for detecting errors early.

- *Schedule-oriented, rather than error detection-oriented.* These ineffective evaluations discuss status of the project with respect to the schedule, rather than where the design is with respect to the technical criteria or maturity. Tradeoff studies, evaluation data, and risk assessments are not presented.

- *Informal, rather than formal.* These ineffective evaluations fail to follow formal procedures for defining roles and viewpoints. They also fail to document results and action items to make sure errors are detected and corrected early. A review team that uses informal procedures risks failure to meet the total system requirements.

- *Superficial, rather than detailed and thorough.* These ineffective evaluations fail because participants prepare too little before the meeting, cover too much material during the meeting at too fast a pace, or use vague or subjective evaluation and acceptance criteria.

Stress technical reasons for reviews and inspections

Have reviews and inspections for any of the following reasons:

- assess maturity level
- identify errors
- detect interface problems
- determine technical risks
- adhere to design for manufacture criteria
- ensure product adheres to standards and design rules

Choose a technically competent team

Choose a team of technically competent people, usually from the designers' peers. Their expertise should be in areas such as testing, producibility, interfaces, and maintenance. This diversity of experience ensures the team will find errors, inconsistent interfaces, and technical risks.

Follow formal procedures

Use the steps identified in this chapter to customize specific procedures for your project. People will then know exactly how to plan, familiarize and prepare before the meeting. They will know how to find problems during the meeting. And, they will know how to resolve errors, follow up after the meeting, and use the results to improve the product and the review and inspection process.

Checklists help designers follow design rules and avoid common errors. Entry and exit criteria ensure standards are met before the material goes to the next development phase. Specific guidelines for individual review and pace increase the chance of detecting errors early when they are easy to fix.

Obtain management commitment and support

Encourage potential reviewers and inspectors to attend courses. Have tutorials and refresher courses within the project to communicate specific project standards and guidelines.

To achieve the maximum benefits from the training, establish a management structure that supports the design review process. Include time and resources for the design reviews in project schedules. Make sure these are not the first to be sacrificed when deadlines approach. Managers should also praise people for finding errors and suggesting ways to improve the processes or products. Results from reviews and inspections should not be part of individual performance appraisals.

Track measures of success

Perhaps the most important way to achieve effective reviews is to *institute and track measures to success.* Use outcome and process measures to see what parts of the process are working and what parts need to be improved. Process measures have no inherent validity or value in themselves but do give more immediate feedback and thus allow fine tuning for continual improvement. As described in Step 7, the usefulness of process measures depends on their correlation with outcome measures.

Using these keys to success, following the procedures in this part, referring to the Templates and *Best Practices*...all will help ensure that effective design reviews are the norm rather than rare events.

3

Application

This chapter is an application of how to use design reviews within the design process. It outlines data packages used to prepare for the reviews and provides examples of key factors and metrics which may be helpful. This chapter is not meant as a prescription for all projects, but rather as a guide.

Reviews occur at many points during the design process to assess design maturity. A design is considered mature when evaluation and verification show it meets all requirements, e.g., performance, reliability, maintainability, and supportability at a particular stage of development. The design process for complex programs may be divided into three stages: initial, intermediate, and final stages. Reviews occur during each of these stages. On government contracts, internal reviews usually precede the contract-mandated reviews.

Design reviews should be conducted as described in the Procedures section. However, each review must be tailored to the topics, materials, and analyses relative to the level of design maturity. Subcontractors may follow the same sequence of design reviews. They may also participate in the prime contractor's design reviews and in contract-mandated reviews.

Approach

Figure 1 provides an overview of a typical design process after the needs and mission profile have been developed. The figure depicts the stages of the design process and the key tasks and activities for each stage. Time is represented from left to right. Internal and contract-mandated reviews are positioned in the figure corresponding to associated key tasks and activities. The figure does not reflect the iterative nature of tasks and reviews within each stage of the design process. However, feedback can occur between any stage or activity.

INITIAL STAGE

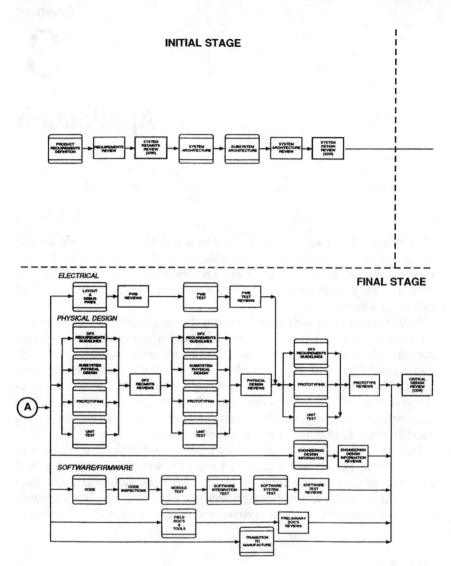

Figure 1

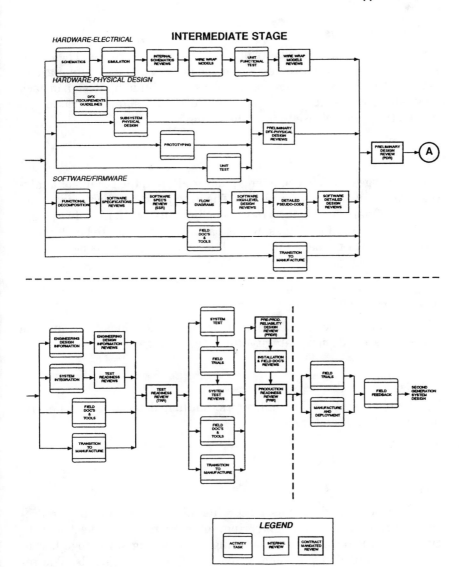

Figure 1 (*Continued*)

Sequence of reviews

The sequence of design reviews parallels the design process. The initial reviews cover the requirements and the system and subsystem architecture. The intermediate reviews cover the preliminary design of each subsystem. The final reviews cover the detailed design, integrated system testing, and release to manufacturing.

Data package

Data packages include documents, materials, and the results of analyses that are examined at a review. The data packages for the intermediate and final stage reviews have been broken down for hardware and software. The data packages in this part are not meant to be all inclusive. Each data package must be adjusted to fit the needs of the individual review.

Key factors

The key factors sections are examples of key questions that must be asked to assess design maturity during each stage.

Metrics

These are examples of key indicators that help determine whether the design process is on target.

Initial Stage

Description

After the needs and the mission profile have been established, representatives from systems engineering, software, firmware, electrical design, physical design, manufacturing, and testing work together to generate detailed requirements that satisfy customer needs. The requirements provide a baseline for evaluating risks and modifications as the design matures.

The architecture is developed after the requirements have been baselined. Representatives from hardware and software generate and explore ideas for allocating the main functions (e.g., control, timing, data processing) to system and subsystems. Design conflicts may be resolved using historical data from similar projects, data from tradeoff studies, and feasibility studies of trial allocation of requirements to hardware and software. When a suitable system architecture emerges, the process is repeated for each subsystem.

Scheduling

Reviews are scheduled when the requirements, system architecture, and subsystem architecture are ready to be compared to entry criteria. Figure 2 and Table 2 show the scheduling of the reviews during the initial stage.

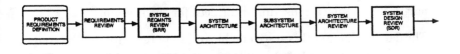

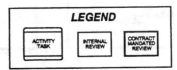

LEGEND

| ACTIVITY TASK | INTERNAL REVIEW | CONTRACT MANDATED REVIEW |

Figure 2   Initial stage.

TABLE 2 Initial Stage Reviews

Review	Type	Participants may include representatives from:	Objectives
System Requirements Reviews	Internal	Systems engineering, program management, hardware, and software	Review requirements and specifications to ensure they are clear, concise, accurate, and verifiable against customer needs.
System Requirements Review (SRR)	Contract-mandated	Systems engineering, program management, hardware, software, and the customer	Review the requirements and specifications with the customer to obtain agreement.
System Architecture Reviews	Internal	Hardware, software, and systems engineering	Review system partitioning and various alternative architectures.
Subsystems Architecture Reviews	Internal	Hardware, software, and systems engineering	Review subsystem partitioning and various alternative architectures.
System Design Review (SDR)	Contract-mandated	Systems engineering, program management, hardware, software, and the customer	Review the high-level system design and requirements allocations to obtain customer agreement.

Data package

The data package at this stage includes:

- functional requirements
- interface requirements
- design requirements
- environment constraints
- partitioning, allocations, and interfaces for the system.

The data package at this stage emphasizes requirements and the system and subsystem architectures, and thus is not broken down into hardware and software as later stages are. Examples are shown in Figures 3 and 4.

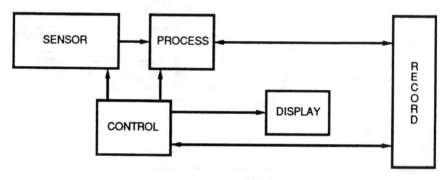

SUBSYSTEM PARTITIONING

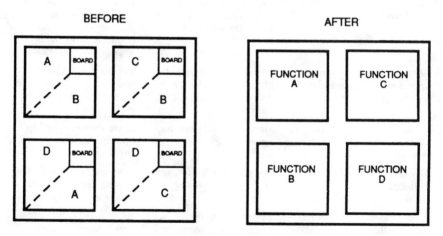

SUBSYSTEM PARTITIONING

Figure 3 Initial stage example: Data package. Illustration of system and subsystem partitioning into functional modules. (Adapted from J. W. Priest, *Engineering Design for Producibility and Reliability*. Reprinted by permission of Marcel Dekker. New York, NY. 1988.)

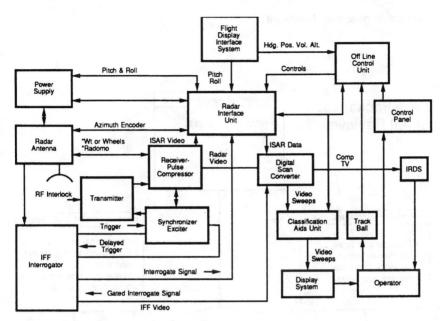

Figure 4 Initial stage example: Data package. Example of a functional block diagram for a radar system. (From J. W. Priest, *Engineering Design for Producibility and Reliability.* Reprinted by permission of Marcel Dekker. New York, NY. 1988.)

Data packages: initial reviews

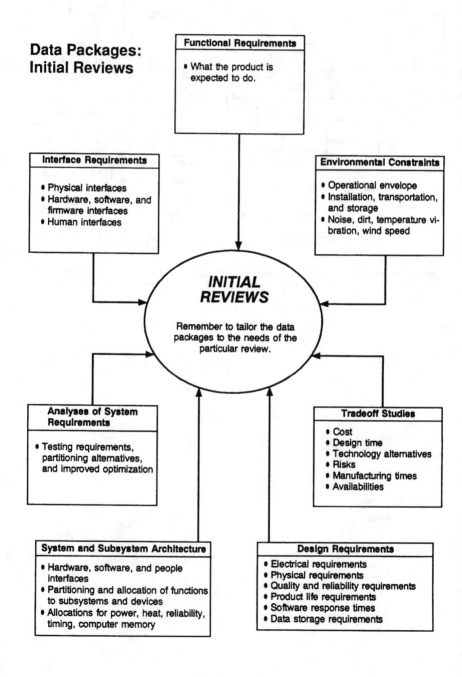

**Data Packages:
Initial Reviews**

Functional Requirements

• What the product is
  expected to do.

Interface Requirements

• Physical interfaces
• Hardware, software, and
  firmware interfaces
• Human interfaces

Environmental Constraints

• Operational envelope
• Installation, transportation,
  and storage
• Noise, dirt, temperature vibration, wind speed

**INITIAL
REVIEWS**

Remember to tailor the data
packages to the needs of the
particular review.

**Analyses of System
Requirements**

• Testing requirements,
  partitioning alternatives,
  and improved optimization

Tradeoff Studies

• Cost
• Design time
• Technology alternatives
• Risks
• Manufacturing times
• Availabilities

System and Subsystem Architecture

• Hardware, software, and people
  interfaces
• Partitioning and allocation of functions
  to subsystems and devices
• Allocations for power, heat, reliability,
  timing, computer memory

Design Requirements

• Electrical requirements
• Physical requirements
• Quality and reliability requirements
• Product life requirements
• Software response times
• Data storage requirements

Key factors—initial stage

■ What percentage of requirements are traceable to the mission profile?

■ What are the results of the environmental envelope study with expected conditions for operations, installation, and support specified, e.g., temperature, pressure, moisture?

■ Do system and subsystem representatives agree on the allocations and interpretations of requirements?

■ Do the tradeoffs and feasibility results indicate the allocations to hardware or software are realistic?

Examples of these key factors are shown in Figure 5.

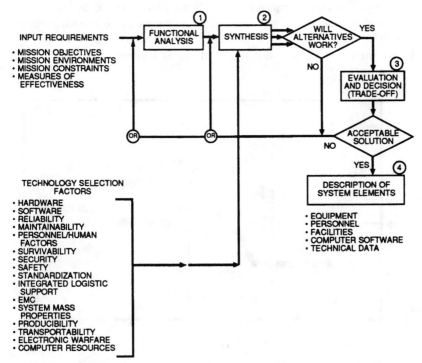

Figure 5 Initial stage example: Key factors. Overview of the systems engineering process showing how system elements are traceable to input requirements and tradeoff decisions. (From Defense System Management College. *System Engineering Management Guide*. MDA 903-82-C-0339 Fort Belvoir, VA. October 3, 1983.)

Metrics—initial stage

- How much development effort is spent before the requirements are baselined?
- How stable are the requirements over time?
- Does actual work spent on defining requirements and architecture agree with the estimates?

Examples of these metrics are shown in Figures 6 and 7.

Intermediate Stage

Description

The outcome of the intermediate stage of the design process is to produce preliminary prototypes of the hardware and software. This stage involves the concurrent development of the hardware and software, based upon the requirements and architecture derived in the initial stage. Coordination is of particular importance due to the simultaneous development of software and hardware. The architecture and interfaces must be carefully documented under change control.

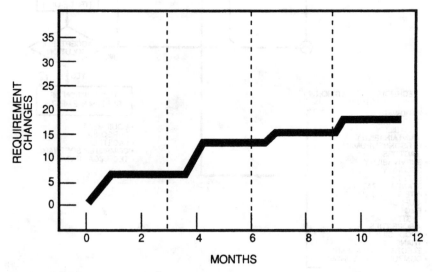

Figure 6 Initial stage example: Metrics. Cumulative changes in requirements over time. (American Telephone and Telegraph. New York, NY. 1989.)

REQUIREMENTS AVAILABILITY

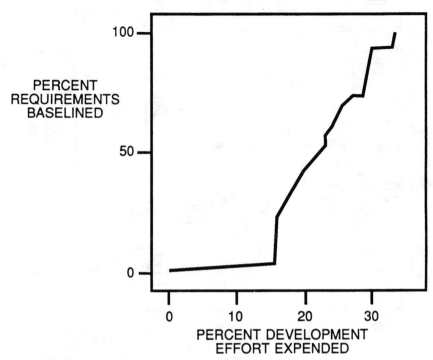

Figure 7 Initial stage example: Metrics. Metric showing the extent to which development effort tracks the baselining of requirements. (American Telephone and Telegraph. New York, NY. 1989.)

Scheduling

The purpose of reviews during this stage of the design process is to verify the preliminary design against requirements. Reviews are held when schematics, sketches, flow diagrams, and early prototypes are ready to be compared against the entry criteria. Figure 8 and Table 3 show reviews typically scheduled in the intermediate stage.

Data package

The data package for reviews during this stage is broken down for hardware and software (see Figures 9, 10, 11, and 12). Remember to tailor the data package to the needs of the particular review.

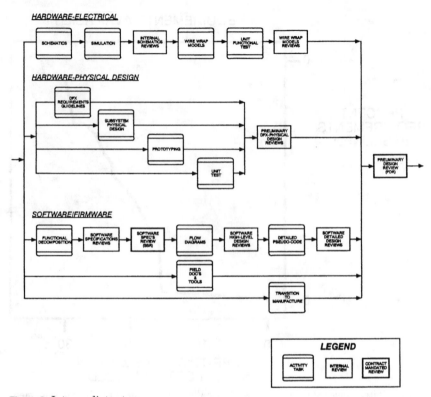

Figure 8 Intermediate stage.

TABLE 3 Intermediate Stage Reviews

Review	Type	Participants may include representatives from:	Objectives
Schematics Reviews	Internal	Hardware, software, systems engineering, and manufacturing	Review the schematics, layouts, routing, and preliminary functional simulations.
Software Specifications Reviews	Internal	Hardware, software, systems engineering, and project management	Review the software requirements and specifications against customer needs.
Software Specifications Review (SSR)	Contract-mandated	Hardware, software, systems engineering, project management, and the customer	Review the software requirements for clarity, consistency, and completeness to obtain customer agreement.
Software High-Level Design Reviews	Internal	Hardware, software, systems engineering, and manufacturing	Review the functional flow, top-level structure, development methodology and tools.

TABLE 3 Intermediate Stage Reviews (Continued)

Review	Type	Participants may include representatives from:	Objectives
Wire Wrap Models Reviews	Internal	Hardware, software, systems engineering, and manufacturing	Review individual signal processing, logic flow chains, preliminary sneak circuit and worst case analyses.
Preliminary DFX*-Physical Design Reviews	Internal	Hardware, software, systems engineering, and manufacturing	Review preliminary plans, prototypes, and guidelines.
Software Detailed Design Reviews	Internal	Hardware, software, systems engineering, and manufacturing	Review detailed design, interfaces, pseudocode, tests, analyses, and draft manuals.
Preliminary Design Review (PDR)	Contract-mandated	Hardware, software, systems engineering, project management, subcontractors, and the customer	Review hardware and software prototypes for functional flow, requirements allocations, thermal analysis, power, packaging, reliability, maintainability, and manufacturability.

*Note: DFX means Design for X where X may stand for the product's manufacturability, testability, safety, installability, etc.

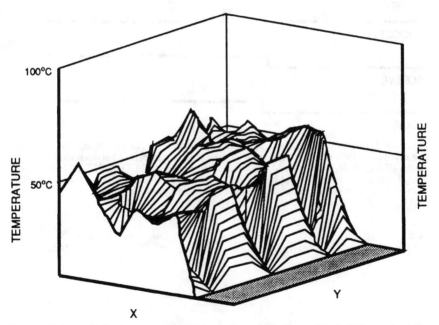

Figure 9 Intermediate stage example: Data package— hardware. Typical thermal profile of a circuit board showing temperatures taken from x-y locations on the board. (From J. W. Priest, *Engineering Design for Producibility and Reliability*. Reprinted by permission of Marcel Dekker. New York, NY. 1988.)

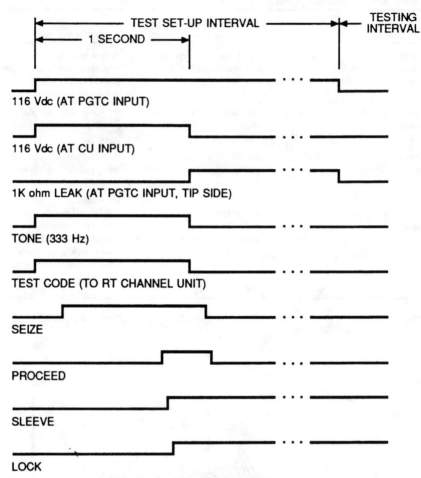

Figure 10   Intermediate stage example: Data package— hardware. Illustration of timing analyses which show the sequence of events at the interface between channel test units. (From M. M. Luniewicz, J. W. Olson, and K. E. Stiefel. "The SLC 96 Subscriber Loop Carrier System: Channel Bank," *AT&T Technical Journal*. Reprinted by permission of American Telephone and Telegraph. New York, NY. December 1984.)

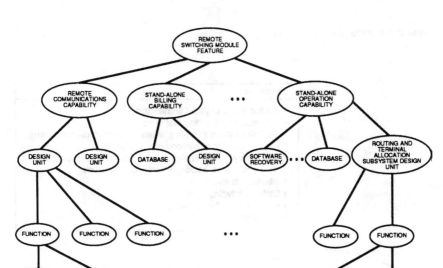

Figure 11 Intermediate stage example: Data package— software. Illustration of a successive decomposition of features into software modules, capabilities, units, functions, and lines of code. (From J. T. Beckett, D. G. Dempsey, E. M. Prell, and A. Villarosa. "5ESS™ Switching System Software: Methods for Managing a Large Software Project," *AT&T Technical Journal*. Reprinted by permission of American Telephone and Telegraph. New York, NY. January 1986.)

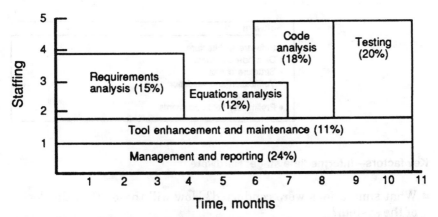

Figure 12 Intermediate stage example: Data package— software. Percentage of staff allocated to various functions to carry out a software verification and validation plan. (From B. W. Boehm. *Software Engineering Economics*. Reprinted by permission of Prentice-Hall, Inc. Englewood Cliffs, NJ. 1981.)

Data packages: intermediate

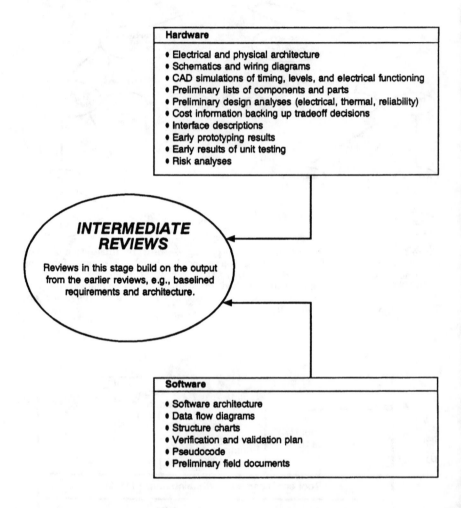

Key factors—Intermediate stage: Hardware

- What simulations were conducted? How will these affect the design of the system?
- What redundancies are required as a result of preliminary prototyping? How will these affect the overall system operation? What precautions were taken to avoid having redundancies substituted for good design?
- What plans have been developed to identify technology risks, production risks, and critical materials?

- What are the results of preliminary tests to verify the interfaces and timing requirements?

Examples of these key factors are shown in Figures 13 and 14.

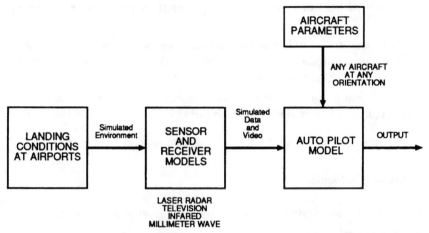

Figure 13 Intermediate stage example: Key factors. Illustration of a simulation model for a landing system. (From J. W. Priest, *Engineering Design for Producibility and Reliability*. Reprinted by permission of Marcel Dekker. New York, NY. 1988.)

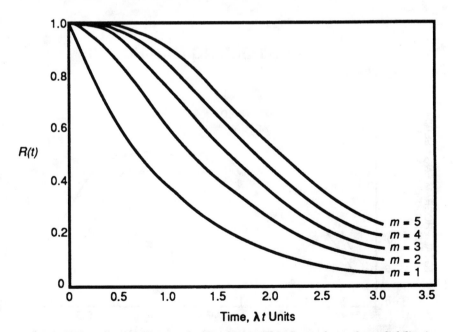

Figure 14 Intermediate stage example: Key factors. This figure shows that reliability increases as the number of redundancies (*m*) increases. (From ARINC Research Corporation. *Reliability Engineering*. Reprinted by permission of Prentice-Hall, Inc. Englewood Cliffs, NJ. 1964.)

Key factors—Intermediate stage: Software

- What exception conditions, violation limits, and capacity limits have been taken into account?
- What modifications are needed as a result of preliminary prototype testing?
- What percentage of the software can be reused from earlier systems? How many standard and off-the-shelf software components are estimated for the final system?
- Do the results of the software system test verify performance, reliability, and other critical requirements?

Examples of these key factors are shown in Figures 13 and 14.

Metrics—Intermediate stage

- Actual versus planned technical hours of design engineering
- Actual design to cost versus goal

Examples of these key factors are shown in Figures 15 and 16.

ENGINEERING MAN-HOURS

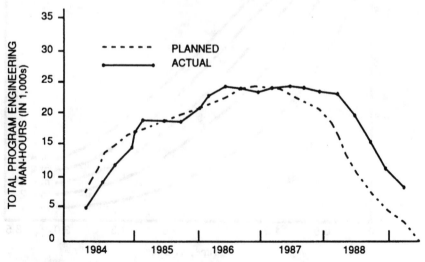

Figure 15 Intermediate stage example: Metrics. Curves showing relationship between actual and planned staff months of development effort. (Honeywell, Inc. Hopkins, MN. 1989.)

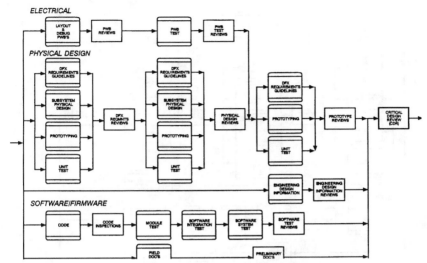

Figure 16 Intermediate stage example: Metrics. Tolerance limits for comparing actual costs with the design-to-cost (DTC) goal. [From H. K. Burbridge. "Life-Cycle Costs." in L. Walsh, R. Wurster, and R. J. Kimber (eds.). *Quality Management Handbook*. Reprinted by permission of Marcel Dekker. New York. New York, NY. 1986.]

Final Stage

Description

At this stage, early models of software and hardware are available. Most of the system's development and validation occurs at this stage. The goal is to develop and test the prototypes and then integrate and evaluate the system and prepare for production. The design should be stable when system integration begins.

System integration verifies the interfaces of the hardware and software of the system being built. Each separate build represents a subset of the entire system; integration continues until a representation of the entire operational system exists.

System testing evaluates the product from the customer's perspective in as realistic an environment as possible. The field trial verifies the technical performance, transportability, installation, and supportability in an actual operational environment. It also tests system documentation (including the installation and troubleshooting procedures).

Scheduling

Figure 17 and Table 4 show the scheduling of reviews during the final stage.

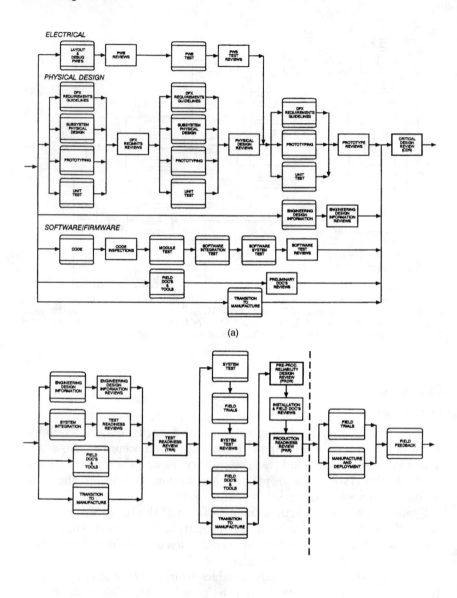

(a)

(b)

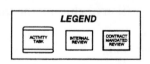

Figure 17 Final stage.

TABLE4 Final Stage Reviews

Review	Type	Participants may include representatives from:	Objectives
DFX Requirements Reviews	Internal	Hardware, software, systems engineering, manufacturing, and project management	Review DFX guidelines and design rules.
Printed Wiring Board (PWB) Reviews	Internal	Hardware, software, systems engineering, and manufacturing	Review individual signal processing or logic flow chains, preliminary sneak circuit analyses, worst case analyses and DFX guidelines.
Software Code Inspections	Internal	Software and possibly subcontractors	Review the code for compliance with requirements, interfaces, standards, error recovery, and exceptional handling procedures.
Printed Wiring Board Test Reviews	Internal	Hardware, software, systems engineering, and manufacturing	Review functional, thermal, electrical, power, reliability analyses, against design requirements and DFX guidelines.
Physical Design Reviews	Internal	Hardware, software, systems engineering, manufacturing, test, and subcontractors	Review assemblies and subassemblies against design requirements and DFX guidelines.
Prototype Reviews	Internal	Hardware, software, systems engineering, manufacturing, test, and subcontractors	Review functional, thermal, electrical, power, and reliability analyses, against design requirements and DFX guidelines.
Engineering Design Information Reviews	Internal	Hardware, software, systems engineering, manufacturing, test, and subcontractors	Review preliminary drawings against DFX guidelines.
Software Test Reviews	Internal	Hardware, software, systems engineering, project management, test, manufacturing, and subcontractors	Review the results of software module testing, integration testing, and system testing for compliance with requirements.
Preliminary Documentation Reviews	Internal	Hardware, software, systems engineering, manufacturing, documentation support group, and subcontractors	Review users' guides, installation manuals, and other field manuals against design requirements.

TABLE 4   Final Stage Reviews (Continued)

Review	Type	Participants may include representatives from:	Objectives
Critical Design Review (CDR)	Contract-mandated	Hardware, software, systems engineering, project management, test, manufacturing, subcontractors, and the customer	Review the working models of hardware and software to assess readiness for release to manufacturing.
Engineering Design Information Reviews	Internal	Hardware, software, systems engineering, project management, test, manufacturing, and subcontractors	Review the final drawings and data, component selection, component ordering, and manufacturing planning.
Test Readiness Reviews	Internal	Hardware, software, project management, systems engineering, test, manufacturing	Review the requirements and design changes, test plans and procedures, integration, and test cases.
Test Readiness Review (TRR)	Contract-mandated	Hardware, software, project management, systems engineering, test, manufacturing, subcontractors, and the customer	Review the requirements and design changes, test plans and procedures, integration, and test cases to obtain customer agreement.
System Test Reviews	Internal	Hardware, software, project management, systems engineering, test, manufacturing, and subcontractors	Review results of system testing and field trials including performance, functionality, and interfaces.
Preproduction Reliability Design Review (PRDR)	Contract-mandated	Hardware, software, project management, systems engineering, test, manufacturing, subcontractors, and customer	Review failure reports, failure analysis reports, corrective actions, and retests to assess whether the system's design is mature enough for transition to production.
Installation and Field Documentation Reviews	Internal	Hardware, software, project management, systems engineering, test, manufacturing, documentation support, subcontractors, and the customer	Review and finalize installation and users' guides based on feedback from integration and field tests.
Production Readiness Review (PRR)	Contract-mandated	Hardware, software, project management, systems engineering, test, manufacturing, subcontractors, and the customer	Assess whether to go ahead with full production. Review the transition plan, manufacturing plan, and list of critical parts, components, and equipment.

Data package

Tailor the data package (Figure 18) to the needs of the particular review.

Specified Transistor, Microcircuit, and Diode Stress Levels for Design			
Number of parts with "junction temperature" (°C)			
Total number of parts	≤110	≤100	≤70
387	387 (100%)	384 (99%)	360 (93%)

Figure 18 Final stage example: Data package. Analysis of the percentage of components achieving various junction temperatures. From J. W. Priest. *Engineering Design for Producibility and Reliability*. Reprinted by permission of Marcel Dekker. New York, N.Y. 1988).

Data packages: Final reviews

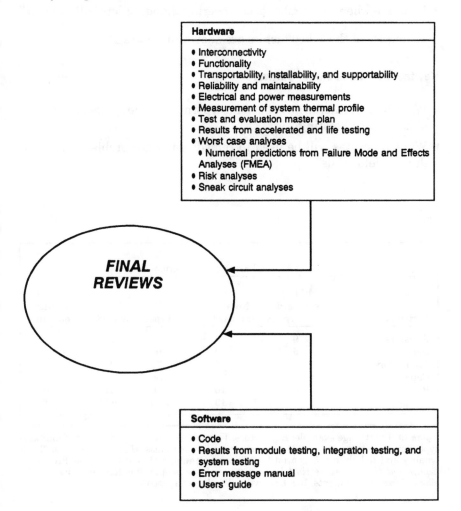

Hardware

- Interconnectivity
- Functionality
- Transportability, installability, and supportability
- Reliability and maintainability
- Electrical and power measurements
- Measurement of system thermal profile
- Test and evaluation master plan
- Results from accelerated and life testing
- Worst case analyses
  - Numerical predictions from Failure Mode and Effects Analyses (FMEA)
- Risk analyses
- Sneak circuit analyses

FINAL REVIEWS

Software

- Code
- Results from module testing, integration testing, and system testing
- Error message manual
- Users' guide

Key factors—Final stage: Hardware

- What are the results of worst case analyses done over the full range of temperature variation and power supply variations?
- What tests and analyses show the product is ready for production?
- What percentage of the components conform to military specifications? What percentage of the components will be obsolete and not available from their suppliers in two years? What percentage will be obsolete in five years?
- What analyses are focusing on high risk/low yield manufacturing processes or materials and on production planning, facilities allocation, producibility, tools, and test equipment?
- Estimate when the number of engineering change orders will stabilize?

Examples of these key factors are shown in Figure 19.

Key factors—Final stage: Software

- What improvements have been incorporated in the system as a result of adversarial testing?
- What processes will be used to remedy software problems detected during field trials?

	Achieved Part Standardization				
Part class	Military standard part types	Not military standard types	% Types	Total quantity	Military quantity
Resistors	9	0	100	993	100
Diodes	5	0	100	115	100
Transistors	7	2	78	106	95
Capacitors	4	1	80	1616	47
IC	0	10	0	166	0
Totals	26	13	66.7	2936	66.9

Figure 19 Final stage example: Key factors. This table shows the percentage of military-type components vs. not military components by part class. (From B. Grimes. "High Quality and Reliability: Essential for Military, Commercial, and Consumer Products." *Equipment Group Engineering Journal.* Texas Instruments, Inc. Reprinted by permission of Texas Instruments, Inc. Lewisville, TX. July-August 1983.)

- What systems are in place to monitor configuration control? Who controls when and how changes will be incorporated into the software?

Examples of these key factors are shown in Figure 19.

Metrics—Final stage

- Root cause analyses of faults
- Percent use of standard components
- Fault densities of errors introduced at various stages
- Results of reliability analyses (Mean Time Between Failures [MTBF], Mean Time to Repair [MTTR], FIT rates) and thermal analyses (maximum junction temperatures)

Examples of these key factors are shown in Figures 20–23.

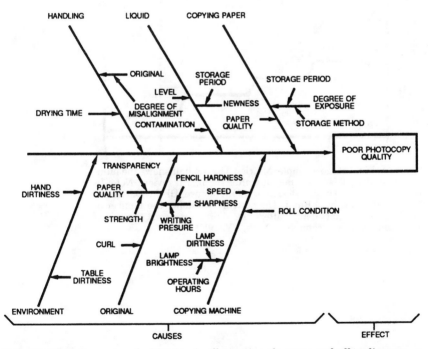

Figure 20 Final stage example: Metrics. An illustration of a cause-and-effect diagram—a tool used to determine the root causes or problems. (From AT&T Quality Steering Committee. Process Quality Management and Improvement Guidelines. Reprinted by permission of American Telephone and Telegraph Bell Laboratories. Indianapolis, IN. 1988.)

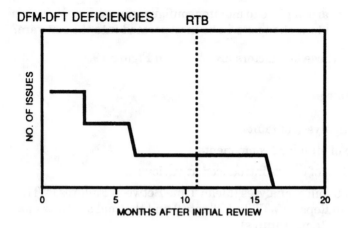

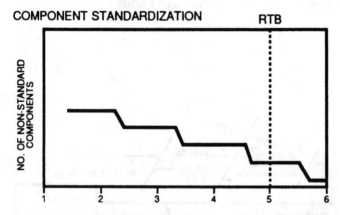

Figure 21 Final stage example: Metrics. The top half shows the number of DFM-DFT issues remaining after the initial review. Note: In the figure above, DFM = Design for Manufacture, DFT = Design for Test, RTB = Ready to Build. The bottom half shows the number of non-standard components after the initial review. (American Telephone and Telegraph. New York, NY. 1989.)

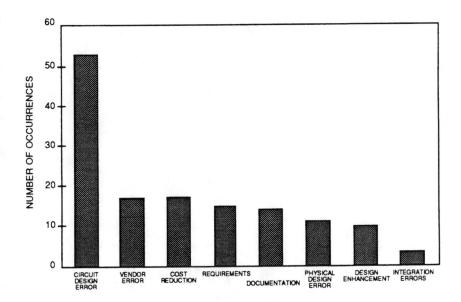

Figure 22 Final stage example: Metrics. Pareto analysis showing the frequency of occurrence of hardware faults. Data could be used in a root cause analysis. (American Telephone and Telegraph. New York, NY. 1989.)

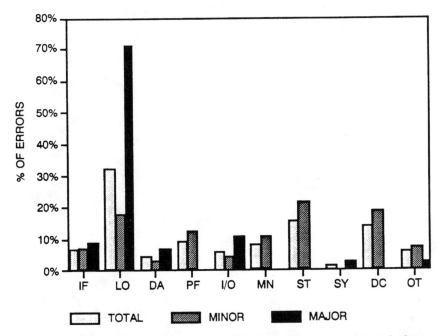

Figure 23 Final stage example: Metrics. Pareto analysis showing frequent types of software errors. Note: In the figure above, IF = Interface, LO = Logic, DA = Data, PF = Performance, I/O = Input/Output, MN = Maintenance, ST = Standards, SY = Syntax, DC = Documentation, OT = Other. (American Telephone and Telegraph. New York, NY. 1989.)

Summary

Design reviews assess design maturity and find problem areas and errors in specifications, requirements, detailed design, schematics, prototypes, lines of codes, etc. They are cost-effective—with savings of ten to one if errors are found in the design stage rather than the system test stage; hundred to one if errors are found in the requirements stage rather than at system test. Other benefits include increased quality, easier maintenance, and more timely deliveries. Employee benefits include improved feedback, less frustration, and increased job satisfaction.

Follow Best Practices

The benefits occur if projects follow the best practices for reviews and inspections. Examples of these best practices are:

- Obtain total management commitment and support for the reviews and inspections in the project schedules.
- Select a review or inspection team with technical expertise, including hardware, software, firmware, as well as test and manufacturing.
- Define roles for team members.
- Define specific entry and exit criteria for each stage of the development process.
- Classify and track errors according to type, class, and severity.
- Summarize results to find errors commonly caught by the team.
- Include in summary reports for management, summaries of the number of errors detected, analysis of errors, suggestions for training, fine tuning, etc.
- Track measures of success using process and outcome measures to fine tune the design review process for continual improvement.

Summary of Procedures

Effective design reviews do not happen by accident: They are the result of careful planning, following the step-by-step procedures outlined in this part, and adhering to the principles in *Best Practices* and the templates in *Transition from Development to Production*. The table below shows a summary of all of the steps presented in the Procedures chapter.

What	Who	Key Points
	Step 1: Plan	
Choose Team Members	Project Management Review Leader	■ Selects review leader. ■ Selects other team members who are technically competent peers of the designer. ■ Distributes a list of team member names to interested parties.
Assign Roles and Tasks	Review Leader	■ Determines the roles for individuals to fulfill during review. E.g.: Reader, recorder, and reviewer; all members except the designer and review leader have review duties as well. ■ Assigns review tasks to reviewers, E.g.: review for compliance with templates, proper interfaces, adequacy of documentation, adherence to standards.
Provide Guidelines	Review Leader	■ Define individual review time, types of analyses to be conducted, meeting times, and meeting pace. Review guidelines may have been previously developed. ■ Publish meeting schedule and agenda.
Set and Use Entry Criteria	Review Leader and Designer	■ Define criteria that must be met *before* review meeting is scheduled.
Set Exit Criteria	Review Leader and Designer	■ Define criteria that must be met for material to be acceptable. ■ Publish and distribute exit criteria to reviewers.

What	Who	Key Points
	Step 2: Familiarize	
Have an Overview Meeting	All Review Team Members	▪ Familiarize review team and management with the design goals, contents, and required effort.
	Step 3: Individual Review	
Examine Material	Reviewers	▪ Adhere to guidelines to review material individually. ▪ Develop a preliminary list of errors and issues to be discussed during the review meeting (Step 4: Team Review).
Prepare Analysis Results	Reviewers	▪ Examine results of analysis tools and simulations. ▪ Prepare results for use during the review meeting.
Track Individual Review Time	Reviewers	▪ Record and submit time spent in individual review to review leader.
	Review Leader	▪ Keep track of total time, project size, and complexity of material for use in process measures (Step 7: Improve).
	Step 4: Team Review	
Facilitate Discussion	Review Leader	▪ Bring review meeting objectives, agenda, guidelines, and material to be reviewed to meeting. ▪ Encourage all team members to participate.
Pace Meeting	Review Leader	▪ Use appropriate pace so that all issues are examined. ▪ Keep meeting on track.
Examine Material	All Reviewers	▪ Examine material at appropriate pace and depth.
Cover Key Issues	Review Leader and Reviewers	▪ Encourage all to contribute; make certain all issues are covered.
Record and Classify Errors and Issues	Recorder	▪ Write down issues raised during meeting; classify errors by type and class; submit to review leader.
	Review Leader	▪ Use list of issues as a starting place for assigning responsibility (Step 5: Resolve).
Judge the Severity	Review Leader and Recorder	▪ Assign *critical, major,* or *minor* severity to each error.

What	Who	Key Points
	Step 5: Resolve	
Assign Responsibility	Review Leader	■ With input from management, assign individuals to resolve errors and open issues. ■ Document and track assignments and resolution of errors and issues.
	Step 6: Follow Up	
Is Another Review Needed?	Review Leader	■ Compare criteria set in Step 1 to meeting outcome; determine if another meeting is required to address design review issues properly; publish meeting notice.
Track Errors and Issues	Review Leader	■ Implement a FRACAS to track errors and issues.
Track Error Trends	Project Management	■ Use automated tools to track trends in errors and report findings to help eliminate problems in future reviews. ■ Emphasize traceability throughout the project.
Issue Reports	Review Leader	■ Prepare reports on individual design review meetings.
	Project Management	■ Prepare summary reports for management and others.
	Step 7: After	
Use Outcome and Process Measures for Fine Tuning	Review Leader or Project Management	■ Use outcome measures to see what parts of the process are working. ■ Use process measures, in conjunction with outcome measures, to gauge the effectiveness of the process.

Summary of Application

The Application chapter of this part discusses initial, intermediate, and final stage design reviews. This section covers scheduling and sequence of reviews in each stage, as well as the required data packages and key factors to be considered.

Initial reviews: Requirements and architecture

- System Requirements Review (Internal)
- System Requirements Review (SRR) (Contract-Mandated)
- System Architecture Reviews (Internal)
- Subsystems Architecture Reviews (Internal)
- System Design Review (SDR) (Contract-Mandated)

Intermediate reviews: Preliminary subsystem design

- Schematic Reviews (Internal)
- Software Specifications Reviews (Internal)
- Software Specifications Review (SRR) (Contract-Mandated)
- Software High-Level Design Reviews (Internal)
- WireWrap Models Reviews (Internal)
- Preliminary DFX-Physical Design Reviews (Internal)
- Software Detailed Design Reviews (Internal)
- Preliminary Design Review (PDR) (Contract-Mandated)

Final reviews: Detailed subsystem design, system integration and test, release to manufacturing

- DFX Requirements Reviews (Internal)
- Printed Wiring Board (PWB) Reviews (Internal)
- Software Code Inspections (Internal)
- Printed Wiring Board Test Reviews (Internal)
- Physical Design Reviews (Internal)
- Prototype Reviews (Internal)
- Engineering Design Information Reviews (Internal)
- Software Test Reviews (Internal)
- Preliminary Documentation Reviews (Internal)

- Critical Design Review (CDR) (Contract-Mandated)
- Engineering Design Information Reviews (Internal)
- Test Readiness Reviews (Internal)
- Test Readiness Review (TRR) (Contract-Mandated)
- System Test Reviews (Internal)
- Preproduction Reliability Design Review (PRDR) (Contract-Mandated)
- Installation and Field Documentation Reviews (Internal)
- Production Readiness Review (PRR) (Contract-Mandated)

Chapter

5

References

Bennett, R. *Preliminary and Critical Design Review: Procedures and Effectiveness.* Masters Thesis: School of Systems and Logistics of the Air Force Institute of Technology Air University, Wright Patterson Air Force Base, Ohio, September, 1987. Presents data from a survey of 271 junior and senior Air Force program managers who evaluated their last Preliminary Design Review or Critical Design Review. On the average, less than half of the respondents gave a favorable assessment of their last design review's reduction of technical risk, interfaces, producibility, timeliness and completion of the data package. Most rated their own preparation times inadequate because of the size and complexity of the materials and the late delivery.

Best Practices: How to Avoid Surprises in the World's Most Complicated Technical Process. Department of the Navy (NAVSO P-6071), March 1986. Discusses how to avoid traps and risks by implementing best practices for 47 areas or templates that include topics in design, test, production, facilities, and management. These templates give program managers and contractors an overview of the key issues and best practices to improve the acquisition life cycle.

Boehm, B. "Software Engineering," *IEEE Transactions on Computers,* Vol. C-25(12), December, 1976. Presents data from studies at IBM, GTE, TRW, and AT&T that shows that the cost to fix software errors increases logarithmically as the errors are detected and corrected in later and later phases.

Boehm, B. *Software Engineering Economics.* Englewood Cliffs, NJ: Prentice-Hall, 1981. Discusses the economic rationale for having reviews and inspections at the end of each developmental phase in order to detect, resolve, and fix errors while the cost is low.

Davis, H. A. "Peer Reviews," *Engineering Manager,* November, 1984. Discusses the rationale and procedures for effective design reviews. Emphasizes the importance of clear roles and responsibilities, management support for schedules and resources, and a climate that rewards rather than punishes error detections.

Defense Systems Management College. *Systems Engineering Management Guide.* MDA 903-82-C-0339. Fort Belvoir, VA: DSMC, October 3, 1983. Describes reviews often mandated in government contracts. Includes a program manager's checklist for design reviews with tips on scheduling, tailoring, personnel, and administration.

Deutsch, M. S. *Software Verification and Validation.* Englewood Cliffs, NJ: Prentice-Hall, 1982. Discusses design reviews (Systems Requirement Reviews, System Design Reviews, Preliminary Design Reviews, Critical Design Reviews) conducted to validate and verify systems.

Dhillon, B. S. *Quality Control, Reliability, and Engineering Design.* New York: Marcel Dekker, 1985. Discusses engineering design reviews: types, subjects discussed, required items, participants on design review teams, and reliability considerations during the design review.

Dhillon, B. S. and Reiche, H. *Reliability and Maintainability Management.* New York: Van Nostrand Reinhold, 1985. Discusses design review objectives, types of design reviews, input and output, and benefits.

Fagan, M. E. "Advances in Software Inspections," *IEEE Transactions on Software Engineering,* July 1986. Discusses the rationale, procedures, and data from software inspections. Gives results of empirical studies at IBM showing that reviews and inspections pay for themselves.

Fagan, M. E. "Design and Code Inspections to Reduce Errors in Program Development," *IBM Systems Journal,* Vol. 15(3), 1976. Discusses the improvements in quality and productivity from formal inspections of software designs and code. Describes procedures for defining roles, entry and exit criteria, and classifying errors. Includes sample forms for reporting results.

Fowler, P. J. "In-Process Inspections of Workproducts at AT&T," *AT&T Technical Journal,* Vol. 65(2), 1986. Describes procedures and data on formal inspections and training courses for reviews and inspections. Includes sample forms: meeting notices, inspection reports, and summaries.

Freedman, D. P. and Weinberg, G. M. *Handbook of Walkthroughs, Inspections, and Technical Reviews.* Boston: Little, Brown, 1982. Discusses trends, techniques, and methods of formal reviews, inspections, and walkthroughs. Gives details for what to do before, during and after review meetings for reviews of different types. Gives tips for review leaders and other team members on how to act during the meetings.

Koontz, W. L. G. "Experience with Software Inspections in the Development of Firmware for the Digital Loop Carrier System," *IEEE International Conference on Communications,* 1986. Discusses procedures and data from firmware inspections.

Landers, R. R. *Reliability and Product Assurance.* Englewood Cliffs, NJ: Prentice-Hall, 1963. Discusses design review objectives, scheduling, and data packages.

Military Standard 785: *Reliability Program for Systems and Equipment Development and Production.* Discusses reliability improvement using Failure Mode Analysis (FMEA), Criticality Analysis, and Test, Analyze, And Fix (TAAF).

Military Standard 2167: *Defense System Software Development.* Describes the content and scheduling of software reviews that are often mandated on government contracts.

Military Standard 1521: *Technical Reviews and Audits for Systems, Equipments, and Computer Software.* Details reviews typically mandated on contracts, e.g., Systems Requirement Review, Preliminary Design Review, Critical Design Review, Functional Configuration Audit, Physical Configuration Audit, Production Readiness Review.

Parnas, D. L. & Weiss, D. M. "Active Design Reviews: Principles and Practices," *8th International Conference on Software Engineering,* 1985. Discusses techniques for effective design reviews with illustrations from the Naval Research Laboratory's projects.

Priest, J. W. *Engineering Design for Producibility and Reliability.* New York: Marcel Dekker, 1988. Discusses techniques for test and evaluation for reliability growth and design maturity including design reviews for verification, test, analyze, and fix, integrated test, thermal and vibration tests, etc.

Transition from Development to Production. Department of Defense (DoD 4245.7-M), September, 1985. Techniques for avoiding technical risks in 47 key areas or templates including funding, design, test, production, facilities, logistics, management, and transition plan. Identifies critical engineering processes and controls for the design, test, and production of low risk products.

Walsh, L., Wurster, R. and Kimber, R. J. (Eds.) *Quality Management Handbook.* New York: Marcel Dekker, 1986. Discusses scheduling and purpose of design reviews, failure reporting, analysis, and corrective systems, failure mode and effects analysis, and design tradeoffs.

Design Release

1

Introduction

To the Reader

This Design template part discusses the Design Release template.

The templates, which reflect engineering fundamentals as well as industry and government experience, were first proposed in the early 1980s by a Defense Science Board task force of industry and government leaders, chaired by Willis J. Willoughby, Jr. The task force sought to improve the effectiveness of the transition from development to production of systems. The task force concluded that most program failures were due to a lack of understanding of the engineering and manufacturing disciplines used in the acquisition process. The task force then focused on identifying engineering processes and control methods that minimize technical risks in both government and industry. It defined these critical events in design, test, and production in terms of templates.

The template methodology and documents

A template specifies:

- areas of technical risk
- fundamental engineering principles and proven procedures to reduce the technical risks

Like classical mechanical templates, these templates identify critical measures and standards. By using the templates, developers are more likely to follow engineering disciplines.

In 1985, the task force published 47 templates in the DoD *Transition from Development to Production* (DoD 4245.7-M) manual.[1] The templates cover design, test, production, management, facilities, and logistics.

In 1986, the Department of the Navy issued the *Best Practices* (NAVSO P-6071) manual,[2] which illuminates DoD practices that increase risks. For each template, the *Best Practices* manual describes:

- potential traps and practices that increase the technical risks
- consequences of failing to reduce the technical risks
- an overview of best practices to reduce the technical risks

The *Best Practices* manual seeks to make practitioners more aware of traps and pitfalls so they can avoid repeating mistakes.

In September 1987, the Army Materiel Command made the templates the foundation for their risk reduction roadmaps for program managers.[3] In February 1991, the templates were incorporated into the DoD 5000.2 document as part of core of a fundamental policies and procedures for acquisition programs.[4]

The templates are the foundation for current educational efforts

In 1988, the government initiated an educational program, "Templates: Professionalizing the Acquisition Work Force." This program includes books, such as this one, that increase awareness and promote the use of good engineering practices.

The key to improving the DoD's acquisition process is recognizing that the process is an industrial process, not an administrative process. This is a change in perspective that implies a change in the skills and technical knowledge of the acquisition work force in government and industry. Many in this work force do not have engineering backgrounds. Those with engineering backgrounds often do not have broad experience in design, test, or production. The work force must understand basic design, test, and production processes and associated technical risks. The basis for this understanding should be the templates which highlight the critical areas of technical risk.

---

[1]Department of Defense. *Transition from Development to Production.* DoD 4245.7-M, September 1985.

[2]Department of the Navy. *Best Practices: How to Avoid Surprises in the World's Most Complicated Technical Process.* NAVSO P-6071, March 1986.

[3]U. S. Army Materiel Command. *Program Management Risk Reduction Roadmaps.* Alexandria, VA: U.S. Army Materiel Command, September 1987.

[4]Department of Defense. *DoD Instruction 5000.2. Defense Acquisition Management Policies and Procedures.* Washington, DC: Department of Defense, February 23, 1991.

The template educational program meets the needs of the acquisition work force. The program consists of a series of technical books The books provide background information for the templates. Each book covers one or more closely related templates.

How the parts relate to the templates. The parts describe:

- the templates, within the context of the overall acquisition process
- risks for each included template
- best commercial practices currently used to reduce the risks
- examples of how these best practices are applied

The books do not discuss government regulations, standards, and specifications, because these topics are well-covered in other documents and courses. Instead, the books stress the technical disciplines and processes required for success.

Clustering several templates in one book makes sense when their best practices are closely related. For example, the best practices for the templates in Part 3, Parts Selection and Defect Control, interrelate and occur iteratively within design and manufacturing. Designers, suppliers, and manufacturers all have important roles.

Courses on the templates. The booksare designed to be used either in courses or as stand-alone documents. The managers should recognize that adherence to engineering discipline is more critical to reducing technical risk than strict adherence to government military standards. They should especially recognize when their actions (or inactions) increase technical risks as well as when their actions reduce technical risks.

The templates are models

The templates defined in DoD 4245.7-M are not the final word on disciplined engineering practices or reducing technical risks. Instead, the templates are references and models that engineers and managers can apply to their industrial processes. Companies should look for high-risk technical areas by examining their past projects, by consulting their experienced engineers, and by considering industry-wide issues. The result of these efforts should be a list of areas of technical risk and best practices which becomes the company's own version of the DoD 4245.7-M and NAVSO P-6071 documents. Companies should tailor the best practices and engineering principles described in the books to suit their particular needs. Several military suppliers have already produced manuals tailored to their processes.

Figure 1 shows where to find more and more details about risks, best practices, and engineering principles. Participants in the acquisition process should refer to these resources.

A major milestone for design organization is the release of a design that is ready for production. What must take place prior to design release to ensure that the design meets the implicit requirements of cost, reliability, maintainability and producibility? In today's environment, many issues are overlooked in the design area that cause redesigns while the product is in manufacturing. Why? There are many reasons: including schedule pressures, cost over-runs, and lack of a sound configuration control system. However, the primary reason is producibility, an issue that is usually overlooked by designers. Many organizations do not involve manufacturing engineers in product design until the design is ready to be released. By the end of the design cycle, a product has acquired more than 75% of its total value thereby making changes more costly and difficult. To avoid schedule delays and additional costs, it is imperative that designers get all organizations and specialists involved throughout the design cycle to optimize the design from both product and process viewpoints.

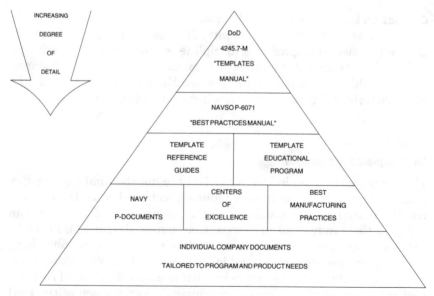

Figure 1 Information about acquisition-process templates.

Designers and manufacturing engineers must be a team and ask each other the correct questions prior to design release so manufacturability, producibility and quality-improvement issues do not impede manufacture. Organizations have guidelines that control the release of a design to manufacturing, so why are there problems? Some reasons are: design reviews are skipped because of schedule pressures; standards are ignored; restrictions force organizations to conform; or lack of planning.

Table 1 identifies risks in the Design Release that the design may encounter unless the project uses good Design Release methods and practices.

TABLE 1   Risks and Consequences Addressed by Design Release

Risks	Consequences
Drawings are not released in a timely manner	Delays occur throughout the manufacturing process
Drawings are released before they are ready	Change orders must be prepared causing unnecessary rework in design and manufacturing
Designers release long lead items last	Delays occur in manufacturing and cause delays in the deployment of the system
"Back planning" from delivery schedules drive design schedules	Unrealistic schedules make the product late
Designers ignore feedback from design reviews	Designs do not reach maturity causing costly rework
Designers ignore feedback from models	Not incorporating feedback causes manufacturing delays and mistakes to occur in manufacturing Project lacks a "lessons learned" database
Project lacks a configuration management system	Producers cannot control manufacture of items shipped to the field
Designers have no common system to transfer designs to vendors	Delays occur in manufacturing and shipment of the system
Designers establish no baseline	Shipment of configured items may not be the latest design
Project lacks one organization with coordination responsibilities	Many important issues will slip through the cracks causing additional delays and costs to the program
Long-lead assemblies and sub-assemblies are released before integration testing is completed	System fails to perform as required resulting in redesign and retesting of the system

Drawings Are Not Released in a Timely Manner

Design organizations usually release drawings as "packages of information." The "packages" usually include stocklists, bills of material, standard parts, etc., i.e, all the items needed to manufacture the product. Without complete information, scheduling is difficult.

Drawings Are Released Prior to Approval

Because of program pressures, drawings are released before they have been reviewed and approved by all individuals or organizations. When this happens, errors in design are likely, and issues such as reliability and producibility are overlooked. This results in producing change orders that result in unnecessary rework.

Long-Lead Items Are Released Last

Assemblies or parts that require extensive design, or parts that require complex tooling, are usually the last parts to be released. Because these items require longer lead times to finalize both the part and the tooling, design priorities must encourage completion and release as early as possible. Otherwise, manufacturing will incur schedule delays.

"Back Planning" from Delivery Schedules

Schedules for the design organization are usually made by working backwards from the delivery milestones. This type of planning does not leave any room for schedule adjustments and adds pressure on the design organization. Consequently, designers rush to make "the schedule," leaving errors in the released designs for the manufacturing organization to correct. This slows the schedule, delays the product, and brings cost over-runs and decreased performance.

Not Using the Feedback from Design Reviews

The design review is supposed to improve the design, based on the entrance and exit criteria established for the review. Any design that fails these criteria must be corrected. Otherwise, issues that surfaced may be addressed in manufacturing or later causing delays and additional costs.

Not Using the Feedback from Models or Simulations

A model or simulation built to provide information on manufacturability, producibility, testability or design intent must be examined by the appropriate individuals and, through change orders, incorporate any characteristics that may impact the design, or the manufacturing processes. Failure to address these issues will result in unnecessary rework and delays.

Lack of a "Lessons Learned" Database

All organizations learn from projects in progress or completed. These lessons should be shared with other organizations within the company. A "lessons-learned database" is the ideal repository for this information. This data base can avoid repeating past mistakes and failures.

Program Lacks a Configuration Management System

A configuration management system applies discipline to functional and physical characteristics at all levels of the program. A program that lacks or has limited configuration management will have difficulty manufacturing, shipping, and/or replacing existing products.

Project Lacks a Common Database

Having a project database allows everyone in the project to see the design information as it is created. As the design changes, all functional organizations have access to the information immediately. This saves time because organizations have immediate access to the design information and do not have to wait for the hard copy to arrive by mail. Lacking this common database will cause delays in schedules and add cost to the program.

No System to Transfer Design Information to Vendors

A method of assuring that vendors and subcontractors obtain the latest revision of a drawing is through (controlled) access to the project database. Working from the project database allows immediate transfer of the design information.

No Design Baseline Established

Failure to establish a system, assembly or document baseline can lead to the shipment of systems or replacement parts that have not been accepted as the final design. Failure to establish a baseline will impact cost, schedules and maintenance of the system.

Lack of Coordination Responsibilities

In today's environment, organizations must partner with other organizations to form a team. However, when organizations depend on each other, one organization must have the responsibility for coordination. With no clear responsibility, organizations may assume some other organization is responsible and let the milestones slip, causing delays and cost over-runs.

Long Lead Items Are Released Before Integrated Testing

Long lead assemblies and sub-assemblies are released to production before they are fully tested. This results in redesigning the interconnecting units, causing schedule slips and rework costs to the project.

Keys to Success

To minimize risks, a documented design release procedure is required. The procedure must require early involvement of designers, manufacturing engineers, program managers and contractors involved in the design process so decisions can be made where they cost the least and can have a significant impact on cost and schedules. Listed are the practices found to be the most effective for a successful design release process.

Integrated configuration management system

The configuration management system for the program must be an integrated system controlled by the engineering or design organization. However, everyone involved in the program must have access to the database and must be able to populate and change that portion of that database that pertains to their area of involvement (i.e., manufacturing controls process information; logistics enters sparing information, etc.)

Using design reviews

Planning of design reviews is the most important part of the design process. By appropriating a sequence in the master schedule, design reviews can be conducted in a manner that is most effective, without rushing. Planning for design reviews allows everyone to prepare and allot their time efficiently. Design reviews must be conducted when adequate progress, in the design, has been made to warrant a review, not when the calendar or schedule says it's required.

A Policy on flow-down of requirements to subcontractors and vendors

Items to include in the contractual flow-down requirement are: prime contractors conducting critical design reviews (CDR) or preliminary design reviews (PDR) at the subcontractor's facilities, design release policy of subcontractors, and transfer mechanisms (electronic or otherwise) between the prime and the subcontractors for rapid dissemination of vital information. Failure to include such a requirement will result in delays and added costs.

A release policy

There must be a corporate policy stating the criteria for design release. Some items to include in the policy are: who is responsible for sign-offs; which documentation is required; and what organizations control the overall process.

Early involvement of the manufacturing engineers in the design

Early involvement of the manufacturing organizations can resolve many of the issues, such as producibility and manufacturability, that may cause schedule slips due to redesigns. Through this early involvement, the development cycle of a system can be reduced significantly.

Use of "Graybeard teams" to assess design maturity prior to release

"Graybeard" is a term used for the most experienced individual in a particular area of expertise. Organizations are finding that teams consisting of graybeards are effective in evaluating a design just prior to release. These teams are unbiased and can quickly analyze a design. Organizations that use these teams find they are cost-effective and usually save time.

All-inclusive design release package

The design-release package must be all-inclusive. The information included in the package must be a checklist of all issues that are closed, process documentation and layouts for manufacturing the product, all tooling information, and a list of vendors who are or will be manufacturing the various assemblies or sub-assemblies. If necessary, checklists may be used as a guide to ensure completeness of the release information.

2

Procedures

Before any organization can implement an effective Design Release Process, two key items must be in place: an involved project management team and an effective configuration management system. The project management team must provide the leadership *and* the resources to meet the program goals and objectives. With an effective configuration management system in place, both the manufacturing and design organizations know the product manufactured will be using the latest design information. In addition, any changes that are required will be implemented in an orderly process through the configuration management system. Developing a large system involves the coordination of information between many organizations. Configuration management, particularly in complex systems, is the key tool to achieve this coordination to assure minimized costs after a design has been released, to minimize risks and to provide significant benefits throughout the product life-cycle.

Lack of a configuration management system has been acknowledged by many organizations to be a major reason for manufacturing problems.

Poor configuration control is a leading cause of increasing program costs and procurement schedules. Lack of a good configuration control system leads to many pitfalls, including an unknown design baseline, excessive production rework, poor spares effort, stock purging rather than stock control and the inability to resolve field problems.[5]

Figure 2 shows the life-cycle phases of a typical DoD project. For example, at the end of the demonstration and validation phase (D&V), 85% (upper curve) of the decisions affecting the life-cycle costs have been made, leaving 22% (lower curve) of the cost reductions opportunities in the D&V phase. With the implementation of cross-functional

---

[5]Transition from Development to Production, DoD 4245.7-M, pp. 3.30-3.32.

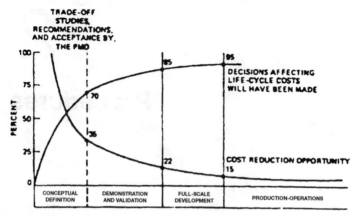

Figure 2 Department of Defense life-cycle phases.

teams, projects should get a larger percentage of the decisions earlier in the concept or D&V phases, reducing the number of changes in the engineering and manufacturing development (EMD) phase.

Lacking control of design information leads to a design-release process that increases problems in the manufacturing process. Additionally, in second-sourcing contracts, a prime developer or manufacturer that lacks a sound configuration management and design release process adds problems for the new contractor. As more contracts are bid using leader-follower teams, the management of the design information becomes crucial in the life-cycle process.

Also as organizations move toward totally automated systems for reducing intervals and improving productivity and quality, the management of both the design and release information becomes crucial to the development process.

Figure 3 shows world-class organizations using product data management (PDM) have significant improvement at reducing the number of design changes at various stages of the development cycle. Using information systems that coordinate the activities at the start of design, manufacturers target a 50% reduction in the time to market along with quantifiable quality and cost improvements.[6] The systems employed are integrated: that is, they use pieces of modeling, configuration management, project management and communications packages to comprise the system from development to manufacturing. By using the elements from different, yet compatible MIS systems, organizations

---

[6]Brown, D. H. *Product Information Management,* Cleveland, OH: Computer-Aided Engineering, May 1991, p. 60.

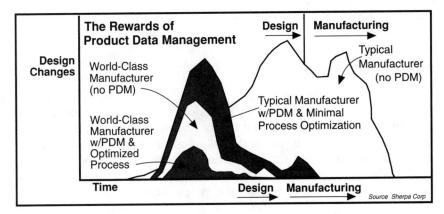

Figure 3 Rewards of product data management.

can customize with little or no programming. A drawback to such customization is that the organization must conform to the practices of the MIS system rather than designing a system around standard corporate practices. For determining the benefits of customization or modularity, see Part 5, Computer-Assisted Technology.

Data Management

A product data package is the data provided by the design organizations to manufacturing that is used to build the product. The product data would include parts lists, drawings, artwork, etc. Two related terms are configuration management and data requirements. Configuration management is concerned with controlling the different versions of the product (different revisions of product data) as well as data needed to manufacture the product such as documentation. Configuration management, according to the government, refers to the management of configuration items, while in the commercial world, the term refers to the management of all data related to the design throughout the product life-cycle.

Product Data Management

Product data management is the management of product data to ensure that it is complete and consistent, i.e., the parts list sent to purchasing is consistent with the drawing sent to the manufacturing engineer, all the parts are on the approved list, etc. Product data management is best done using CAD/CAM systems and electronic tools that are interconnected and use a common data base. Product data management is similar to configuration management but is primarily focused on content and delivery of product documentation.

Process Optimization

Process Optimization means that you have taken a TQM approach to defining, continuously improving, and documenting your processes. In particular, process optimization means that you have eliminated non-value added activities by using the latest quality tools and methods. In particular for design transfer, it means using electronic tools and CAD/CAM systems to develop the product data and transfer the product data electronically to the factory, vendors and subcontractors.

Step 1: Establish Project Schedule

All projects, whether a month or several years in duration, require a formalized plan for allocation of resources and for estimating the project cost. This planning must be done by the program manager from the project's inception through the final life-cycle phase. The plans must cover objectives, policies, procedures, milestones, tracking cost and tracking resources, and contingency plans. The plans must be tailored to the project, but flexible, and contain the input from all organizations. As the program changes, plans and schedules must be modified to meet the changing conditions. All the work must be defined through the work breakdown structure. If the tasks are well-defined and understood at the beginning of the project, then there are fewer changes, with increased performance, during the various phases of the program. Without the proper "up-front planning," no program can succeed. Each program must have its own goals and milestones; basing a program on previous goals and objectives will lead to poor use of resources, fire-fighting, and continuous crisis management. The program objectives must be defined in a documented plan that is communicated to all participants in the program. Without the proper planning, the program will have:

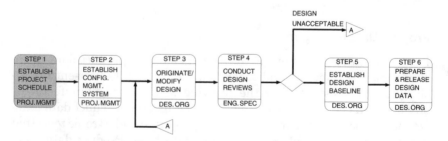

Figure 4 Steps in design release.

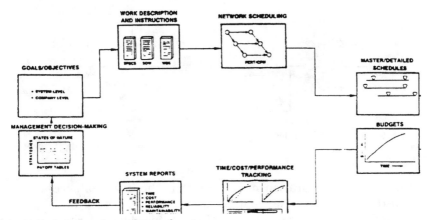

Figure 5 Project planning and control system.

- difficulty getting started
- a constant state of uncertainty
- difficulty monitoring the organizational or functional responsibilities
- difficulty improving the program

Developing and updating a schedule is a disciplined process with continuous adjustments required to keep the project on track.

Figure 5 depicts the various planning and control elements of a project.[7] Project Management's responsibility is to perform the functions and provide feedback to the organizations on progress toward the goals and objectives of the program.

Step 2: Establish the Configuration Management System

Early in the program the configuration management (CM) system must be implemented, otherwise both product integrity and product stability will be lost. There are costs to develop and implement a configuration management system but the loss of productivity, increased design and manufacturing costs, and schedule delays are consequences that out-weigh the costs of implementation. In today's environment, companies must increase quality while lowering costs and maintaining

---

[7]Kerzner, H. *Project Management A Systems Approach to Planning, Scheduling, and Controlling.* New York:Van Nostrand Reinhold, 1989, p. 577.

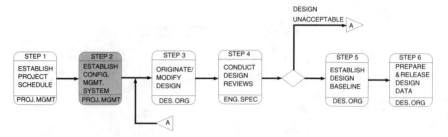

Figure 6 Steps in design release.

schedules. Configuration management is the key to a successful design-release process that controls costs.

Configuration management is often confused with a change control or an accounting system. Configuration management prevents unnecessary changes and improves the effectiveness of all organizations while reducing costs and improving schedules. Proper planning and application of a CM system can provide the technical professionals with a tool to accomplish these goals.

Management must address the configuration management and design-release issue at the beginning of the project to derive full benefits. Failure is evidenced by the following examples.

- A defense contractor *traditionally* plans for a 300-percent rate of design change.

- An aircraft manufacturer inflates its fabrication lead times by 50 percent to *allow* for anticipated uncontrolled design changes.[8]

These are examples that demonstrate lack of program tailoring. This type of planning builds program failures by adding additional delays to the schedule that are unnecessary.

There are four key principles to CM: definition, communication, control and incorporation. Case histories have demonstrated ways to reduce costs, avoid expenses and compress manufacturing cycles by 50%.[9]

More details on configuration management can be obtained by consulting Part 7, Configuration Management Control.

---

[8]"Configuration Management—The Next Engineering Management Revolution," R. G. Boznak. 1989, 2nd International Engineering Management Conference Proceedings, pp. 207-210.

[9]Configuration Management–The Next Engineering Management Revolution," R. G. Boznak. 1989 2nd International Engineering Management Conference Proceedings, pp. 207-210.

Step 3: Originate or Modify Design

Some organizations are reaping the benefits of computer aided design (CAD), computer aided manufacturing (CAM), and computer integrated manufacturing (CIM) systems and paperless factories, while others are wasting valuable time and money generating paper drawings, doing revisions manually and maintaining drawing files for paper copies. Additional costs may be incurred by engineers redrawing lost drawings or using the wrong revision to update drawings. One study estimates that 3% of all corporate drawings fall into the lost drawing category.[10] Corporations committed to the management of design data have little trouble justifying the expenditures for integrated systems. However, organizations that are considering the expenditure for CAD systems should track the number of revisions per year, the number of lost drawings, and any other metric that could influence the purchase. Figure 8 shows the cost of automated systems versus manual. The benefits of automated systems are:

- improved company efficiency

- improved security

- generation of production information

Figure 8 illustrates how US engineers spend their time in the development process in a non-automated environment. Clearly organizations need to improve the productivity of their technical workforce to remain competitive.

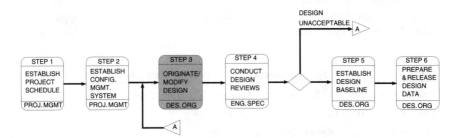

Figure 7 Steps in design release.

[10]Dvorak, P. "Taking Care of Technical Drawings." *Machine Design* March 7, 1991, pp. 34-40.

The cost of redrawing and revising drawings			
	Cost/drawing time ($/hr)	Drawing time (Hr)	Cost ($)
Making revisions			
Manually	40	10	400
Using CAD	65	3	195
Data capture methods			
Redraw with CAD	65	10	650
CAD redigitize	65	8	520
Scan and vectorize	Service bureau	—	250
Scan and hybrid store	100	0.25	25

Source Intergraph

Figure 8 Redrawing and revising.

As Table 2 shows, and surveys indicate, US engineers spend 10 to 20% of their time on true engineering activities while Japanese engineers spent 40% in the design and analysis functions.[11] Using automated tools will increase the productivity of the US engineer during the design and release processes. However, many US engineers still lack problem solving skills, teamwork, are poor communicators and lack the proper planning and organizational skills. If the US is to gain a competitive advantage, then US management must invest the time in training so these deficiencies will no longer exist: the rewards would be repaid many times.

In some organizations, studies show that engineers spend approximately 40% of their time waiting for other organizations' feedback on their designs in automated environments.[12] Organizations must reduce the time engineers spend waiting for feedback from support services to reduce cycle time.

TABLE 2　Time Spent by US Engineers in a Non-automated Environment.

10% - Design (Thinking, Theorizing) - Analysis (Modeling, Simulation)

10% - Testing (Data Reduction, Prototypes)

25% - Documenting (Memos, Drawings, Sketches)

15% - Planning (Gantt or Pert)

40% - Communicating (Meetings, Sharing Information, Talking, Listening)

---

[11]From the course conducted by the University of Lowell titled "Concurrent Engineering and Design for Manufacture", June 17-19, 1991.

[12]From the course conducted by the University of Lowell titled "Concurrent Engineering and Design for Manufacture", June 17-19, 1991.

Step 4: Conduct Design Reviews

Design reviews must be scheduled and conducted on a regular basis to assess design maturity. Timely design reviews eliminate the surprises of cost, schedule and delivery that result when significant changes are required. Conducting effective design reviews is outlined in Part 8, Design Reviews. Some key points for conducting successful design reviews are:

- provide guidelines for conducting reviews
- choose the appropriate team members
- set entry and exit criteria
- provide information prior to review
- issue the appropriate reports
- use the data to refine the process

Step 5: Establish Baseline

Baseline management involves several baselines that are established throughout the product life-cycle; some are contractual while others

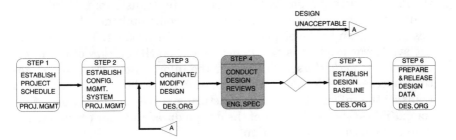

Figure 9 Steps in design release.

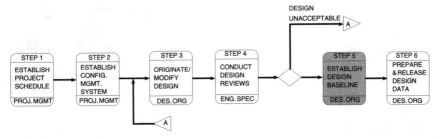

Figure 10 Steps in design release.

are program-specific. The final baseline is the design baseline. The establishment of this final baseline signals that all updates of information have been integrated into the design, and the information is ready to be transferred to manufacturing. The product baseline defines all approved or conditionally approved documentation that determines the product that will be manufactured. Included are engineering models, prototypes, testing information, and any feedback from field trials or demonstrations. The product baseline occurs at the end of EMD and *after* the functional and physical configuration audits. Changes to the product baseline are made through engineering change proposals (ECP's) that are accepted by the customer. A program may have multiple baselines consisting of individual configuration items (CIs) and even multiple baselines for each CI depending on the model variations. For instance, the design of a plane may consist of separate baselines for its avionics, subsystems, armament and propulsion systems. Each of these major areas will have subsystem components that must be integrated to form an overall system baseline that meets the project goals.

Step 6: Prepare and Release Design Data

This is the final step before manufacturing the product. The design organization exists to transfer the technology, the design information, to the manufacturing organization. A successful transfer of this information implies the engineering knowledge is completely developed so the design data can be developed into a system or product. The success of the system or product is *totally* dependent on the *quality* of the information and *method* used to transfer the information. Many organizations successfully develop the information in the laboratory but fail to transfer the information to manufacturing successfully because of incompatible systems, formats of the design data, and poor configuration management of the information.

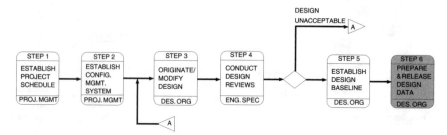

Figure 11 Steps in design release.

Chapter

3

Application

In this chapter are examples of the use of the principles and processes discussed in the Procedures chapter. The examples include:

- Martin Marietta
- GTE Government Systems
- GE Aircraft Engines
- Raytheon Missile Division
- AT&T Bell Laboratories, Merrimack Valley
- Textron Defense Systems, Wilmington, MA.

Martin Marietta

A technology that can track activities and manage the release of information to manufacturing is applications software. This technology uses the computing work-group architectures, allowing individuals and teams to communicate with each other on ideas, designs, and changes. In addition, this custom designed software can plan manufacturing processes, program numerical control (NC) equipment, estimate the product cost, and assist in the documentation process, and ensure the product can be built as designed.

The process starts with the development of a solid model representing the product to be built. The process continues with further enhancements of the design, including design rules and relationships that define the product. Some of the benefits from this technology are:

- encourages concurrent engineering practices
- develops prototypes on the first iteration
- increases productivity in the design and manufacturing processes

A prototype system to handle shop-floor travelers (STFs) was installed at the Martin Marietta facility in Waterton, CO. This system was designed to eliminate the STFs from manually being handled by 300 production workers. Based on the prototype savings, the system is expected to increase productivity by 200%, and save $300,000 in paper and reproduction costs.

GTE Government Systems

GTE Government Systems Command, Control, and Communications Systems Sector has developed a process for design-release that uses the Defense Science Board Templates in its Producibility Involvement Design/Manufacturing Process.[13] Each program develops a set of templates that is tailored to the program.

Figure 12 (Courtesy of GTE) is a modified version of its product development cycle showing the relationship between the templates and the phase of the program.

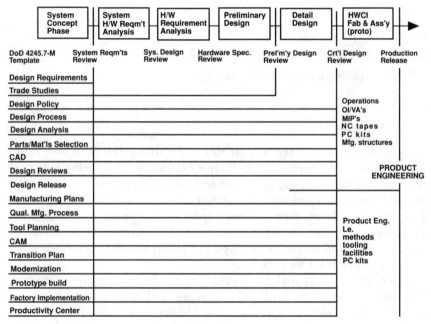

Figure 12   Producibility in design/manufacturing process.

---

[13]Courtesy of GTE Government Systems Corp, Waltham, MA.

The key elements of this process are:

- The Program Manager develops a master schedule with all disciplines providing input on the items to be designed and the priority of the design-releases.

- All disciplines have influence during the design phase minimizing surprises during design reviews and at drawing release sign-offs.

- Schedules are developed for purchasing components and fabrication of all configuration items. Any conflicts in the schedule are resolved at weekly status meetings. Issues that cannot be resolved are brought to the attention of upper management for resolution. All disciplines "buy-in" to the schedule with full agreement.

Some of the benefits GTE has realized from this process are:

- Concurrent engineering concepts are mandated by GTE management on all programs

- Configuration management systems are employed earlier

- More comprehensive design reviews are conducted

- Tradeoff studies are completed earlier in the design process

General Electric's ENACT System

The need for a new system to control configuration implementation, development assembly, and stores and tracking was recognized by the General Electric engineering community after a study of the engineering functions was completed in 1983. The General Electric Aircraft Engine Division[14] facility in Lynn, Mass., installed the ENACT system in 1986 that supports a jet engine from the development through the production phases. ENACT covers engineering, assembly, configuration management and test organizations. Prior to the development and installation of this system, information on the development, testing, material and assembly of a jet engine was stored in various forms, in several locations and with several organizations maintaining the same information. In this form, retrieval of the information was performed manually, was time consuming and was often outdated or incorrect when the information was received.

Figure 13 shows the organizations that can use the system through the various stages of a program.

---

[14]Courtesy of General Electric Aircraft Engine Division, Lynn, MA.

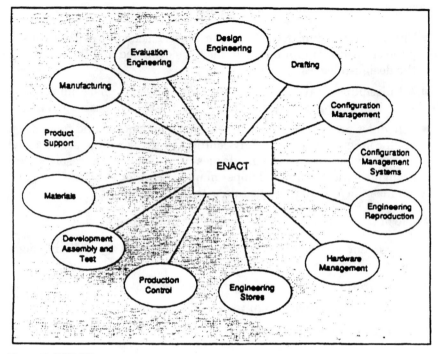

Figure 13 ENACT users.

The ENACT system supports the following features through the stages of the development of a jet engine.

Establish the product stage

During this phase, all organizations work together to determine the product configuration and a list of parts. The key elements are:

- Define the test plan

  All test plans are created, maintained, and updated using the system and

- Prepare product definition

  All phases of the design are coordinated through the system including configuration management, parts lists, design reviews, part weights and specifications.

- Make design changes

  All changes and approvals are coordinated through the system

- Distribute drawings

Acquire hardware stage

During this phase the hardware is purchased, inspected, stored and distributed for assembly. The key elements during this phase are:

- Develop requirements

  This deals with determining the dates and quantities of final assemblies required.

- Receive hardware

  As material is received, it is inspected, and disbursed to the assembly operations.

- Tracking hardware

  Material is tracked to determine the availability of parts and determine parts shortages.

Develop product stage

During this phase, the engine is assembled, tested and disassembled. The key elements are:

- Scheduling
- Define assembly procedures
- Assembly
- Breakdown of engine

The system has all information stored in a central location for easy access. The advantages of the ENACT system are:

- compatibility with other systems—the system was designed to be compatible with other systems at the Lynn facility.

- easy accessibility–all the information is available either on-line or in batch reports to all users who are authorized.

- consistency–because all the information is electronically stored in one area all users see the same information.

- timeliness–all information is stored immediately, therefore information can be quickly retrieved.

- visibility–because of quick access decisions can be made immediately.

- anyone can access the system provided they have need for the information. The system can:

- create documents, changes in designs or bills of materials
- review and approve parts lists and changes in designs
- perform on-line inquiries and printout reports

GE is using process improvement teams to make improvements to the system. Through the use of process improvement teams and the ENACT system, the benefits of the ENACT realized are:

- shortened the change in design process by 50%
- reduced the configuration management workforce by 50%
- improved tracking of all parts, subassemblies and assemblies through manufacturing
- reduced the vendors request for design changes by 75%
- turn around approximately 75% of the producibility design changes requested in less than 30 days

The use of an automated data storage and retrieval system, an electronic drawing storage system coupled with ENACT provides the benefit of only the latest drawings can be used to record changes or manufacture a product.

Raytheon Missile Systems Division and Missile Systems Laboratories

The Raytheon Missile Systems Laboratory has the responsibility for establishing and maintaining the configuration management, design release, and the tracking systems for all products developed within the laboratory. Recently this laboratory had the responsibility for developing and transitioning, into production, the Patriot and Hawk missiles.

The implementation of the configuration management, design release process, and the tracking systems is the responsibility of the program manager. Using corporate and divisional policies and procedures, the program manager tailors the program's processes in conjunction with a transition to production plan and checklist.

Figure 14 shows the Raytheon Missile Systems design release process. Some of the key features of this process are:

- design reviews are used throughout the process with both engineering and manufacturing teams involved
- engineering performs changes to the data using engineering change orders (ECOs) through a CCB that includes manufacturing

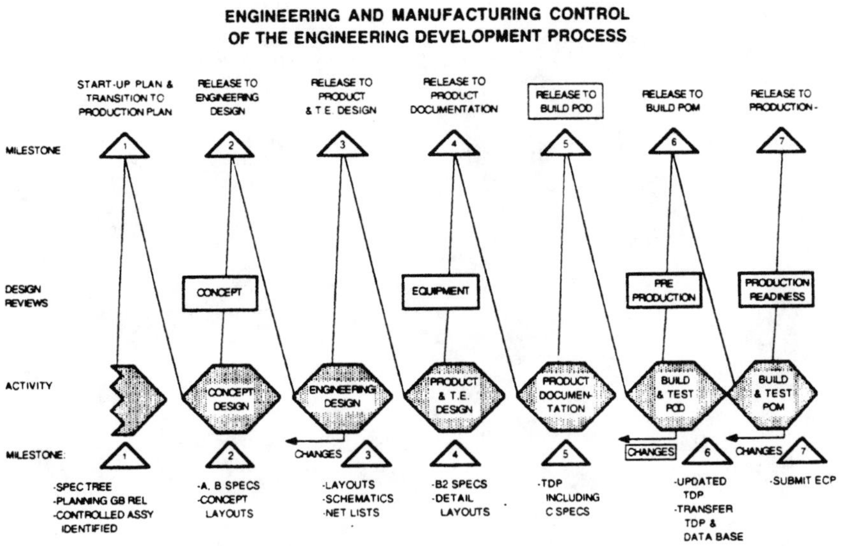

ENGINEERING AND MANUFACTURING CONTROL
OF THE ENGINEERING DEVELOPMENT PROCESS

Figure 14 Control of engineering development process.

- manufacturing engineers receive information updates as engineering changes are approved
- engineering with laboratory CM controls the configuration and the CAD database prior to production release, with manufacturing having read only permission to the CAD database
- after the production release, engineering with manufacturing CM controls the database.

AT&T Bell Laboratories, Merrimack Valley Works

Within AT&T Bell Laboratories Merrimack Valley, there is an organization that is responsible for converting the circuit designer's requirements into a circuit design that meets the factory guidelines. This organization is responsible for coordinating the design reviews and correcting the design data. Figure 15 is a high-level diagram of that process.

Some of the key features of this process are:

- use of subject matter experts to continually improve the design release process

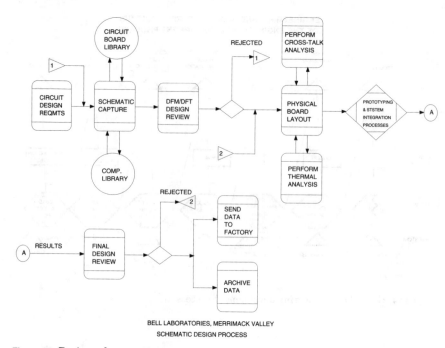

BELL LABORATORIES, MERRIMACK VALLEY
SCHEMATIC DESIGN PROCESS

Figure 15 Design release process.

- use of mentors to train new people
- regularly scheduled user's meetings to exchange information
- both automatic and manual audits are conducted with the errors continuously tracked and distributed to the design organizations
- all inspections must be performed in the design process because the factory has no tools to perform the re-inspections
- configuration management is the key to successful design transfer
- a project component database is established early in design process from which the designers must select components
- a circuit board database established by the circuit manufacturing organization
- early in the project, the manufacturing location and line is established opening communications early in the process
- sign-off forms are used through-out the design process to obtain agreement on the information prior to releasing the design data. Figures 16 and 17 are examples of sign-off forms

DETAIL AUDIT SHEET

CODE: VERSION: DATE:

IN-CIRCUIT TEST					
AUDITS	CLASS				SEE NOTE
	1	2	3	4	
MISSING SITES					
UNUSED PINS					
BOARD EDGE CLEAR					
TOTALS					

Figure 16 In-circuit test audit sheet.

DETAIL AUDIT SHEET

CODE: VERSION: DATE:

PART ASSEMBLY					
AUDITS	CLASS				SEE NOTE
	1	2	3	4	
SPAN					
ORIENTATION					
MOUNT LAYER					
ASSEMBLY CLEARANCE					
PART in VOID					
TOTALS					

Figure 17 Part assembly audit sheet.

- design reviews are used throughout the design process and in the design release process. For example, the final design review includes results from model fabrication, functional testing and system integration.

Textron Defense Systems

At the beginning of a contract, it is the Program Manager's responsibility to coordinate the development of a realistic schedule that meets all the organizational requirements and delegate the appropriate level of authorization for releasing design information to the various organizations for manufacture or purchase. Design release direction is provided via Program Directives that either contain the required tailoring of the approval process by each organization so it is consistent with the project needs or identifies a System Engineering Management Plan (SEMP) and/or a Configuration Management Plan developed to guide the program. For instance, various organizations may require different levels of authorizations for various stages of the program.

Computer aided tools are the key to the early definition of product configuration, the development of a solid model representing the product to be manufactured, and the generation of process plans and instructions for the fabrication of the product. Distribution of the materials, to be reviewed for release, can be done electronically or through "hard copies". It is the responsibility of Program Management to coordinate with Purchasing the transfer of the latest design information to all sub-contractors and vendors in a consistent and timely manner as well as scheduling timely reviews with sub-contractors.

Key to a successful design release is the timeliness of reviews by key individuals. Problem areas are identified prior to release, risks determined, trade-offs negotiated and plans are put in place to mitigate the risks as the design matures. Any information that can impact a design becomes costly to incorporate as the design matures.

4

Summary

The previous pages have explained the best practices associated with a design release process. The best practice is the key to minimizing the technical risks in the design release process. The release process is where designers, manufacturing engineers, planners, purchasing representatives and subcontractors must agree the data and information is of the highest quality and will produce a product that meets or exceeds the customer's expectations.

The following are the key best practices. When implemented, they will make an effective design release process possible.

Keys to Success

To minimize risks, a documented design release procedure is required. The procedure must require designers, manufacturing engineers, program managers and contractors involved in the design process so decisions can be made where they cost the least and can have a significant impact on cost and schedules. Listed are the practices found to be the most effective for a successful design release process.

Integrated configuration management system

The configuration management system for the program must be an integrated system controlled by the engineering or design organization. However, everyone involved in the program must have access to the database and must be able to populate and change that portion of that database that pertains to their area of involvement (i.e., manufacturing controls process information; logistics enters sparing information, etc.).

Using design reviews

Planning of design reviews is the most important part of the design process. By appropriating a sequence in the master schedule, design reviews can be conducted in a manner that is most effective, without rushing. Planning for design reviews allows everyone to prepare and allot their time efficiently. Design reviews must be conducted when adequate progress, in the design, has been made to warrant a review, not when the calendar or schedule says it's required.

A policy on flow-down of requirements to subcontractors and vendors

Items to include in the contractual flow-down requirement are: prime contractors conducting critical design reviews (CDR) or preliminary design reviews (PDR) at the subcontractor's facilities, design release policy of subcontractors, and transfer mechanisms (electronic or otherwise) between the prime and the subcontractors for rapid dissemination of vital information. Failure to include such a requirement or a policy will result in delays and added costs.

A release policy

There must be a corporate policy stating the criteria for design release. Some items to include in the policy are: who is responsible for sign-offs; which documentation is required; and what organizations control the overall process.

Early involvement of the manufacturing engineers in the design

Early involvement of the manufacturing organizations can resolve many of the issues, such as producibility and manufacturability, that may cause schedule slips due to redesigns. Through this early involvement, the development cycle of a system can be reduced significantly.

Use of "Graybeard teams" to assess design maturity prior to release

"Graybeard" is a term used for the most experienced individual in a particular area of expertise. Organizations are finding that teams consisting of graybeards are effective in evaluating a design just prior to release. These teams are unbiased and can quickly analyze a design.

Organizations that use these teams find they are cost-effective and usually save time.

All-inclusive design release package

The design-release package must be all-inclusive. The information included in the package must be a checklist of all issues that are closed, process documentation and layouts for manufacturing the product, all tooling information, and a list of vendors who are or will be manufacturing the various assemblies or sub-assemblies. If necessary, checklists may be used as a guide to ensure completeness of the release information.

Table 3 outlines procedures and supporting activities for the Design Release template:

TABLE 3   Summary of Design-Release Procedures

Procedures	Supporting Activities
Establish Project Schedule	■ Plans must be tailored to the program with input from all organizations ■ As the program matures plans and schedules must be revised to reflect the program maturity ■ All tasks must be defined through the WBS ■ All program plan must be communicated to all program participants ■ Plan and budget appropriately in order to prevent schedule slippage and delay in the design release process
Configuration management must be implemented early in the program	■ Control of the system is maintained through the product life-cycle with reduced costs ■ Product integrity is maintained ■ The configuration management system must be tailored to the program ■ Audits must be constantly performed as a check on the system and progress of the program
Resources must be used effectively to reduce waste in the program	■ The design process must be tailored to the program to eliminate poor utilization of people and equipment ■ Systems must be designed into the process ■ The configuration control board is the key to effective configuration management

TABLE 3   Summary of Design-Release Procedures (Continued)

Procedures	Supporting Activities
Effective use of design reviews	■ Use to evaluate design maturity ■ Select the team members appropriately ■ Set entrance and exit criteria for effective reviews ■ Provide feedback to improve the design ■ Provide reports to management on action items
Establish Baselines	■ System maturity can be assessed through the establishment of the various baselines ■ Imposes efficiency and discipline to the process ■ Increases design integrity
Successful release of the final design data depends upon	■ Frequent use of design reviews ■ Early involvement of the manufacturing engineers in the design process ■ Design release guidelines established early in the program

Boznak, R. G., "Configuration Management the Next Engineering Management Proceedings Revolution." Second of the International Conference on Engineering Management, Toronto, September, 1989. pp. 207—210. This paper cites the principles and benefits of establishing a configuration management system.

Configuration and CIM SME White Paper, Society of Manufacturing Engineers, 1987. This document provides a good overview of the philosophy, basic principles and accepted practices of Configuration Management in the CIM environment.

Department of Defense, *Defense Acquisition Management Policies and Procedures.* DoD 5000.2, February 23, 1991. This document establishes a frame work for translating mission needs into programs that meet the user's needs and a management process for acquiring quality products with aggressive risk management by both government and industry.

Department of Defense, *Transition from Development to Production.* DoD 4245.7-M, September 1985. Describes techniques for avoiding technical risks in 48 key areas or templates in funding, design, test, production, facilities, logistics, management, and transition plan.

Department of the Navy, *Best Practices: How to Avoid Surprises in the World's Most Complicated Technical Process.* NAVSO P-6071. March 1986. Discusses how to avoid traps and risks by implementing best practices for 48 areas or templates including parts and materials selection, piece part control, defect control, and manufacturing screening.

Department of the Navy Memorandum, A-12 Administrative Inquiry, Nov. 28, 1990. This memorandum describes the findings, conclusions and recommendations based on an in-depth inquiry of the A-12 in the FSD phase. This memorandum describes the background, failures of the government program manager and recommendations to improve FSD programs.

Kerzner, H., *Project Management.* A Systems Approach to Planning, Scheduling and Controlling, 3rd ed. New York, New York, Van Nostrand Reinhold, 1989. This book is an excellent resource for managers at all levels who must plan and provide support for projects. Included are questions and case studies derived from the author's experiences.

Krogh, L. C., "Measuring and Improving Laboratory Productivity/Quality." Research Management Nov.-Dec. 1987, pp. 22-24. This article states the product of a laboratory is information and the customers for this information are marketing and manufacturing. It describes how 3M used this strategy to implement a quality improvement program within the R&D laboratory.

MIL-HDBK-727, Military Handbook Design Guidance for Producibility. This document provides the design engineer with information that will aid in reducing and eliminating features of a design that would make the design difficult to produce. Included are sections that deal with different materials and their specific producibility considerations. This document provides an excellent set of checklists that one could use to check the producibility of a design.

Mills, R. "Linking Design and Manufacturing," *Computer-Aided Engineering* Mar. 1991, pp. 42-48. Discuss how organizations can use product data management to reduce the number of design changes and provide the missing link between design and manufacturing.

Nichols, K., "Getting Engineering Changes Under Control," *Journal of Engineering Design*, vol. 1, no. 1, pp. 5-15.

US Army Material Command. *Program Management Risk Reduction Roadmaps.* September 1987. The roadmap the US Army will use to reduce risk in programs.

Index